CHEMICAL AND BIOCHEMICAL ASPECTS
OF SUPEROXIDE AND SUPEROXIDE DISMUTASE

DEVELOPMENTS IN BIOCHEMISTRY

Volume 1—Mechanisms of Oxidizing Enzymes, edited by Thomas P. Singer and Raul N. Ondarza, 1978

Volume 2—Electrophoresis '78, edited by Nicholas Catsimpoolas, 1978

Volume 3—Physical Aspects of Protein Interactions, edited by Nicholas Catsimpoolas, 1978

Volume 4—Chemistry and Biology of Pteridines, edited by Roy L. Kisliuk and Gene M. Brown, 1979

Volume 5—Cytochrome Oxidase, edited by Tsoo E. King, Yutaka Orii, Britton Chance and Kazuo Okunuki, 1979

Volume 6—Drug Action and Design: Mechanism-Based Enzyme Inhibitors, edited by Thomas I. Kalman, 1979

Volume 7—Electrofocus/78, edited by Herman Haglund, John G. Westerfeld and Jack T. Ball, Jr., 1979

Volume 8—The Regulation of Coagulation, edited by Kenneth G. Mann and Fletcher B. Taylor, Jr., 1980

Volume 9—Red Blood Cell and Lens Metabolism, edited by Satish K. Srivastava, 1980

Volume 10—Frontiers in Protein Chemistry, edited by Teh-Yung Liu, Gunji Mamiya and Kerry T. Yasunobu, 1980

Volume 11A–Chemical and Biochemical Aspects of Superoxide and Superoxide Dismutase, edited by J.V. Bannister and H.A.O. Hill, 1980

Volume 11B–Biological and Clinical Aspects of Superoxide and Superoxide Dismutase, edited by W.H. Bannister and J.V. Bannister, 1980

CHEMICAL AND BIOCHEMICAL ASPECTS OF SUPEROXIDE AND SUPEROXIDE DISMUTASE

Proceedings of the Federation of European Biochemical Societies
Symposium No. 62

Editors:

J.V. BANNISTER
Inorganic Chemistry Department, University of Oxford, Oxford, England

H.A.O. HILL
Inorganic Chemistry Department, University of Oxford, Oxford, England

Organizing Committee:

J.V. Bannister (England), B. Babior (U.S.A.), E.M. Fielden (England),
I. Fridovich (U.S.A.), J. McCord (U.S.A.), A.M. Michelson (France),
G. Rotilio (Italy)

ELSEVIER/NORTH-HOLLAND
NEW YORK • AMSTERDAM • OXFORD

Published by:

Elsevier North Holland, Inc.
52 Vanderbilt Avenue, New York, New York 10017

Sole distributors outside USA and Canada:

Elsevier/North-Holland Biomedical Press
335 Jan van Galenstraat, PO Box 211
Amsterdam, The Netherlands

Library of Congress Cataloging in Publication Data

International Symposium on Superoxide and Superoxide Dismutases, 2d, Malta, 1979.
Chemical and biochemical aspects of superoxide and superoxide dismutase.
(Developments in biochemistry; 11A ISSN 0165-1714)
(Proceedings of the Federation of European Biochemical Societies symposium; no. 62)
Sponsored by the Federation of European Biochemical Societies.

Bibliography: p.
Includes index.
1. Superoxide dismutase—Congresses. 2. Superoxide — Metabolism — Congresses. I. Bannister, Joe V., 1945- II. Hill, Hugh Allen Oliver. III. Federation of European Biochemical Societies. IV. Title. V. Series. VI. Series: Federation of European Biochemical Societies. FEBS symposium; no. 62.
QP603.S94I57 1979 612'.015 80-24010
ISBN 0-444-00442-4

Manufactured in the United States of America

Contents

Preface

This book and its companion volume represent the second collection of articles to be published after an international meeting on superoxide and superoxide dismutases. The subject matter of the articles was communicated to the Second International Symposium on Superoxide and Superoxide Dismutases held in Malta, 1st-5th October, 1979. The book is concerned with chemical and biochemical aspects of the subject; the companion volume deals with biological and clinical aspects. The subdivision of the material has been dictated to some extent by the exigencies of two volume publication.

Since the pioneering discovery of enzymatic catalysis of superoxide dismutation by McCord and Fridovich in 1968-69, work in this exciting area of scientific activity has involved the chemistry of the superoxide ion (and other oxygen-derived radicals), the biochemistry of the superoxide dismutases (which proved to be a family of metallo-enzymes), the biological implications of superoxide and superoxide dismutases, and the involvement of superoxide (and other active oxygen species) in pathological processes. The latter provides the rationale for therapeutic applications of superoxide dismutase. The work cuts across the disciplines of chemistry, biochemistry, biology and medicine and has one unifying theme - the interaction of living organisms and their constitutents with the products of one-electron reduction of oxygen. Curiously superoxide dismutase activity was first discovered in a copper protein, erythrocuprein, which had independently been tested for anti-inflammatory activity (as orgotein of Huber and co-workers).

The discovery of superoxide dismutase linked two fields previously somewhat considered separate: the chemistry of superoxide and the biochemistry of molecular oxygen. The study of superoxide dismutase has drawn together workers from the physical and life sciences in a way that few proteins or enzymes can be said to have done. The wide-ranging interest it has created has been astounding.

Oxford - March, 1980 J.V. Bannister - H.A.O. Hill

Acknowledgments

The papers published in this book were presented at an International Symposium on Superoxide and Superoxide Dismutases held in Malta between 1st and 5th October 1979. The meeting was sponsored by the Federation of European Biochemical Societies. The help received from Professors G. Bernardi (Paris), P.N. Campbell (London), S.P. Datta (London) and T.W. Goodwin, FRS (Liverpool) in getting this meeting funded and organised is gratefully acknowledged.

American participation at the meeting and also support for the meeting was made possible through sponsorship from the Office of Naval Research, National Cancer Institute, National Heart Lung and Blood Institute, National Institute of Environmental Health Sciences and Fogarty International Center.

Further support was obtained from Diagnostic Data, Inc. of Mountain View, California, USA and from Lederle Laboratories, A Division of the American Cyanamid Company, Pearl River, New York, USA. Thanks are also due to Professors A.M. Michelson, J. McCord and I. Fridovich for generously donating royalties towards the meeting.

The editorial assistance of Professor W.H. Bannister in producing this volume is gratefully acknowledged. Professor W.H. Bannister and Dr.J.V. Bannister gratefully acknowledge the Wellcome Trust, London, England for Research Fellowships. This publication would not have been possible without the support of the Wellcome Trust.

REDOX REACTION CHEMISTRY OF SUPEROXIDE ION

DONALD T. SAWYER, DAVID T. RICHENS, EDWARD J. NANNI, JR. AND MARTIN D. STALLINGS
Department of Chemistry, University of California, Riverside,
California 92521, USA

INTRODUCTION

Although the electrochemical reduction of oxygen to superoxide ion radical (O_2^-) in aprotic media has been known since 1965 (Refs. 1-6), the oxidation-reduction reactivity of O_2^- in such media has not been addressed to the extent that its nucleophilic properties have been studied.[23] Electrochemistry continues to provide a convenient means to prepare stable solutions of O_2^- in aprotic solvents with known concentrations by means of controlled potential coulometry.[2,7]

In 1968 McCord and Fridovich discovered that O_2^- is a respiratory intermediate of aerobic organisms,[8] and a year later observed[9] that hemocuprein, a protein whose function had been unknown since 1938, (Ref. 10) is involved in the dismutation of this intermediate. Since then, two other types of superoxide dismutase[11,12] (SOD) enzymes have been discovered; the three types generally are classified as copper-zinc (hemocuprein), manganese, or iron SODs. These discoveries have prompted widespread and intense interest among chemists and biochemists in the chemical properties and reactivity of O_2^- with organic functional groups and metal ion complexes.

In the solid state the O-O bond length of O_2^- is 1.33Å, which corresponds to a bond order of 1.5 (Ref. 13). The IR stretching frequency for O_2^- is 1145 cm^{-1} in contrast to 1556 cm^{-1} for O_2 and $\sim$770 cm^{-1} for O_2^{2-} (Refs. 14, 15). In acetonitrile with 0.1 M tetrapropylammonium perchlorate, O_2^- has a single absorption band with λ_{max} at 255 nm (ε, 1460 $M^{-1}cm^{-1}$) (Ref. 16). ESR spectra of this same solution at 77°K yield for O_2^- $g_\perp$ = 2.008 and $g_\parallel$ = 2.083 (Ref. 16). In dimethyl sulfoxide, O_2^- from KO_2 plus 18:6-crown ether has an absorption maxium at 250 nm (ε = 2086 $M^{-1}cm^{-1}$) (Ref. 17).

Much of the reactivity of O_2^- can be explained in terms of its Bronsted basicity, its nucleophilicity, and its redox activity. Also, the reduction potentials for the O_2/O_2^-, O_2^-/H_2O_2, and O_2/H_2O_2 couples are extremely pH and solvent dependent (see Table 1). Although the one-electron processes that are summarized in Table 1 for aqueous media involve unstable species and are not observed normally by electrochemical methods, these couples can be calculated from pulse radiolysis data and thermodynamic parameters.[18,20] Based on the kinetics and thermodynamics for a number of fundamental reactions of O_2^- and $HO_2^{\cdot}$ in water by the pulse radiolysis technique,[21,22]

$$H^+ + O_2^- \rightleftharpoons HO_2^\bullet \qquad pK_a = 4.88 \qquad (1)$$

$$O_2^- + HO_2^\bullet \rightleftharpoons O_2 + HO_2^- \qquad k_2 = 8.5 \times 10^7 M^{-1}s^{-1} \qquad (2)$$

$$HO_2^\bullet + HO_2^\bullet \rightarrow O_2 + H_2O_2 \qquad k_3 = 7.6 \times 10^5 M^{-1}s^{-1} \qquad (3)$$

$$O_2^- + O_2^- \xrightarrow{H_2O} O_2 + HO_2^- \qquad k_4 < 100\ M^{-1}s^{-1} \qquad (4)$$

$$O_2^- + {}^\bullet OH \rightarrow O_2 + OH^- \qquad k_5 = 1x\ 10^{10} M^{-1}s^{-1} \qquad (5)$$

$$^\bullet OH + H_2O_2 \rightarrow HO_2^\bullet \qquad k_6 = 3 \times 10^7 M^{-1}x^{-1} \qquad (6)$$

the reduction of O_2 in aqueous media can be explained by a one-electron transfer followed by a rapid dismutation of the O_2^- to H_2O_2 and O_2.

TABLE 1

REDOX POTENTIALS FOR SUPEROXIDE IONS IN AQUEOUS AND APROTIC MEDIA

		E^o, V *vs* NHE
A.	Aqueous Media, pH 7	
	$O_2 + e^- \rightarrow O_2^-$	-0.33
	$O_2^- + 2H^+ + e^- \rightarrow H_2O_2$	+0.87
	$O_2 + 2H^+ + 2e^- \rightarrow H_2O_2$	+0.27
B.	Aqueous Media, pH 0	
	$O_2 + H^+ + e^- \rightarrow HO_2^\bullet$	-0.04
	$HO_2^\bullet + H^+ + e^- \rightarrow H_2O_2$	+1.41
	$O_2 + 2H^+ + 2e^- \rightarrow H_2O_2$	+0.68
C.	Aqueous Media, pH 14	
	$O_2 + e^- \rightarrow O_2^-$	-0.33
	$O_2^- + H_2O + e^- \rightarrow HO_2^- + OH^-$	+0.17
	$O_2 + H_2O + 2e^- \rightarrow HO_2^- + OH^-$	-0.08
D.	Aprotic Media, dimethylformamide (0.1 *M* TEAP)	
	$O_2 + e^- \rightarrow O_2^-$	-0.55
	$O_2^- + HSol + e^- \rightarrow HO_2^- + Sol^-$	-1.75
	$O_2 + HSol + 2e^- \rightarrow HO_2^- + Sol^-$	-1.15
	$O_2 + H^+ + e^- \rightarrow HO_2^\bullet$	-0.04

The Bronsted basicity of O_2^- is exemplified by its capacity to effect proton removal from substrates and solvents.[23] The pK_a for $HO_2^{\bullet}$ in water is 4.88 which infers that O_2^- is a weak base.[21] However, combining the reduction half-reaction for O_2^- in water[24]

$$O_2^- + H_2O + e^- \rightarrow HO_2^- + OH^- \qquad E^o = +0.17 \text{ V } \underline{vs} \text{ NHE} \tag{7}$$

with the oxidation half-reaction for O_2^- in water,[20]

$$O_2^- \rightleftharpoons O_2 + e^- \qquad E^{o\prime} = -0.33 \text{ V } \underline{vs} \text{ NHE} \tag{8}$$

yields the net expression

$$2O_2^- + H_2O \rightleftharpoons O_2 + HO_2^- + OH^- \qquad K = 2.5 \times 10^8 \tag{9}$$

at pH 14. Sawyer and Gibian[25] conclude that the value of the equilibrium constant indicates that O_2^- can promote proton transfer from substrates and solvents to an extent equivalent to that of the conjugate base of an acid with a pK_a value of approximately 23. Recent studies establish that a number of weakly acidic organic compounds are deprotonated efficiently in the presence of O_2^- (Refs. 26,27).

The first reports of nucleophilic substitution by O_2^- with alkyl halides in aprotic solvents appeared in 1970 (Refs. 28, 29). Because the reaction rates followed the order $1^o > 2^o >> 3^o$ for alkyl halides and tosylates and the attack by O_2^- results in inversion of configuration, the O_2^--RK reactions proceed by an S_N2 mechanism.[30,31] The reaction O_2^- with esters yields carboxylic acids and alcohols,[32,33] and with acyl halides yields diacylperoxides.[34] Acyl oxygen scission occurs for the reaction of esters with O_2^- on the basis of complete retention of configuration in the alcohol moiety from an optically active ester.[32] Further, the order of reactivity for different n-octanoate esters follows typical acyl oxygen scission patterns.[32] Therefore, the intial reaction is proposed to occur via the usual nucleophilic addition-eliminations at the carbonyl carbon (eqs. 10 and 11).

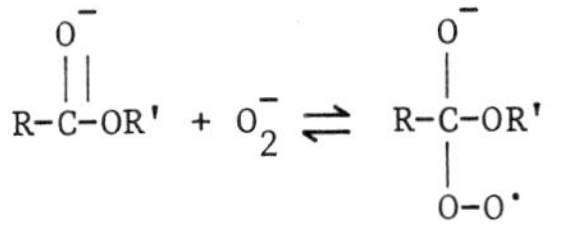
(10)

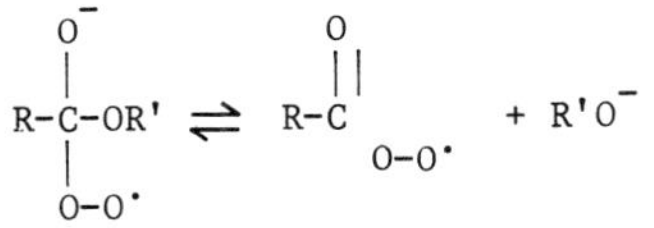
(11)

Numerous groups have concluded that O_2^- is an effective reducing[35-41] and oxidizing agent.[42-50] However, the direct transfer of an electron to O_2^- is an unreasonable characteristic for solution conditions where O_2^- is stable, i.e. aprotic media.[26] Indeed as Table 1 illustrates, the oxidation of O_2^- to O_2 is thermodynamically favorable in aprotic media and is roughly equivalent to dithionite ion (E^o = -0.46 V vs NHE) (Ref. 51). Conversely, the reduction of O_2^- to HO_2^- is difficult and virtually impossible in the absence of proton sources. In their presence O_2^- dismutases rapidly to HO_2^- and O_2, which then may be involved in the cited oxidation of organic molecules by O_2^-. Similarly, metal ions such as Mn^{2+} (Ref. 8) or Fe^{2+} are Lewis acids and can promote the dismutation of O_2^-.

This report summarizes a detailed study of the reactivity of O_2^- with a number of oxidants and acidic reductants. The results indicate that O_2^- effectively reduce oxidants such as quinones, sulfur dioxide, and transition metals by a one-electron mechanism. With acidic reductants such as phenol, hydroquinone, catechol, α-tocopherol, and ascorbic acid there is not a direct electron transfer to O_2^-, but rather a proton induced dismutation of O_2^- to O_2 and HO_2^- followed by oxidation of the substrate anion by O_2.

MATERIALS AND METHODS

Equipment. The cyclic voltammetric experiments were accomplished with either a three-electrode potentiostat-amperiostat constructed with operational amplifiers[52] or a Princeton Applied Research Model 173/179 potentiostat-galvanostat. The voltammograms were recorded with a Houston Instruments 2000 XY recorder. A Princeton Applied Research Model 179 digital coulometer was utilized for the controlled potential electrolysis experiments. Kinetic studies were performed with a Beckman Model 1855 rotating disc electrode in conjunction with a potentiostat and a Hewlett Packard Model 7030A XY recorder. All air-sensitive electrochemical experiments were conducted inside a Vacuum Atmosphere Corporation Dri-Train HE 193 Dr-Lab glove box in the presence of purified nitrogen.

The electrochemical cell consisted of a Leeds and Northrup olyethylene electrochemical cell top and a 100-ml electrolytic beaker. The cell top included holes to accommodate the electrodes, the auxiliary compartment, and the introduction and removal of samples and gases.

The working electrode for the cyclic voltammetric experiments was a Beckman platinum-inlay electrode (surface area 0.23 cm^2); a cylindrical platinum mesh electrode was used for controlled potential electrolysis; and a platinum-inlay electrode (surface area 0.34 cm^2) was used in the rotating disc experiments. The auxiliary electrode was a platinum flag electrode which was isolated from the bulk solution by a medium porosity glass frit. The reference electrode was a Ag/AgCl (aqueous tetramethylammonium chloride) cracked glass-bead electrode which was adjusted to 0.000 V vs SCE (Ref. 53). The reference electrode was placed inside a luggin capillary in the cell assembly.

UV-visible spectra were recorded with a Cary 17D or a Cary 219 spectrophotometer, with the concentration of the supporting electrolyte in the reference and the sample cells the same as in the electrochemical experiments (0.1 M TEAP). ESR spectra were recorded with a Varian 4502 spectrometer equipped with standard quartz cells. Low temperature measurements were made by use of liquid nitrogen in a

Wilmad quartz Dewar flask. The field was standardized with 1,1-di-phenyl-2-picryl-hydrazyl (Eastman Kodak Co.) (g = 2.0037).

Chemicals and Reagents. Burdick and Jackson "distilled in glass" dimethylformamide (DMF) and dimethyl sulfoxide (DMSO) were obtained in quart bottles to minimize water contamination. The water content as specified by the manufacturer was 0.029% for DMSO and 0.008% for DMF.

Tetraethylammonium perchlorate (TEAP) from G. Frederick Smith Chemical Co. was used as the supporting electrolyte (0.1 M TEAP) in the electrochemical experiments. High purity oxygen was obtained from the Chemetron Corporation. Superoxide was generated electrochemically by reducing a solution through which oxygen was bubbled continuously. Upon completion of electrolysis the superoxide solution was degassed with argon to remove any residual oxygen. Tetraethylammonium hydroxide (TEAOH) was obtained from Eastman Kodak Co. as a 25% methanol solution. 3,5-Di-*tert*-butylcatechol (DTBC) and *p*-benzoquinone were obtained from the Aldrich Chemical Co., and hydroquinone and L-(+)-ascorbic acid from Matheson Coleman and Bell. α-Tocopherol was supplied by Calbiochem. The remaining chemicals were reagent grade and were used without further purification.

RESULTS

Dismutation of superoxide ion. Dismutation of O_2^- is induced by charge neutralization from protons or Lewis acids. In aqueous media the dismutation of O_2^- is pH dependent,[54] but highly favored under all conditions. The equilibrium constants at various pH values can be calculated from the data of Table 1;

at pH 14

$$2O_2^- + H_2O \rightarrow O_2 + HO_2^- + OH^- \qquad K = 3 \times 10^8 \qquad (12)$$

at pH 7

$$2O_2^- + 2H^+ \rightarrow O_2 + H_2O_2 \qquad K = 2 \times 10^{20} \qquad (13)$$

at pH 0

$$2HO_2^\cdot \rightarrow O_2 + H_2O_2 \qquad K = 3 \times 10^{28} \qquad (14)$$

Reference to the pulse radiolysis data of eqs. 1-4 confirms that the rate for the dismutation process increases as the pH is lowered. Recent data confirm that in the limit of highest pH the direct unprotonated dismutation in water

$$O_2^- + O_2^- \xrightarrow{H_2O} HO_2^- + O_2 \qquad (15)$$

has a second-order rate constant no greater than 0.3 $M^{-1}s^{-1}$ (Ref. 55).

In aprotic media the dismutation of O_2^- also requires a proton or Lewis acid source. Figure 1 illustrates the effect of a proton source (NH_4^+) on the reduction of oxygen to O_2^- in dimethylformamide (DMF). With increasing concentrations of NH_4^+ the reduction is shifted from a one-electron process to a two-electron process at more positive potentials (due to the proton induced dismutation of O_2^- to H_2O_2 and O_2). Table 2 (part A) provides additional data for the effect of several acids on the reduction potential and overall electron stoichiometry for oxygen in DMF.

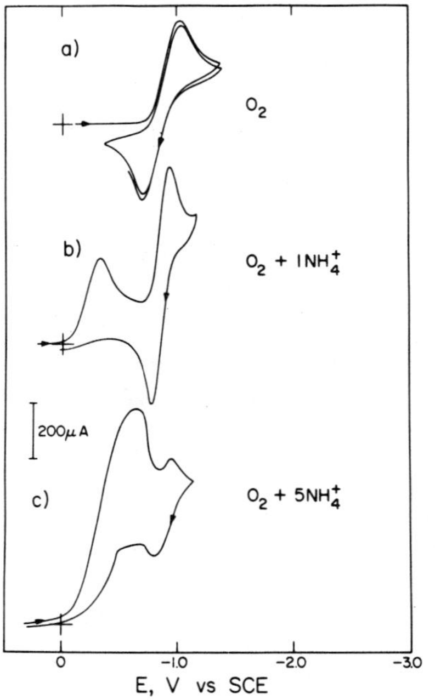

Fig. 1. Cyclic voltammograms in dimethylformamide (0.1 M tetraethylammonium perchlorate) of (a) 4.8 mM O_2, (b) 4.8 mM O_2 plus 4.8 mM NH_4ClO_4, and (c) 4.8 mM O_2 plus 24 mM NH_4ClO_4. Measurements were made with a platinum electrodes (area, 0.23 cm^2) at a scan rate of 0.1 V s^{-1}; temperature, 25°C. E_{NHE}=E_{SCE} + 0.25 V.

The dismutation of O_2^- by acidic substrates has been investigated by cyclic voltammetry and with a Pt rotating disc electrode in DMF and DMSO. After the addition of a proton source to an O_2^- solution the rest potential shifts positively, O_2 is produced, and in the case of phenol and α-tocopherol, an anodic peak due to their anions is observed. For slow reactions the decrease of the O_2^- concentration has been monitored by cyclic voltammetry. With more acidic substrates the rate of disappearance of O_2^- has been monitored with a Pt rotating disc electrode. Because of efficient conventive mass transfer with such electrodes, a steady-state current is observed;[56] i.e. the current for a constant concentration is independent of time. Hence, rates of reaction can be determined for rapid processes by monitoring

TABLE 2

THERMODYNAMICS AND KINETICS FOR THE PROTON-INDUCED DISMUTATION OF SUPEROXIDE ION IN DIMETHYLFORMAMIDE

A. Potentials for the reduction of O_2 in the presence of a five-fold excess of three acids

Acid	$(E_{pc})_{\frac{1}{2}}$, V vs NHE	$n(e^-/O_2)$
HCl	+0.12	2
NH_4ClO_4	-0.05	2
phenol	-0.63	2
-	-0.65	1

B. Observed rate constants for the dismutation of superoxide anion in the presence of excess substrates at 25°C;

$$\frac{-d(O_2^-)}{dt} = k[HA][O_2^-]$$

Substrate (HA)	k, $M^{-1}s^{-1}$ (d)
NH_4ClO_4 (a)	$(2.0 \pm 0.2) \times 10^3$
phenol (a)	$(1.6 \pm 0.2) \times 10^3$
ascorbate anion^{-1} (b)	$(1.9 \pm 0.2) \times 10^2$
α-tocopherol (a)	$(7.8 \pm 0.8) \times 10^1$
H_2O (c)	$(1.8 \pm 0.2) \times 10^{-3}$
n-butanol (c)	$(1.5 \pm 0.2) \times 10^{-3}$

(a) The rates of these substrates were measured with a rotating disc electrode at 0.00 V vs SCE.
(b) Measured by rapid scan cyclic voltammetry.
(c) Measured by cyclic voltammetry.
(d) $k = 2k_{20}$; see DISCUSSION. Mixing times limit measurements to rate constants that are less than $10^3 M^{-1}s^{-1}$.

the current as a function of time. At potentials more positive than -0.62 V vs SCE, O_2^- in DMF is oxidized to O_2 and provides a measure of the O_2^- concentration.

Figure 2 illustrates the decay of O_2^- concentrations as a function of time for five different protic substrates in DMF. Clearly, the rate of dismutation is highly dependent upon substrate acidity. Because the semilog plots are linear, the loss of O_2^- in the presence of excess substrate appears to follow pseudo

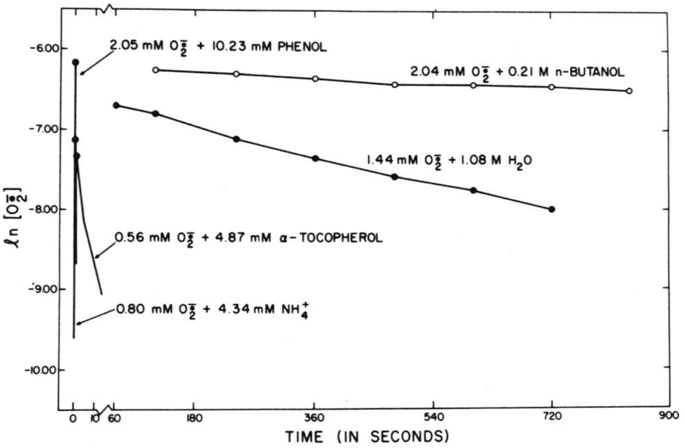

Fig. 2. Rate of disappearance of $O_2^{\bar{\cdot}}$ in the presence of various protic substrates in dimethylformamide (0.1 M tetraethylammonium perchlorate) at 25°C. Concentration of O_2^- were monitored by a rotating platinum disc electrode at 0.0 V vs SCE or by cylic voltammetry.

first-order kinetics. However, the dismutation reaction is followed by a number of secondary processes, especially with the more acidic substrate (NH_4^+ and phenol). Hence, the initial rates have been used to evaluate the rate constants which are summarized in Table 2, part B. The data indicate that the rate of dismutation roughly correlates with the acidity of the substrate.

Metal ions may be classified as Lewis acids and, like the Bronsted acids HCl, NH_4ClO_4 and phenol, affect the reduction of O_2 and the dismutation of O_2^-. Table 3 summarizes the reduction potentials of O_2 in DMSO in the presence of $Mn(ClO_4)_2$, $Co(ClO_4)_2$, $Fe(ClO_4)_2$, $Zn(ClO_4)_2$, $Cd(ClO_4)_2$, $Ni(ClO_4)_2$, $LiClO_4$ at a Pt electrode and the overall electron stoichiometry. These metal ions cause the reduction of O_2 to O_2^- to occur at more positive potentials in the order: $Fe^{2+} > Mn^{2+} \gtrsim Zn^{2+} \sim Co^{2+} > Cd^{2+} > Ni^{2+} > Li^+ \sim TEA^+$ (where TEA^+ is tetraethylammonium ion, the cation commonly employed as a supporting electrolyte). The overall electron stoichiometry in the presence of Mn^{2+} and Co^{2+} is four electrons per O_2 whereas in the presence of Fe^{2+}, Zn^{2+}, Cd^{2+}, and Ni^{2+} it is two electrons per O_2. When Li^+ is present, only one electron is consumed in the process. Combination of $Mn^{II}(ClO_4)_2$ and O_2^- in acetonitrile yields O_2 and Mn^{3+} when monitored by cyclic voltammetry and spectroscopy, which is consistent with the conclusion that metal cations act as Lewis acids to induce the dismutation of O_2^-.

Superoxide ion as a reducing agent. Reference to Figure 3 illustrates the cyclic voltammograms for 3,5-di-*tert*-butyl-*o*-quinone (DTBQ), O_2^-, and the products of their 1:1 combination in DMF. The first one-electron reduction of DTBQ is

TABLE 3

METAL-CATALYZED REDUCTION OF O_2 IN DMSO (0.1 M TEAP) AT A Pt ELECTRODE[a]

Metal	$E_{p,c}$, V vs SCE	$n(e^-/O_2)$
-	-0.80	1
Li^+	-0.79	1
Zn^{2+}	-0.67	2
Cd^{2+}	-0.67	2
Fe^{2+}	-0.52	2
Mn^{2+}	-0.61	4
Co^{2+}	-0.66	4

[a]Scan rate 0.02 V s^{-1} and $[M^{n+}]/[O_2]$ = 10 → 30

reversible ($E^{o\prime}$ = -0.24 V vs NHE), but the second one-electron transfer at -1.08 V vs NHE yields an unstable dianion and is an irreversible process.

Controlled potential electrolysis of DTBQ at -0.74 V vs NHE confirms that the first reduction is a one-electron process that yields the anodic couple with a positive voltammetric scan. On the basis of the ESR (Ref. 57) and UV-visible spectra[43,58] the blue product species is 3,5-di-*tert*-butyl-o-semiquinone ($DTBQ^{\overline{\cdot}}$). Combination of O_2^- and DTBQ yields a blue solution with the electrochemistry that is illustrated by Figure 3, trace c). When scanned in the positive direction, a small anodic peak at -0.50 V vs NHE is obtained that is due to some unreacted superoxide. The next peak at -0.15 V vs NHE is due to the oxidation of $DTBQ^{\overline{\cdot}}$ to DTBQ. When the voltage is scanned initially in the negative direction a reduction peak is observed at -0.80 V vs NHE which can be eliminated by first purging the solution with argon; hence, it is due to the reduction of molecular oxygen. The next reduction peak at -1.23 V vs NHE is due to the reduction of $DTBQ^{\overline{\cdot}}$ to the dianion. When the scan direction is reversed, an anodic peak due to the oxidation of $DTBQ^{2-}$ to $DTBQ^{\overline{\cdot}}$ is observed at -0.99 V vs NHE, a peak due to the oxidation of O_2^- to O_2 at -0.50 V vs NHE, and a peak due to the oxidation of $DTBQ^{\overline{\cdot}}$ to DTBQ at -0.15 V vs NHE.

Figure 4 illustrates the electrochemical behavior of O_2^-, sulfur dioxide, and the products of their 1:1 combination in DMF. Addition of SO_2 to a solution of O_2^- initially results in a blue solution that becomes colorless in about 30 seconds. The cyclic voltammogram for this solution after it is purged with argon (Figure 4) is identical to electrochemically generated $S_2O_4^{2-}$ (Ref. 51). The ESR spectrum for this solution has a g-value of 2.006 which is identical to that for the $SO_2^{\overline{\cdot}}$ radical anion[58-60] and is consistent with the dissociation of $S_2O_4^{2-}$. If additional SO_2 is added to this solution it turns bright blue (λ_{max} at 580 nm) and gives a spectrum characteristic of the $(SO_2)_2SO_2^{\overline{\cdot}}$ complex.[59-65]

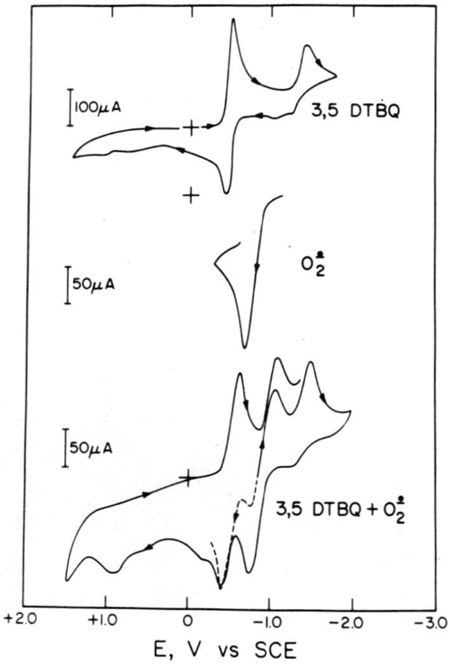

Fig. 3. Cyclic voltammograms in dimethylformamde (0.1 M tetraethylammonium perchlorate) of 5.3 mM 3,5-di-tert-butyl-o-quinone (3,5 DTBQ), of 4.0 mM O_2^-, and of 5.3 mM 3,5 DTBQ plus 4.0 mM O_2^-. Measurements were made with a platinum electrode (area, 0.23 cm^2) at a scan rate of 0.1 V s^{-1}; temperature, 25°C.

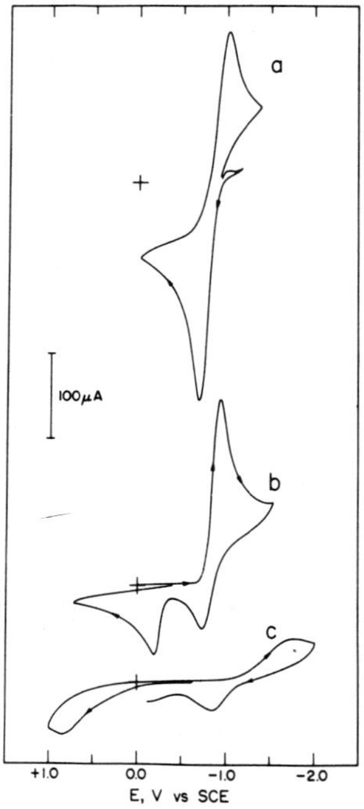

Fig. 4. Cyclic voltammograms at a platinum electrode (area, 0.23 cm^2) in dimethylformamide (0.1 M tetraethylammonium perchlorate); scan rate, 0.1 V s^{-1}; temperature, 25°C. (a), 5 mM O_2^-; (b), 5 mM SO_2; (c), product solution from the comination of 5 mM O_2^- and 5 mM (after deaeration with argon). $E_{NHE} = E_{SCE} + 0.25$ V.

Superoxide is also an effective one-electron reducing agent of transition metal complexes. Studies in progress indicate that O_2^- is capable of reducing Fe^{3+}, ferricinium ion, $Mn_2^{IV}O_2(o\text{-phen})_4^{4+}$, $Co^{III}(o\text{-phen})_3^{3+}$, and $Ir^{IV}Cl_6^{2-}$. The biological significance of the reducing capacity of O_2^- is illustrated by its ability to reduce ferricytochrome c (Refs. 66-68).

Oxidation of superoxide ion to singlet oxygen. Formation of singlet oxygen ($^1O_2(^1\Delta g)$) has been reported to result from the oxidation of O_2^- with ferricinium ion[69] and diacyl peroxides,[70] and from the dismutation of O_2^- by proton sources.[71] Recent studies in our laboratory support the formation of $^1O_2$ by the oxidation of O_2^- with ferricinium ion in DMF.[72] However, there is little evidence for its formation from the proton-induced dismutation of O_2^- in DMF. Relative to the production of $^1O_2$ by the reaction of 0.01 M NaOCl with 12.7 M H_2O_2 in H_2O, half as much $^1O_2$ results from the combination of 0.01 M ferricinium perchlorate with 0.01 M O_2^- in DMF on the basis of photon counting experiments. In contrast, combination of 0.01 M $Fe(ClO_4)_3$ with 0.01 M O_2^- and of 0.01 M $HClO_4$ with 0.01 M O_2^- in DMF only results in a 2% and 3% relative yield of $^1O_2$, respectively.

Conversion of $^3O_2$ to $^1O_2$ is a highly endothermic process:[73]

$$^3O_2 \rightarrow {}^1O_2 \qquad \Delta G = +91\ KJ$$

On the basis of the relationship;

$$\Delta G_{rxn} = nF\Delta E^o_{rxn},$$

where ΔG_{rxn} = 91 KJ, n = 1, and F = 96,487 C, the value for ΔE^o_{rxn} is -0.94 V. Since in DMF the reversible reduction of $^3O_2$ occurs at E^o = -0.55 V vs NHE (Table 1), combination of the appropriate redox potentials yields:

$$^1O_2 + e^- \rightarrow O_2^- \qquad E^o = +0.39\ V\ vs\ NHE$$

Hence, from thermodynamic considerations the oxidation of O_2^- to $^1O_2$ requires an oxidant with a redox potential of at least +0.39 V vs NHE. Because the reversible reduction potential for ferricinium ion occurs at +0.69 V vs NHE, oxidation of O_2^- to $^1O_2$ is thermodynamically favored. Although the reduction potential for $Fe(ClO_4)_3$ is +0.64 V vs NHE, a low yield of $^1O_2$ results. This probably is due to the difference in the coordination chemistry around the iron atom which results in different transition states for the O_2^--oxidant reaction. Other possible oxidants with redox potentials greater than +0.39 V and with the appropriate coordination chemistry include $Mn^{III}(o\text{-terpy})_2^{3+}$, $(Mn^{IV}(o\text{-phen})_2O)_2^{4+}$ (Goodwin's complex), CoIII $(o\text{-phen})_3^{3+}$, and $Ir^{IV}Cl_6^{2-}$.

Superoxide ion as an oxidizing agent. Under aprotic conditions O_2^- is no more effective as an oxidant than sodium ion (E^o_{red} = -1.75 V vs NHE). However, recent reports have concluded that it directly oxidizes a number of biologically important molecules such as vitamin E (α-tocopherol)[48,49] and vitamin C (L-(+)ascorbic acid).[50] These conclusions were based on studies with O_2^- generated in situ by reduction of O_2 with xanthine-xanthine oxidase in aqueous media or by use of KO_2 in tetrahydrofuran.

Careful stoichiometric and kinetic studies on the reactivity of O_2^-, O_2 and H_2O_2 with these substrates as well as with phenol, hydroquinone, and 3,5-di-*tert*-butylcatechol in several non-aqueous solvents have been undertaken to determine the mechanism for the net oxidation process. The reactions have been monitored by use of cyclic voltammetry, absorption spectroscopy, ESR spectroscopy, and HPLC. Initial rates have been measured by use of a rotated disc electrode and by rapid scan voltammetry.

Reaction with catechols and hydroquinones. The results for phenol, which have been discussed in a preceding section, confirm that it reacts with one or two equivalents of O_2^- in DMF to produce a stoichiometric yield of phenoxide ion, O_2 and H_2O_2. A detailed study of the reaction of O_2^- with 3,5-di-*tert*-butylcatechol in DMF illustrates the results for catechols and hydroquinones. This reaction has been followed both in a sealed electrochemical cell, previously purged with argon (Figure 5, trace a) and in an open cell continuously purged with argon during and after the addition of the reactants (Figure 5, trace b). In each case a rapid oxidation occurs to give a blue solution of the semiquinone radical anion (λ_{max} at 313, 345 and 380 nm; E^o_{ox} = -0.15 V *vs* NHE, and E^o_{red} = -1.2 V *vs* NHE). The yield of semiquinone during continuous argon purging is only about 66% of that obtained with a sealed cell. The catechol monoanion (λ_{max} at 305 nm, ε = 10,000 M^{-1} cm^{-1}; E^o_{ox} = +0.05 V *vs* NHE) is an additional product. When purging with argon is continued for about ten minutes after combination of the reactants, it is the dominant product. The cyclic voltammogram of Figure 5 (trace a) clearly shows that semiquinone anion, catechol monoanion, molecular oxygen, and residual O_2^- are all present, and in an apparent equilibrium which can be shifted by purging with either O_2 or argon. Further addition of O_2^- results in increasing yields of semiquinone anion and the evolution of O_2.

An analogous reaction occurs when hydroquinone and O_2^- are combined to produce the semiquinone radical anion (λ_{max} at 453 nm, ε = 7,000 M^{-1} cm^{-1}; E^o_{ox} = -0.15 V *vs* NHE), but it requires less than two equivalents of O_2^- for a stoichiometric yield (in contrast to at least three equivalents for 3,5-di-*tert*-butylcatechol). The yields of semiquinone have been calibrated with standard preparations from combination of hydroquinone or catechol with the corresponding benzoquinone in the presence of excess TEAOH.

The results provide strong support for the conclusion that the actual oxidation step for the combination of O_2^- and hydroquinone or catechol is electron transfer from the anion of the substrate to the O_2 from the proton induced dismutation of O_2^-.

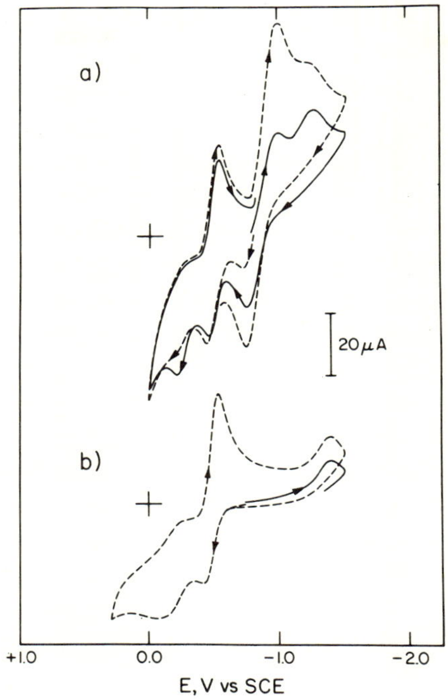

Fig. 5. Cyclic voltammograms of 2.0 mM O_2^- plus 1.0 mM 3,5-di-*tert*- butylcatechol at a platinum electrode (area 0.23 cm^2) in dimethyl formamide (0.1 *M* tetraethylammonium perchlorate); scan rate, 0.1 V s^{-1}; temperature, 25°C. (a), sealed cell; (b), continuous purging of the reaction mixture with argon. Solid lines for initial negative scans and dashed lines for initial positive scans. $E_{NHE} = E_{SCE}$ + 0.25 V.

Reaction with α-tocopherol (vitamin E). Combination of α-tocopherol with one or two equivalents of O_2^- in DMF, DMSO, or CH_3CN results in a golden-orange color (λ_{max} at 385 nm and 415 nm, ε = 1,500 M^{-1} cm^{-1}) due to oxidation of the substrate. Almost identical spectral features are obtained from reaction of the anion with a stoichiometric quantity of O_2 in DMF, which provides further support that this is the actual oxidation step. The slower reaction of O_2^- with α-tocopherol in DMSO is consistent with the observed slower reaction of substrate anion with O_2 in this solvent. Furthermore, vigorously purging a solution of α-tocopherol and two equivalents of O_2^- with argon results in almost quantitative production (>75%) of the substrate anion (λ_{max} at 327 nm, ε = 5,000 M^{-1} cm^{-1} in DMF; E^{o}_{ox} = 0.15 V *vs* NHE) without significant substrate oxidation.

The nature of the oxidation products from the combination of O_2^- with α-tocopherol or of O_2 with its anion is solvent dependent. If the product solution in DMF is neutralized with one or two equivalents of HCl, and absorption maximum at 288 nm (ε = 2,700 M^{-1} cm^{-1}) is observed. However, in DMSO or CH_3CN, a comparable addition of HCl yields only an absorption maximum at 268 nm (ε = 5,700 M^{-1} cm^{-1} for CH_3CN and ε = 2,400 M^{-1} cm^{-1} for DMSO), which is consistent with α-tocopherol quinone[74] as one of the oxidation products in these solvents. HPLC experiments have confirmed that several oxidation products are present and that their distribution is dependent both on the solvent used and on the amount of O_2 introduced. A number of oxidation products of α-tocopherol have been identified when $FeCl_3$, $AgNO_3$, alkaline $Fe(CN)_6^{3-}$, and p-benzoquinone are used as oxidants.[74-77] However, these products do not appear to be the same as those that result from the combination of O_2^- and α-tocopherol.

The kinetics of the initial reaction of α-tocopherol with O_2^- in DMF have been followed with a rotating Pt-disc electrode. For a constant α-tocopherol concentration of 5 mM, the O_2^- concentration has been varied between 0.5 and 1.0 mM under pseudo first-order conditions. The decay of O_2^- concentration obeys first-order kinetics with respect to the initial rate, with an apparent rate constant of $(3.78 \pm 0.4) \times 10^{-1} sec^{-1}$. Doubling the α-tocopherol concentration causes the initial rate to double, which confirms that the reaction also has pseudo first-order dependence on α-tocopherol. The calculated overall second-order rate constant is listed in Table 2 (part B) and is discussed in a later section.

Reaction with L-(+)ascorbic acid (vitamin C). Combination of O_2^- with one equivalent of L-(+)ascorbic acid in DMF results in the rapid production of a pale orange solution that contains the ascorbate monoanion (λ_{max} at 274 nm, ε = 10,000 M^{-1} cm^{-1}; E^o_{ox} = +0.25 V *vs* NHE) as well as a number of oxidation products. As with α-tocopherol and 3,5-di-*tert*-butylcatechol, continuous purging of the solution with argon during the addition of reactants results in increased yields of the monoanion (about 50%) relative to semiquinone. Superoxide ion also reacts with the monoanion of ascorbic acid, but at a much slower rate to give a pale orange solution with an absorption maximum at 285 mm (ε = 1,000 M^{-1} cm^{-1}) in DMF and oxidation peaks at +0.65 V and +1.1 V *vs* NHE. Almost identical features are obtained from the combination of L-(+)ascorbic acid with two equivalents of O_2^-. The addition of excess TEAOH to these solutions produces a pale yellow solution with an absorption maximum at 295 nm (ε = 1,8000 M^{-1} cm^{-1}), which slowly turns pink on prolonged standing. A similar solution is obtained from the combination of L-(+)ascorbic acid, excess TEAOH, O_2, and H_2O_2 or by the addition of H_2O_2 to solutions of L-(+)-dehydroascorbic acid in the presence of excess TEAOH in DMF. In contrast, when L-(+)ascorbic acid, TEAOH, and H_2O_2 are combined in the absence of O_2, there is no appreciable reaction unless a large excess of TEAOH is present, which promotes the decomposition of H_2O_2 to O_2^- and $\cdot OH$ (Ref. 78).

L-(+)ascorbic acid does not react with O_2 in these solvents, but addition of excess TEAOH leads to rapid oxidation of the dianion of L-(+)ascorbic acid[79]

(λ_{max} at 305 nm; E^{o}_{ox} = -0.65 V vs NHE). The monoanion reacts with one equivalent of O_2 in DMF to produce a yellow solution of the anion of L-(+)2,3-diketogulonic acid (λ_{max} at 356 nm, ε = 1,100 $M^{-1}cm^{-1}$), the ring-opened δ-lactone of L-(+)-dehydroascorbic acid. Addition of excess TEAOH and one equivalent of H_2O_2 to L-(+)-ascorbic acid yields the species with the 295 nm absorption band. Controlled potential oxidation of the ascorbate dianion in DMF at -0.8 V vs SCE in the presence of excess TEAOH and H_2O_2 also yields the species with the 295 nm band. The ε values for this band are similar to those for the $(-(CHOH)_2CO_2^-)$ group (e.g. for glycerate ($CH_2(OH)CH(OH)CO_2^-$; λ_{max} at 296 nm, ε = 2,000 M^{-1} cm^{-1}).

The initial rate for the reaction of L-(+)ascorbic acid with O_2^- in DMF is rapid on the basis of ongoing experiments with the rotating-disc electrode. However, the reaction of the monoanion of ascorbic acid with O_2^- is much slower and has been followed by rapid scan voltammetry. For an initial ascorbate concentration of 0.97 mM, the O_2^- concentration has been varied between 0.15 mM and 1.0 mM. The decay of the O_2^- concentration obeys first-order kinetics during the initial two seconds of reaction with a rate constant of (1.85 ± 0.2) x $10^{-1}s^{-1}$. Variation of the monoanion concentration also indicates a first-order process. The calculated overall second-order rate constant is listed in Table 2 (part B).

Reduction of oxygen to superoxide ion. That reduced xanthine oxidase reacts with O_2 to produce O_2^- has been known since the 1960's (Ref. 81). Reduced flavins (FMN, FAD, riboflavin) also yield O_2^- when combined with O_2 (Refs. 82,83).

On the basis of recent experiments, O_2^- can be produced in good yield by reduction of O_2 by the dianions of hydroquinone and ascorbic acid in aprotic media. Figure 6 illustrates the voltammogram of the dianion of hydroquinone in DMF that is formed by the addition of two equivalents of TEAOH. The voltammogram (scanned in the positive direction initially) exhibits two oxidation peaks, one at -0.70 V vs NHE that is due to the oxidation of the dianion to semiquinone, and a second at -0.10 V vs NHE that is due to the oxidation of semiquinone to quinone. When an equimolar amount of O_2 is added to the dianion solution and the voltammogram is scanned in the positive direction, an O_2^- oxidation peak is observed as a shoulder of the semiquinone oxidation peak at -0.10 V vs NHE; the combined peak current is much larger than the corresponding reduction peak for the quinone at -0.21 V vs NHE on the reverse scan. The latter is followed by a reduction peak for O_2 at -0.67 V vs NHE. When the solution is scanned in the negative direction from the rest potential, the voltammogram does not exhibit a peak for O_2, but a reduction wave at -0.95 V vs NHE is observed for reduction of semiquinone to hydroquinone.

Superoxide from the combination of O_2 and hydroquinone dianion as well as from combination of O_2 and ascorbate dianion has been detected by ESR spectroscopy. However, in the case of ascorbate, a ten-fold excess of TEAOH is required to obtain the maximum yield of O_2^-. Voltammetry, UV-visible spectroscopy, and ESR spectro-

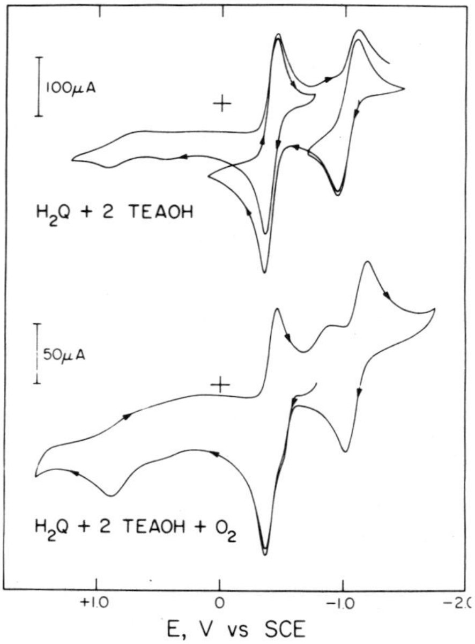

Fig. 6. Cyclic voltammograms at a platinum electrode (area, 0.23 cm^2) in dimethylformamide (0.1 M tetraethylammonium perchlorate) of 4.8 mM hydroquinone dianion and of 2.4 mM hydroquinone dianion plus 2.4 mM O_2; scan rate, 0.1 V s^{-1}; temperature, 25°C.

scopy confirm that one-half of an equivalent of O_2^- is produced per ascorbate. Apparently, only one-half of the ascorbate is converted to the dianion because of the high pK_a value for the second proton of ascorbic acid (pK_2 = 11.5; Ref. 84).

DISCUSSION

Dismutation of superoxide. Aqueous thermodynamic data indicate that reduction of O_2 to H_2O_2 is favored over the reduction of O_2 to O_2^- by +0.60 V (see Table 1 and Ref. 85). Similarly, electrochemical experiments confirm that O_2 is reduced at more positive potentials in acidified DMF and these potentials are dependent upon the degree of acidity (Table 2, part A). When NH_4^+ is added to an O_2-containing solution, a new, broad reduction peak appears at the expense of the aprotic peak (Figure 1); the latter peak is eliminated when five equivalents of NH_4^+ have been added and the overall electron stoichiometry for the first peak becomes two (Table 2, part A). These observations are consistent with the conclusion that protons induce the dismutation of the O_2^- that results from the electrochemical reduction of O_2 (in Figure 1 the more positive peak represents an overall reduction to peroxide and the second peak is due to a one-electron reduction to O_2^-). However, both reductions occur via one-electron processes at the electrode surface. In the presence of protic substrates the product species dismutates to O_2 and H_2O_2, which is confirmed by the broad irreversible peak of Figure 1, trace c). A reasonable mechanism for O_2 reduction in the presence of excess protons is represented by eqs. 16 and 17

$$O_2 + H^+ + e^- \rightarrow HO_2^{\bullet} \quad (16)$$

$$HO_2^{\bullet} + HO_2^{\bullet} \rightarrow O_2 + H_2O_2 \quad (17)$$

Numerous workers[21,85-88] have studied the mechanism of the dismutation of O_2^- in water and observe second-order kinetics with respect to O_2^-. However, for alkaline solutions a first-order component is present during the initial reaction stage. Bielski and Allen[88] attribute this to the presence of catalytic impurities, namely, trace metals. The kinetic data for O_2^- dismutation in DMF in the presence of NH_4^+, phenol, ascorbic acid monoanion, α-tocopherol, water and n-butanol also indicates a pseudo first-order process with respect to O_2^- (see Figure 2). In our view this is not due to the presence of catalytic impurities, but indicates that the transfer of protons is rate limiting in aprotic solvents. Even in aqueous alkaline media the rate of proton transfer is so rapid that it cannot be measured.[89]

Under proton-limited conditions, HO_2^- is an initial dismutation product. It is extremely unstable in aprotic solvents and reacts with H_2O_2 to yield O_2^-, $^{\bullet}OH$, and H_2O (Ref. 78):

$$H_2O_2 + HO_2^- \rightarrow O_2^- + {}^{\bullet}OH + H_2O \quad (18)$$

This reaction complicates the kinetics and the reaction profiles (current vs. time) that are obtained with the rotating disc electrode. Moreover, the relative acidities of HO_2^- (pK_a = 11.75) and $HO_2^{\bullet}$ (pK_a = 4.88) indicate that the reaction:

$$HO_2^- + HO_2^{\bullet} \rightarrow H_2O_2 + O_2^- \quad (19)$$

is favored and will further complicate the kinetics after the initial stage. The rate law for the proton-induced decay of O_2^- in aprotic solvents can be developed from a set of sequential reactions:

$$HA + O_2^- \underset{k_{-20}}{\overset{k_{20}}{\rightleftharpoons}} HO_2^{\bullet} + A^- \quad (20)$$

$$HO_2^{\bullet} + O_2^- \xrightarrow{k_{21} = 9\times10^7 M^{-1}s^{-1}} HO_2^- + O_2 \quad (21)$$

$$HO_2^- + HA \xrightarrow{k_{22}} H_2O_2 + A^- \qquad (22)$$

$$HO_2^- + H_2O_2 \xrightarrow{k_{23}} O_2^- + {}^{\bullet}OH + H_2O \qquad (23)$$

$$HO_2^- + HO_2^{\bullet} \xrightarrow{k_{24}} H_2O_2 + O_2^- \qquad (24)$$

The rate of loss for $[O_2^-]$ is given by

$$\frac{-d[O_2^-]}{dt} = k_{20}[O_2^-][HA] + k_{21}[HO_2^{\bullet}][O_2^-] - k_{-20}[HO_2^{\bullet}][A^-] - k_{23}[HO_2^-][H_2O_2] - k_{24}[HO_2^-][HO_2^{\bullet}] \qquad (25)$$

Invoking the steady-state approximation for $[HO_2^{\bullet}]$ gives

$$\frac{d[HO_2^{\bullet}]}{dt} = 0 = k_{20}[HA][O_2^-] - k_{-20}[HO_2^{\bullet}][A^-] - k_{21}[HO_2^{\bullet}][O_2^-] - k_{24}[HO_2^-][HO_2^{\bullet}] \qquad (26)$$

which can be solved for $[HO_2^{\bullet}]$

$$[HO_2^{\bullet}] = \frac{k_{20}[HA][O_2^-]}{k_{-20}[A^-] + k_{21}[O_2^-] + k_{24}[HO_2^-]} \qquad (27)$$

Substitution into eq. 25 yields:

$$\frac{-d[O_2^-]}{dt} = \frac{\{2k_{20}k_{21}[HA][O_2^-]^2 - k_{23}k_{24}[H_2O_2][HO_2^-]^2 - k_{23}k_{21}[H_2O_2][HO_2^-][O_2^-] - k_{23}k_{-20}[H_2O_2][HO_2^-][A^-]\}}{k_{-20}[A^-] + k_{21}[O_2^-] + k_{24}[HO_2^-]} \qquad (28)$$

Because $[A^-] = [HO_2^-] = [H_2O_2] = 0$ at the beginning of the reaction, the last three terms in the numerator and the first and third terms in the denominator of eq. 28 can be neglected. Thus, the rate-law for the initial stage of the reaction can be expressed as

$$\frac{-d[O_2^-]}{dt} = 2k_{20}[HA][O_2^-] \qquad (29)$$

Eq.28 indicates the complexity of the overall kinetics for the loss of O_2^- concentration that is induced by protic reagents in DMF. Measurement of initial rates from data such as illustrated by Figure 2 provides the means to evaluate k_{-20}; the values are summarized in Table 2, part B. Evidently the rate determining step for the proton-induced dismutation of O_2^- in DMF is reaction (20), and is due to the much slower rate of proton transfer in this solvent. The data of Table 2, part B indicate that the rate of O_2^- dismutation in DMF is roughly proportional to substrate acidity (and rate of proton transfer). In water, the rate of proton transfer is extremely rapid and the rate determining step is reaction (21) with $k_{21} = 9 \times 10^7 M^{-1}s^{-1}$ (Ref. 88). We envisage that reaction (21) involves a triplet transition state complex with O_2^- acting as an electron-pair ligand (Ref. 2):

$$O_2^- + HO_2^{\cdot} \rightarrow (^{\cdot}O_2\text{-H-}O_2^{\cdot})^-_{tr} \rightarrow {}^{\cdot}O_2^{\cdot} + HO_2^-$$

The reduction of O_2 in the presence of the metal ions Mn^{2+}, Co^{2+}, Fe^{2+}, Zn^{2+} (Ref. 4), Cd^{2+} (Ref. 4) and Ni^{2+} (Lewis acids) occurs at more positive potentials than in their absence (Table 3) and results in either a two-or four-electron overall reduction. As with protic substrates the first O_2 reduction wave is irreversible and excess metal ion must be added before the second O_2 reduction wave is suppressed. These observations indicate that O_2 reduction is dependent upon the Lewis acid and that either two- or four-electron reductions of O_2 occur via dismutation reactions after the initial one-electron reduction to O_2^-. As with protons, the O_2^- acts as an electron-pair donor (Lewis base) to the metal ion. In fact, some of the earliest electrochemical studies of O_2^- in aprotic media have demonstrated this with protons, alkali metal ions, and alkaline earth ions.[2,4,90] These superoxy complexes are only transiently stable and decompose to give O_2 and the corresponding peroxide. However, Li^+ ion appears to form a stable $Li(:O_2^{\cdot})$ complex (Table 3). ESR and kinetic measurements demonstrate that O_2^- forms coordination complexes with Ca(II) and Ba(II) in aqueous media as well as DMSO solutions.[91]

The results of Table 3 indicate that O_2^- also acts as an electron-pair donor for Zn(II), Cd(II), Mn(II), Fe(II), Co(II) and Ni(II). When Zn^{2+} ion and O_2^- are combined in aprotic solvents, a two-step reaction occurs with an initial transiently stable bis-coordination complex that decomposes to give O_2 and zinc peroxide.[4,90] Again, we envisage this to involve a triplet transition-state complex

$$O_2^- + Zn^{2+} \rightarrow Zn(:O_2^{\cdot})^+$$

$$O_2^- + Zn(:O_2^{\cdot})^+ \rightarrow (^{\cdot}O_2:)Zn(:O_2^{\cdot})_{tr} \rightarrow ZnO_2 + {}^{\cdot}O_2^{\cdot}$$

Analogous mechanisms are proposed for Cd(II), Fe(II), Mn(II), Co(II) and Ni(II). A recent report[92] indicates that O_2^- produced by pulse radiolysis in water forms a transiently stable electron-pair donor complex with Mn(II). The combination of electrogenerated O_2^- in acetonitrile with $Mn^{II}(ClO_4)_2$ produces similar results. The complex decomposes to yield O_2 and Mn(III). Based on the overall electron stoichiometry (Table 3) the peroxides of Mn(II) and Co(II) are unstable and dismutate. A possible mechanism for the decomposition of O_2^- by Mn(II) or Co(II) can be represented by a sequence of dismutation reactions.

$$Mn^{2+} + O_2^- \rightarrow Mn^{II}(:O_2^{\cdot})^+ \quad (30)$$

$$Mn^{II}(:O_2^{\cdot})^{+} + O_2^{-} \rightarrow (^{\cdot}O_2:Mn^{II}:O_2^{\cdot}) \rightarrow O_2 + Mn^{II}O_2 \tag{31}$$

$$Mn^{II}O_2 + Mn^{II}O_2 \rightarrow 2Mn^{II}O + O_2 \tag{32}$$

$$Mn^{II} + O_2 \rightarrow Mn^{III}(:O_2^{\cdot})^{2+} \tag{33}$$

$$Mn^{III}(:O_2^{\cdot})^{2+} + Mn^{II} \rightarrow (Mn^{III}\text{-O-O-}Mn^{III})^{4+} \tag{34}$$

$$(Mn^{III}\text{-O-O-}Mn^{III})^{4+} + Mn^{II} \rightarrow (Mn^{III}\text{-O-}Mn^{III})^{4+} + (Mn^{III}O\cdot)^{2+} \tag{35}$$

$$(Mn^{III}O\cdot)^{2+} + Mn^{II} \rightarrow (Mn^{III}\text{-O-}Mn^{III})^{4+} \tag{36}$$

$$(Mn^{III}O\cdot)^{2+} + (Mn^{III}\text{-O-O-}Mn^{III})^{4+} \rightarrow (Mn^{III}\text{-O-}Mn^{III})^{4+} + (Mn^{III}(:O_2^{\cdot}))^{2+} \tag{37}$$

Superoxide as one-electron reducing agent. The voltammetric data of Figure 3 clearly indicate that O_2 is evolved when O_2^- is combined with a quinone (DTBQ) and that the major product is 3,5-di-tert-butyl-o-semiquinone anion radical (λ_{max} at 383 nm, ε = 2,250 M^{-1} cm^{-1}) (Ref. 58). The relevant redox potentials in DMF are:

$$O_2 + e^- \rightarrow O_2^- \qquad E^{o\prime} = -0.62 \text{ V } \underline{vs} \text{ NHE} \tag{38}$$

$$DTBQ + e^- \rightarrow DTBQ^{\overline{\cdot}} \qquad E^{o\prime} = -0.25 \text{ V } \underline{vs} \text{ NHE} \tag{39}$$

which indicate that the equilibrium constant for the reaction of O_2^- with DTBQ,

$$O_2^- + DTBQ \rightleftharpoons DTBQ^{\overline{\cdot}} + O_2, \tag{40}$$

has a value of 2.8×10^6. Analogous results are obtained for the reduction of p-benzoquinone by O_2^- in DMF, with a stoichiometric yield of the semiquinone anion radical on the basis of the ESR spectrum.[43]

Previous studies of sulfur dioxide[59-65] and molecular oxygen,[2,67] as well as the data in Figure 4, trace a) indicate that O_2^- has a sufficiently negative potential (eq. 38) to reduce SO_2 to a limited extent:

$$SO_2 + e^- \rightleftharpoons SO_2^{\overline{\cdot}} \qquad E' = -0.58 \text{ V } \underline{vs} \text{ NHE} \tag{41}$$

On the basis of these redox potentials the equilibrium constant of the reaction of SO_2 with O_2^-,

$$SO_2 + O_2^- \rightleftharpoons O_2 + SO_2^{\overline{\cdot}}, \tag{42}$$

has a value of 4.8. However, the reaction goes to completion (see Figure 4, trace c)) because of complexation of $SO_2^{\overline{\cdot}}$ by SO_2 and its dimerization to dithionite ion.[63-66]

In summary, electrochemical, spectroscopic, and chemical evidence confirms that O_2^- is a stronger and more effective reducing agent than dithionite ion in aprotic solvents. This capacity of O_2^- to act as a strong one-electron reducing agent ($E^{o\prime}$ = -0.57 V vs NHE) may represent its most serious biological hazard.

For example, the reduction of ferricytochrome c by O_2^-,

$$\text{cyt c}(Fe^{3+}) + O_2^- \rightleftharpoons \text{cyt c}(Fe^{2+}) + O_2, \quad (43)$$

has an equilibrium constant with a value of 3.67 x 10^4 in aqueous media.[68]

Formation of singlet oxygen from superoxide oxidation. The results of photon counting experiments indicate that both proton induced dismutation ($HClO_4$ + O_2^-) and transition-metal ion oxidation ($Fe^{III}(ClO_4)_3$ + O_2^-) of O_2^- yield predominantly triplet oxygen.[72] This is believed to be the result of a triplet transition state complex for the dismutation process, $((\cdot O_2{:})H({:}O_2^{\cdot}))_{tr}$, and a triplet state for the redox reaction $(\cdot Fe^{III}({:}O_2^{\cdot}))_{tr}$. Wavefunctions for the ground state of O_2^- have been calculated by use of unrestricted open shell LCAO-SCF functions that are based on the INDO formalism.[72] The calculations reveal that removal of the highest energy electron from O_2^- will produce a triplet state. The calculations also show that a triplet transition state leads to triplet O_2, while a singlet transition state leads to $^1\Delta g$ O_2. Because these reactions yield $^3O_2$, triplet transition state complexes are plausible.

The photon counting experiments for the oxidation of O_2^- by ferricinium perchlorate indicate a significant yield of $^1O_2$. While both $Fe^{III}(C\ell O_4)_3$ and ferricinium ion are thermodynamically strong enough one-electron oxidants ($E^{o\prime}$> +0.40 V vs NHE) to oxidize O_2^- to $^1O_2$, only ferricinium ion does so. This can be rationalized on the basis that ferricinium ion is coordinately saturated and may form a singlet transition state complex with superoxide, $((cp)_2Fe\text{-}{}^{\cdot}O_2{:})^{2+}_{tr}$.

Oxidation of acidic substrates by superoxide ion. Combination of O_2^- with protic substrates (α-tocopherol, L-(+)ascorbic acid, hydroquinone, and 3,5,-di-tert-butylcatechol) yields products that are consistent with an apparent one-electron oxidation of the substrate and the production of H_2O_2. Previously, either direct electron-transfer[26,42,43] or proton induced dismutation followed by O_2 oxidation of substrate anion appeared equally viable. However, the present studies provide clear evidence for the dismutation mechanism and appear to clearly rule out direct electron transfer to O_2^- from the substrate via a "cage" mechanism, even as a competitive pathway. Thus, by continuously purging the O_2 that results from dismutation of O_2^- from the solution with argon, quantitative yields of substrate anion are obtained without significant oxidation. This is especially evident for the reaction of O_2^- with 3,5-di-tert-butylcatechol in DMF (Figure 5) and with α-tocopherol in DMSO. Furthermore, the presence of residual O_2, O_2^-, substrate anion, and semiquinone anion in the reaction solutions (Figure 5, trace a)) is inconsistent with a direct electron-transfer mechanism. These components must result from an initial proton-induced dismutation of O_2^-.

The reported oxidation of hydrazines,[45,46,48] thiols,[29,47] and alcohols[29] by O_2^- also presumably involve initial proton transfer to O_2^- and dismutation prior to the oxidation process (dialkyl sulfides, ethers, and alkoxides are inert).

Furthermore, the oxidation of hydrazines by O_2^- has been shown to be catalyzed by protons or metal cations.

Proposed mechanisms for the reaction of O_2^- with acidic substrates are outlined by eqs. 44–55, where (HA) represents α-tocopherol or ascorbate anion and (H_2B) hydroquinone, 3,5-di-tert-butylcatechol, or L-(+)ascorbic acid.

For HA:

$$HA + O_2^- \underset{k_{-44}}{\overset{k_{44}}{\rightleftharpoons}} HO_2^{\bullet} + A^- \tag{44}$$

$$HO_2^{\bullet} + O_2^- \xrightarrow{k_{45} = 10^8 M^{-1}s^{-1}} O_2 + HO_2^- \tag{45}$$

$$HA + HO_2^- \underset{k_{-46}}{\overset{k_{46}}{\rightleftharpoons}} A^- + H_2O_2 \tag{46}$$

$$A^- + O_2 \xrightarrow{k_{47}} A^{\bullet} + O_2^- \tag{47}$$

$$n\,A^{\bullet} \xrightarrow[(O_2,\ O_2^-,\ H_2O_2)]{k_{48}} \text{dimeric, dismutation, and redox products} \tag{48}$$

For H_2B:

$$H_2B + O_2^- \underset{k_{-49}}{\overset{k_{49}}{\rightleftharpoons}} HB^- + HO_2^{\bullet} \tag{49}$$

$$HO_2^{\bullet} + O_2^- \xrightarrow{k_{50} = 10^8 M^{-1}s^{-1}} O_2 + HO_2^- \tag{50}$$

$$H_2B + HO_2^- \underset{k_{-51}}{\overset{k_{51}}{\rightleftharpoons}} HB^- + H_2O_2 \tag{51}$$

$$2HB^- \underset{k_{-52}}{\overset{k_{52}}{\rightleftharpoons}} H_2B + B^{2-} \tag{52}$$

$$B^{2-} + O_2 \xrightarrow{k_{53}} B^{\bar{\bullet}} + O_2^- \tag{53}$$

$$B^{\bar{\bullet}} + H_2B \longrightarrow HB^{\bullet} + HB^- \tag{54}$$

$$n\,HB^{\bullet} \xrightarrow[(O_2,\ O_2^-,\ H_2O_2)]{k_{55}} \text{dimeric, dismutation, and redox products} \tag{55}$$

When O_2^- is combined with diprotic substrate (H_2B) such as catechol or ascorbic acid, several additional reactions are likely:

$$HB^- + O_2 \underset{k_{-56}}{\overset{k_{56}}{\rightleftharpoons}} HB^{\bullet} + O_2^- \tag{56}$$

$$2HB^{\bullet} \underset{k_{-57}}{\overset{k_{57}}{\rightleftharpoons}} H_2B + B \tag{57}$$

$$B + O_2^- \xrightarrow{k_{58}} B^{\cdot} + O_2 \quad (58)$$

If reactions (49) through (53) are dominant, then the overall stoichiometry becomes

$$O_2^- + H_2B \rightarrow B^{\bar{\cdot}} + H_2O_2 \quad (59)$$

If reactions (49) through (54) plus (56), (57) and (58) are dominant, then the overall stoichiometry can be represented by

$$3O_2^- + 3H_2B \rightleftharpoons B^{\bar{\cdot}} + 2HB^- + 2H_2O_2 + O_2 \quad (60)$$

The initial rate determining step is proton-transfer from substrate to O_2^- (eqs. 44 and 49). The initial rate data for phenol and NH_4^+ in DMF are approximately first-order in O_2^- and in HA (or H_2B), as illustrated by Figure 2. The observed overall second-order rate constants are equal to $2k_{44}$ (or $2k_{49}$) and have the values summarized in Table 2, part B.

Reference to Table 2, part B indicates that the value of k_{44} for α-tocopherol is an order of magnitude lower than the value for phenol. This is consistent with the expected lower acidity of the hydroxyl proton of α-tocopherol that results from its greater electron density in the aromatic ring. Because of this, E^o_{ox} for α-tocopherol anion is 0.5 V more negative than E^o_{ox} for phenoxide ion in DMF. The alkyl substituents on the ring also should enhance free radical formation.[93] For these reasons O_2 oxidizes α-tocopherol anion but not phenoxide ion. Further work has shown that the observed stability of α-tocopherol anion in MeOH toward O_2 oxidation is due to a positive shift of 0.5 V in the E^o_{ox} for the anion in this solvent.

In the case of L-(+)ascorbic acid, corresponding experiments with D-glucitol and D-glucono-δ-lactone indicate that they react with O_2^- at a rate similar to that observed for n-butanol (Table 2, part B). Hence, the enhanced reactivity for L-(+)ascorbic acid is due to the higher acidity of its α and β protons (in H_2O pK_1 = 4.76 and pK_2 = 11.56) (Ref.84), relative to aliphatic hydroxyl protons (for n-butanol in H_2O pK_a = 18). The rate of oxidation of L-(+)ascorbic acid also may be enhanced by its ability to form relatively stable free radicals,[94,95] particularly when complexed to metal ions, e.g. Tl(III) (Ref. 95) and Cu(II) (Ref. 96).

Reference to Table 2, part B confirms that the initial-rate constants for the reaction of O_2^- with protic substrates correlates with their acidity. The rate constant for reaction of L-(+)ascorbic acid with O_2^- has not been determined, but preliminary work indicates that it is much greater than those for phenol and NH_4^+. The initial rate-constants follow the order: L-(+)ascorbic acid > phenol ~ NH_4^+ > ascorbate monoanion > α-tocopherol > H_2O > n-butanol ~ D-glucono-δ-lactone ~ D-glucitol.

In summary, pseudo first-order kinetic behavior for the initial rate of reaction of O_2^- with protic substrates indicates that the primary step is proton transfer from the substrate to O_2^- and that its rate is directly related to the acidity of the substrate. The rate of subsequent oxidation is governed by the equilibrium constant for reaction of the substrate anion with O_2 and is, therefore, related to the E^o_{ox} value for the anion and the stability of the radical product.

AKNOWLEDGMENT

This work was supported by the U.S.D.A. Forest Service Forest Products Laboratory, Madison, Wisconsin, under Agreement No. 12-75. We thank Mary Minick, NSF-URP Fellow, for her assistance with the kinetic measurements.

REFERENCES

1. Maricle, D.L. and Hodgson, W.G. (1965) Anal. Chem. 37, 1562-1565.
2. Sawyer, D.T. and Roberts, J.L., Jr., (1966) J. Electroanal. Chem. 12, 90-101.
3. Peover, M.E. and White, B.S. (1966) Electrochim. Acta, 11, 1061-1067.
4. Johnson, E.J., Pool, K.H. and Hamm, R.E. (1966) Anal. Chem. 38, 183-185.
5. Goolsby, A.D. and Sawyer, D.T. (1968) Anal. Chem. 40, 83-86.
6. Bauer, D. and Beck, J.P. (1972) J. Electroanal. Chem. Interfacial Electrochem. 40, 233-254.
7. Fee, J.A. and Hildenbrand, P.G. (1974) FEBS Lett. 39, 79-82.
8. (a) McCord, J.M. and Fridovich, I. (1968) J. Biol. Chem. 243, 5753-5760. (b) Fridovich, I. (1972) Account. Chem. Res. 5, 321-326.
9. McCord, J.M. and Fridovich, I. (1969) J. Biol. Chem. 244, 6049-6055.
10. Mann, T. and Keilin, D. (1938) Proc. Roy. Soc. London, B126, 303-308.
11. Keele, B.B., Jr., McCord, J.M. and Fridovich, I. (1970) J. Biol. Chem. 245, 6176-6181.
12. Yost, F.J. and Fridovich, I. (1973) J. Biol. Chem. 248, 4905-4908.
13. Vaska, L. (1976) Account. Chem. Res. 9, 175-191.
14. Andrews, L. (1968) J. Amer. Chem. Soc. 90, 7368-7370.
15. Evans, J.C. (1969) J. Chem. Soc. Chem. Commun., 682-683.
16. Ozawa, T., Hanaki, A. and Yamamoto, H. (1977) FEBS Lett. 74, 99-102.
17. Kim, S., DiCosino, R. and San Filippo, J., Jr., (1979) Anal. Chem. 31, 679-681.
18. Parsons, R. (1959) Handbook of Electrochemical Constants, Buttersworths, London.
19. Fee, J.A. and Valentine, J.S. (1977) in Superoxide and Superoxide Dismutases, Michelson, A., McCord, J.M. and Fridovich, I., eds., Academic Press, New York, pp. 19-60.
20. Ilan, Y.A., Meisel, D. and Czapski, G. (1974) Israel J. Chem. 12, 891-895.
21. Behar, D., Czapski, G., Rabani, J., Dorfman, L.M. and Schwartz, H.A. (1970) J. Phys. Chem. 74, 3209-3213.
22. Dorfman, L.M. and Adams, G.E. (1973) Reactivity of the Hydroxyl Radical in Aqueous Solutions, NSRDS-NBS 46 (SD Catalog No. C13. 48:46), U.S. Government Printing Office, Washington, D.C., pp. 35, 38, 39.
23. Tezuka, M. Hamada, H., Ohkatsu, Y. and Osa, T. (1976) Denki Kogaku, 44, 17-21.
24. Ilan, Y.A., Czapski, G. and Meisel, D. (1976) Biochim. Biophys. Acta, 430, 209-224.
25. Sawyer, D.T. and Gibian, M.J. (1979) Tetrahedron, 35, 1471-1481.
26. Sawyer, D.T., Gibian, M.J., Morrison, M.M. and Seo, E.T. (1978) J. Amer. Chem. Soc. 100, 627-628.
27. Gibian, M.J. Sawyer, D.T., Ungermann, T., Tangpoonphlvivat, R. and Morrison, M.M. (1979) J. Amer. Chem. Soc. 101, 640-644.
28. Merritt, M.V. and Sawyer, D.T. (1970) J. Org. Chem. 35, 2157-2159.
29. Dietz, R., Forno, A.EJ., Larcombe, B.E. and Peover, M.E. (1970) J. Chem. Soc. B, 816-820.
30. San Fillipo, J., Jr., Chern, C.I. and Valentine, J.S. (1975) J. Org. Chem. 40, 1678-1680.

31. Johnson, R.A. and Nidy, E.S. (1975) J. Org. Chem. 40, 1680-1681.
32. San Fillip, J. Jr., Ramano, L.J., Chern, C.I. and Valentine, J.S. (1976) J. Org. Chem. 41, 586-588.
33. Magno, F. and Bontempelli, G. (1976) J. Electroanal. Chem. 68, 337-344.
34. Johnson, R.A. (1976) Tetrahedron Lett., 331-332.
35. Stallings, M.D. and Sawyer, D.T. (1979) J. Chem. Soc. Chem. Commun., 340-341.
36. (a) Roupko, R. and Rosenthal, I. (1973) J. Phys. Chem. 77, 1722-1724.
(b) Rosenthal, I. and Frimer, A. (1976) Tetrahedron Lett., 2805-2809.
37. Valentine, J.S. and Curtis, A.B. (1976) J. Amer. Chem. Soc. 97, 224-226.
38. Klug-Roth, D. and Rabani, J. (1976) J. Phys. Chem. 80, 588-591.
39. Valentine, J.S. and Quinn, A.E. (1976) Inorg. Chem. 15, 1997-1999.
40. Howie, J.K., Morrison, M.M. and Sawyer, D.T. (1977) ACS Symp. Ser. 38, 97.
41. Halliwell, B. (1975) FEBS Lett. 56, 34-38.
42. Moro-oka, Y. and Foote, C.S. (1976) J. Amer. Chem. Soc. 98, 1510-1514.
43. Lee-Ruff, E., Lever, A.B.P. and Rigaudy, J. (1976) Can. J. Chem. 54, 1837-1840.
44. Afanes'ev, I.B. and Polozova, H.I. (1976) Zh. Org. Khim. SSSR, 12, 1833.
45. Misra, H.P. and Fridovich, I. (1976) Biochemistry, 15, 681-687.
46. Chern, C.V. and San Fillipo, J., Jr., (1977) J. Org. Chem. 42, 178-180.
47. Asada, K. and Kanematsu, S. (1976) Agri. Biol. Chem. Tokyo, 40, 1891-1892.
48. Nishikimi, M. and Machlin, L.J. (1975) Arch. Biochem. Biophys. 170, 684-689.
49. Matsumoto, S. and Matsuo, M. (1977) Tetrahedron Lett., 1999-2006.
50. Nishikimi, N. (1975) Biochem. Biophys. Res. Commun., 63, 463-468.
51. Martin, R.P. and Sawyer, D.T. (1972) Inorg. Chem. 11, 2644-2646.
52. Goolsby, A.D. and Sawyer, D.T. (1967) Anal. Chem. 39, 411-414.
53. Sawyer, D.T. and Roberts, J.L., Jr., (1974) Experimental Electrochemistry for Chemists, Wiley-Interscience, New York.
54. Czapski, G. (1971) Ann. Rev. Phys. Chem. 22, 171-208.
55. Bielski, B.H.J. and Allen, A.O. (1977) J. Phys. Chem. 81, 1048-1054.
56. Piekarski, S. and Adams, R.N. (1971) in Physical Methods of Chemistry, Part IIA Electrochemical Methods, Weissberger, A. and Rossiter, B.W., eds., Wiley-Interscience, New York, pp. 567.
57. EPR at Work, No. 28, Varian Associates Analytical Instruments, Palo Alto, California.
58. Stallings, M.D. and Sawyer, D.T. (1978) unpublished results.
59. Rinker, R.G. and Lynn, S. (1969) Ind. Eng. Chem. 8, 339-341.
60. Dinse, K.P. and Mobius, K. (1962) Z. Naturforsch. 23a, 695.
61. Bonnaterre, R. and Cauquis, G. (1971) J. Electroanal. Chem. 32, 215-218.
62. Martin, R.P. (1973), Ph.D. Dissertation, University of California, Riverside.
63. Magno, F., Mazzochin, G.A. and Bontempelli, G. (1974) J. Electroanal. Chem. Interfacial Electrochem. 57, 89-96.
64. James, J. and Hambright, P. (1969) J. Coord. Chem. 8, 339.
65. Rinker, R.G. and Lynn, S. (1968) J. Phys. Chem. 72, 4706-4707.
66. Nishinaga, A., Itahara, T. and Matsuura, T. (1974) Bull. Chem. Soc. Japan, 47, 1811-1812.
67. Lambeth, D.O. and Palmer, G. (1973) J. Biol. Chem. 248, 6095-6101.
68. Rao, R.S. and Hayon, E. (1975) J. Phys. Chem. 79, 397-402.
69. Mayeda, E.A. and Bard, A.J. (1973) J. Amer. Chem. Soc. 95, 6223-6226.
70. Dauen, W.C. and Arudi, R.S. (1978) J. Amer. Chem. Soc. 100, 3944-3945.
71. Mayeda, E.A. and Bard, A.J. (1974) J. Amer. Chem. Soc. 96, 4023-4024.
72. Birge, R.R., Hubbard, L.M., Morrison, M.M. and Sawyer, D.T. (1979) submitted for publication.
73. Guiraud, H.J. and Foote, C.A. (1976) J. Amer. Chem. Soc. 98, 1984-1986.
74. Skinner, W.A. and Alaupovic, P. (1963) J. Org. Chem. 28, 2854-2858.
75. Skinner, W.A. (1964) Biochem. Biophys. Res. Commun. 15, 469-472.
76. Nelson, D.R. and Robeson, C.D. (1962) J. Amer. Chem. Soc. 84, 2963-2965.
77. Nilsson, J.L.G., Daves, G.D. and Folers, K. (1968) Acta. Chem. Scand. 22, 207-218.

78. Roberts, J.L., Jr., Morrison, M.M. and Sawyer, D.T. (1978) J. Amer. Chem. Soc. 100, 329-330.
79. Weissberger, A., LuValle, J.E. and Thomas, O.S., Jr., (1943) J. Amer. Chem. Soc. 65, 1934-1939.
80. Kenyon, J. and Munro, J. (1948) J. Chem. Soc. 158-161.
81. Bray, R.C. (1961) J. Biochem. 81, 189-195.
82. Michelson, A.M., McCord, J.M. and Fridovich, I. (1977) Superoxide and Superoxide Dismutases, Academic Press, New York.
83. See Ref. 9.
84. The Merck Index, 9th ed. (1976), Merck and Co., Rahway, New Jersey.
85. George, P. (1965) in Oxidases and Related Redox Systems, King, T.E., Mason, M.S. and Morrison, M., eds., John Wiley and Sons, New York, pp. 3-36.
86. Rabani, J. and Nielsen, S.O. (1969) J. Phys. Chem. 73, 3736-3744.
87. Bielski, B.H.J. and Chan, P.C. (1976) J. Biol. Chem. 251, 3841-3844.
88. See Ref. 55.
89. Eigen, M. and DeMaeyer, L. (1958) Proc. Roy. Soc. London, 247A, 505-515.
90. Johnson, E.J., Pool, K.H. and Hamm, R.E. (1969) Anal. Chem. 39, 888-891.
91. Bray, R.C., Mautner, G.N., Fielden, E.M. and Carle, C.J. (1977) in Superoxide and Superoxide Dismutases, Michelson, A.M., McCord, J.M. and Fridovich, I., eds., Academic Press, New York, pp. 61-75.
92. Bielski, B.H.J. and Chan, P.C. (1978) J. Amer. Chem. Soc. 100, 1920-1921.
93. Svanholm, U., Bechgaard, K. and Parker, V.D. (1974) J. Amer. Chem. Soc. 96, 2409-2413.
94. Green, R.C. and O'Brien, P.J. Biochim. Biophys. Acta, 293, 334-338.
95. Stegmann, H.B.,Scheffler, K. and Schuler, P. (1978) Angew. Chem. Int. Ed. 17, 365.
96. Weissberger, A. and LuValle, J.E. (1944) J. Amer. Chem. Soc. 66, 700-705.

FENTON'S REAGENT - ONCE MORE REVISITED

GERALD COHEN[+] AND PIERRE M. SINET[++]
+ Department of Neurology, Mount Sinai School of Medicine of the City University of New York, New York, USA; ++ Laboratoire Biochemie Genetique, Hôpital Necker, Paris, France

The title of this article has been borrowed from Cheves Walling's[1] review "Fenton's Reagent Revisited". Walling explored aspects of the Fenton reaction from an organic chemist's point of view. In this article, we explore aspects of the Fenton reaction primarily from a biologist's point of view.

The Fenton reaction is shown in equation (1):

$$Fe^{2+} + H_2O_2 \longrightarrow Fe^{3+} + OH^- + {}^{\bullet}OH \qquad (1)$$

Ever since the observations of Fenton[2,3] concerning the oxidation of organic substrates by mixtures of H_2O_2 and reduced iron, the Fenton reagent has attained a position of prominence and great interest was evoked in the reaction mechanism. In the 1930's, the radical ($OH^{\bullet}$; equation (1)) was the reactive species mediating the oxidation of organic substrates. Today, this concept is widely accepted. Recent electron spin resonance studies with spin-trapping agents[5] have verified the presence of OH in reaction (1).

In general, most studies with the Fenton reagent have been concerned with relatively concentrated solutions of Fe^{2+} and H_2O_2. This is true both for studies with added organic substrates and for studies of the iron-catalyzed decomposition of H_2O_2 in the absence of substrates. In general, experiments have been carried out at acid pH (pH 3 or less) in order to avoid precipitation of ferric hydroxide. It has been shown that certain iron chelates, such as iron-EDTA, can participate in Fenton-type reactions[1,6] as follows:

$$Fe^{2+}\text{-chelate} + H_2O_2 \longrightarrow Fe^{3+}\text{-chelate} + OH^- + OH^{\bullet} \qquad (2)$$

A special value to the use of iron chelates is that they are more soluble than free ferrous ions at or near neutral pH. A second value from a biologist's point of view is that they may serve as models for biological reactions: Iron within cells is not "free", but is chelated with ubiquitous cellular substituents (e.g., amino acids) or specific proteins (e.g., cytochrome P-450).

For the purpose of our experiments, we define a biologic system as one at neutral pH with relatively dilute reactants. Our experiments were carried out at pH 7.0 (50 mM phosphate buffer) with 10-50 μM Fe^{2+} and 10 μM - 1.0 mM H_2O_2. Two chelating agents were used, namely, ethylenediaminetetraacetic acid (EDTA) and diethylenetriaminepentaacetic acid (DTPA or DETAPAC).

We wish to make two main points about the Fenton reaction as it may occur in biologic systems:

(1). The addition of ferrous chelates or ferrous salts to biologic media will produce both H_2O_2 and $OH^{\cdot}$ from the spontaneous reactions between ferrous iron and oxygen. That is, $OH^{\cdot}$ will be produced even when H_2O_2 is not added or generated by other endogenous sources.

(2). Reaction (2) consumes oxygen. The equation, as written, is not quite correct because it fails to show the simultaneous consumption of oxygen.

Hydroxyl Radical Production by Ferrous Chelates With and Without Added H_2O_2

(A) Reaction Mechanisms. Some of the reactions that will be discussed are shown below:

$$Fe^{2+} + O_2 \longrightarrow Fe^{3+} + O_2^- \quad (3)$$

$$Fe^{2+} + O_2^- + 2H^+ \longrightarrow Fe^{3+} + H_2O_2 \quad (4)$$

$$Fe^{2+} + H_2O_2 \longrightarrow Fe^{3+} + OH^- + OH^{\cdot} \quad (5)$$

$$\text{KTBA} + OH^{\cdot} \longrightarrow CH_2{=}CH_2 + \text{other products} \quad (6)$$

where O_2^- is the superoxide radical anion, and KTBA is 2-keto-4-thiomethylbutyric acid. Reactions (3), (4) and (5) show a sequential oxidation of ferrous ions by oxygen, O_2^- and H_2O_2 to form the $OH^{\cdot}$ radical. Reaction (6) shows a reaction between $OH^{\cdot}$ and KTBA to form ethylene gas as a product. The formation of ethylene from methional or KTBA has gained acceptance in recent years as a simple way to measure the production of $OH^{\cdot}$ radicals.[7,8] We used the reaction to measure $OH^{\cdot}$ radical production duirng the autoxidation of Fe^{2+} and during the reaction of Fe^{2+} with added H_2O_2 (i.e., the Fenton reaction).

(B) Autoxidation Rates for Fe^{2+}-EDTA and Fe^{2+}-DTPA. Preliminary studies evaluated the autoxidation rate for ferrous ions chelated with either EDTA or DTPA at pH 7.0 (Figure 1). The appearance of ferric ions was measured with acidic thiocyanate reagent. We observed that the autoxidation of 100 μM Fe^{2+}-EDTA at pH 7.0 proceeds very rapidly with a half-life of less than 15 seconds. In contrast, Fe^{2+}-DTPA is comparatively stable with half-life in the range of 15 minutes.

(C) Ethylene Production During the Autoxidation of Fe^{2+}-EDTA or Fe^{2+}-DTPA. The autoxidation reaction of 50 μM Fe^{2+}-EDTA or Fe^{2+}-DTPA in the presence of 10 mM KTBA evolved ethylene gas (reaction (6); Tables 1 and 2, line 1). Experiments were carried out in 25 ml Erlenmeyer flasks that were sealed with pliable rubber stoppers; the reaction was initiated by injection of ferrous ammonium sulfate through the stopper and samples of gas phase were taken sequentially for analysis. The reaction of Fe^{2+}-EDTA appeared to be relatively rapid (Table 1, line 1); any apparent difference between the rate of autoxidation (Figure 1) and the rate of appearance of ethylene in the gas phase can probably be attributed

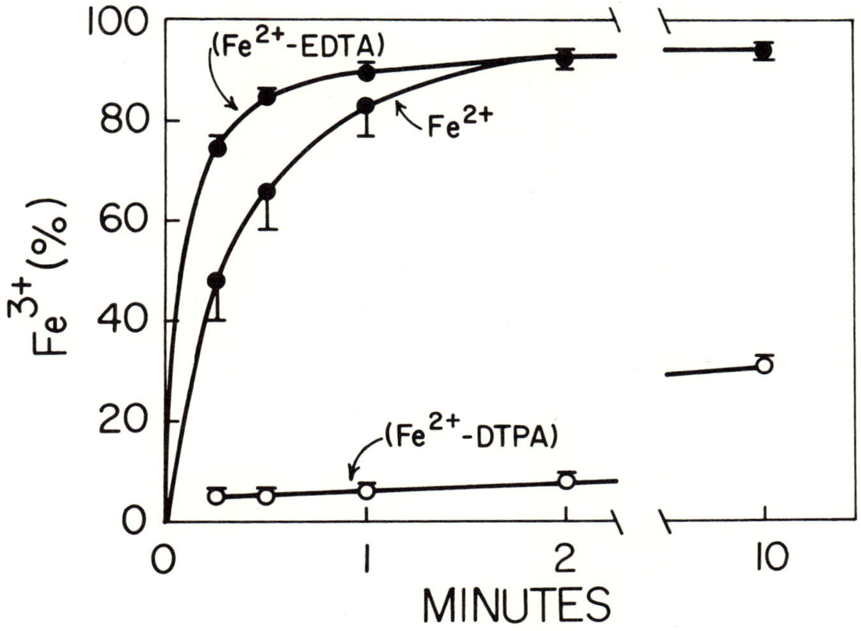

Fig. 1. Autoxidation rates for 100 µM Fe^{2+}-EDTA, Fe^{2+}-DTPA and Fe^{2+} (viz., Fe^{2+}-phosphate) in 50 mM potassium phosphate buffer at pH 7.0. Reactions were initiated by the addition of a stock solution of 2M $Fe(NH_4)(SO_4)$ to the reaction medium which contained twice the molar quantity of chelating agent. The appearance of Fe^{3+} was measured with acidic thiocyanate reagent.

to the time required for distribution of ethylene between the aqueous and gas phases. The rate of ethylene production by Fe^{2+}-DPTA was considerably slower (Table 2, line 1) and compared favourably to the very much slower rate of autoxidation of Fe^{2+}-DPTA (Figure 1).

The ethylene-forming reaction was blocked when helium gas was substituted for air in the flasks (Tables 1 and 2, line 2). Therefore, the formation of ethylene cannot be attributed to a reaction between Fe^{2+} and KBTA. No ethylene was formed when ferric sulphate was injected in place of the ferrous salt (not shown); therefore Fe^{3+} was unreactive in the system. Moreover, when the addition of KTBA was delayed for several minutes to allow for complete oxidation of Fe^{2+} to Fe^{3+}, no ethylene was formed. It can be concluded that ethylene formation requires the reduction of oxygen by the ferrous chelate. Catalase blocked production of ethylene (Tables 1 and 2, line 3); therefore, a requirement for H_2O_2 is established. Boiled catalase was without effect. Superoxide dismutase did not have any major action on the production of ethylene (Tables 1 and 2, line 4). The formation of ethylene is best explained by the generation of $OH^{\cdot}$ via a Fenton-type reaction between generated H_2O_2 and the Fe^{2+}-EDTA chelate (equations (5) and (6).

TABLE 1

ETHYLENE FORMATION FROM KTBA DURING THE OXIDATION OF Fe^{2+}-EDTA BY OXYGEN OR HYDROGEN PEROXIDE[a]

Condition	nmoles Ethylene ± S.D. (N) 5 Minutes	20 Minutes
1. Air	31.0 ± 1.3 (4)	44.6 ± 6.0 (5)
2. Helium	0.4 ± 0.3 (7)	2.1 ± 1.1 (7)
3. Air + Catalase	2.4 ± 0.7 (4)	3.9 ± 1.7 (7)
4. Air + SOD	32.1 ± 4.7 (4)	38.3 ± 4.1 (6)
5. Air + 1 mM H_2O_2	84.7 ± 18.6 (3)	89.5 ± 17.2 (6)

[a]Samples consisted of 4 ml of 50 mM potassium phosphate buffer, pH 7.0, containing 50 μM ferrous ammonium sulfate, 100 μM EDTA, and 10 mM KTBA at room temperature in sealed flasks. Where indicated, the following additions were made: catalase (125 μg/ml), superoxide dismutase (SOD, 10 μg/ml).

TABLE 2

ETHYLENE FORMATION FROM KTBA DURING THE OXIDATION OF Fe^{2+}-DTPA BY OXYGEN OR HYDROGEN PEROXIDE[a]

Condition	nmoles Ethylene ± S.D. (N) 5 Minutes	20 Minutes	20 Hours
1. Air	2.6 ± 0.6 (6)	7.9 ± 1.7 (4)	33.9 ± 11.9 (6)
2. Helium	0	0.2 ± 0.2 (6)	3.2 ± 0.3 (4)
3. Air + Catalase	0.9 ± 0.4 (7)	-	2.8 ± 0.5 (7)
4. Air + SOD	2.1 ± 0.5 (4)	6.1 ± 0.7 (4)	30.5 ± 5.2 (6)
5. Air + 1 mM H_2O_2	179 ± 13 (6)	203 ± 22 (4)	153 ± 30 (5)

[a]Experimental conditions are the same as in Table 1 except that DTPA was substituted for EDTA.

(D) Ethylene Production by Fe^{2+}-EDTA or Fe^{2+}-DTPA with excess H_2O_2. A direct study of the Fenton reaction between Fe^{2+}-EDTA or Fe^{2+}-DTPA and added H_2O_2 was also carried out (Tables 1 and 2, line 5). The H_2O_2 was added immediately (within 5 seconds) after the injection of the ferrous ammonium sulfate. H_2O_2 stimulated the production of ethylene (lines 5 compared to lines 1). The reaction appeared essentially complete at the early sampling time (5 minutes). In experiments with Fe^{2+}-DTPA, no greater yields of ethylene were observed at either higher or lower concentrations of added H_2O_2 (not shown). The yield of ethylene from Fe^{2+}-EDTA was doubled by the addition of H_2O_2. It would appear that increased yields would be attained just as readily by adding more Fe^{2+} as by adding H_2O_2. We had expected that the yield of $OH^{\cdot}$ (i.e., ethylene) would be greatly more increased by adding a large excess of H_2O_2. This would have been true if the cyclical reduction of Fe^{3+} to Fe^{2+} (Refs.4,9) had occurred. Our data with bith Fe^{2+}-EDTA and Fe^{2+}-DTPA are in keeping with a relatively fast reaction with H_2O_2, optimizing the yield of $OH^{\cdot}$ based on equation (5) and minimizing a "wasteful" loss of Fe^{2+} equations (3) and (4).

<u>(E) Discussion of Hydroxyl Radical Production by Ferrous Chelates</u>. These studies indicate that both H_2O_2 and $OH^{\cdot}$ are formed during the autoxidation of dilute solutions of Fe^{2+}-EDTA (fast reaction) or Fe^{2+}-DPTA (slow reaction) at physiological pH. The Fenton reactions as measured by ethylene production (equations (2) and (6) for Fe^{2+}-EDTA and Fe^{2+}-DTPA are both relatively fast. The corresponding Fe^{3+}-chelates are unreactive in the system. A number of conclusions seem apparent:

(1) The addition of Fe^{2+}-EDTA to biologic systems, for example, to tissue homogenates or preparations of subcellular organelles, would be expected to evoke oxidative reactions dependent upon the formation of reactive $OH^{\cdot}$ radicals. This action can probably be augmented when small amounts of reducing agent are present to recycle the Fe^{3+} back to Fe^{2+}. It is not clear to what extent such reactions may explain oxidative effects noted for Fe^{2+} or Fe^{2+}-chelates in various reported experiments.

(2) Fe^{2+}-DTPA can provoke formation of $OH^{\cdot}$ radicals in systems where H_2O_2 is added or generated. Although Fe^{2+}-EDTA exerts a similar action, it can appear comparatively less effective as a reactant with added H_2O_2 (compare Tables 1 and 2, lines 1 and 5 at both 5 minutes and 20 minutes). The reason for this effect is the very rapid autoxidation of Fe^{2+}-EDTA, resulting in its transformation to the inactive Fe^{3+}-chelate and, at the same time, producing considerable $OH^{\cdot}$ in the absence of added H_2O_2.

(3) If care is not taken in the preparation of Fe^{2+}-EDTA stock solutions (e.g., preparation in an anaerobic environment), investigators may find that they are working with the ferric-chelate rather than the ferrous-chelate.

<u>(F) Ancillary Studies With Iron-Chelates</u>. Ferrous ions chelated with ADP are rapidly oxidized to Fe^{3+} (not shown). Ferrous ions added to phosphate buffer (i.e., Fe^{2+}-phosphate) also undergo a rapid autoxidation (Figure 1). Although ethylene production was not studied, it seems likely that both Fe^{2+}-ADP and Fe^{2+}-phosphate would behave similarly to Fe^{2+}-EDTA in biologic systems.

In contrast, the Fe^{2+}-chelate with o-phenanthroline was relatively stable to oxidation by either oxygen or H_2O_2. Therefore, the Fe^{2+}-o-phenanthroline chelate does not readily participate in Fenton-type reactions.

<u>The Fenton Reaction: Stoichiometry and Consumption of Oxygen</u>

<u>(A) Reaction Mechanisms</u>. In the absence of added substrate (e.g., KTBA), the reaction sequence for the Fenton reagent is given by equation (5) followed by equation (7):

$$Fe^{2+} + H_2O_2 \longrightarrow Fe^{3+} + OH^- + OH^{\cdot} \quad (5)$$

$$Fe^{2+} + OH^{\cdot} \longrightarrow Fe^{3+} + OH^- \quad (7)$$

$$2Fe^{2+} + H_2O_2 + 2H^+ \longrightarrow 2Fe^{3+} + 2H_2O \quad (8)$$

or

$$2[Fe^{2+}\text{-chelate}] + H_2O_2 + 2H^+ \longrightarrow 2[Fe^{3+}\text{-chelate}] + 2H_2O \quad (9)$$

That is, the $OH^{\bullet}$ radical that is produced (equation (5)) is subsequently consumed by the oxidation of a second molecule of ferrous ion (equation (7)). The sum of equations (5) and (7) written to show formation of water as a product is given by equation (8) or (9).

Similarly, when exogenous H_2O_2 is omitted and the autoxidation of a Fe^{2+}-chelate is studied, the reaction sequence can be given by equations (3) to (5), followed by equation (7). The sum, written to show formation of water, is equation (10) or (11).

$$Fe^{2+} + O_2 \longrightarrow Fe^{3+} + O_2^- \quad (3)$$

$$Fe^{2+} + O_2^- + 2H^+ \longrightarrow Fe^{3+} + H_2O_2 \quad (4)$$

$$Fe^{2+} + H_2O_2 \longrightarrow Fe^{3+} + OH^- + OH^{\bullet} \quad (5)$$

$$Fe^{2+} + OH^{\bullet} \longrightarrow Fe^{3+} + OH^- \quad (7)$$

$$4Fe^{2+} + O_2 + 4H^+ \longrightarrow 4Fe^{3+} + 2H_2O \quad (10)$$

$$4[Fe^{2+}\text{-chelate}] + O_2 + 4H^+ \longrightarrow 4[Fe^{3+}\text{-chelate}] + 2H_2O \quad (11)$$

The experimental predictions that can be made for the consumption of Fe^{2+} based on equation (9) or (11) are:

(1) In the presence of H_2O_2 (i.e., a typical Fenton reaction), two moles of Fe^{2+} should be consumed per mole of H_2O_2 added (see equation (9)).

(2) In the absence of added H_2O_2 (i.e., an autoxidation reaction), four moles of Fe^{2+} should be consumed per mole of oxygen (see equation (11)).

One potential complication would be the continued decomposition of H_2O_2 in the presence of ferric chelates accompanied by recycling of the Fe^{3+} to Fe^{2+} (Refs.4,6,9). However, under our experimental conditions (viz., neutral pH and dilute reactants), no decomposition of H_2O_2 by Fe^{3+}-chelates was seen.

(B) Observed Stoichiometry Between the Disappearance of H_2O_2 and the Appearance of Fe^{3+}. As noted above, the expected stoichiometry between Fe^{3+} formed and H_2O_2 lost during a Fenton reaction is 2:1 (equation (9)). However, when 50 μM H_2O_2 was mixed with 50-200 μM Fe^{2+}-DTPA, we observed a ratio of 3.3-3.7 molecules of Fe^{3+} formed per mole of H_2O_2 lost (Table 3). Similarly, in experiments with Fe^{2+}-EDTA (Table 3), ratios greater than 2:1 (theoretical) were observed. Experiments with Fe^{2+}-EDTA are complicated because the fast rate of autoxidation of Fe^{2+}-EDTA (Figure 1) makes it difficult to ascertain whether or not the observed stoichiometry (Table 3) reflects complex interaction or simply independent reactions of chelated Fe^{2+} with oxygen or H_2O_2. However, the results with Fe^{2+}-DTPA can be clearly interpreted: Fe^{2+}-DTPA is relatively stable to autoxidation (Figure 1). And yet more Fe^{3+} is formed than in predicted by equation (9). Therefore, another factor must be added to account for the extra Fe^{3+}. One such factor could be participation of oxygen in the Fenton reaction.

TABLE 3

STOICHIOMETRY BETWEEN THE DISAPPEARANCE OF H_2O_2 AND THE APPEARANCE OF FERRIC IONS[a]

Iron-Chelate Added	H_2O_2 Lost	Fe^{3+} Formed	Ratio Fe^{3+}/H_2O_2
	(µmoles ± S.E.M.)		
50 µM Fe^{2+}-DTPA	12.3 ± 1.3	45.3 ± 0.9	3.7
200 µM Fe^{2+}-DTPA	42.9 ± 2.4	157.0 ± 4.7	3.7
50 µM Fe^{2+}-EDTA	14.6 ± 0.9	45.3 ± 1.5	3.1
200 µM Fe^{2+}-EDTA	44.7 ± 3.1	185.1 ± 3.0	4.1

[a]The reaction medium was air-equilibrated 50 mM potassium phosphate buffer, pH 7.0, containing 50 µM H_2O_2 and 200 µM DTPA or EDTA. Ferrous ammonium sulfate was added to a final concentration of 50 or 200 µM. After two minutes, samples were acidified and Fe^{3+} and H_2O_2 were separately measured with thiocyanate reagent. Results are from five experiments with DTPA and 6 experiments with EDTA.

(C) Observed Stoichiometry Between Fe^{2+} Oxidized and H_2O_2 Added in the Presence and Absence of Oxygen. Experiments were conducted in sealed flasks under helium at pH 7.0. Results with Fe^{2+}-DTPA in excess (100 µM) and with smaller quantities of added H_2O_2 (12.5 µM) are shown in Table 4. The Fe^{2+} was measured colorimetrically with o-phenanthroline. Under helium, the ratio between Fe^{2+} lost and H_2O_2 added was 1.7 (Table 4, line 2) which is close to the expected stoichiometry of 2.0 (equation (9)). In the presence of oxygen, some autoxidation of the Fe^{2+}-DTPA occurred (line 3) and the observed loss of Fe^{2+} (line 4) was even greater. When line 4 is corrected for the autoxidation rate (line 3), the observed ratio (Fe^{2+} lost/H_2O_2 added) is 3.4. This value is much greater than the theoretical value of 2.0 and it is double that seen in the absence of oxygen. It appears, once again, that a new factor must be added to account for the extra Fe^{2+} that is lost. The factor is clearly dependent upon the presence of oxygen (line 4). Therefore, it seems probable that oxygen is consumed during the reaction.

(D) Direct Observation of Oxygen Consumption. The Fenton reaction between Fe^{2+}-DTPA and H_2O_2 was studied with an oxygen electrode. The addition of H_2O_2 to final concentrations of either 20 µM or 40 µM resulted in the consumption of oxygen (Figure 2). In these and other experiments, the rate of oxygen consumption was proportional to the quantity of H_2O_2 added over the range studied (10-80 µM H_2O_2). The use of the relatively stable chelate, Fe^{2+}-DPTA, was crucial for these observations to be made. Because this chelate is stable to autoxidation, the stimulation of oxygen consumption upon addition of H_2O_2 becomes readily apparent.

TABLE 4

LOSS OF FERROUS IONS DURING THE REACTION OF H_2O_2 WITH EXCESS Fe^{2+}-DPTA IN THE ABSENCE AND PRESENCE OF OXYGEN[a]

Condition	H_2O_2 Added (μM)	Fe^{2+} Lost μM ± SEM	(μM Corr)	Ratio Fe^{2+} Lost/H_2O_2 Added
Helium	0	0		
Helium	12.5	21.0 ± 1.0	(21.0)	1.7
Helium + Oxygen	0	16.5 ± 1.1		
Helium + Oxygen	12.5	58.7 ± 1.8	(42.2)	3.4

[a]Reactions were carried out in sealed 25 ml Erlenmeyer flasks in 4.0 ml of 50 mM potassium phosphate buffer, pH 7.0, containing 200 μM DTPA (Fe^{2+}-DTPA, 100 μM). Where indicated, 2 ml of oxygen were substituted for 2 ml of helium. Results are from 8-13 determinations in three experiments.

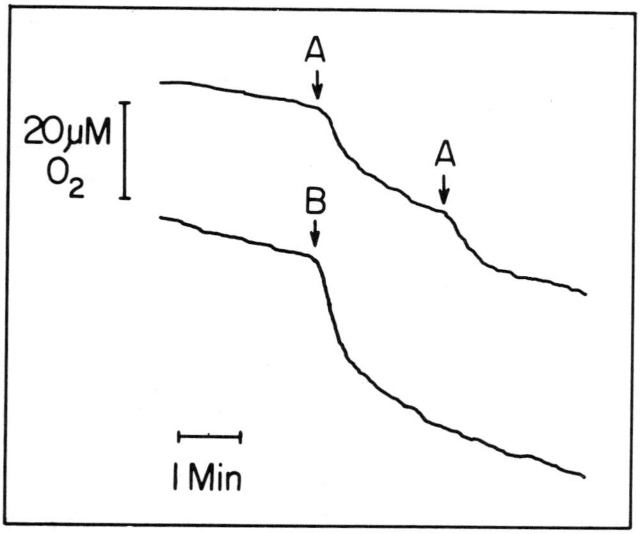

Fig. 2. Oxygen consumption by Fe^{2+}-DTPA. The reaction well contained 4.0 ml of 50 mM potassium phosphate buffer (pH 7.0). Fe^{2+} solution (50 μl) was added to give a final concentration of 200 μM. Where indicated by arrow, 20 μl (A) or 40 μl (B) of 4 mM H_2O_2 was added to give final concentrations of 20 or 40 μM H_2O_2.

With Fe^{2+}-EDTA, on the other hand. the very rapid consumption of oxygen during the autoxidation reaction obscured the effect of adding H_2O_2 to the system. Nonetheless, when the sequence of additions was reversed (H_2O_2 first, Fe^{2+} last to initiate the reaction) it was observed that oxygen consumption was still present. Therefore both oxygen and H_2O_2 are consumed during a Fenton reaction between dilute H_2O_2 and Fe^{2+}- EDTA at neutral pH.

(E) Discussion of Oxygen Consumption and Stoichiometry in the Fenton Reaction. Both oxygen and H_2O_2 are consumed during the Fenton reaction between H_2O_2 and Fe^{2+}-DTPA or Fe^{2+}-EDTA in dilute solution at neutral pH. With Fe^{2+}-DTPA, oxygen consumption is markedly stimulated upon the addition of H_2O_2 (Figure 2). It should be emphasized that the consumption of oxygen, rather than its production, is unusual. Earlier studies of the interaction of iron salts with H_2O_2 had been concerned mainly with oxygen formation. Most earlier studies with the Fenton reagent were performed with much more concentrated solutions of reagents. Under these conditions, the oxygen content of the solution is limited: The concentration of oxygen is only 0.20–0.25 mM for air-equilibrated aqueous solutions. Moreover, at high cocnentrations of reactants, other reaction pathways, such as equation (12) become favoured:

$$Fe^{3+} + O_2^- \longrightarrow Fe^{2+} + O_2 \qquad (12)$$

Some studies reported by Barb et al.[9] are of interest because they were conducted in dilute solution in our concentration range and lower (but, at acid pH). These authors noted an "anomaly" in kinetics under these conditions, associated with "a corresponding anomaly in stoichiometry" in which more than two molecules of Fe^{2+} were consumed per mole of H_2O_2 that disappeared. The ratio tended to a maximum of 4.0 as further dilutions were made to the range of 10 μM Fe^{2+} and 1 μM H_2O_2. The authors considered the results to be abnormal and they noted that the ratio returned to the "normal value of 2.0" as the concentration of reactants was raised. They attributed these results to a presumed presence of trace organic impurities that may have produced a chain oxidation of Fe^{2+}. We suggest, to the contrary, that the stimulated consumption of oxygen may be a characteristic of the reactions of ferrous ions or ferrous chelates with H_2O_2 and that this effect is observed mainly in dilute solution.

A criticism that might be raised to our experimental observations is that the samples may be contaminated with chemicals that can act as acceptors for $OH^\cdot$ to produce a chain oxidation of ferrous ions. The contaminants are the chelating agents themselves, EDTA or DTPA, and the buffer salts. Therefore, we repeated the oxygen electrode experiment in the absence of chelating agents and buffer salts. Under these conditions, the ferrous salt (200 μM $FeSO_4$) was relatively stable to autoxidation, more so than Fe^{2+}-DTPA, and the pH was neutral. When H_2O_2 was added, a stimulated consumption of oxygen was observed. We conclude, therefore, that oxygen consumption is a characteristic of the reaction between Fe^{2+} and H_2O_2.

The consequence of these observations is that oxygen consumption may be an integral part of Fe^{2+}/H_2O_2 reactions in biological systems. It will be of interest to determine if the presence of H_2O_2 can result in the "activation" of oxygen for biologic oxidations. Of particular interest are enzymatic reaction mechanisms in which both H_2O_2 and oxygen have been implicated, e.g., the reactions of cytochrome P-450 in liver cells. An active complex between the Fe-chelate, H_2O_2 and oxygen seems a distinct possibility. Such complexes have been invoked by others for reactions of superoxide, ferric salt, and H_2O_2 (Ref.10). At least one enzymatic reaction, that of tryptophan dioxygenase, proceeds through an intermediate that can be formed either from the ferrous state with oxygen or from the ferric state with superoxide, and may simultaneously involve H_2O_2 (Ref.11). It remains to be seen if our observations with model iron-chelates indicate a more general phenomenon in biologic reactions.

SUMMARY

(1) The spontaneous reactions of Fe^{2+}-EDTA (fast reaction) or Fe^{2+}-DPTA (slow reaction) with oxygen produces $OH^{\cdot}$ radicals. These radicals arise via an H_2O_2-dependent (i.e., catalase-sensitive) mechanism. The required H_2O_2 is generated by the reduction of oxygen during the autoxidation process.

(2) The Fenton reaction between relatively dilute H_2O_2 and ferrous chelates at neutral pH consumes oxygen and produces a greater yield of oxidized iron than can be accounted for by theory. In one sense, this observation indicates an "activation" of molecular oxygen during the reaction.

(3) It seems reasonable to expect that the Fenton reaction on living cells between endogenous iron-chelates and endogenously-generated H_2O_2 can serve as a stress on cellular vitality. Or, alternatively, the reaction may serve a useful purpose in cellular metabolism.

ACKNOWLEDGEMENT

Parts of these studies were supported by U.S. Public Health Service Grant Ns-05184 (Centre for the Study of Parkinson's and Allied Diseases), by the Centre Nationale de la Recherche Scientifique, France (ERA 47) and by a NATO Fellowship to P.M.S.

REFERENCES

1. Walling, C. (1975) Account. Chem. Res. 8, 125-131.
2. Fenton, H.J.H. (1894) J. Chem. Soc. 65, 899-910.
3. Fenton, H.J.H. (1899) Proc. Chem. Soc. 15, 224.
4. Haber, F. and Weiss, J. (1934) Proc. Roy. Soc. A, 147, 332-351.
5. Lai, C.-S. and Piette, L.H. (1979) Tetrahedron Lett. 9, 775-778.
6. Walling, C., Partch, R.W. and Weil, T. (1975) Proc. Nat. Acad. Sci. USA, 72, 140-142.

7. Beauchamp, C. and Fridovich, I. (1970) J. Biol. Chem. 245, 4641-4646.
8. Cohen, G., Heikkila, R.E., Allis, B., Cabbat, F., Dembiec, D., MacNamee, D., Mytilineou, C. and Winston, B. (1976) J. Pharmacol. Exp. Ther. 199, 336.
9. Barb, W.G., Baxendale, J.H., George, P. and Hargrave, K.R. (1951) Trans. Faraday Soc. 47, 462-468.
10. Czapski, C. and Ilan, Y. (1978) Photochem. Photobiol. 28, 651-653.
11. Hayaishi, O. (1980) in Proc. 4th Int. Symp. Microsomes and Drug Oxidations. Coon, M.J., ed., Academic Press, New York (in press).

THE BLEACHING OF CROCIN BY OXYGEN RADICALS

MANFRED SARAN, CHRISTA MICHEL AND WOLF BORS
Abteilung für Strahlenbiologie, Institut für Biologie Gesellschaft für Strahlen- und Umweltforschung, 8042 Neuherberg, W.Germany.

INTRODUCTION

There are growing doubts on the direct involvement of oxygen radicals proper (O_2^-, $^{\bullet}OH$) in biological reactions.[1] Among the alternative possibilities, we discussed the involvement of organic oxygen radicals as put forward primarily by Pryor[2] and Borg et al.[3] In this context we mentioned the analytical difficulties of differentiating between the various oxygen radicals by spectroscopic methods.

In a recent completed study on the bleaching of the water-soluble carotenoid crocin (i) during co-oxidation of linoleic acid by lipoxygenase, (ii) by photobleaching with isolated chloroplasts, and (iii) after generating $^{\bullet}OH$, O_2^- and hydrated electrons (e_{aq}^-) radiolytically, we found evidence for secondary reactions involving peroxy radicals.[4] This is supported by the findings that steady-state irradiation resulted in higher bleaching efficiency than application of similar radiolytic doses by pulse radiolysis and that N_2O/O_2-saturated solutions gave the highest bleaching rate after cumulative reactions of $^{\bullet}OH$ and O_2. To verify this hypothesis we investigated several systems for their ability to generate alkoxy ($RO^{\bullet}$) or peroxy ($ROO^{\bullet}$) radicals, using the bleaching of crocin as an assay method, which can be evaluated kinetically.

MATERIALS AND METHODS

Crocin was isolated from saffron as described previously.[5] 13-Hydroperoxy linoleic acid (13-LOOH) was prepared enzymatically.[6] The substances were gifts of Dr.E.F. Elstner and Dr.W. Grosch, respectively. In addition, *tert*-butyl hydroperoxide (*t*-BOOH) and di-*tert*-butyl peroxide (BOOB; Merck) were used as easily available peroxide compounds.

Alkoxy radicals were produced from the hydroperoxides by reaction with radiolytically generated e_{aq}^- in nitrogenerated solutions with *tert*-butanol as scavenger of $^{\bullet}OH$ radicals[7] as well as by photolysis of t-BOOH and BOOB.[8] Peroxy radicals were formed after reaction of radiolytically generated $^{\bullet}OH$ radicals with the respective molecule and subsequent attachment of oxygen in solutions saturated with gas mixtures of N_2O and O_2.[9] Table 1 lists the various radiolytic systems which were used as radical sources in these investigations

TABLE 1

AQUEOUS RADIOLYTIC SYSTEMS FOR THE GENERATION OF RADICALS

Gas	Scavenger/Converter[a]	Solutes[b]	Reactive Radicals
N_2O	-	-	$^{\bullet}OH$
N_2	H_2O_2	-	$^{\bullet}OH$
N_2O/O_2	-	-	$^{\bullet}OH + O_2^-$
O_2	-	-	$^{\bullet}OH + O_2^- + O_2$
N_2	t-butanol	-	e_{aq}^-
O_2	$HCOO^-$	-	O_2^-
N_2	t-butanol	t-BOOH	t-BO$^{\bullet}$
N_2	"	13-LOOH	13-LO$^{\bullet}$
O_2	$HCOO^-$	t-BOOH	O_2^- (no RO$^{\bullet}$)
O_2	"	13-LOOH	"
N_2O/O_2	-	t-butanol	t-BOO$^{\bullet}$

[a]Concentration of scavengers/converters generally 10 mM

[b]Solutes 10 μM (13-LOOH) and 0.5-1 mM (t-BOOH)

RESULTS

To demonstrate the very rapid processes involved in the radical-induced bleaching of crocin, the time traces up to 100 msec after pulse-radiolytic generation of various radical species are depicted in Figure 1. The curves were obtained as a composite of a number of pulses and different observation periods. It is evident that $^{\bullet}OH$ radicals are most efficient, followed by alkoxy radicals derived from 13-LOOH and t-BOOH. The highest sequential bleaching is observed in oxygenated solutions after initial attack of $^{\bullet}OH$ and presumably chain propagation via peroxy radicals derived from crocin itself. Rather slow and incomplete bleaching is observed with e_{aq}^- and the peroxy radical derived from t-butanol - $(CH_3)_2$-C(OH)-$CH_2OO^{\bullet}$ - and minimal effect for all systems containing O_2^-.

Even the most powerful initial bleaching by $^{\bullet}OH$ radicals relfects only a fraction of the total radical concentation. This evidence for slow secondary processes being primarily responsible for the overall reaction is substantiated by the almost complete disappearance of the crocin absorption after steady-state X-irradiation (Figure 2).

These radiolytical conditions do not allow a continuous observation of the reaction. Nevertheless the prominent effect of secondary reactions involving

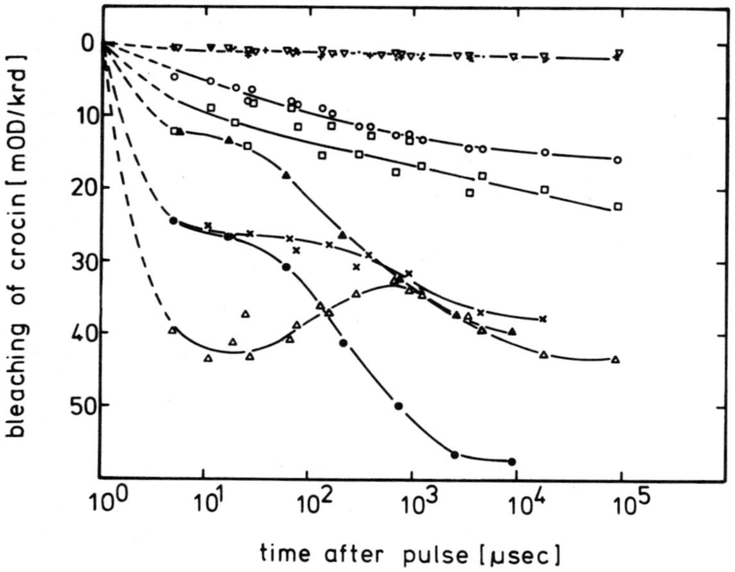

Fig. 1. Kinetics of the bleaching of crocin at 440 nm after pulse-radiolytic generation of various radicals. Concentration of crocin (ε = 1.335 x 10^5 M^{-1} cm^{-1} at 440 nm) varied between 8.3 - 12.5 μM in neutral solution except for 13-LOOH to improve its solubility as ammonium salt (pH 8.5); the start of the reactions was arbitrarily fixed at 1 μsec after the pulse (the end of the pulse is around 100 nsec and the first observation point at 5 μsec). The respective radical species generated in the various systems are listed in Table 1. (●) N_2O-saturated solutions; (▲) O_2-saturated solutions; (△) 7.8 μM 13-LOOH, 10 mM t-butanol, N_2; (X) 0.5 mM t-BOOH, 10 mM t-butanol, N_2; (□) 10 mM t-butanol, N_2; (O) 10 mM t-butanol, N_2O/O_2 gas mixtures; (•) 10 mM $HCOO^-$, O_2; (▽) 7.8 μM 13-LOOH, 10 mM $HCOO^-$, O_2; (+) 0.5 mM t-BOOH, 10 mM $HCOO^-$, O_2.

organic oxygen radicals, partly derived from crocin itself, is obvious. As proof that alkoxy radicals were indeed produced during radiolysis and were causing the depletion of crocin, we used the standard procedure of UV-irradiation of organic peroxides[8] (Figure 3).

As expected, the dialkyl peroxide was more stable towards UV-exposure, yet it too generated alkoxy radicals. The photolysis of the hydroperoxide, on the other hand, results in homolytic dissociation with the concomitant formation of both RO˙ and ˙OH radicals.[9] A discrimination of these two species would only be possible by the use of specific scavengers - or kinetic observation as we did in pulse radiolysis. The antioxidant nor-dihydroguaiaretic acid (NDGA) was found to inhibit the enzyme-catalyzed bleaching of crocin, whereas sorbic acid

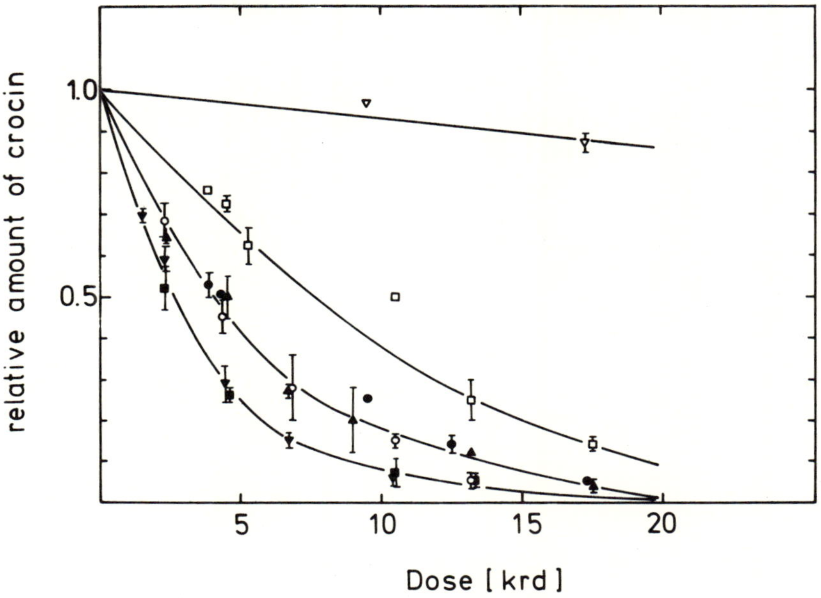

Fig. 2. Bleaching of crocin at 440 nm in X-irradiated solutions. Concentration of crocin 9 µM, pH 5.9; dose rate 9.5 krd/min. (▽) 10 mM $HCOO^-$, O_2; (□) 10 mM t-butanol, N_2; (●) N_2O-saturated solutions; (▲) O_2-saturated solutions; (○) 10 mM H_2O_2, N_2; solutions saturated with N_2O/O_2 gas mixtures: (■) 9:1, (▼) 3:2

TABLE 2

INHIBITORY EFFECTS OF ANTIOXIDANTS ON THE BLEACHING OF CROCIN BY THE UV-PHOTOLYSIS OF PEROXIDES.[a]

Antioxidant	Concentration (mM)	Change of net bleaching (%)	
		t-BOOH[b]	BOOB
Sorbic acid	0.01	-20	+15[c]
	0.1	-37	-145[d]
	1.0	-96	n.d[e]
nor-dihydro-	0.01	-	+12.5[c]
guaiaretic	0.1	- 6	-114[d]
acid	1.0	-41	n.d

[a]Concentration of tert-butyl hydroperoxide (t-BOOH) and di-tert-butyl peroxide (BOOB): 10 mM; crocin: 10 µM (neutral solutions). Control: net bleaching rate (without additives and minus basal bleaching of 0.24 µM/min); t-BOOH - 1.15 µM/min; BOOB - 0.21 µM/min.

[b]Inhibition

[c]Stimulation

[d]Diminution also of basal bleaching

[e]Not determined

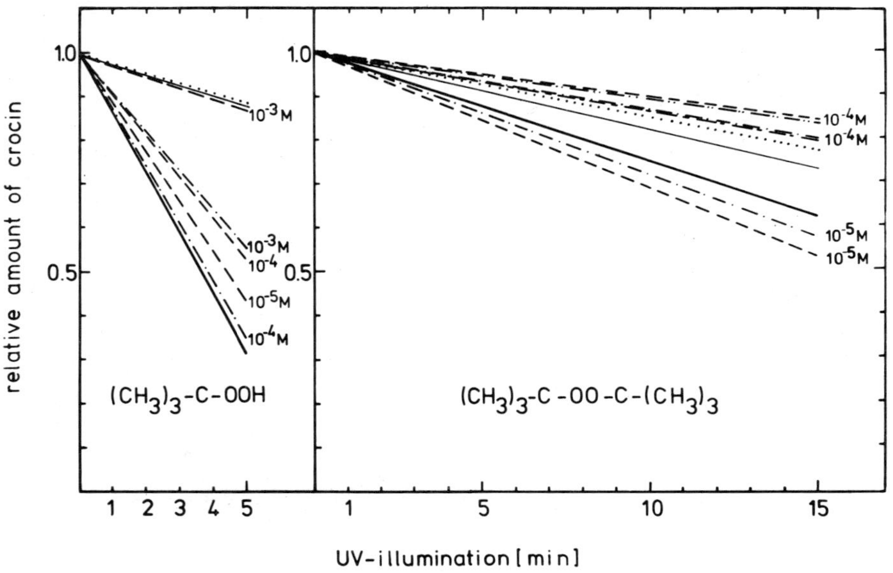

Fig. 3. Bleaching of crocin at 440 nm after photolytic generation of alkoxy radicals and effect of antioxidants. Original concentration of crocin 10 µM in neutral solution. (•••••••) photobleaching of crocin by 254 nm-illumination without additives; (———) photobleaching in the presence of 0.1 mM of either t-butyl hydroperoxide (t-BOOH) or di-t-butyl peroxide (BOOB); (———) 10 mM of t-BOOH or BOOB present; (—·—) 10 mM of peroxides with different concentrations of nor-dihydroguaiaretic acid; (— — —) 10 mM of peroxides with different concentrations of sorbic acid; (—··—) nor-dihydroguaiaretic acid without peroxides; (— - —) sorbic acid without peroxides.

had no effect (manuscript in preparation). These two agents were therefore compared in the photolytic model systems for generating alkoxy and/or ˙OH radicals (Table 2, see also Figure 3).

The results are far more ambiguous than expected. First of all, rather high concentrations of the organic peroxides (10 mM) were necessary to stimulate a minor background photobleaching of crocin. Duplication of these experiments with 13-LOOH was not possible. Only in the hydroperoxide solution, a clear-cut inhibition is observed - with sorbic acid being considerably more effective than NDGA, relfected also by the difference in the rate constants with ˙OH radicals of these diverse substances:

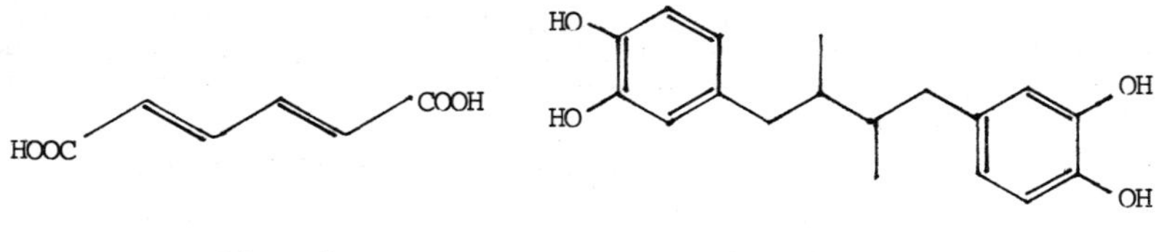

sorbic acid nor-dihydroguaiaretic acid

Competitive bleaching of p-nitrosodimethylaniline after $^{\cdot}OH$-generation in X-irradiated N_2O-saturated solutions resulted in the tentative values of $3x10^9$ $M^{-1}s^{-1}$ for sorbic acid and $1.5x10^8$ $M^{-1}s^{-1}$ for NDGA. The inhibitory effects of sorbic aicd and NDGA are less clear in the presence of the dialkyl peroxide: low concentrations of the antioxidants (10 μM) stimulate the bleaching, while 100 μM diminish even the base photobleaching of crocin in the absence of the dialkyl peroxide.

DISCUSSION

The intent of the these investigations was to evaluate the bleaching of crocin as a kinetic assay for the differentiation of various types of radicals - except for O_2^- which does not react at all.[4] That this is indeed possible was demonstrated primarily by the results from pulse radiolysis (Figure 1). During steady-state X-radiolysis competitive scavenging reactions are still a problem for the exclusive generation of alkoxy radicals (Figure 2). This is even more the case during photolysis of di-*tert*-butyl peroxide as potential source of $RO^{\cdot}$ radicals, where the stimulation with 10 μM of the antioxidants might be attributed to their involvement in a chain-propagating step. As both compounds do not absorb in the region of crocin, the diminishing of the base bleaching could be best explained by energy transfer from an excited crocin.

The comparison of the various systems established furthermore the versatility of radiolytic methods to generate the desired radicals selectively. Thus alkoxy radicals were generated by the following reactions:

$$ROOH + e^-_{aq} \longrightarrow RO^{\cdot} + OH^- \quad (1)$$

$$ROOH \text{ or } ROOR \xrightarrow{h\nu} RO^{\cdot} + {}^{\cdot}OH \text{ or } 2\ RO^{\cdot} \quad (2)$$

whereas the reaction with O_2^-

$$ROOH + O_2^- \not\longrightarrow RO^{\cdot} + OH^- + O_2 \quad (3)$$

does not lead to this species[7] as previously reported.[2] Both irradiation systems are artificial and can only serve as models for biological genration of alkoxy radicals. The most likely reaction under the latter conditions would be the so-called "organic Fenton reaction":

$$ROOH + Me^{n+} \longrightarrow RO^{\cdot} + OH^- + Me^{(n+1)+} \quad (4)$$

The reaction has already been demonstrated by ESR experiments to yield alkoxy radicals[11] and is presently under further investigation in our laboratory, uisng highly specific chelators for ferrous or ferric iron.

Still amother method of generating $RO^{\cdot}$ radicals was applied by Schieberle et al.,[12] i.e. the thermolytic decomposition of *tert*-butyl peroxalate. This alkoxy radical was previously shown to preferentially abstract hydrogen atoms from hydroperoxy groups.[13] It was thus possible to selectively generate peroxy radicals, as Scheiberle et al.[12] did from linoleic acid hydroperoxide. The multitude of degradation products under oxic conditions demonstrated that such peroxy radicals are similarly capable to promote autoxidation chain processes.

Radiolytic generation of peroxy radicals is less clear-cut, as it depends on the indiscriminate attack of ˙OH radicals and subsequent attachment of oxygen:

$$RH + {}^{\cdot}OH \longrightarrow R^{\cdot} + H_2O \qquad (5)$$

$$R^{\cdot} + O_2 \longrightarrow ROO^{\cdot} \qquad (6)$$

While the detailed structure of these peroxy radicals may thus remain obscure, they are produced unambiguously in solutions saturated with N_2O/O_2 gas mixtures. Radiolysis of purely oxygenated solutions, to the contrary, lead to O_2^- in addition to ˙OH and ROO˙ - with the chance of O_2^- reacting with the peroxy radical, thus complicating kinetic observations.

In conclusion the investigation of the mechanism of the bleaching of the cartenoid crocin demonstrated the versatility of this compound as an assay for organic oxygen radicals, excluding O_2^-. Assuming that free ˙OH radicals are unlikely to be involved in most biochemical oxidation processes[1,14] such reactions with strongly absorbing, yet generally unreactive substances, may present an important tool for the detection respectively differentiation of such organic oxygen radicals.

REFERENCES

1. Bors, W., Saran, M. and Czapski, G., this publication, Vol.2.
2. Pryor, W.A. (1978) Photochem. Photobiol. 28, 787-801.
3. Borg, D.C., Schaich, K.M., Elmore, J.J. and Bell, J.A. (1978) Photochem. Photobiol. 28, 887-907.
4. Bors, W., Saran, M., Michel, C., Elstner, E.F., Grosch, W. and Laskawy, G. (1979), submitted for publication.
5. Friend, J. and Mayer, A.M. (1960) Biochim. Biophys. Acta, 41, 422-429.
6. Garssen, G.J., Vliegenthart, J.F.G. and Boldingh, J. (1971) Biochem. J. 122, 327-332.
7. Bors, W., Michel, C. and Saran, M. (1979) FEBS Lett. in press.
8. Howard, J.A. (1973) in Free Radicals, Vol.II, Kochi, J.K., ed., Wiley-Interscience, New York, p.9.
9. Hiatt, R, (1971) in Organic Peroxides, Vol.II, Swern, D., ed., Wiley-Interscience, New York, pp. 1-56.
10. Ilan, Y., Rabani, J. and Henglein, A. (1976) J. Phys. Chem. 80, 1558-1562.
11. Gilbert, B.C., Holmes, R.G.G., Laue, H.A.H. and Norman, R.O.C. (1976) J. Chem. Soc., Perkin Trans.II, 1047-1052.
12. Schieberle, P., Tsoukalas, B. and Grosch, W. (1979) Z. Lebensm. Unters. Forsch. 168, 448-456.
13. Porter, N.A., Funk, M.O., Gilmore, D., Isaac, R. and Nixon, J. (1976) J. Amer. Chem. Soc. 98, 6000-6005.
14. Bors, W., Michel, C. and Saran, M. (1979) Eur. J. Biochem. 95, 621-627.

PRODUCTION OF SUPEROXIDE BY OSMOCHROMES

J.W. BUCHLER[+], J. BILLECKE[++] AND W. KOKISCH[++]
[+]Fachbereich Anorganische Chemie und Kernchemie, Technische Hochschule Darmstadt, D-6100 Darmstadt, W. Germany; [++]Institut für Anorganische Chemie, Technische Hochschule Aachen, D-5100 Aachen, W. Germany

ABSTRACT

A substitutionally inert "osmochrome", bis(1-methylimidazole)-octaethylporphinatoosmium(II), produces appreciable amounts of superoxide by direct outer-sphere electron transfer to dioxygen in pyridine-water (99:1) as evidenced by electron paramagnetic resonance. The subsequent dismutation of superoxide to dioxygen and hydrogen peroxide (detected as the peroxotitanyl complex) constitutes the "acid-induced autoxidation of osmochromes" which are air-stable in aprotic solvents because they do not offer a coordination site for the dioxygen molecule. This reaction sequence is suggested for the autoxidation of hemochromes in pyridine/acid or water, or of cytochromes b. The stoichiometry of the autoxidation is followed by respirometry. In the absence of water, superoxide reduces osmichrome salts to osmochromes.

INTRODUCTION

During the last decade it has been shown that autoxidation of hemoglobins to methemoglobins can produce superoxide, O_2^-.[22,26,29,30] This reduction of molecular oxygen by a heme protein is usually explained as follows:[9] The dioxygen molecule is first coordinated to Fe^{II}. The resulting dioxygen complex then dissociates yielding Fe^{III} and O_2^-. The prerequisite for such a reduction of O_2 seems to be the availability of a free coordination site at the heme iron (inner sphere electron transfer). Nevertheless, some cytochromes (especially those having a rather low redox potential, e.g. cytochrome b_5) are oxidized by molecular oxygen although they cannot have a free coordination site as evidenced by their inertness towards carbon monoxide.[21]

The possibility of a direct electron transfer between a hemochrome and dioxygen cannot be investigated because hemochromes are labile in solution. Due to this lability a coordination site is offered to the dioxygen molecule which is then degraded to oxide by excess heme.[9] Recently, a kinetic study of hemochromes, e.g. bis(pyridine)octaethylporphinatoiron(II) ($Fe(OEP)Py_2$)(**1A**), under quasi-inert conditions in a large excess of ligands has led Chu et al.[14] to the assumption that an acid-induced autoxidation to hemichromes could take place starting from a

TABLE 1

SOME PROPERTIES OF HEMOCHROMES, OSMOCHROMES, AND RUTHENOCHROMES

Compound No.[a]	M	L[b]	Redox potentials [V][c] E_1	E_2	E_3	O_2^- concentration by e.p.r.[d] [%]
1A	Fe	Py	-0.15[e]	-0.000	-	n.d.[g]
1B	Fe	1-MeIm	-	-0.41[f]	-	n.d.[g]
2A	Os	Py	-0.37	-0.17	-0.17	0
2B	Os	1-MeIm	-0.63	-0.49	-0.49	3.5 ± 1
2C	Os	NMe_3	-0.56	-	-	0.02
2D	Os	γ-Pic	-0.42	-	-	0
2E	Os	THT	-0.26	-	-	0
2F	Os	$P(OMe)_3$	-0.08	-	-	0
3A	Ru	Py	-0.02	+0.18	+0.18	0
3B	Ru	1-MeIm	-0.32	-	-	0.02
4A	Os	Py	-0.17	-	-	0
4B	Os	1-MeIm	-0.36	-	-	0.02

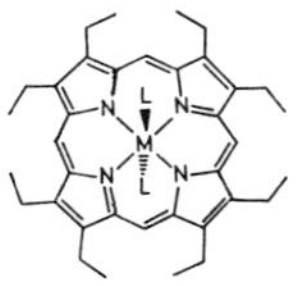

1: M= Fe

2: M= Ru

3: M- Os

[a] 1 - 3: derived from octaethylporphyrin; 4: derived from meso-tetra(p-tolyl) porphyrin.

[b] Axial ligands L: Py, pyridine; 1-MeIm, 1-methylimidazole; NMe_3, trimethylamine; γ-Pic, γ-picoline; THE, tetrahydrothiophene; $P(OMe)_3$, trimethylphosphite.

[c] Metal (II/III) redox potentials have been measured in the following solvents: E_1, CH_2Cl_2; E_2, pyridine; E_3, pyridine/H_2O (99:1).

[d] See Table 3 and Text; percentages below 0.02% have been estimated from the e.p.r. signal amplitude.

[e] Measured in CH_2Cl_2/Py (99:1).

[f] Measured in 1-MeIm.

[g] n.d., not determined (interference of unidentified Fe(III) signals precluded O_2^- determination.

charge transfer complex between hemochrome and dioxygen which is protonated; the resulting hydroperoxy radical then disproportionates. However, the hydrogen peroxide thus formed cannot be detected because it reacts with the heme under oxidation of the porphyrin ring.[14,23] In another recent investigation of the autoxidation of a hemochrome, bis(imidazole)(tetra-p-sulfonatophenylporphinato)-iron(II), in phosphate-buffered water Fleischer and Ferra[18] also postulated the formation of hydrogen peroxide. They explain their results by an inner sphere electron transfer because of the retarding effect of excess imidazole.

Bis(pyridine)octaethylporphinatoosmium(II), (**2A**, $Os(OEP)Py_2$), was obtained in the course of the investigation of the first dinitrogen carrying metalloporphyrin, $Os(OEP)N_2(THF)$ (THF= tetrahydrofuran) (Ref.6). A variety of other bis(ligand) species, e.g. **2B** - **2F**, have been prepared[1,2,7-9] (see Table 1). These compounds were named "osmochromes" because of their similarity to hemochromes. We observed that in the presence of acids HX the osmochromes, $Os(OEP)L_2$, can undergo a slow autoxidation to osmichrome salts, e.g. $[Os(OE)L_2]^+ X^-$. Here, we wish to present evidence that the steps (1) and (2) lead to the total acid induced autoxidation (3):

$$2Os(OEP)L_2 + 2O_2 \rightleftharpoons 2[Os(OEP)L_2]^+ + 2O_2^- \quad (1)$$

$$2O_2^- + 2HX \longrightarrow O_2 + H_2O_2 + 2X^- \quad (2)$$

$$2O_s(OEP)L_2 + O_2 + 2HX \rightarrow 2[Os(OEP)L_2]X + H_2O_2 \quad (3)$$

A prerequisite for the operation of the equilibrium (1) - an outer sphere electron transfer process - is the kinetic inertness of both the osmochrome and the osmichrome cation. This is dealt with below. The total reaction (3) is also treated and our knowledge on the equilibrium (1) is elaborated.

Contrary to expectation porphyrin complexes containing transition metal ions such as Co^{III}, Cr^{III}, Ru^{II}, whose complexes are normally inert, are observed to be substitutionally labile.[15,17] In the absence of proton sources, molecular oxygen does not react with osmochromes in dilute solutions. If the osmochromes were labile, as is e.g. $Os(OEP)N_2(THF)$, autoxidation should give the trans-dioxo osmium(VI) complex, $OsO_2(OEP)$ (Ref. 6).

MATERIALS AND METHODS

The Osmochromes were prepared and purified as previously described.[1,2,7-12] Ruthenochromes and hemochromes were obtained by standard procedures (see Ref. 12). Dry solvents were stored over molecular sieves. Pyridine was distilled from calcium hydride prior to storage. Potassium superoxide was purchased from EGA-Chemie, D-7294 Steinheim; dibenzo-18-crown(6) from Strem Chemicals, Newburyport, MA, U.S.A. Argon was purified by an Oxisorb Cartouche (Messer-Griesheim, Germany).

Redox potentials were determined by cyclic voltammetry (see Figure 1 for an example) with Princeton Applied Research equipment described elsewhere.[11] A Gibson instrument was used as respirometer, with modifications.[3] Weighed samples were introduced in sealed capillaries and cracked by magnetic stirring within the "submarine unit". A high-boiling organic solvent was used as the manometer fluid to avoid distillation. Optical measurements were done with a Unicam SP 1750 spectrophotometer. Oxygen detection in an argon stream was achieved with a solid state electrochemical detector. End-O-Mess L8 (ZrO_2 electrolyte) purchased from Friedrichsfeld GmbH, D-68 Mannheim 71. E.p.r. spectra were taken at 77K with a Bruker X-band spectrometer (type BER 420) fitted with a Nicolet computer (sweep width 1000 Gauss, microwave energy control at 200 μA, field modulation amplitude set at 1 Gauss, time constant 1s).

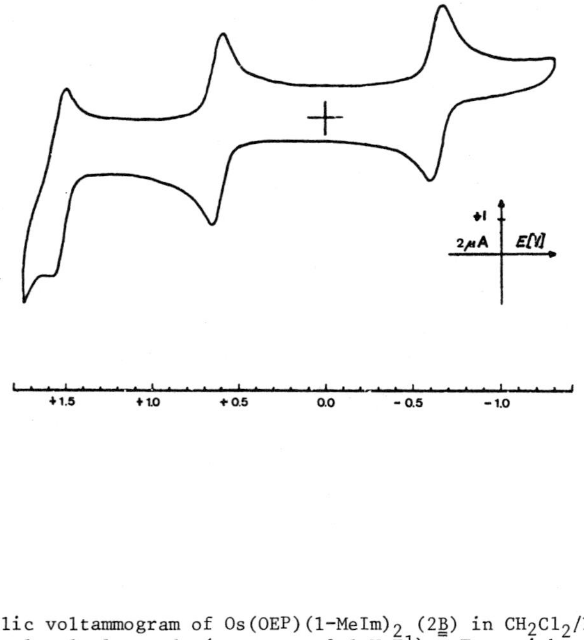

Fig. 1. Cyclic voltammogram of Os(OEP)(1-MeIm)$_2$ (2B) in CH_2Cl_2/NBu_4PF_6 _versus_ saturated NaCl-calomel electrode (scan rate 0.1 Vs^{-1}). From right to left: Os^{II}/Os^{III}-pair, Os^{III}/Os^{IV}-pair, porphyrin ring oxidation.

RESULTS

Acid-induced autoxidation of osmochromes. The acid-induced autoxidation of Os(OEP)Py_2 (2A) was monitored by a Gibson respirometer which was modified for non-aqueous solvents and calibrated with hemochromes.[3] With this instrument the oxygen uptake of 5 to 10 mg of a metalloporphyrin could be followed (about 7 to 12 μmole). An oxygen consumption of 0.5 ± 0.05 moles per mole of osmochrome was found using dimethylacetamide as solvent and benzenesulfonic acid as the inducing proton source

HX. Other runs in benzonitrile yielded similar results. Thus, the oxygen consumption agrees with eq. (3).

The osmichrome salts produced were reduced with sodium amalgam in tetrahydrofuran and the products isolated. The complete recovery of the osmochromes was checked by optical and ^1H-NMR spectra, cyclic voltammetry, or thin layer chromatography. Typical spectral data are given in Ref. 8.

For a complete establishment of eq. (3) at least the identification of the second product, hydrogen peroxide, was necessary. Application of several H_2O_2 assay methods[27] in dimethylformamide, acetonitrile or acetic acid did not give conclusive results. The negative test after autoxidation in glacial acetic acid can be explained by assuming that H_2O_2 acts as an oxidant in acetic acid as well (eq. (4), X = OAc).

$$Os(P)L_2 + H_2O_2 + 2HX \longrightarrow 2\ [Os(P)L_2]X + 2H_2O \qquad (4)$$

Three observations corroborate this assumption:

(a) H_2O_2 is a stronger oxidant than O_2 in aqueous solutions.

(b) In the absence of O_2, osmochromes are rapidly oxidized in H_2O_2/H_2SO_4 (eq. (4), $X = 0.5SO_4$).

(c) In glacial acetic acid, the O_2 uptake is only ~0.35moles O_2 per mole of $Os(OEP)Py_2$, and ~0.40 mole O_2 per mole of $Os(OEP)$-$(1\text{-MeIm})_2$. Thus, the hydrogen peroxide formed according to eq. (3) reacts with excess osmochrome, the more reactive bisimidazole osmochrome having a smaller chance of a further reaction with H_2O_2 than the bispyridine osmochrome.

Positive tests for H_2O_2 were obtained using pyridine as solvent. In a typical experiment, the osmochrome (10^{-5} mole) is dissolved in 1 ml dry, air-saturated pyridine. 5 ml H_2O_2-reagent (1 g $TiOSO_4$ in 500 ml 20% aq.K_2SO_2) was added. The osmochrome salt immediately precipitated. Ten extractions with 5 ml portions of CH_2Cl_2 yielded a porphyrin-free aqueous phase the optical density of which was measured at 407 nm, the absorption maximum of the peroxotitanyl cation. The procedure was run several times at various standardized H_2O_2 concentrations without porphyrin in order to calibrate the system. Also, weighed samples of KO_2 were disproportionated in the same system.

The results are shown in Table 2 and prove that H_2O_2 amounts are generated by osmochromes in an order of magnitude according with eq. (3). The fact that the yields are not 100% and depend on the concentration in the bis(imidazole) series can be explained by interference of eq. (3) which becomes more important at higher osmochrome concentrations.

The osmochrome/dioxygen-osmichrome/superoxide equilibrium. The acid-induced autoxidation of osmochromes is explained above as consisting of the equilibrium (1) which is driven to the right by the acid-induced disproportionation of superoxide according to reaction (2). From the electrochemical oxidation reduction potentials, $E_o[O_2^- \rightleftharpoons O_2] = -0.9$ V in pyridine (Ref. 24), and

TABLE 2

HYDROGEN PEROXIDE DETERMINATIONS WITH THE TITANYL REAGENT

H_2O_2-source (10^5 . mmol)	ℓg E[a]	H_2O_2-content (10^5 . mmol)	Yield (%)
KO_2			
2.67	2.16	1.38	97
3.09	2.12	1.34	87
Os(OEP) $(1\text{-MeIm})_2$			
0.68	0.362	0.24	71
1.17	0.303	0.19	32
1.50	0.480	0.31	41
1.75	0.330	0.21	42
5.72	0.693	0.50	18
$Os(OEP)Py_2$			
0.99	0.117	0.08	16
1.50	0.165	0.12	16

[a]Measured with a Unicam SP 1750 spectrophotometer in a 1 cm quartz cuvette at 407 nm.

$E_o[Os(P)L_2 \rightleftharpoons Os(P)L_2^+] = -0.5 \ldots. -0.2V$ in pyridine (see Table 1), it can be taken that equilibrium (1) lies far to the left. Therefore, if superoxide is added to an osmichrome salt, oxygen should be evolved. This is indeed the case.

To a solution of an osmichrome salt derived from 2A or 2B ($c \approx 5 \times 10^{-5}$ mol/l) was added an excess of a saturated solution of KO_2/dibenzo-18-crown-6 in pyridine under argon. The complete reduction of the osmichrome salt was monitored spectrophotometrically. The oxygen evolved in the same manner from more concentrated osmichrome solutions ($c \approx 2 \times 10^{-3}$ mol/l) was identified by purging the system with argon which then passed a cooling trap retaining the solvent vapours and an electrochemical oxygen detector operating in the ppm-range. These reductions did not proceed instantaneously. At $c \approx 0.5 \times 10^{-5}$ mol/l, $[Os(OEP)Py_2]^+$ had a half-life of ~4min and $[Os(OEP)(1\text{-MeIm})_2]^+$ a half-life of ~20 min, indicating that the reaction rate may be proportional to the difference of the redox potentials (and hence, the free enthalpy of the reaction, see Table 1). Thus the reverse of eq. (1) was established. In this respect, osmochromes behave like cytochromes.

The final proof for the forward reaction (1) came from electron paramagnetic resonance spectroscopy. The osmichrome cations are e.p.r.-silent. They do not interfere with the well-known O_2^- signals.[5] Initial experiments in dry pyridine

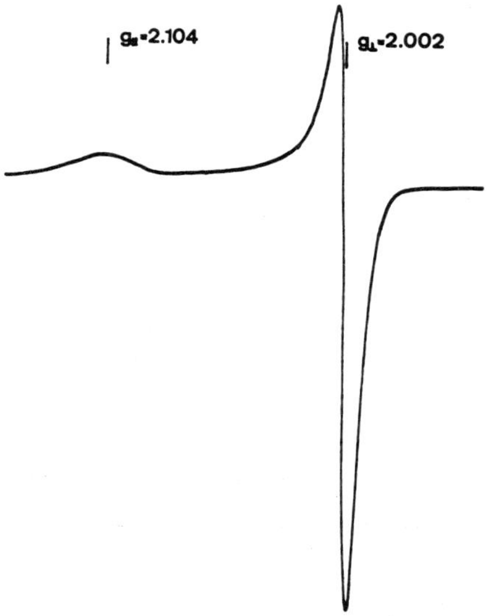

Fig. 2. E.p.r. spectrum of superoxide formed from Os(OEP)(1-MeIm)$_2$ (2B) and dioxygen in pyridine/water (99:1) and frozen at 77K.

and various other dry solvents did not show an e.p.r. signal. Following an accidental observation, wet pyridine was found to be a very suitable solvent for the recognition of O_2^- by its e.p.r. spectrum (Figure 2), an addition of 1% water (by volume) to pyridine being most favourable. Larger amounts of water cause a more rapid decay of the superoxide according eq. (2) and hence less intense e.p.r. signals. The half-lifetime of Os(OEP)(1-MeIm)$_2$ in Py/H_2O (99:1) was about 45 h, in Py/H_2O (9:1) only 1 min. Concentrations of superoxide were determined by computer-assisted integration versus external $CuSO_4 \cdot 5H_2O$/HCl/$NaClO_4$ ($10^{-3}/10^{-2}/2$ mol/l in water) as standard and were found to be in the percent range of the initial osmochrome concentrations ($0.6...2x10^{-3}$ mol/1), indicating a notable effect of the water addition.

This obvious increase in superoxide concentration from nondetectable to the percent range is probably due to hydration according eq. (5) which provides an additional driving force to eq. (1). The hydration of O_2^- is well-known (see Ref. 16 for a review) and reflected in the fact that O_2^- has a more positive redox potential in water than in nonaqueous solvents, i.e. O_2 is a stronger oxidant in water. The addition of 1% water to pyridine does not show a notable effect on the redox potential of osmochromes (see Table 1).

$$O_2^- + H_2O \rightleftharpoons [O_2 \cdot H_2O]^- \qquad (5)$$

The superoxide is clearly a transient. This is seen from a typical experiment : 40 µl of water were injected into a solution of 22.9 mg Os(OEP)(1-MeIm)$_2$ in 4 ml dry air-saturated pyridine and 300 µl samples withdrawn in the intervals given in Table 3. The samples were immediately frozen (77K) and e.p.r. spectra taken. Similar experiments showed that after 0.5-1 min 3.5 ± 1% superoxide was present (related to 100% initial osmochrome); after 2h $\leqslant$1.9% was found.

TABLE 3

SUPEROXIDE FORMATION BY BIS(1-METHYLIMIDAZOLE)OCTAETHYLOSMOCHROME

Sample No.	t (min)	$10^8 \cdot n(O_2^-)$[a] (mol)	$\frac{n.(O_2^-)}{n.(Os)} \times 100$[b] (%)
1	0	-	-
2	1	5.52	2.85
3	2	4.06	2.09
4	3	4.65	2.40
5	4	4.43	2.28
6	10	4.81	2.48
7	20	4.11	2.12
8	30	4.02	2.07
9	59	4.01	2.07
10	119	3.63	1.87

[a] Determined by integration of the e.p.r. signal.
[b] Mole percent superoxide formed per mole of initial osmochrome; $n(Os) = 1.94 \times 10^{-6}$ mol.

Only two other osmochromes, 2C and 4B, and a ruthenochrome, 3B, produced very small amounts of superoxide (see Table 3). The ruthenochrome 3B and a meso-tetra-p(tolyl)porphyrin osmochrome, give minor amounts of superoxide because their redox potentials are also rather low. The rather strong reductant, $Os(OEP)(NMe_3)_2$, contrary to expectation, gives also a low transitory O_2^--concentration. It seems that both an unsaturated axial ligand and a strongly negative potential facilitate superoxide production. Therefore, $Os(OEP)(1\text{-MeIm})_2$ is the most efficient osmochrome in this process. The hemochrome 1B is also a good candidate for O_2^- generation, but the e.p.r. measurements were inconclusive (see Table 1).

The above mentioned oxidation of osmochromes with hydrogen peroxide in acidic media can be reversed in alkaline media (eq. (6)), e.g. in pyridine containing KOH or NBu_4OH.

The reaction

$$[Os(OEP)L_2]^+ \xrightarrow[-O_2]{H_2O_2/OH^-} Os(OEP)L_2 \qquad (6)$$

was followed spectrophotometrically or by gas analysis, the oxygen evolved being detected electrochemically as described above. However, the stoichiometry of the reaction was not studied. Roberts et al.[25] have described the decomposition of H_2O_2 in pyridine and base according to the following equations:

$$H_2O_2 + OH^- \longrightarrow HO_2^- + H_2O \qquad (7)$$

$$HO_2^- + H_2O_2 \longrightarrow OH^{\bullet} + O_2^- + H_2O \qquad (8)$$

While these authors did not explore the fate of the hydroxyl radical, which probably reacts with pyridine, they identified the O_2^- ion electrochemically. The formation of superoxide then explains the reduction of osmichromes according the reverse of eq. (1). Only the bispyridine osmichrome, being the stronger oxidant, is reduced by 30% H_2O_2 in dilute pyridine solutions ($c \approx 10^{-5}$ mol/l) without hydroxide added.

DISCUSSION

The experiments described above establish the possibility of an efficient, direct, outer-sphere electron transfer between the dioxygen molecule and a metalloporphyrin carrying unsaturated axial ligands and having a highly negative redox potential. The reaction can occur in a basic organic solvent when traces of water solvate the superoxide ion and lower its activity. As these conditions also exist in the hydrophobic pockets of cytochromes, it is suggested that the autoxidation of certain strongly reducing cytochromes (e.g. cytochrome b_5) consists in an outer-sphere electron transfer to dioxygen under formation of hydrated superoxide. This leaves the hydrophobic region and then disproportionates in regions of higher water concentrations. Therefore, cytochromes b are also candidates for causing the oxygen consumption due to carbon monoxide- and cyanide-insensitive respiration which leads to H_2O_2.[13]

A novel cytochrome b appears[28] to be involved in the oxidase system being important for the microbicidal processes[19,20] performed by stimulated neutrophilic polymorphonuclear leukocytes (neutrophils). Following our arguments, it could well be that this cytochrome is capable of directly producing the superoxide which is formed when the neutrophils are stimulated by phorbol myristate acetate.[28] Babior,[4] however, has proposed that superoxide synthetase is a flavoprotein.

As regards the work of Chu et al.[14] it seems possible that the intermediate in his study is not the charge-transfer complex $Fe(OEP)L_2...O_2$, but rather the hemichrome superoxide ion pair. This does not affect the kinetic results. His conclusion that both inner sphere and outer sphere electron transfer is possible between ferrous prophyrins and oxygen is corroborated by our observations with osmium porphyrins. We now postulate the autoxidation of hemochromes consisting of the consecutive steps (9) below and (2) above.

$$Fe(P)L_2 + O_2 \longrightarrow [Fe(P)L_2]^+ + O_2^- \qquad (9)$$

The experiments of Fleischer and Ferra[18] do not at all rule out an outer-sphere process. These workers do not discuss the sequence of steps (9) and (2) although their kinetic results do not exclude it. The imidazole amounts added exceed the amounts of phosphate buffer. Thus the imidazole could cause a pH increase which retards the acid-induced dismutation of superoxide.

ACKNOWLEDGEMENTS

We are very grateful to Professor K. Gersonde (Aachen) for the use of his Bruker e.p.r. spectrometer, to Dr. H.A.O. Hill (Oxford) for the communication of unpublished results and illuminating discussions, and to Dr. H. Twilfer (Aachen) for advice on the use of the e.p.r. spectrometer. We thank the Fonds der Chemischen Industrie and the Deutsche Forschungsgemeinschaft for financial support and the Degussa Wolfgang for a gift of osmium tetroxide.

REFERENCES

1. Antipas, A., Buchler, J.W., Gouterman, M. and Smith, P.D. (1978) J. Amer. Chem. Soc. 100, 3015-3024.
2. Antipas, A., Buchler, J.W., Gouterman, M. and Smith, P.D. (1979) J. Amer. Chem. Soc., in press.
3. Billecke, J. and Buchler, J.W. (1977): To be published; and Billecke, J. (1977) Diplomarbeit, Technische Hochschule Aachen.
4. Babior, B.M. (1977) in Superoxide and Superoxide Dismutases, Michelson, A.M., McCord, J.M. and Fridovich, I., eds., Academic Press, New York, pp. 271-278.
5. Bray, R.C., Mautner, G.N., Fielden, E.M. and Carle, C.I. (1977) in Superoxide and Superoxide Dismutases, Michelson, A.M., McCord, J.M. and Fridovich, I., eds., Academic Press, New York, pp. 61-75.
6. Buchler, J.W. and Smith, P.D. (1974) Angew. Chem. Int. Ed. 13, 745-746.
7. Buchler, J.W., Kokisch, W. and Smith, P.D. (1976): To be published; and Smith, P.D. (1976) Doctoral Thesis, Technische Hochschule Aachen; Kokisch, W. (1976) Diplomarbeit, Technische Hochschule Aachen.

8. Buchler, J.W., Kokisch, W. and Smith, P.D. (1978) Struct. Bonding 34, 79-134.
9. Buchler, J.W. (1978) Angew. Chem. Int. Ed. 17, 407-423.
10. Buchler, J.W. (1978) Hoppe-Seylers Z. Physiol. Chem. 359, 1065-1066.
11. Buchler, J.W., Kokisch, W., Smith, P.D. and Tonn, B. (1978) Z. Naturforsch. 33b, 1371-1380.
12. Buchler, J.W. (1978) in The Porphyrins, Vol. IA, Dolphin, D., ed., Academic Press, New York, pp. 389-483.
13. Chance, B., Sies, H. and Boveris, A. (1979) Physiol. Rev. 59, 527-605.
14. Chu, M.M.L., Castor, C.E. and Hathaway, G.M. (1978) Biochemistry, 17, 481-486.
15. Eaton, S.S., Eaton, G.R. and Holm, R.H. (1972) J. Organometal. Chem. 39, 179-195.
16. Fee, J.A. and Valentine, J.S. (1977) in Superoxide and Superoxide Dismutases, Michelson, A.M, McCord, J.M. and Fridovich, I., eds., Academic Press, New York, pp. 19-60.
17. Fleischer, E.B., Jacobs, S. and Mesticelli, L. (1968) J. Amer. Chem. Soc. 90, 2527-2531.
18. Fleischer, E.B. and Ferra, R.V. (1979) Bioinorg. Chem. 10, 91-100.
19. Green, M.R., Hill, H.A.O., Okolow-Zubkowska, M.J. and Segal, A.W. (1979) FEBS Lett. 100, 23-27.
20. Hill, H.A.O. (1978): Personal communication.
21. Lemberg, R. and Barret, J. (1973) Cytochromes, Academic Press, London.
22. Misra, H.P. and Fridovich, I. (1972) J. Biol. Chem. 247, 6960-6962.
23. Pasternack, R.F. and Halliwell, B. (1979) J. Amer. Chem. Soc. 101, 1026-1031.
24. Peover, M.E., and White, B.S. (1966) Electrochim. Acta, 11, 1061-1067.
25. Roberts, Jr., J.L., Morrison, M.M. and Sawyer, D.T. (1978) J. Amer. Chem. Soc. 100, 329-330.
26. Rotilio, G., Fioretti, E., Falcioni, G. and Brunori, M. (1977) in Superoxide and Superoxide Dismutases, Michelson, A.M., McCord, J.M. and Fridovich, I., eds., Academic Press, New York, pp. 239-244.
27. Schumb, W.C., Satterfield, C.N. and Wentworth, R.L. (1955) Hydrogen Peroxide, Chapman and Hall, London.
28. Segal, A.W. and Jones, O.T.G. (1979) Biochim. Biophys. Res. Commun. 88, 130-134, see also this publication, Vol.2.
29. Wallace, W.J., Maxwell, J.C. and Caughey, W.S. (1974) Biochim. Biophys. Res. Commun. 57, 1104-1110.
30. Wever, R., Oudega, B. and van Gelder, B.F. (1973) Biochim. Biophys. Acta, 302, 475-478.

KINETIC FACTORS AFFECTING THE STEADY STATE CONCENTRATION OF SUPEROXIDE IN BIOLOGICAL SYSTEMS AND THEIR RELATION TO THE PRODUCTION OF HYDROXYL RADICALS BY METAL IONS AND METALLOPROTEINS AND TO THE ROLE OF SUPEROXIDE DISMUTASE.

ADELIO RIGO[+] AND GIUSEPPE ROTILIO[++]
+ Institute of Physical Chemistry, University of Venice, Venice, Italy;
++ Institute of Biological Chemistry, University of Rome, Rome, Italy

The superoxide anion radical and hydrogen peroxide appear to be present in biological systems as intermediates in the biological reduction of molecular oxygen to water,[1] Though they are relatively unreactive species a great deal of experimental evidence shows that the very reactive hydroxyl radical is formed in biochemical systems in the presence of H_2O_2 and O_2^- (Refs.2,3).

Direct reaction between O_2^- and H_2O_2, the so-called Haber-Weiss reaction,[4] although thermodynamically possible ($\Delta G^o \sim -16$ kcal/mole) occurs at a very low rate, In fact a kinetic rate constant value $<10^{-4}\ M^{-1}\ sec^{-1}$ has been measured for this reaction[5] which excludes that in biological systems the direct reaction between O_2^- and H_2O_2 be significant in comparison with the typical reactions of these species, even in the most favourable conditions. As a consequence the formation of $OH^\cdot$ radicals from O_2^- and H_2O_2 must proceed catalytically. A catalytic cycle involving the reduction, via O_2^-, of suitable metal ion complexes followed by their reoxidation, via H_2O_2 (Fenton-type reaction[6]) appears to be possible in view of the kinetic rate constants of these reactions.

Though it was demonstrated that such a mechanism is operative in the case of Fe-EDTA chelates,[7,8] no experimental evidence has been reported showing that formation of $OH^\cdot$ radicals from H_2O_2 and O_2^- is a quite general reaction in the presence of metal ion complexes, and in particular that metalloproteins could be involved in the catalytic cycle. A discussion of the kinetic aspects of this reaction and of its feasibility is the object of this communication.

Steady state concentration of O_2^- in aqueous solution in the absence of external catalysts. Superoxide is unstable in aquoeus solutions, even in the absence of any catalyst except protons, since its protonated form, $HO_2^\cdot$, reacts rapidly with O_2^- (Ref.9):

$$(1)\quad HO_2^\cdot + O_2^- \xrightarrow[k_1]{H^+} H_2O_2 + O_2 \qquad k_1 = 7.8 \times 10^7\ M^{-1}sec^{-1}$$

Therefore the rate of this dismutation is a pH dependent process, with a maximum at pH 4.8, which corresponds to the pK of $HO_2^\cdot$ (Ref.9). If a flux of O_2^- is generated at a constant rate, R_o, in a buffered aqueous solution, a steady state concentration of O_2^-, $[O_2^-]_s$ is reached:

$$(2) \qquad [O_2^-]_s = \frac{10^{\frac{1}{2}(4.8-pH)}}{k_1^{\frac{1}{2}}} R_o^{\frac{1}{2}}.$$

According to this equation, $[O_2^-]_s$ vs R_o is reported in Figure 1 at three different pH values. It must be pointed out that the R_o values that have been considered in Figure 1, 10^{-9} - 10^{-7} M sec^{-1}, represent a very small fraction of cellular oxygen consumption rates, which can reach values as high as 10^{-4} M sec^{-1} in some tissues of aerobic organisms.[10]

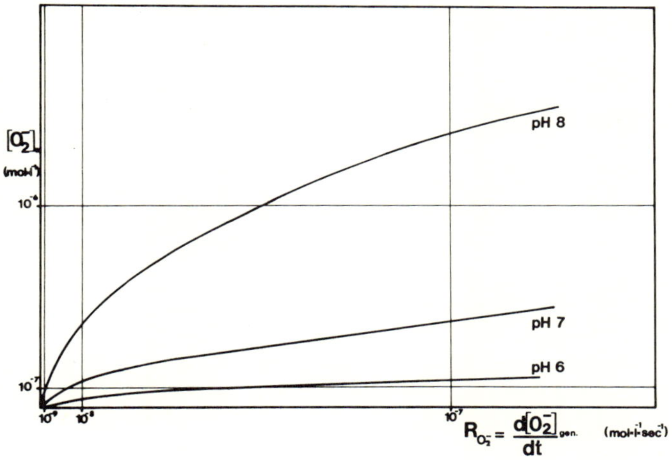

Fig. 1. H^+-controlled steady-state concentrations of O_2^- as function of the rate of O_2^- production at different pH values.

<u>Reduction of metal ions by O_2^-, and mechanisms of reoxidation of the reduced metal ion</u>. In the presence of suitable metal ions, such as Cu(II) and Fe(III), additional mechanisms of reactions are operative besides the H^+-catalyzed dismutation (1). In fact O_2^-, according to its redox properties, $E^{o\prime} = -0.33$ V (Ref.11), can donate one electron to the oxidized form of these metal ions:

$$(3) \qquad M^{n+} + O_2^- \xrightarrow[k_3]{} M^{(n-1)+} + O_2$$

$M^{(n-1)+}$ can be reoxidized in different ways. There can be one electron donation to O_2^-:

(4) $$M^{(n-1)+} + O_2^- \xrightarrow[k_4]{H^+} M^{n+} + H_2O_2$$

The sum of reactions (3) and (4) is the metal-catalyzed O_2^- dismutation, and the kinetic rate constant for the overall catalytic process is:

$$k_c = \frac{2\,k_3k_4}{k_3 + k_4} .$$

Dismutation of O_2^- by the metalloproteins referred to as copper-zinc, manganese and iron superoxide dismutases typifies this case.[12-14] $M^{(n-1)+}$ can react with H_2O_2:

(5) $$M^{(n-1)+} + H_2O_2 \xrightarrow[k_5]{} M^{n+} + HO^{\cdot} + HO^-$$

which is the well known Fenton reaction.[15] Reactions (3) and (5) may be defined as metal-catalyzed Haber-Weiss reaction or superoxide-driven Fenton reaction.[6] Finally, $M^{(n-1)+}$ can react with molecular oxygen. In inorganic complexes, the rate of reoxidation of $M^{(n-1)+}$ by O_2 is usually lower[16-18] than that of reactions (4) and (5). In bioinorganic complexes, reoxidation by O_2 can be efficient but is however a multi-electron process leading to formation of H_2O. The reaction:

(6) $$M^{(n-1)+} + O_2 \xrightarrow[k_6]{} M^{n+} + O_2^-$$

which leads to production of free O_2^- in solution, occurs, for instance in iron-sulphur proteins,[19] which however transfer electrons much more efficiently to other acceptors than oxygen. Therefore this reaction will not be considered here in connection with its possible effects on the steady state concentration of O_2^- in biological systems.

<u>Steady state concentration of reduced metal ions in the presence of O_2^- and H_2O_2</u>. From the considerations given above it appears that $[O_2^-]_s$ can be controlled by proton and/or metal ion catalyzed processes, reaction (1) and reactions (3), (4), respectively. In particular at pH $\geq$ 7 and for $k_c \times [M]_o \leq 1$ sec^{-1}, where $[M]_o = [M^{(n-1)+}] + [M^{n+}]$, it results

$$[O_2^-]_H >> [O_2^-]_M = \frac{R_o}{k_c[M]_o} ,$$

where $[O_2^-]_H$ and $[O_2^-]_M$ are the steady state concentrations of O_2^- when its dismutation is controlled by H^+ or by metal ions, respectively. If these conditions are met and the reduced metal ion is not part of a specific oxidase or oxygenase, the steady state concnetration of $M^{(n-1)+}$ ions (M^{red}) in the reaction system is essentially established by metal reduction *via* O_2^- and metal reoxidation *via* O_2^-

or H_2O_2. Therefore in these conditions the production of $OH^{\cdot}$ via oxidation of M^{red} by H_2O_2 will depend on the competition for M^{red} between O_2^- and H_2O_2, reoxidation by O_2 being negligible.

Three cases should be considered: (A) $k_4\ [O_2^-]_s \gg k_5\ [H_2O_2]$: M^{red} reacts with O_2^- much faster than with H_2O_2. In these conditions the $OH^{\cdot}$ yield, η_{OH}, is given by

$$(7) \qquad \eta_{OH} = \frac{d[OH^{\cdot}]/dt}{R_o} = \frac{k_5\ [M^{red}]\ [H_2O_2]}{R_o}$$

Since at steady state $k_3[M^{ox}] = k_4[M^{red}]$, equation (7) can be rearranged as follows:

$$(8) \qquad \eta_{OH} = \frac{k_3k_5}{k_3 + k_4} \times \frac{[M]_o\ [H_2O_2]}{R_o}$$

According to equation (7) the $OH^{\cdot}$ yield is independent of the absolute values of k_3 and k_4 (it depends on their ratio only) and increases by increasing the H_2O_2 and total metal ion concentrations, while being inversely proportional to the rate of O_2^- production (see also the discussion of case (B)).

From an analysis of equation (8) it results that $\eta_{OH} \geqslant 0.1$ could be obtained in conditions that are likely to occur in biological systems. For instance, $\eta_{OH} = 0.1$ for $[M]_o = 10^{-6}M$, $[H_2O_2] \sim 10^{-7}M$, (Ref.20), $R_o \sim 10^{-10}\ M\ sec^{-1}$, (Ref.21) and Figure 1, and $k_5 = 100\ M^{-1}sec^{-1}$. The value of $k_5 = 10^2\ M^{-1}sec^{-1}$ appears to be justified on the basis of the comparable or higher values that have been found with Cu and Fe complexes (Table 1). The corresponding rate of $OH^{\cdot}$ formation, R_{OH}, is $10^{-10}\ M\ sec^{-1}$ and, at first glance, this rate may look insignificant.

TABLE 1

KINETIC RATE CONSTANT VALUES FOR THE REOXIDATION OF REDUCED METAL COMPLEXES BY HYDROGEN PEROXIDE

M^{red}		pH	$k_{H_2O_2 + M^{red}}$ (k_5) $(M^{-1}\ sec^{-1})$	References
Fe^{2+}	(aqua)	2.0	85	(17)
Fe^{2+}	(EDTA)	6.5	1.1×10^4	(17)
Cu^+	(aqua)	4.6	10	(23)
Cu^+	(pyr)	7.0	200	(18)
Cu^+	(dipy)	4.6	850	(24)

However, even at this low rate a significant amount of biomolecules may be the preferential target of $OH^{\bullet}$ radicals. For instance, DNA and proteins may scavenge $OH^{\bullet}$ radicals very effectively on account of their relatively high concentration and of the high reactivity of their constituents. In fact the second order rate constants of the reaction between $OH^{\bullet}$ radicals and pyrimidine bases or amino acids are in most cases near the diffusion limit (Ref.22).

(B) $k_5\ [H_2O_2] \gg k_4\ [O_2^-]_s$: M^{red} reacts with H_2O_2 much faster than with O_2^-. In these conditions one $OH^{\bullet}$ is produced for each O_2^- generated in the system if the catalytic cycle of oxidation and reduction of metal ions, via reactions (3) and (5), is operative. The corresponding steady state concentration of O_2^- is

$$[O_2^-]_s = \frac{R_o}{k_3[M^{ox}]} \ .$$

Cytochrome c, $k_3 = 1.3 \times 10^6\ M^{-1}sec^{-1}$ at pH 7.0 (Ref.25), is a good example of this case, that it $k_5 \gg K_4$ (Ref.26). According to these data, $\eta_{OH} = 1$ was calculated for $[H_2O_2] = 10^{-5}$ M and $[M]_o = 10^{-6}$ M when $R_o < 10^{-8}$ M sec^{-1}. If R_o increase above 10^{-8} M sec^{-1}, η_{OH} decrease because of the dismutation of O_2^- by H^+. These conclusions may support the hypothesis that O_2^--dismutating systems are beneficial in mitochondria because of the high concentration of cytochrome c and the low fluxes of O_2^- (Ref.21). However, since dismutation leads to increase of H_2O_2 and cytochromes are continuously reduced by the electron flow, only the high efficiency of the cytochrome oxidase reaction minimizes the probability of reaction of reduced cytochromes with H_2O_2. Furthermore similar considerations can offer a rationale for the presence of cytochrome c peroxidase in yeast mitochondria (Ref.27). It is also noteworthy that hypoxic conditions would be more favourable to the formation of $OH^{\bullet}$ radicals in these systems by lowering R_o, provided that other molecules can reduce M^{ox} at a rate comparable with k_4.

(C) $k_5\ [H_2O_2] \sim K_4\ [O_2]_s$: H_2O_2 and O_2^- react with M^{red} at comparable rates. In this case the $OH^{\bullet}$ yields can be calculated from equation (7), where M^{red} is obtained from the solution of the following kinetic equations relative to the reactions (3), (5), assuming steady state conditions for O_2^- and M^{red}:

$$R_o - k_3\ [M^{ox}]\ [O_2^-] - k_4\ [M^{red}]\ [O_2^-] = 0$$

$$- k_3\ [M^{ox}]\ [O_2^-] + k_4\ [M^{red}]\ [O_2^-] + k_5\ [M^{red}]\ [H_2O_2] = 0$$

$$[M]_o = [M^{red}] + [M^{ox}]$$

The solution of this system is particularly simple in the case of $k_3 = k_4$. The $OH^{\bullet}$ yields, calculated according to this hypothesis for different values of R_o, are reported as a function of $[M]_o$ in Figure 2. It is again apparent that significant flows of $OH^{\cdot}$ are generated from O_2^-, via metal ion catalysis, in conditions which can occur in buochemical systems. It is of relevance to note again that η_{OH} increase when R_o decreases.

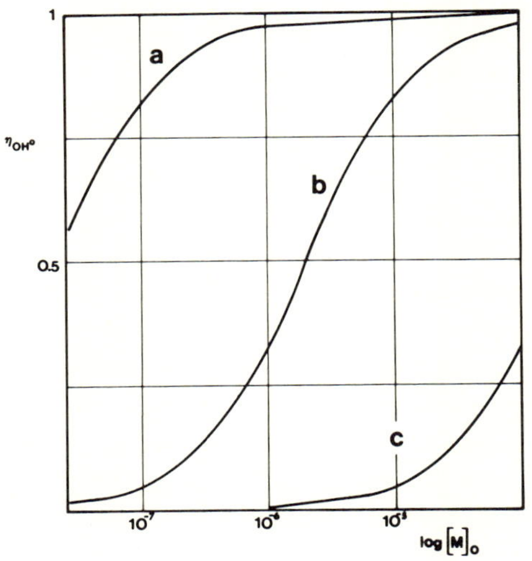

Fig. 2. Fraction of O_2^- converted to $OH^{\cdot}$ radicals, via M-ion catalysis, as function of the total concentration of M ions, M_o, assuming that the kinetic rate constant values for the reduction and for the oxidation of M ions by O_2^- are equal. Additional assumptions are:
(a) $k_5 \times [H_2O_2] = 10^{-2} sec^{-1}$; $R_o = 10^{-10} M\ sec^{-1}$;
(b) $k_5 \times [H_2O_2] = 10^{-3} sec^{-1}$; $R_o = 10^{-9} M\ sec^{-1}$;
(c) $k_5 \times [H_2O_2] = 10^{-3} sec^{-1}$; $R_o = 10^{-7} M\ sec^{-1}$.

The situations discussed above are not so straightforward _in vivo_ because of metalloproteins which can lower the O_2^- and H_2O_2 concentrations by catalytic reactions (superoxide dismutases, catalase, peroxidases), and or reducing system different from O_2^- which can affect the steady state concentration of M^{red}. However, if a biological role related to cellular defence against reactive oxygen derivatives is to be assigned to such enzymes on the basis of the present considerations, it may be safely stated that they prevent the accumulation of significant local concentrations of O_2^- and H_2O_2 and thus interfere with the systems described above. It is also clear that, to fulfill this role efficiently, the structure of their metal binding sites must be such as to kinetically disfavor their participation in reaction (5). Nevertheless, at least in some conditions, they can form OH$^\cdot$ radicals, through this reaction, and this has been verified for Cu-Zn superoxide dismutase.[28] Furthermore, it should be note that the kinetic considerations given here confirm that superoxide dismutation cannot be a beneficial process if it does not kinetically cope with an associated system competent for H_2O_2 removal[29,30] as it alone would lower O_2^- and increase H_2O_2 and thus favor, as discussed under cases (A), (B) and (C), OH$^\cdot$ production.

Experimental requirements. From the above discussion it appears that biochemically attainable fluxes of OH$^\cdot$ radicals are obtained at very low rates of O_2^- generation. This conclusion has experimental relevance as well as the already mentioned biological implications. In fact low R_o values for O_2^- are not easy to obtain *in vitro* because of the instability of its aqueous solutions. In addition, detecting the corresponding low fluxes of OH$^\cdot$ demands a very sensitive OH$^\cdot$ scavenger which must not interfere with the O_2^- generating system. A further complication, in particular when metalloenzymes are used in the catalytic cycle of OH$^\cdot$ radicals with the metal-binding molecule itself. This instance was in fact verified to occur in the simple system Fe-EDTA (Ref.31) and it is obviously more likely with the bulky protein moiety of metalloenzymes. Different sources of O_2^-, the characteristics of which are reported in Table 2, have been proposed; however, none appears to satisfy the tight experimental demands emerging from the kinetic analysis of the system.

TABLE 2

SYSTEMS FOR SUPEROXIDE PRODUCTION *IN VITRO*

System	Minimum rate of O_2^- production ($M\ sec^{-1}$)	Possible interferences of the O_2^- producing system	
		OH$^\cdot$ scavenger	$M^{ox} - M^{red}$ couple
Xanthine-xanthine oxidase (Ref.32)	10^{-10}	yes	yes
KO_2-crown ether in DMSO (Ref.33)	10^{-6}	yes	no
Strongly alkaline aqueous solutions of KO_2 (Refs.5,34)	10^{-6}	no	no
Electrochemical (Refs.35,36)	10^{-10}	yes	yes

Conclusions. More work is still needed to test experimentally the formation of $OH^{\cdot}$ radical from O_2^- and H_2O_2 via metal-ion catalysis, which appears to be the most reasonable way of explaining O_2^- and H_2O_2 cytotoxicity.[6] In particular new systems for O_2^- production should be set up and highly sensitive $OH^{\cdot}$ scavengers should be provided. However, even purely theoretical consideration cast some doubt on the effectiveness of superoxide dismutase *per se* as an enzyme that combats oxidative damages. More appropriately, it can at present be inferred that the high catalytic rate of O_2^- dismutation displayed by some metalloproteins and several metal complexes as well is related to the better competition displayed by O_2^- than by H_2O_2 for the reduced form of metal systems with certain coordination geometries. In this context, the biological merit of having superoxide dismutases instead of low-molecular weight complexes may find more credibility if the relative efficiency of the reduced state of either system in being reoxidized by H_2O_2 would favor the enzyme as an evolutionary solution to segregate metal ions in environments that are less prone to Fenton chemistry. Nonetheless, where other reduced metal systems that are capable of being reoxidized by H_2O_2 more efficiently than by O_2^- are present, the production of $OH^{\cdot}$, the actually oxidizing species, will be counteracted only by catalytic decomposition of H_2O_2.

REFERENCES

1. Chance, B., Sies, H. and Boveris, A. (1979) Physiol. Ref. 59, 527-605.
2. Beauchamp, C. and Fridovich, I. (1970) J. Biol. 245, 4641-4646.
3. Fridovich, I. (1974) Adv. Enzymol. 41, 35-97.
4. Haber, F. and Weiss, J. (1934) Proc. Roy. Soc. A, 147, 332-335.
5. Rigo, A., Stevanato, R., Finazzi-Agro, A. and Rotilio, G. (1977) FEBS Lett. 80, 130-132.
6. Fee, J.A. (1980) 3rd ISOX Proceedings, in press.
7. McCord, J.M. and Day, D. (1978) FEBS Lett. 86, 139-142.
8. Halliwell, B. (1978) FEBS Lett. 92, 321-326.
9. Czapski, G. and Dorfman, L.M. (1964) J. Phys. Chem. 68, 1169-1177.
10. Netter, H. (1969) Theoretical Biochemistry, Oliver and Boyd, Edinburgh, pp. 432-436.
11. Fee, J.A. and Valentine, J.S. (1977) in Superoxide and Superoxide Dismutases, Michelson, A.M., Mcord, J.M. and Fridovich, I., eds., Academic Press, New York, pp. 19-60.
12. Fielden, E.M., Roberts, P.B., Bray, R.C., Lowe, D.J., Mautner, G.N., Rotilio, G. and Calabrese, L. (1974) Biochem. J. 139, 49-60.
13. Lavelle, F., McAdam, M.E., Fielden, E.M., Roberts, P.B., Puget, R. and Michelson, A.M. (1977) Biochem. J. 161, 3-11.
14. McAdam, M.E., Fox, R.A., Lavelle, F. and Fielden, E.M. (1977) Biochem. J. 165, 71-79.
15. Fenton, H.J.M. (1894) J. Chem. Soc. 65, 899-910.
16. George, P. (1954) J. Chem. Soc., 4349-4359.
17. Borgaard, O.K., Farrer, D. and Anderse, V.S. (1971) Acta Chem. Scand. 25, 3541-3543.
18. Rigo, A., Tomat, R. and Viglino, P. (1979) Euchem. Conference, Venice.
19. Misra, H.P. and Fridovich, I. (1971) J. Biol. Chem. 246, 6886-6890.
20. Oshino, N., Chance, B., Sies, H. and Bucher, T. (1973) Arch. Biochem. Biophys. 154, 117-131.
21. Tyler, D.D. (1974) Biochem. J. 147, 493-504.

22. Scoles, G., Shaw, P. and Willson, R.L. (1964) in Pulse Radiolysis, Ebert, M., Keene, J.P., Swallow, A.J. and Baxendale, J.H., eds., Academic Press, New York, pp. 151-164.
23. Milyavskii and Khusid, L.B. (1974) Zh. Anal. Chim. 29, 2359-2364.
24. Pecht, I. and Anbar, M. (1968) J. Chem. Soc., 1902-1904.
25. Koppenol, W.H., Van Buuren, K.J., Butler, J. and Braams, R. (1976) Biochim. Biophys. Acta, 449, 157-178.
26. Rigo, A., in preparation.
27. Yonetani, T. (1978) in The Enzymes, Vol. 13, Boyer, P.D., ed., Academic Press, New York, pp. 345-361.
28. Bray, R.C., Cockle, S.A., Fielden, E.M., Roberts, P.B., Rotilio, G. and Calabrese, L. (1974) Biochem. J. 139, 43-48.
29. Rotilio, G., Calabrese, L., Finazzi Agro, A., Argento-Ceru, M.P., Antinori, F. and Mondovi, B. (1973) Biochim. Biophys. Acta, 321, 98-102.
30. Bozzi, A., Mavelli, I., Finazzi Agro, A., Strom, R., Wolf, A.M., Mondovi, B. and Rotilio, G. (1976) Mol. Cell. Biochem. 10, 11-16.
31. Kaneko, H., Nozaki, K. and Ozawa, T. (1978) J. Electroanal. Chem. 87, 149-153.
32. Fridovich, I. (1970) J. Biol. Chem. 245, 4053-4057.
33. Valentine, J.S. and Curtis, A.B. (1975) J. Amer. Chem. Soc. 97, 224-226.
34. Marklund, S. (1976) J. Biol. Chem. 251, 7504-7507.
35. McCord, J. and Fridovich, I. (1968) J. Biol. Chem. 243, 5763-5770.
36. Rigo, A., Viglino, P. and Rotilio, G. (1975) Anal. Biochem. 68, 1-8.

METABOLIC ROLE OF XANTHINE OXIDASE AS SOURCE OF SUPEROXIDE RADICALS AND HYDROGEN PEROXIDE

RAINER FRIED,
Department of Biochemistry, Creighton University, Medical School, Omaha, NE 68178, U.S.A.

INTRODUCTION

Superoxide radicals are reactive products of univalent reduction of molecular oxygen; they are inactivated by superoxide dismutase (E.C.1.15.1.), which is widely distributed throughout animals, plants and microorganisms. Superoxide radicals and superoxide dismutase participate in many reactions and reaction sequences, such as biological oxidations, photosynthesis, phagocytosis, mechanism of action of pollutants and several antibiotics, the process of senescence, as well as in a number of diseases.[1-11]

The superoxide radicals are mainly end products of a reaction sequence or are oxidative radical chain propagators. However, these are not the only reactions in which superoxide radicals can participate. I have proposed[1,12] that superoxide radicals are generated in a coupled enzyme system with xanthine oxidase and are used for biochemical oxidation or oxygenation reactions, and that this synthetic oxidative pathway is terminated by superoxide dismutase; xanthine oxidase and superoxide dismutase thus would act as a coupled enzyme system in biosynthetic pathways ("Unified Theory of Xanthine Oxidase").

In the present report I shall discuss older work which lead to the development of this theory and modern reports dealing with the synthetic or participatory role of superoxide radicals. The last section deals with the mechanism of auto-inactivation of xanthine oxidase in relation to superoxide radicals and hydrogen peroxide.

Metabolic Role of Xanthine Oxidase. Xanthine oxidase (E.C.1.2.3.2) is a metalloflavoprotein[13-15] recently defined as: Molybdenum iron-sulfur flavin hydroxylase. This enzyme has been identified more than a hundred years ago and has been extensively studied, yet it is rather surprising that its biological function is not yet fully understood. The enzyme acts on a wide range of substrates of many different chemical classes, as well as with numerous electron acceptors.

The main function of xanthine oxidase is considered to be the irreversible oxidation of purines, oxidizing hypoxanthine and xanthine to form urate. In most animals, urate is further oxidized to allantoin and subsequent metabolites. However, urate oxidase (E.C.1.7.3.3) is a deleted enzyme in humans and in primates; hence urate accumulates and is excreted. Thus, in man, relatively soluble and harmless compounds, hypoxanthine and xanthine, are converted into urate which is quite insoluble and causes many health problems, due to precipitation of uric acid or sodium urate. Xanthine oxidase is the terminal enzyme in the catabolism of nucleic acids in humans. The teleological function of this strange reaction system, which also requires a large amount of oxygen, is hard to understand, and may be considered a "practical joke of evolution."

I and other investigators have been puzzled by this reaction. I have proposed[1,12] that the oxidation of purines to urate may not be the only, indeed not the main function of xanthine oxidase, and that xanthine oxidase serves as a ubiquitous source of endogenous hydrogen peroxide and superoxide radicals. These biological oxidants, in turn, would be utilized for oxidative reaction sequences, and would serve as a general tool for oxidations.

The substrates of xanthine oxidase are readily available to most cells: purines are formed by total biosynthesis from "non-essential" amino-acids which themselves can be readily formed from other precursors. This is not the only pathway of formation of hypoxanthine and xanthine; they also derive from catabolism of adenine and guanine coming from nucleic acids and compounds such as ATP. Indeed, xanthine oxidase is considered to be the rate-controlling enzyme in the catabolism of nucleic acids.[16-17]

The aerobic oxidation of purines also forms reduced metabolites of molecular oxygen: two very active oxidants, hydrogen peroxide and the superoxide radical. Their inactivation can be carried out by enzymes (superoxide dismutase and catalase or peroxidase respectively); they can also be inactivated by protective compounds, which include thiols such as cysteine or glutathione, ascorbic acid, and tocopherol. In addition, many other compounds can act as antioxidant protective compounds or scavengers of free radicals. The interpretation of what occurs is, of course, a matter of perspective. The "inactivation" of superoxide radicals and hydrogen peroxide must be accompanied by the oxidation of the compounds which react with them. Depending on the point of view, this interaction can be either the "defensive objective" of the cell, or it could have a "synthetic purpose", where the emphasis would be directed to the anabolism or "useful" transformation of the reducing compounds.

Two comments should be made here: (a) I am strongly inclined to teleology, and feel that biochemical reactions occurring in living systems either have a biological function or had such a function during earlier stages of evolution. In the preceding paragraph this teleological function may have been over-emphasized, but the use of this terminology is a convenient method to express my point of view of the synthetic function of metabolites of xanthine oxidase;

(b) The present discussion does not enter into the specificity of superoxide radicals or hydrogen peroxide as oxidative agents in a reaction sequence; the concept would apply equally to other compounds such as the hydroxyl radical and singlet oxygen. Nor is it necessary to mention, that xanthine oxidase is not the only enzyme which forms oxidizing metabolites, which are formed from many other enzyme sources, and would all contribute to the pool of "active forms of oxygen" available for endogenous "useful" oxidations.

I select xanthine oxidase, because of its wide distribution and the prominence of its role in oxygen consumption of mammalian organs, based on *in vitro* studies. The older literature indicates that a major share of the oxygen consumption of liver preparations is the oxidation of purines stemming from degradation of nucleic acids. "Endogenous respiration" without added substrate can last for more than one hour, and is mainly due to xanthine oxidase. An interesting sidelight is the high specific activity of xanthine oxidase in the intestinal mucosa, where the special function of this enzyme has not been explained. The total amount of xanthine oxidase in the mammalian body is highest in the liver, but the specific activity is ten times higher in the intestinal mucosa.[18]

From the teleological point of view, the negative aspect of xanthine oxidase has already been mentioned. I find it hard to conceive that xanthine oxidase be solely responsible for terminal oxidation of nucleic acid derivatives, leading -- in human -- to the accumulation and excretion of uric acid. I cannot perceive any "survival value" of the unchecked formation of uric acid, which is a paradox if we consider man the apex of evolution. It would be easier to understand if the biosynthesis of uric acid, unbalanced by its catabolism, were a "bargain", where its potentially damaging effects (e.g., gout and other metabolic diseases[19,20] could be compensated by a "beneficial" function of its enzymatic formation. Thus, xanthine oxidase would be a "useful tool" whose beneficial effects outweigh its drawbacks.

Possible Participation of Xanthine Oxidase in Synthetic Reactions. A cursory examination of older literature indicates several reactions in which xanthine oxidase may participate; a systematic survey may reveal yet others. The evidence for the participation is based on several experimental findings: (1) In the majority of cases metabolic reactions were carried out in crude preparations, e.g., liver homogenate, which includes xanthine oxidase among many other enzymes; (2) One finds mention of hypoxanthine or other purines as "co-factors" for biochemical oxidations, such as the oxidation of free triglycerides or phospholipids, and the oxidations catalyzed by monamineoxidase [21-26]. It is interesting here, that sulfite oxidation is also catalyzed by hypoxanthine; in this reaction sequence superoxide radicals play a prominent role[27,28]; (3) The third indication of participation of xanthine oxidase in a given reaction, is its inhibition by allopurinol, a potent specific inhibitor of xanthine oxidase.[17,29] Such is the case, for example with tryptophan oxygenase, where a complex situation prevails, and several mechanisms of oxygenation seem to exist.[30,31] Allopurinol inhibits the oxidation of several drugs in vivo[32] which involves hydroxylation of aromatic ring systems.

Xanthine oxidase reduces molecular oxygen both to hydrogen peroxide and the superoxide radical. Which one of these agents participates in a sequencial oxidation can be determined only indirectly, by testing whether the reaction is inhibited by further addition of exogenous catalase or superoxide dismutase. Thus we would have a system of two enzymes with opposing effects: I. Xanthine oxidase releases hydrogen peroxide or superoxide radicals, and makes them available as oxidants of other compounds; II. Excess of these oxidants, beyond the "synthetic requirements", would be transformed into less active metabolites and the hydrogen peroxide or superoxide radical flux would be controlled by the two inactivating enzymes.

The xanthine oxidase system is frequently used to generate superoxide radicals, in vitro; so far attention has been directed mainly towards the "damaging" effect; to oxidise catecholamines or compounds like pyrogallol, and to induce lipid peroxidation. It is to be expected that the "synthetic" effect of "active" oxygen metabolites will gain more prominence in the next period of investigation.

Oxidation of Catecholamines. Catecholamines, e.g., epinephrine, are unstable in solution and readily autoxidize to colored compounds. This reaction is carried out by superoxide radicals, which act both as electron acceptors and as chain propagators. The reaction rate is enhanced by xanthine oxidase; originally the mechanism had been considered a co-oxidation of xanthine and epinephrine.[33-36] The reaction is inhibited by superoxide dismutase; this is still used as an assay of this enzyme, whose activity is inversely proportional to color formation. The neurotoxic compound 6-hydroxy-dopamine is also oxidized by a similar mechanism.[37-39]

Iodination of Proteins. The iodination of tyrosine, as well as the condensation of tyrosine molecules to form thyroxine both require the participation of hydrogen peroxide. Thyroid xanthine oxidase has been proposed as a source for hydrogen peroxide; on the other hand, formation of iodinated tyrosines and of thyroxine is inhibited by allopurinol.[40,41] The possible participation of the superoxide ion in this reaction has not been discussed by the cited authors.

Activation of Guanylcyclase.[42] Partially purified guanyl-cyclase is activated by addition of superoxide dismutase and further activated by nitrate reductase. The stimulation caused by superoxide dismutase is inhibited by cyanide and glutathione and also by catalase. The interpretation given for these reactions is that hydrogen peroxide is generated by superoxide dismutase from superoxide radicals, said to be formed during guanyl-cyclase activity due to contaminating flavoproteins (not further identified). It is proposed that the hydroxyl radical formed from superoxide radical and hydrogen peroxide be the active stimulating agent of guanyl-cyclase. One deals here with a complex system, where the activation of cGMP formation is enhanced by some form of "activated oxygen" generated in situ, but the mechanism requires further clarification. Addition of hydrogen peroxide alone did not activate guanyl-cyclase. Scavengers of superoxide radicals also inhibit this reaction; conflicting reports indicate either no effect or inhibition of activation by scavengers of hydroxyl radicals. The presence of xanthine oxidase in the guanyl-cyclase preparation has not been tested.

Phagocytosis.[43] Regulation of guanyl-cyclase activity, and hence the level of intercellular cGMP may also be connected with the overall process of phagocytosis, where superoxide radicals play yet another prominent role. Phagocytosis includes rapid formation of microfilaments which promote and stabilize vacuoles formed from the leucocyte membrane; possibly this formation of microfilaments is catalyzed by the action of cGMP. Cell membranes of phagocytosing leucocytes secrete a stream of superoxide radicals, which starts about thirty seconds after the attachemnt of an antigenic particle to the leucocyte membrane. The superoxide radicals may play multiple roles: Weismann[43] proposes a "destructive effect" of superoxide radicals and its metabolites on membrane lipids, and goes on to state: "However there is reason to suspect that superoxide also participates in other, more 'constructive' activities (though some of the latter may have destructive consequences as well). ...One of the metabolic pathways or one of its successor compounds lead to the formation of the compound HETE (12-L-hydroxy-5,8,10,14-eicosatetraenoic acid, a metabolite of arachidonic acid), which is a chemotactic substance that attracts additional PMN (leucocytes) to the invasion site, in effect, bringing in more troops to deal with the enemy." In addition to superoxide radicals participating in the formation of HETE, superoxide radicals, catalyzed by the enzyme cyclo-oxygenase, lead to the formation of a series of prostaglandins and thromboxanes, which participate in the phagocytosis sequence in different periods and with different physiological functions.[43]

Microsomal Oxygenations Catalyzed by Superoxide Radicals. Superoxide radicals have been invoked for oxygenation of the sulfur atom of thioethers resulting in sulfoxide formation. This reaction is a common step of biotransformations involving sulfur derivatives. Molecular oxygen can be introduced by a system in which superoxide radicals are generated; this is partially inhibited by addition of superoxide dismutase.[44] Similarly, superoxide radicals are intermediates in biotransformations of amines promoted by mixed function amine oxidases.[45] Superoxide served as source of hydrogen peroxide in a reconstituted monoxygenase system,[46] the reaction could be inhibited by addition of superoxide dismutase.

In a similar reconstituted system, aromatic hydroxylation was blocked by addition of superoxide dismutase; the part of the enzyme system involved in the generation of superoxide radicals, could be replaced *in vitro* by xanthine plus xanthine oxidase.[47] Inhibition of the hydroxylation by superoxide dismutase and replacement of part of the enzyme system by xanthine oxidase could be observed only in the reconstituted system, but not in intact microsomes.

This paper[47] is one of the first in which the xanthine oxidase-xanthine system is used as a generator of superoxide radicals, and the first where xanthine oxidase is used to promote hydroylation of aromatic ring systems, beyond the inherent function of introducing oxygen into purine and pteridine rings.

A more recent example of xanthine oxidase participating in oxidative biotransformations is the oxidation of hydroxylamine *in vitro*,[45] which is inhibited by further addition of superoxide dismutase. The xanthine oxidase system, replacing the more specific enzyme, mixed function amine oxidase, liberates superoxide radicals which are essential oxidants in this reaction.

Carboxylation of Glutamate in Prothrombin.[48] This reaction, an important step in blood coagulation, also involved the participation of superoxide radicals, probably due to superoxide radical formation during the oxidation of metabolites of vitamin K. Both the carboxylation of bound glutamate and the formation of vitamin-K epoxide are inhibited by very high levels of superoxide dismutase. On the other hand, addition of vitamin K to a microsomal enzyme system containing NADH greatly enhances the rate of formation of superoxide, which can be blocked by the customary low levels of superoxide dismutase.

Platelet Aggregation and Neurotransmitter Release. When platelets obtained from healthy donors were incubated in a medium containing xanthine oxidase plus xanthine, platelets aggregated and serotonin was released.[49] Both effects could be blocked by addition of low levels of superoxide dismutase. Aggregation and serotonin release caused by low levels of thrombin were greatly enhanced in platelets which had been preincubated with the xanthine oxidase-xanthine system at low xanthine oxidase concentration, i.e. platelets which had been exposed to a low level of superoxide radical flux. This pretreatment lowered the threshold to thrombin more than ten-fold.

Similar results were obtained with rat peritoneal mast cells, incubated in a medium containing xanthine oxidase plus xanthine. This treatment caused release *in vitro* of histamine; the release of histamine was inhibited by catalase but not by superoxide dismutase or mannitol.[50] Low levels of hydrogen peroxide also caused histamine release.

Lipid Peroxidation. The autoxidation of unsaturated lipid has been intensively studied. In a model system, microsomal lipid from rat liver was incubated with xanthine oxidase-xanthine system, in presence of ferric salts. The peroxidation is inhibited by superoxide dismutase or scavengers of singlet oxygen. A different mechanism of peroxide seems to prevail in presence of ascorbate as promoter.[51] In a recent paper,[52] the oxidation of lineolic acid by the xanthine oxidase-xanthine system was studied as a model for lipid peroxidation. Superoxide radicals are intermediates in this reaction, which requires catalytic amount of linoleic acid hydroperoxides. A new mechanism for the reaction has been proposed, which entails generation of alkoxy radicals by interaction of the superoxide radicals with hydroperoxides, and that these alkoxy radicals in turn promote lipid peroxidation.

EXPERIMENTAL: MECHANISM OF AUTO-INACTIVATION OF XANTHINE OXIDASE

The xanthine oxidase-xanthine system has also been used to inactivate other enzymes (e.g. NaK-ATPase of bovine erthrocytes and rat brain synaptosomes,[53,54] yeast ribonuclease and lysine-tRNA ligase[55]). It has been known for a long time[56-58] that xanthine oxidase inactivates itself during its catalytic activity; and it has become customary to add exogenous catalase (E.C.1.11.1.6) to the assay system to stabilize xanthine oxidase, especially during prolonged incubations. Of course, these experiments had been carried out long before the existence of superoxide dismutase or superoxide radicals had even been suspected. It is now known that both hydrogen peroxide and the superoxide radical are reaction products of enzymatic oxidation of xanthine, both of which are powerful oxidants. We therefore decided to test the role of superoxide dismutase alone or in conjunction with catalase on the auto-incubation of xanthine oxidase to verify whether superoxide dismutase would contribute to the stability of the enzyme.

The experiments were carried out in three phases: (a) preincubation of high levels of purified cream xanthine oxidase in presence and absence of xanthine and appropriate supplements; (b) dialysis; (c) addition of fresh substrate to an aliquot of the pre-incubated enzyme system, and spectrophotometric assay.[59] Pre-incubation with xanthine caused irreversible inactivation of xanthine oxidase, assayed both as oxidase and as dehydrogenase activity, but pre-incubation without xanthine had no effect. Superoxide dismutase, present in the pre-incubation phase, did not prevent auto-inactivation of xanthine oxidase (Table 1). On the other hand pre-incubation with catalase protected xanthine oxidase against auto-inactivation. Similar results were observed for pre-incubation in presence of allopurinol, which prevents enzymic activity; however, cysteine did not protect the enzyme. At the levels tested, catalase plus superoxide dismutase did not provide additional protection or activation

TABLE 1

EFFECTS OF PRE-INCUBATION WITH XANTHINE AND SUPEROXIDE DISMUTASE (SOD) ON AUTO-INACTIVATION OF CREAM XANTHINE OXIDASE (XO)

Pre-incubation System	Xanthine	Enzyme Activity Dehydrogenase (A_{540})	Oxidase (A_{298})
XO (1:20; 1.0ml)	-	1.72	1.72
XO	20 μMoles	0.04	0.03
XO + SOD (30 μg)	-	1.21	1.84
XO + SOD	20 μMoles	0.05	0.04

Cream xanthine oxidase (Sigma) was diluted from stock 1:20 with Na-phosphate, 0.1 M, pH 7.8 ("Phosphate"). The diluted enzyme was incubated for one hour at 37° and pH 9.45 with 20 μmoles of xanthine (0.01 M Na-xanthine, dissolved in 0.01 M NaOH) and supplements indicated; assays containing 0.01 M NaOH without substrate were used as controls. Supplements were added during pre-incubation and the volume completed to 6.0 ml with phosphate. After pre-incubation, the mixtures were dialyzed at room temperature against three changes of phosphate during 1-2 hrs. in a Durrum dialyzer. Aliquots were removed, and enzyme activity was assayed conventionally with a fresh supplement of 0.5 μmoles of xanthine in a total volume of 3.0 ml in a Beckman DU Spectrophotometer, at 25° and pH 7.8. All systems were assayed for xanthine oxidase activity at 298 nm (Ref.60) and xanthine dehydrogenase activity at 540 nm (Ref.15). All experiments were carried out in duplicate and repeated at least once. Superoxide dismutase (beef erythrocytes, Sigma) and catalase (beef liver, Sigma) and other reagents were dissolved in phosphate.

TABLE 2

EFFECT OF CATALASE AND SUPEROXIDE DISMUTASE (SOD) ON AUTO-INACTIVATION OF XANTHINE OXIDASE (XO)

Pre-incubation System	Xanthine	Enzyme Activity Dehydrogenase A_{540}/10 min.	Oxidase A_{298}/10 min.
XO (1:20; 1.0 ml.)	-	0.85	0.85
XO	+	0.00	0.04
XO + catalase (100 μg)	-	0.85	0.85
XO + catalase	+	0.92	0.89
XO + catalase + SOD (30 μg)	-	0.34	0.96
XO + catalase + SOD	+	0.28	0.89

Experimental conditions, as in Table 1

of xanthine oxidase. It is possible that at lower levels these two protective enzymes would show additive effects (Table 2 and 3). We conclude that hydrogen peroxide, but not the superoxide radical, is responsible for auto-inactivation of xanthine oxidase, confirming earlier results.[56-58]

TABLE 3

EFFECT OF ALLOPURINOL AND CYSTEINE ON AUTO-INACTIVATION OF XANTHINE OXIDASE (XO)

Pre-incubation System	Xanthine	Enzyme Activity Dehydrogenase A_{540}/10 min.	Oxidase A_{298}/10 min.
XO	-	0.81	0.61
XO	+	0.03	0.04
+ allopurinol, 5 μmoles	-	0.64	0.34
+ allopurinol, 5 μmoles	+	0.63	0.46
+ cysteine, 10 μmoles	-	0.71	0.55
+ cysteine, 10 μmoles	+	0.02	0.04

Experimental conditions, as in Table 1

COMMENTS AND SUMMARY

The present discussion deals with the possibility of an extended biological function of xanthine oxidase beyond the oxidation of purines, pteridines and aldehydes. The reaction products of aerobic activity of xanthine oxidase include the superoxide radical and hydrogen peroxide. It is proposed that these compounds be generated in vivo by the catalytic action of xanthine oxidase, and that these "active forms of oxygen" be used for introduction of oxygen into other compounds (oxygenation or hydroxylation reactions) or to trigger more complex biological processes such as platelet aggregation or release of neurotransmitters.

A survey of the older literature, as well as some recent publications, indicate participation of xanthine oxidase in coupled system in which oxygen is used. These reports deal with: a) reaction sequences carried out by crude systems, such as whole liver homogenate; b) a "catalytic" effect of purines, which enhance the reaction rate by unexplained mechanisms; and c) inhibition of a reaction sequence by exogenous allopurinol, which could be explained by the inhibition of xanthine oxidase present in the assay system.

The following section dealt with selected reactions or complex biological processes, which involve the participation of superoxide radicals, hydrogen peroxide or their metabolites. Evidence is accumulating that these "active forms of oxygen" can participate in a large number of reactions. In many cases the source of these endogenous "active forms of oxygen" is not yet known; given the wide distribution of xanthine oxidase and its substrate, this enzyme is quite possibly an important or one of several sources of superoxide radicals or hydrogen peroxide.

It is interesting to note here, that in model experiments in vitro, the xanthine oxidase system (either with hypoxanthine and xanthine or with acetaldehyde as substrate) is frequently used to generate superoxide radicals, which then participate in "harmful reactions", such as lipid peroxidation and inactivation of enzymes. It is possible that endogenous xanthine oxidase promotes the same type of reactions as are produced when the enzyme is added in model systems. The oxidative inactivation of enzymes, including xanthine oxidase itself, may serve as an additional regulatory mechanism.

ACKNOWLEDGEMENTS

This project was supported in part by BRSG Grant SO7RRO5390 awarded by the Biomedical Research Support Grant Program, Division of Research Resources, National Institutes of Health (U.S.A.). The interest and valuable comments of Dr. D.R. Babin are gratefully acknowledged.

REFERENCES

1. Fried, R. (1977) Acta Univ.Lodz. Folia Biochim. et Biophys. ser. II, 11, 273-295.
2. Fried, R. (1979) J.Neurosci.Res. (in press).
3. Fridovich, I. (1973) Account. Chem.Res. 5, 321-326.
4. Fridovich, I. (1976) in Free Radicals in Biology, Vol. 1, Pryor, W.A. ed., Academic Press, New York, pp. 239-277.
5. Malmstrom, B.G., Andreasson, L.-E. and Reinhammar, B. (1975) in The Enzymes, Vol. 12B, Boyer, P.D. ed., Academic Press, New York, pp. 507-579.
6. Halliwell, B. (1978) Cell Biol.Int.Rep. 2, 113-128.
7. Michelson, A.M. (1977) C.R.Soc.Biol. 170, 1137-1146.
8. Michelson, A.M., McCord, J.M. and Fridovich, I. eds. (1977) Superoxide and Superoxide Dismutases. Academic Press, New York, pp. 1-568.
9. Weser, U. (1973) in Structure and Bonding, Vol. 17, Dunitz, J.D., Hemmerich, P., Ibers, J.A., Jorgensen, C.K., Neilands, J.B., Reinen, D., Williams, R.J.P. eds., Springer Verlag, Heidelberg, pp. 1-65.
10. Fridovich, I. (1975) Annu.Rep.Med.Chem. 10, 257-264.
11. Bors, W., Saran, M., Lengfelder, E., Spottl, R. and Michel, C. (1974) Curr.Top.Rad.Res.Quart. 9, 247-309.
12. Fried, R., Fried, L.W. and Babin, D.R. (1973) Eur.J.Biochem. 33, 439-445.
13. Bray, R.C. (1975) in The Enzymes, Vol. 12B, Boyer, P.D. ed., Academic Press, New York, pp. 299-419.
14. Bray, R.C. (1963) in The Enzymes, Vol. 7, Boyer, P.D., Lardy, H. and Myrback, K. eds., Academic Press, New York, pp. 533-556.
15. Fried, R. and Fried, L.W. (1974) in Methods of Enzymatic Analysis, Bergmeyer, H.U. ed., Academic Press, New York, pp. 644-649.
16. Pomales, R., Bieber, S., Friedman, R. and Hitchings, G.H. (1963) Biochim. Biophys.Acta, 72, 119-120.
17. Hitchings, G.H. (1969) in Biochemical Aspects of Antimetabolites and of Drug Hydroxylation, Shugar, D. ed., Academic Presss, New York, pp. 11-22.
18. Prajda, N., Morris, H. and Weber, G. (1976) Cander Res. 36, 4639-4646.
19. Wyngaarden, J.B. (1978) in The Metabolic Basis of Inherited Disease, Stanbury, J.B., Wyngaarden, J.B. and Fredrickson, D.S. eds., Academic Press, New York, pp. 918-1010.
20. Wyngaarden, J.B. (1978) in The Metabolic Basis of Inherited Disease, Stanbury, J.B., Wyngaarden, J.B. and Fredrickson, D.S., eds., Academic Press, New York, pp. 1037-1044.
21. Annau, E., Eperjessy, A. and Felszeghy, O. (1942) Hoppe-Seylers Z.Physiol. Chem. 277, 58-65.
22. Annau, E., Eperjessy, A. and Zanthuresczky, Z.L. (1943) Hoppe-Seylers Z. Physiol.Chem. 279, 66-72.

23. Jacob, A. (1956) C.R.Acad.Sci. 242, 2180-2182.
24. Burton, K. (1948) Nature, 161, 606.
25. Lagnado, J.R. and Sourkes, T.L. (1956) Can.J.Biochem.Physiol. 34, 1095-1106.
26. Lagnado, J.R. and Sourkes, T.L. (1958) Can.J. Biochem.Physiol. 36, 587-594.
27. Fridovich, I. and Handler, P. (1956) J.Biol.Chem. 221, 323-331.
28. Fridovich, I. and Handler, P. (1961) J.Biol.Chem. 236, 1836-1840.
29. Spector, T. (1977) Biochem.Pharmacol. 26, 355-358.
30. Chytil, F. (1968) J.Biol.Chem. 213, 893-899.
31. Hirata, F. and Hayaishi, O. (1977) in Superoxide and Superoxide Dismutases, Michelson, A.M., McCord, J.M. and Fridovich, I. eds., Academic Press, New York, pp. 395-406.
32. Vesell, E.S., Passananti, G.T., and Greene, F.E. (1970) N.Engl.J.Med. 283, 1484-1488.
33. Valerino, D.M. and McCormack, J.J. (1971) Biochem.Pharmacol. 20, 47-55.
34. Misra, H.P., and Fridovich, I. (1972) J.Biol.Chem. 247, 3170-3175.
35. Sun, M. and Zigman, S. (1978) Anal.Biochem. 90, 81-89.
36. Bors, W., Michel, C., Saran, M. and Lengfelder, E. (1978) Biochim.Biophys. Acta, 540, 162-172.
37. Heikkila, R.E. and Cohen, G. (1973) Science, 181, 456-457.
38. Heikkila, R.E. and Cabbat, F.S. (1976) Anal.Biochem. 75, 356-362.
39. Heikkila, R.E. and Cabbat, F.S. (1977) Res. Commun.Chem.Pathol.Pharm. 17, 649-662.
40. Lee, H.-S., Carson, J.D., McMahon, T.P., Moyer, T.P. and Fischer, A.G. (1977) Life Sci. 20, 453-458.
41. Fischer, A.G. and Lee, H. (1973) Life Sci. 12, 267-275.
42. Mittal, C.K. and Murad, F. (1977) Proc.Natl.Acad.Sci. USA, 74, 4360-4364.
43. Weissmann, G. (1978) Hospital Practice, Sept. 1978, 53-62.
44. Prema, K. and Gopinathan, K.P. (1974) Biochem.J. 137, 119-121.
45. Rauckman, E.J., Rosen, G.M. and Kitchell, B.B. (1979) Mol.Pharmacol. 15, 131-137.
46. Kuthan, H., Tsuji, H., Graf, H., Ullrich, V., Werringloer, J. and Estabrook, R.W. (1978) FEBS Lett. 91, 343-345.
47. Strobel, H.W. and Coon, M.J. (1971) J.Biol.Chem. 246, 7826-7829.
48. Esnouf, M.P., Green, M.R., Hill, H.A.O., Irvine, G.G. and Walter, S.J. (1978) Biochem.J. 174, 345-348.
49. Handin, R.I., Karabin, R. and Boxer, G.J. (1977) J.Clin.Invest. 59, 959-956.
50. Ohmori, H., Komoriya, K., Azuma, A., Kurozumi, S. and Hashimoto, Y. (1979) Biochem.Pharmacol. 28, 333-334.
51. Pederson, T.C. and Aust, S.D. (1973) Biochem.Biophys.Res.Commun. 52, 1071-1078.
52. Thomas, M.J., Mehl, K.S. and Pryor, W.A. (1978) Biochem.Biophys.Res. Commun. 83, 927-932.
53. Bartosz, G., Fried, R., Grzelinska and Leyko, W. (1977) Eur. J. Biochem. 73, 261-264.
54. Hexum, T.D. and Fried, R. (1979) Neurochem.Res. 4, 73-82.
55. Lavelle, F., Michelson, A.M. and Dimitrijevic, L. (1973) Biochem.Biophys. Res.Commun. 55, 350-7.
56. Dixon, M. (1925) Biochem.J. 19, 507-512.
57. Siegel, S.M. and Weintraub, R.L. (1952) Biochem.J. 52, 324-326.
58. Bergel, F. and Bray, R.C. (1959) Biochem.J. 73, 182-192.
59. Fried, R. (1977) in 20th West-Central States Biochemical Conference, Manhattan, KS, Nov. 4-5, Abs. C-10.
60. Kalckar, H.M. (1947) J.Biol.Chem. 167, 429-443 and 461-465.

THE ROLE OF UBISEMIQUINONE IN SUPEROXIDE PRODUCTION BY DIHYDROOROTATE DEHYDROGENASE: A PROPOSED MECHANISM

HENRY JAY FORMAN
Department of Physiology, A-201 Richards Bldg., G-4, University of Pennsylvania, Philadelphia, PA 19174, USA.

ABSTRACT

This paper is the result of a new interpretation of data obtained with mammalian dihydroorotate dehydrogenase made possible through new information obtained in studies, by others, of the related enzyme, succinate dehydrogenase. Trumpower and Simmons[11] suggested that in the succinate dehydrogenase complex, ubiquinone is reduced in two single electron transfer steps (ubiquinone-ubisemiquinone-ubiquinol). A non-heme iron center was proposed as an intermediate carrier in the transfer of the second electron. Binding of an inhibitor to this iron-sulfur center inhibited the reduction of ubisemiquinone but did not affect superoxide production presumably because ubisemiquinone was the source of the O_2^-. As the older data obtained with dihydroorotate dehydrogenase is quite similar to that obtained with succinate dehydrogenase, an analogous role for ubisemiquinone in the mechanism of mammalian dihydroorotate dehydrogenase is proposed.

INTRODUCTION

Dihydroorotate dehydrogenase is an integral component of the inner mitochondrial membrane of mammals.[1] Previous studies of this enzyme from rat liver and brain have concerned the catalytic mechanism, purification and characterization of the "primary" enzyme and its interactions with detergents and salts.[2-7]

In the early papers in this series a mechanism was presented in which the "primary" dehydrogenase reduced ubiquinone and cytochrome b and produced superoxide, O_2^- through an unidentified cofactor.[3,4] Further studies with an isolated preparation indicated iron as the possible cofactor.[7] However, differences in catalytic properties between the isolated and membrane bound forms of the enzyme made the assignment of the O_2^- producing function to iron less certain.[7]

Since completion of those studies, investigations of related mitochondrial complexes, specifically the ubiquinone and iron-sulfur interactions, in the succinate dehydrogenase system[8-11] have provided insight into the mechanism O_2^- generation in that system which may be applicable to the dihydroorotate dehydrogenase complex in mammalian mitochondria.

Review of previous work on mammalian dihydroorotate dehydrogenase. Studies of membrane bound preparations of dihydroorotate dehydrogenase indicated that the enzyme could reduce ubiquinone[12,13] and or cytochrome b.[14] Forman and Kennedy[2] showed that superoxide could also be generated by this enzyme when the electron transport chain was blocked by cyanide or antimycin A. Thenoyltrifluoroacetone (TTFA), antimycin A and superoxide dismutase all partially inhibit the reduction of dichlorophenol indophenol (DCIP) by submitochondrial particles to which dihydroorotate had been added (Table 1).[3] When ubiquinone was added to particles an increased rate of DCIP reduction was observed. This increase, however, could be completely abolished by addition of TTFA (Table 2).[3] These results were used as supporting evidence upon which a model was based in which the site of superoxide production was proposed to be before the antimycin A block (which is between cytochrome b and c) and independent of ubiquinone reduction (Figure 1).[3] Further studies of partially purified enzyme from which cytochrome b was absent showed that the site of O_2^- production was not cytochrome b (Table 3).[4]

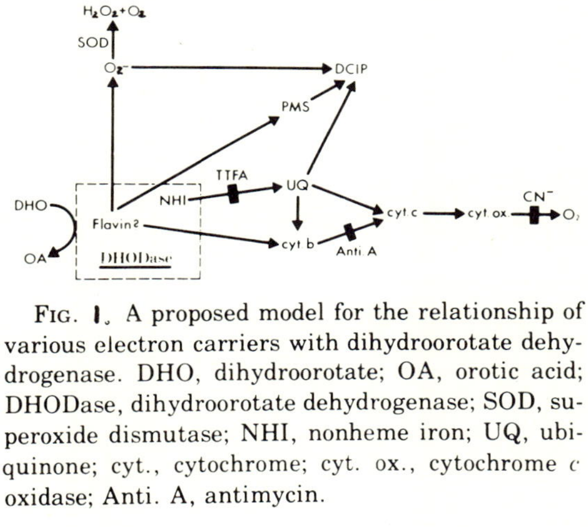

FIG. 1. A proposed model for the relationship of various electron carriers with dihydroorotate dehydrogenase. DHO, dihydroorotate; OA, orotic acid; DHODase, dihydroorotate dehydrogenase; SOD, superoxide dismutase; NHI, nonheme iron; UQ, ubiquinone; cyt., cytochrome; cyt. ox., cytochrome *c* oxidase; Anti. A, antimycin.

Reprinted with permission.[4]

It was also demonstrated that submitochondrial particles could not reduce DCIP directly and that phenazine methosulfate (PMS) could replace O_2^- as the electron carrying intermediate in DCIP reduction.[4] At the time this mechanism was proposed we believed that the enzyme probably contained both a flavin and nonheme iron centers. This was based on data obtained by Miller et al. on partially purified enzyme from bovine liver mitochonria[13] and <u>Neurospora</u> mitochondria.[15]

TABLE 1

EFFECTS OF ANTIMYCIN A, THENOYLTRIFLUOROACETONE, AND SUPEROXIDE DISMUTASE ON DCIP REDUCTION OF DIHYDROOROTATE DEHYDROGENASE

The assay system contained 100 mM Tris-HCl buffer (pH 8.5), 1 mM EDTA, 0.043 mM DCIP, 1mM KCN, 0.17 mM sodium dihydroorotate, and 1.1 mg of mitochondrial protein in a final volume of 3 ml. When the following compondents were present, their final concentrations were: antimycin A, 0.017 mg/ml; thenoyltrifluoroacetone, 1mM; and superoxide dismutase, 1 x 10^{-4}mM. Addition of high concentrations of these inhibitors gave identical results. Since changing the order of addition of inhibitors and substrate was without effect, the reactions were routinely started by addition of dihydroorotate. The values for DCIP reduction are averages of two to five measurements with variations of up to ± 10%.

Addition	DCIP reduced
	nmol/min
None	0.90
Antimycin A	0.90
Thenoyltrifluoroacetone	0.35
Superoxide dismutase	0.56
Thenoyltrifluoroacetone + superoxide dismutase	0

Reprinted with permission.[3]

TABLE 2

EFFECT OF ADDED UBIQUINONE-50 ON DIHYDROOROTATE DEHYDROGENASE ACTIVITY IN AGED RAT LIVER MITOCHONDRIAL PARTICLES

Assay conditions were the same as those described in the legend for Table 1. When present, the concentration of added ubiquinone-50 was 0.1 mg/ml.

Addition	DCIP reduced	
	Ubiquinone not added	Ubiquinone added
	nmol/min	
None	0.54	1.19
Thenoyltrifluoroacetone	0.36	0.34
Superoxide dismutase	0.19	0.84
Superoxide dismutase + thenoyltrifluoroacetone	0	0

Reprinted with permission.[3]

After several years a solubilized and apparently homogenous preparation was made of an enzyme which could reduce DCIP when dihydroorotate was added; however, this preparation could reduce DCIP in the absence of oxygen (while the membrane bound preparation could not) and was completely inhibitable by TTFA.[7] This purified enzyme which we considered to be the "primary" dehydrogenase contained non-sulfur-bound iron as the only detectable reducible prosthetic group (Table 4).[7] Although the solubilized enzyme could produce a small amount of O_2^- in the absence of DCIP, the addition of DCIP greatly enhanced the rate of O_2^- production (Table 5).[7] A mechanism was proposed for a co-reduction of DCIP and O_2 (Figure 2).

TABLE 3

EFFECT OF REMOVAL OF CYTOCHROME b ON SUPEROXIDE PRODUCTION[a]

	Cytochromes				DCIP reduced	
	a	a_3	b	c	Total	Superoxide-dependent
	(μM)				(nmol/min)	
Before centrifugation						
Reduced by dihydroorotate	0	58	230	47	0.39	0.16
Total cytochromes	312	165	450	368		
144,000g supernatant						
Reduced by dihydroorotate	0	0	0	0	0.29	0.18
Total cytochromes	0	0	5	326		

[a] Enzyme assays and determinations of spectra were carried out before and after separation of rat liver dihydroorotate dehydrogenase from cytochrome b. The latter was sedimented by centrifugation at 144,000g for 60 min (see Methods). Protein concentrations before centrifugation and in the 144,000g supernatant were 3.62 and 0.93 mg/ml, respectively. Cytochrome concentrations were determined by the method of Chance and Williams (Ref. 17). Spectra were measured using 2 ml of the enzyme preparations in a total volume of 3 ml, containing 67 mM potassium phosphate buffer, pH 7.8, 50 μg of antimycin A, and in the presence and absence of 0.67 mM sodium dihydroorotate. For determination of total cytochrome concentrations, sodium dithionite and KCN were added in excess.

Reprinted with permission.[4]

$$EH_2 + DCIP \longrightarrow EH^{\bullet} + DCIPH^{\bullet}$$

$$EH^{\bullet} + O_2 \longrightarrow E + O_2^{\bar{\bullet}} + H^+$$

$$DCIP + O_2^{\bar{\bullet}} + H^+ \rightleftharpoons DCIPH^{\bullet} + O_2$$

$$DCIPH^{\bullet} + O_2^{\bar{\bullet}} + H^+ \longrightarrow DCIPH_2 + O_2$$

$$2\,O_2^{\bar{\bullet}} + 2\,H^+ \xrightarrow{SOD} H_2O_2 + O_2$$

Fig. 2. Reprinted with permission.[7]

TABLE 4

EFFECTS OF SUPEROXIDE DISMUTASE AND ANAEROBIOSIS ON PURIFIED DIHYDROOROTATE DEHYDROGENASE ACTIVITY[a]

Sample	DCIP	Air	Superoxide dismutase	DCIP reduced	Orotate produced
				nmol/min	nmol/min
1	+	+	−	2.00	1.89
2	+	+	+	0.05	1.93
3	+	−	−	1.78	1.99
4	+	−	+	1.74[b]	2.09[b]
5	−	+	−		0.18
6	−	+	+		0.16
7	−	−	−		0.00

[a]Assays for DCIP reduction (600 nm) or orotate production (280 nm) at 23° C contained 0.1 M potassium phosphate buffer, pH 7.1, 0.05 M $KClO_4$, 6.7×10^{-4} M sodium dihydrorotate, 0.02% Triton X-100, and, when present, 8.6×10^{-5} M DCIP and/or 3.3×10^{-7} M superoxide dismutase (SOD) in 3 ml. The cuvette used was a slightly modified version of that designed by Hodgson et al. (Ref. 18). Samples were flushed with highly purified N_2 (research grade, Matheson) for 20-30 min.
[b]When air was allowed to re-enter number 4, the rate slowed gradually (to 0.25 after 20 min) for DCIP reduction but continued at the same rate for orotate production.

Reprinted with permission.[7]

TABLE 5

COFACTOR ANALYSES

Cofactor	Mol/mol subunit
Flavin	<0.018[a]
Labile sulfide	<0.012[a]
Zinc	2.17[b] ± 0.43[c]
Iron	8.60 ± 1.90[c]
Ubiquinone	<0.2[a]
Copper	<0.2[a]

[a]Limit of method and/or available material.
[b]Multiple analyses were performed on two separately purified enzyme samples.
[c]SD.

Reprinted with permission.[7]

It seemed reasonable to assume that some of these differences in catalytic properties were the result of changes in accessability of the active site of the enzyme once it was freed from the membrane. This was further supported by the observed inhibition o-phenanthroline of the soluble but not the membrane bound form of the enzyme (Ref. 7 and unpublished observations). At that time many aspects of the catalytic mechanism, particularly of the native membrane-bound enzyme complex, remained unsolved.

Related research on superoxide production in mitochondria. Boveris[15] has recently reviewed the literature on studies of production of O_2^- in mitochondria. It appears that most of the generation of this radical results from the autooxidation of ubisemiquinone.[16] In their most recent work, Trumpower and Simmons[11] have suggested that TTFA inhibits the reduction of ubisemiquinone (QH˙) to ubiquinol (QH_2) but does not inhibit the reduction of ubiquinone (Q) to ubisemiquinone (QH˙). Singer and co-workers have shown that TTFA interacts with an iron-sulfur center in the succinate dehydrogenase complex.[8] Ubiquinone reduction seems to occur in two one electron reduction steps which involved this iron-sulfur center (S-3).[9,10] This led Trumpower and Simmons to suggest that ubisemiquinone may then be the source of O_2^- production in the succinate dehydrogenase system.[11]

Trumpower and Simmons' results[11] with the succinate dehydrogenase system are remarkably similar to what was reported earlier for the dihydroorotate dehydrogenase system.[2-7] In both systems, an increase in cytochrome c reduction due to addition of antimycin A could be inhibited by superoxide dismutase but not by TTFA.[2-11] The inhibition of succinate dependent DCIP reduction by TTFA in the presence or absence of antimycin A and/or superoxide dismutase parallels the results obtained with dihydroorotate (Table 1).[3] Thus the two systems most likely have some common features in their catalytic mechanisms.

A new proposed model for the mechanism of dihydroorotate dehydrogenase. In the original proposed mechanism we assumed that ubiquinone could not be the site of O_2^- production in the dihydroorotate dehydrogenase complex. This conclusion was based on the results of the effects of added ubiquinone, TTFA and superoxide dismutase on the rate of DCIP reduction by the membrane-bound complex (Table 2).[3] While added ubiquinone increased the rate of DCIP reduction, none of the increase was due to O_2^- production. TTFA abolished the increase in DCIP reduction caused by added ubiquinone but had no effect on O_2^- mediated DCIP reduction. What we did not consider was the possible involvement of separate pools of ubiquinone or a two-step process in the reduction of ubiquinone to ubiquinol in which the second step is TTFA inhibitable.

It seems reasonable that ubisemiquinone would be produced in the dihydroorotate dehydrogenase complex as interaction of ubiquinone with an iron in the "primary" enzyme is likely to involve only a single electron transfer. Although the source of the electron which reduces ubiquinone to the semiquinone is probably different in the two systems, it is possible that the source of O_2^- production in the membrane-bound form of dihydroorotate dehydrogenase is the same as for succinate dehydrogenase. In many ways the results obtained with antimycin A, TTFA and superoxide dismutase by Trumpower and Simmons[11] for succinate dehydrogenase resemble the earlier reported observations with dehydroorotate dehydrogenase.[2-4,7] A mechanism which is consistent with the data from earlier work on dehydroorotate dehydrogenase and incorporates new information from the parallel work on succinate dehydrogenase is presented in Figure 3.

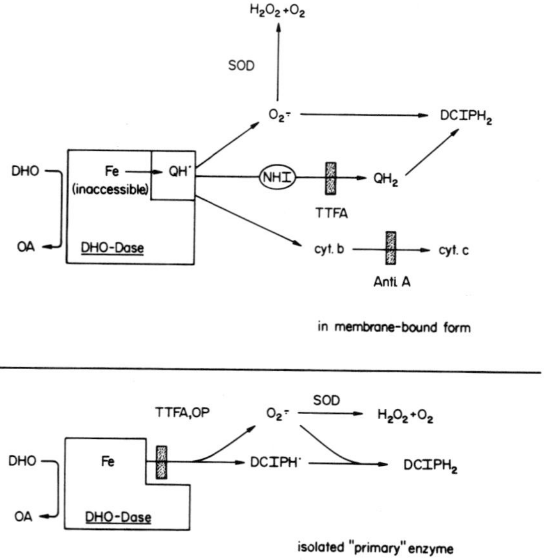

Fig. 3

In the proposed mechanism, two pools of ubiquinone can interact with the complex; one is an integral part of the enzyme while the other reacts with a TTFA inhibitable iron-sulfur center (as with S-3 in succinate dehydrogenase). Alternatively, the mechanism could involve one pool of ubiquinone with a non-heme iron center intermediate in the electron transfer process for the second electron; however, results of the studies with added ubiquinone (Table 2) are more consistent with the first mechanism. Differences in the effects of inhibitors and the catalytic activity of the "primary" dehydrogenase and membrane-bound forms may be the result of removal of the constituitive ubiquinone from the primary dehydrogenase which would allow access to the iron cofactor by inhibitors, DCIP and O_2. The reactions of ubiquinone and O_2 with the membrane-bound complex should then parallel the co-reduction of DCIP and O_2 seen with the isolated enzyme (Figure 2).

REFERENCES

1. Chen, J.J. and Jones, M.E. (1976) Arch.Biochem.Biophys. 176, 82-90.
2. Forman, H.J. and Kennedy, J. (1974) Biochem.Biophys.Res.Commun. 60, 1044-1050.
3. Forman, H.J. and Kennedy, J. (1975) J.Biol.Chem. 250, 4322-4326.
4. Forman, H.J. and Kennedy, J. (1976) Arch.Biochem.Biophys. 173, 219-224.
5. Forman, H.J. and Kennedy, J. (1977) Prep.Biochem. 7, 345-355.
6. Forman, H.J. and Kennedy, J. (1977) J.Biol.Chem. 252, 3379-3387.
7. Forman, H.J. and Kennedy, J. (1978) Arch.Biochem.Biophys. 191, 23-31.
8. Mowery, P.C., Steenkamp, D.J., Ackrell, B.A.C., Singer, T.P. and White, G.A. (1977) Arch.Biochem.Biophys. 178, 495-506.
9. Ruzicka, F.J., Beinert, H., Schlepter, K.L., Dunhasse, W.K. and Sands, R.H. (1975) Proc.Nat.Acad.Sci. USA, 72, 2886-2890.
10. Ingledew, W.J. and Ohnishi, T. (1977) Biochem.J. 164, 617-620.
11. Trumpower, B.L. and Simmons, Z. (1979) J.Biol.Chem. 254, 4608-4616.
12. Kuligowska, E. and Erecinska, M. (1976) Bull.Acad.Pol.Sci.Biol. 15, 187-190.
13. Miller, R.W., Kerr, C.T. and Curry, J.R. (1968) Can.J.Biochem. 46, 1099-1106.
14. Miller, R.W. and Curry, J.R. (1969) Can.J.Biochem. 47, 725-734.
15. Boveris, A. (1977) in Tissue Hypoxia and Ischemia, Reivich, M., Coburn, R., Lahiri, S. and Chance, B., eds., Plenum Publishing, New York, pp. 67-82.
16. Boveris, A., Cadenas, E. and Stoppani, A.O.M. (1976) Biochem.J. 156, 435-444.
17. Chance, B. and Williams, G.R. (1955) J.Biol.Chem. 217, 409-427.
18. Hodgson, E.K., McCord, J.M. and Fridovich, I. (1973) Anal.Biochem. 5, 470-473.

PRODUCTION OF SUPEROXIDE ANION BY THE NADH-DEHYDROGENASE OF MAMMALIAN MITROCHONDRIA

ALBERTO BOVERIS AND JULIO F. TURRENS
Instituto de Quimica Biologica, Facultad de Medicina, Universidad de Buenos Aires, Paraguay 2155, 1121 Buenos Aires, Argentina

ABSTRACT

Beef heart submitochondrial particles (SMP) supplemented with succinate and cyanide produce, upon addition of ATP, 0.6-0.8 nmol O_2^-/min per mg of protein. This O_2^- production supported by reversed electron transfer, is inhibited by rotenone, the uncoupler FCCP, oligomycin and Mn-superoxide dismutase. The amounts of rotenone, FCCP and oligomycin required to inhibit O_2^- production were similar to those needed to inhibit the ATP-dependent succinate-NAD reductase. NADH and rotenone-supplemented SMP generated 0.8 nmol O_2^-/min per mg of protein. Cyanide did not affect this activity, at variance with the marked inhibition produced by cyanide on the succinate and antimycin-supported production of O_2^-. The NADH-dehydrogenase segment of the respiratory chain is quantitatively less important (20-30 %) than the ubiquinone-cytochrome b area in terms of O_2^- production. Autoxidation of the flavin semiquinone of NADH-dehydrogenase could account for the observed O_2^- generation.

INTRODUCTION

Almost a decade ago mitochondria were recognized as physiological sources of H_2O_2 in the laboratories of the Johnson Research Foundation.[1-4] Determination of the mitochondrial rates of H_2O_2 production was approached by different assays: Chance and Oshino[1] used the catalase intermediate of the mitochondrial-peroxisomal fraction of rat liver as indicator; Loschen et al.[2] utilized the coupled oxidation of a fluorescent hydrogen donor (scopoletin); and Boveris et al.[3] and Boveris and Chance[4] trapped H_2O_2 as enzyme-substrate complex with peroxidases. Hydrogen peroxide production, that depends on the mitochondrial metabolic state, could be estimated to account for about 1-2 % of mitochondrial oxygen uptake under physiological conditions.[5-6] The effects of specific mitochondrial inhibitors on H_2O_2 production indicate a reduced member of the respiratory chain, located between the rotenone- and the antimycin-sensitive sites, as the most important autoxidizable component generating H_2O_2 in a side reaction to the main electron flow in the mitochondrial membranes.[2-4] Production of O_2^- by mitochondria was first reported by Loschen et al.[7], and subsequently, Boveris and Cadenas[8] and Dionisi et al.[9] showed that O_2^- is an almost stoichiometric precursor of H_2O_2 in mitochondrial membranes after determining ratios of $O_2^-:H_2O_2$ generation rates of about 2.0. Thereafter, Boveris et al.[10] and Cadenas et al.[11] proposed ubisemiquinone as the main univalent reductant of oxygen in mitochondria, on the basis of: (a) the linear relationship between succinate-reducible ubiquinone and H_2O_2 (and O_2^-) production in ubiquinone-depleted and ubiquinone-reconstituted membranes[10]; (b) the similar rate of O_2^- and H_2O_2 generation in Complexes I and III, which only have ubiquinone as a major common component[11]; (c) the rotenone sensitivity of H_2O_2 production by Complex I[11]; (d) the generation of O_2^- and H_2O_2 in the model reactions of quinol oxidation[11-12]; and (e) the relative stability of the ubisemiquinone radical in mitochondrial membranes[13] (see Figure 1). Trumpower and Simmons[14] have recently related O_2^- production by isolated succinate-cytochrome c reductase to the autoxidation of ubisemiquinone.

In this report we present evidence for a second site of O_2^- production in mammalian mitochondria: the flavoprotein NADH-dehydrogenase.

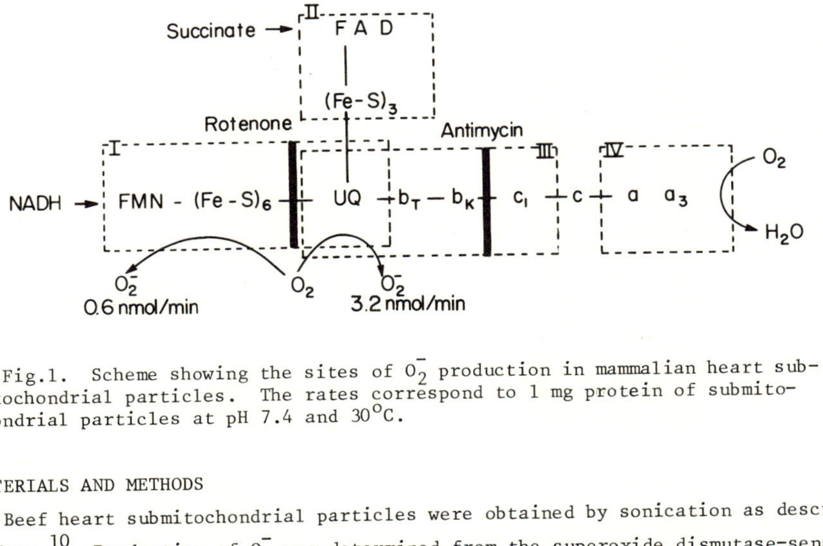

Fig.1. Scheme showing the sites of O_2^- production in mammalian heart submitochondrial particles. The rates correspond to 1 mg protein of submitochondrial particles at pH 7.4 and 30°C.

MATERIALS AND METHODS

Beef heart submitochondrial particles were obtained by sonication as described before.[10] Production of O_2^- was determined from the superoxide dismutase-sensitive rate of adrenochrome formation[15] measuring the absorption change at 480-575 nm in a double beam spectrophotometer as described before.[11] Mn-superoxide dismutase was prepared from rat liver mitochondria according to the procedure of Salin et al.[16] up to stage 4 (90% $(NH_4)_2SO_4$). Other experimental details have been previously described.[10-11] Temperature of the assays was 30°C.

RESULTS

Addition of NADH to submitochondrial particles supplemented with adrenaline

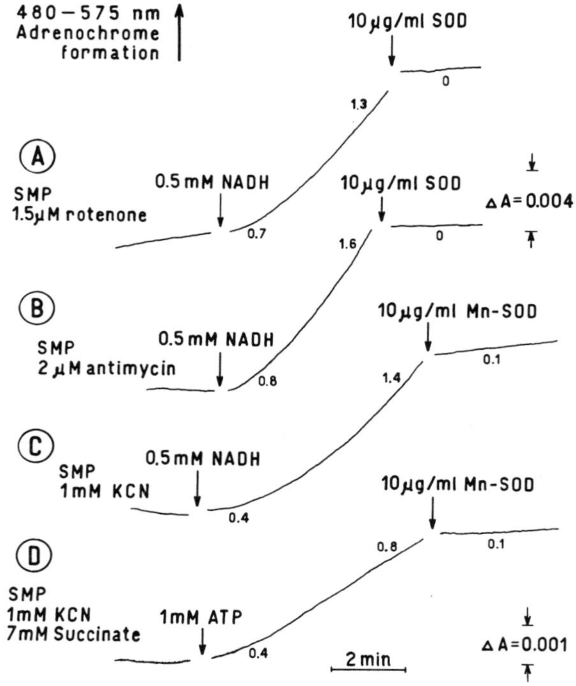

Fig.2. Production of O_2^- by beef heart submitochondrial particles. A, B and C: 0.72 mg protein/ml; D: 0.27 mg protein/ml. Reaction medium was 0.23 M mannitol, 0.07 M sucrose, 0.5 mM EDTA, 0.2 µM catalase heme, 30 mM Tris-HCl, pH 7.4. In trace D, 2.5 mM $MgCl_2$ were added. The numbers near the traces indicate nmol O_2^-/min per mg of protein.

and respiratory inhibitors such as rotenone, antimycin and cyanide, brought about adrenochrome formation at rates of 0.4-1.6 nmol/min per mg of protein (Figure 2, traces A, B and C). Superoxide dismutases were able to inhibit adrenochrome formation indicating O_2^- involvement. However, the rates of adrenochrome production increased linearly with time, i.e. adrenochrome was also being generated, in addition to the O_2^--dependent co-oxidation reaction, in a NADH and O_2^- (or H_2O_2) dependent process, similar to the H_2O_2 and NADH-dependent and O_2^--mediated reduction of cytochrome c reported by Misra and Fridovich.[17] Specificity of the adrenochrome assay is rather poor,[18] and consequently, initial rates (30-60 sec, not different from the values obtained from zero time extrapolation of the plots of adrenochrome formation rates versus time) were considered representing true rates of O_2^- generation. For rate calculations, adrenochrome formation and O_2^- generation were considered in a 1:1 molar ratio.[11]

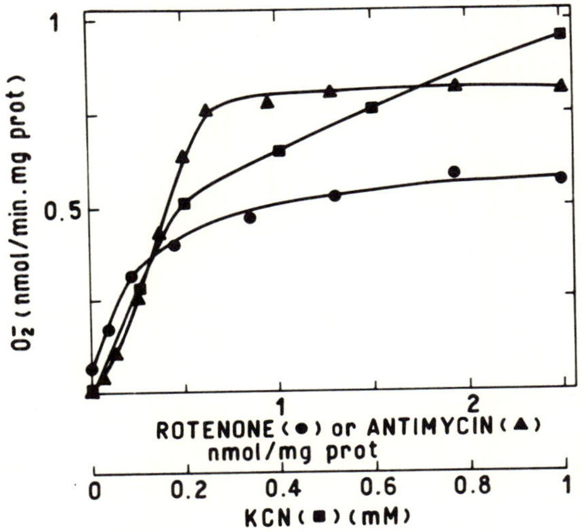

Fig.3. Effects of rotenone, antimycin and cyanide on O_2^- production by NADH-supplemented submitochondrial particles. Experimental conditions as in Figure 2, traces A, B and C.

The amounts of rotenone and antimycin required to produce a maximal effect on O_2^- production (0.36 and 0.60 nmol/mg of protein, respectively; Figure 3) indicate that their action was exerted through an inhibition of the main electron transport system. Higher rates of O_2^- generation were observed in the presence of antimycin, where oxidation of the reduced forms of ubiquinone is possible, than in the presence of rotenone, where only autoxidation of the components located on the substrate side of the rotenone-sensitive site was kinetically important. The profile of cyanide effect was biphasic (Figure 3); more than one site was involved in cyanide action and residual Cu-Zn superoxide dismutase trapped inside submitochondrial particles is likely one of the targets.

Submitochondrial particles in the presence of cyanide, succinate and adrenaline showed, upon addition of ATP, an adrenochrome formation, at a maximal rate of 0.8 nmol/min per mg of protein that was inhibitable by Mn-superoxide dismutase (Figure 2, trace D). An initial lag phase of about 30 sec, presumably due to membrane energization, was followed by a linear rate of adrenochrome (O_2^-) formation. No autocatalytic reaction (no NADH was present) was observed. Addition of ATP produced a reversed electron transfer in which reducing equivalents from the lower potential pool (succinate dehydrogenase, ubiquinone and cytochromes b; about 0 mV midpotential) were transferred to more electronegative components (NADH-dehydrogenase; about -290 mV midpotential) in an energy-dependent reaction. The ATP-dependent production of O_2^- was inhibited by rotenone, the uncoupler FCCP, and oligomycin with titers similar to those required to inhibit the energy-dependent succinate-NAD reductase (0.28 and 0.20 nmol/mg of protein; 0.07 and 0.22 µM; and 0.77 and 0.86 nmol/mg of protein, respectively; Figure 4, A, B and C).

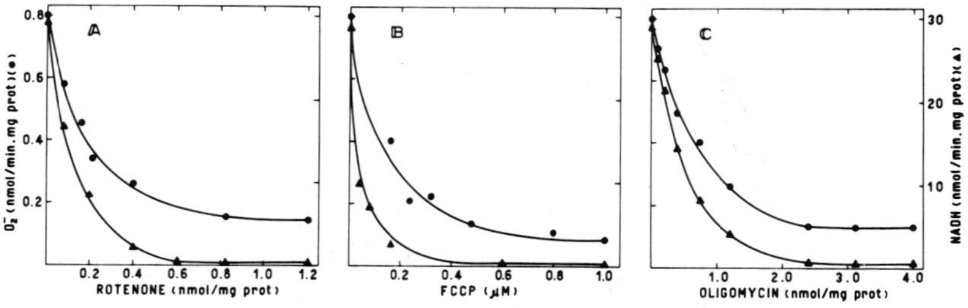

Fig.4. Effects of rotenone, FCCP (carbonyl cyanide p-trifluoromethoxyphenyl-hydrazone) and oligomycin on ATP-dependent O_2^- production and succinate-NAD reductase. Experimental conditions as in Figure 2, trace D.

The effect of cyanide on O_2^- production by submitochondrial particles clearly distinguished between the two segments of the respiratory chain able to generate O_2^- (Figure 5). Cyanide inhibited markedly, by 98 %, the production of O_2^- by submitochondrial particles in the presence of succinate and antimycin, with a half maximal effect at about 0.03 nM.[19] Advantage was taken of this fact to assay O_2^- production by NADH-dehydrogenase; the flavoprotein was reduced by reversed electron flow in the presence of cyanide to reduce background O_2^- production as it is shown in Figure 2, trace D. Contrarily to the marked inhibition produced on the O_2^- generated in the ubiquinone-cytochrome b region of the respiratory chain, cyanide slightly stimulated O_2^- formation in the presence of NADH and rotenone. This slight stimulation may be better interpreted as an inhibition of residual Cu-Zn-superoxide dismutase in the submitochondrial particles. Cyanide had no effect on the O_2^- production of submitochondrial particles supplemented with NADH and antimycin, where inhibition of O_2^- formation by the reduced forms of ubiquinone coexists with inhibition of superoxide dismutase.

DISCUSSION

The data presented here add NADH dehydrogenase to ubiquinone as components of the mitochondrial respiratory chain able to produce measurable amounts of O_2^- upon autoxidation. The production of O_2^- supported by reversed electron transfer clearly indicates that a low potential component, the flavoprotein NADH-dehydro-

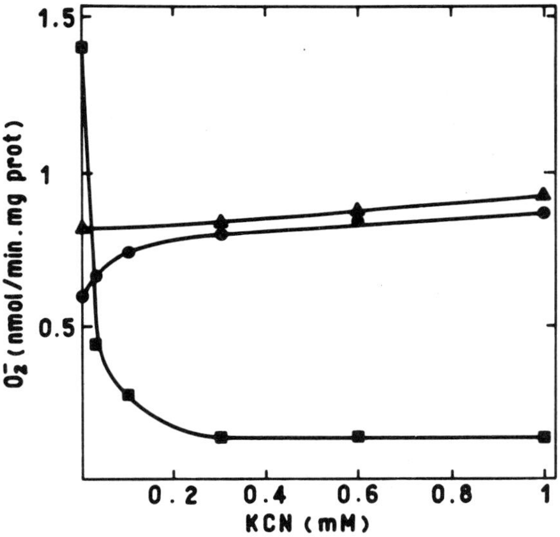

Fig.5. Effects of cyanide on O_2^- production by submitochondrial particles in the presence of (■) 6 mM succinate and 3 μM antimycin; (●) 0.5 mM NADH and 2 μM rotenone; and (▲) 0.5 mM NADH and 3 μM antimycin. Experimental conditions as in Figure 2, traces A and B.

genase, is responsible for O_2^- generation. Indeed, the experimental design is similar to the one used by Hinkle et al.[20] when H_2O_2 formation was measured to locate the first site of energy conservation with respect to the NADH-dehydrogenase flavin.

Previous observations indicated that O_2^- (and H_2O_2) were generated by a low potential component located on the substrate side of the rotenone-sensitive site of the respiratory chain. These observations were: (a) the energy-dependent and reversed electron transfer-driven formation of H_2O_2 by autoxidation of the NADH-dehydrogenase flavin reported by Hinkle et al.[20]; (b) the slight stimulation (about 1 nmol/min per mg of protein) produced by rotenone in the H_2O_2 production of pigeon heart mitochondria supplemented with malate-glutamate and uncoupler (Figure 2 in Ref. 4); and (c) the measured production of 0.6 nmol O_2^-/min per mg of protein in submitochondrial particles added to NADH and rotenone, reported by Thayer (Table 1 in Ref. 21). Plant mitochondria also show an O_2^- production by the autoxidation of the NADH-dehydrogenase.[22-24]

The NADH-dehydrogenase appears quantitatively less important than ubiquinone reduced forms as sources of O_2^- (see Figure 1). Production of O_2^- by NADH-dehydrogenase and ubiquinone reduced forms are less than additive, as it is shown in Figure 3, which could be argued as a kinetic modality of the ubiquinone cycle.[25] However, cyanide effect discriminates between the two sites of O_2^- generation, as it has been pointed out by Rich and Bonner[22] for their measurements of O_2^- production by potato mitochondria.

Generation of O_2^- by flavoprotein autoxidation is well documented[26-27]; flavoproteins acting as electron transferases and dehydrogenases appear likely to act in one-electron steps and to react with oxygen yielding O_2^- rather than H_2O_2 by formation of a stable semiquinone.[28] Anyhow, O_2^- formation from iron-sulfur centers is almost equally possible[29] and a clear distinction between the flavin semiquinone and iron-sulfur center moities of NADH-dehydrogenase as O_2^- source seems unlikely since electron transfer rates between intramolecular redox groups exceed the autoxidation rate by a factor of 10^4.

Mitochondrial membranes may be considered, with 20-60 mg of mitochondrial protein/g of tissue and rates up to 2 nmol O_2^-/min per mg of mitochondrial protein, as the most important loci of O_2^- production in animal cells, with the exception of the liver, where endoplasmic reticulum seems equally effective. It is no wonder Nature has provided mitochondria with their own and special superoxide dismutase.[27]

ACKNOWLEDGMENTS

This work was supported by grants from Consejo Nacional de Investigaciones Cientificas y Tecnicas, Argentina. A.B. and J.F.T. are Career Investigator and Fellow of the same institution.

REFERENCES

1. Chance, B. and Oshino, N. (1971) Biochem.J. 122, 225-233.
2. Loschen, G., Flohe, L. and Chance, B. (1971) FEBS Lett. 18, 261-264.
3. Boveris, A., Oshino, N. and Chance, B. (1972) Biochem.J. 128, 617-630.
4. Boveris, A, and Chance, B. (1973) Biochem.J. 134, 707-716.
5. Chance, B., Boveris, A., Oshino, N. and Loschen, G. (1973) in Oxidases and Related Redox Systems, King, T.E., Mason, H.S. and Morrison, M. eds., University Park Press, Baltimore, pp. 350-353.
6. Chance, B., Sies, H. and Boveris, A. (1979) Physiol.Rev. 59, 527-605.
7. Loschen, G., Azzi, A., Richter, C. and Flohe, L. (1974) FEBS Lett. 42, 68-72.
8. Boveris, A. and Cadenas, E. (1975) FEBS Lett. 54, 311-315.
9. Dionisi, O., Galeotti, T., Terranova, T. and Azzi, A. (1975) Biochim. Biophys. Acta, 403, 292-301.
10. Boveris, A., Cadenas, E. and Stoppani, A.O.M. (1976) Biochem.J. 156, 435-444.
11. Cadenas, E., Boveris, A., Ragan, C.I. and Stoppani, A.O.M. (1977) Arch. Biochem.Biophys. 180, 248-257.
12. Misra, H. and Fridovich, I. (1972) J.Biol.Chem. 247, 188-192.
13. Ingledew, W.J., Salerno, J. and Ohnishi, T. (1976) Arch. Biochem.Biophys. 177, 176-184.
14. Trumpower, B.L. and Simmons, Z. (1979) J.Biol.Chem. 254, 4608-4616.

15. Misra, H. and Fridovich, I. (1972) J.Biol.Chem. 247, 3170-3171.
16. Salin, M.L., Day, E.D.Jr. and Crapo, J.D. (1978) Arch.Biochem.Biophys. 187, 223-228.
17. Misra, H.P. and Fridovich, I. (1973) Biochim.Biophys.Acta, 292, 815-824.
18. Bors, W., Saran, M., Lengfelder, E., Michel, M., Fuchs, C. and Frenzel, C. (1978) Photochem.Photobiol. 28, 629-638.
19. Boveris, A. (1977) Adv.Exptl.Med.Biol. 78, 67-82.
20. Hinkle, P.C., Butow, R.A., Racker, E. and Chance, B. (1976) J.Biol.Chem. 242, 5169-5173.
21. Thayer, W.S. (1977) Chem.Biol.Interact. 19, 265-278.
22. Rich, P.R. and Bonner, W.D.Jr. (1978) Arch.Biochem.Biophys. 188, 206-213.
23. Hug, S. and Palmer, J.M. (1978) Plant Sci.Lett. 11, 351-356.
24. Boveris, A., Turrens, J.F., Sanchez, R.A. and Stoppani, A.O.M. (1979) Abstr. XI Intern.Congress Biochem., Toronto, 06-3-R94.
25. Mitchell, P. (1976) J.Theor.Biol. 62, 327-367.
26. Massey, V., Strickland, S., Mayhew, S.G., Howell, L.G., Engel, P.C., Matthews, R.G., Schuman, M. and Sullivan, P.A. (1969) Biochem.Biophys.Res. Commun. 36, 891-897.
27. Fridovich, I. (1974) Adv.Enzymol. 41, 35-97.
28. Massey, V. and Hemmerich, P. (1979) in Oxidase and Related Redox Systems, King, T.E., Mason, H.S. and Morrison, M. eds., University Park Press, Baltimore (in press).
29. Misra, H.P. and Fridovich, I. (1971) J.Biol.Chem. 246, 6686-6890.

HYDROPEROXIDE-DEPENDENT CHEMILUMINESCENCE OF SUBMITOCHONDRIAL PARTICLES AND ITS RELATIONSHIP TO SUPEROXIDE ANION AND OTHER OXYGEN RADICALS

ENRIQUE CADENAS, ALBERTO BOVERIS AND BRITTON CHANCE
Johnson Research Foundation, School of Medicine, G4, University of Pennsylvania, Philadelphia, PA 19104, USA.

ABSTRACT

Chemiluminescence measures integratively short-life free radicals and excited states derived from side reactions of the lipid peroxidation process. We offer evidence that singlet oxygen is generated in hydroperoxide supplemented submitochondrial particles detected by chemiluminescence methods. Light emission of hydroperoxide supplemented submitochondrial particles is enhanced by antimycin and rotenone and inhibited by cyanide and azide. Chemiluminescence of submitochondrial particles is sensitive to O_2^- and H_2O_2 scavengers (superoxide dismutase and catalase) as well as to singlet oxygen quenchers and enhancers (β-carotene and DABCO (1,4-diazabicyclo-(2,2,2)-octane) respectively). Ferricytochrome c is a main contributor to hydroperoxide-induced chemiluminescence of antimycin-inhibited submitochondrial particles; on the other hand, isolated ferricytochrome c is a source of light emission by catalyzing the homolytic rupture of the hydroperoxide. Hydroperoxide-induced chemiluminescence of submitochondrial particles and ferricytochrome c share a similar mechanism because of their similar sensitivity to β-carotene.

INTRODUCTION

The aerobic environment involves a potential threat for biological membranes which depends on the content of polyunsaturated lipids, the oxygen tension and the presence of catalysts able to initiate the free radical sequence of lipid peroxidation and, on the other hand, on the effectiveness of the enzymatic and non-enzymatic defenses: i.e., superoxide dismutase, catalase, glutathione peroxidase, vitamins A and E, etc.

Different oxygen species such as O_2, O_2^-, H_2O_2, $HO^\cdot$ and singlet oxygen ($^1O_2$) have been proposed as initiators or as rate limiting in the process of lipid peroxidation.[1-6] It seems sensible because of up to date research, that lipoperoxide formation does not depend on the appearance of a single oxidizing species but rather will include all of them in variable proportions in different systems. At any rate, there is substantial agreement that lipid peroxidation involves oxidative radical reactions that could lead to about complete breakdown of polyunsaturated fatty acids and disruption of biological membranes. Stable by-products, generated in the free radical chain reactions are lipid hydroperoxide, malonaldehyde, alkanes and alkenes.[7-8]

The reports of light emission during microsomal peroxidation [9-10] set expectations on chemiluminescence as an integrative measurement of short-life free radicals and excited states derived from side reactions of the lipid peroxidation process. In the present report, we offer evidence that $^1O_2$ is generated in hydroperoxide supplemented submitochondrial particles, detected by chemiluminescence methods. O_2^- participates in some of the modalities of the free radical chain reactions that lead to light emission by submitochondrial particles supplemented with organic hydroperoxides.

MATERIALS AND METHODS

Beef heart submitochondrial particles were obtained by sonication as described before.[11] β-Carotene was incorporated into submitochondrial particles by sonicating beef heart mitochondria in the presence of various amounts of β-carotene (20-50 nmol/mg protein) dissolved in chloroform and its concentration in the membranes was estimated at 494 nm (ε = 58 litre. $mmol^{-1}$. cm^{-1}). β-Carotene was also added to the suspensions as a methanolic solution (ε_{450} = 13.8 litre. $mmol^{-1}$. cm^{-1}). Chemiluminescence was measured with a photon counter as described before[12] in a 35 x 25 x 5 mm cuvette with constant air bubbling to assure a constant oxygen concentration. Chemiluminescence is expressed in the figures as counts/sec or photons/sec, both with identical meaning; these readings are taken over the background level of phototube dark current (350-400 counts/sec). The assays were carried out either in 50 mM-potassium phosphate buffer or in 0.23 M-mannitol/0.07 M-sucrose/20 mM-Tris/HCl buffer, pH 7.4. Temperature was maintained at 37°C.

RESULTS

Hydroperoxide induced chemiluminescence of submitochondrial particles. Beef heart submitochondrial particles showed light emission upon addition of an organic hydroperoxide. In the absence of added hydroperoxide, spontaneous chemiluminescence of submitochondrial particles was about 50 counts/sec; addition of hydroperoxide was necessary to detect substantial chemiluminescence yields. Antimycin-inhibited submitochondrial particles supplemented with t-butyl hydroperoxide showed light emission (Figure 1A): the reactions yielding light emission slowly increased their velocity at a rate of about 400 counts/sec per minute reaching their maximal chemiluminescence of 1500 counts/sec in about 8-10 minutes after the addition of the hydroperoxide and remained constant for about 10 minutes.

Submitochondrial particles showed a 50-60% enhancement of light emission when

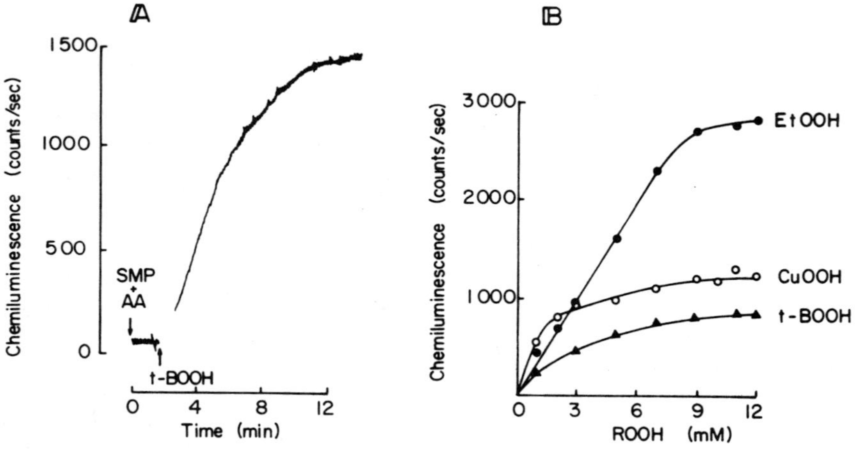

Fig. 1. Chemiluminescence of hydroperoxide supplemented submitochondrial particles. (A) 1.3 mg protein/ml as submitochondrial particles and 10 μM-antimycin in 50 mM-potassium phosphate buffer, pH 7.3 were added to 5mM-t-butylhydroperoxide. (B) the same conditions as in (A) (0.9 mg protein/ml) with different amounts of added hydroperoxide. t-BOOH: t-butylhydroperoxide; EtOOH: ethyl hydroperoxide; CuOOH: cumene hydroperoxide (α,α'-dimethylbenzoylhydroperoxide).

they were supplemented with either rotenone or antimycin and a marked inhibition (about 80-90%) in the presence of cyanide and azide (Table 1; Ref. 13). Addition of substrate to the particles, succinate or NADH, did not show any marked change on the hydroperoxide-dependent chemiluminescence (Table 1). About twice higher chemiluminescence yields were observed with phosphate buffer than with mannitol/sucrose/Tris buffer, presumably because of the reported inhibitory effect of Tris on chemiluminescence[14] or the scavenging effect of mannitol on hydroxyl radicals.[15] The primary hydroperoxide, ethyl hydroperoxide, was more effective in producing light emission than the tertiary hydroperoxides (cumene and t-butylhydroperoxides) (Figure 1B).

Generation of superoxide and chemiluminescence of submitochondrial particles. The rate of production of O_2^- by submitochondrial particles increased by successive washings of the particles whereas the content of superoxide dismutase decreased parallely. Chemiluminescence assayed under these conditions did not vary throughout the washings (Figure 2). The reproducibility of the results shown in Figure 2, mainly depends on the age of the mitochonrial membranes: we have been able to observe that recently prepared submitochondrial particles do increase their O_2^- generation by subsequent washings. In this way, the effect of added superoxide dismutase on chemiluminescence of submitochondrial particles might depend on the endogenous superoxide dismutase activity of the particles that seems to decrease with aging and washing.

TABLE 1

HYDROPEROXIDE-INDUCED CHEMILUMINESCENCE OF SUBMITOCHONDRIAL PARTICLES

1.65 mg protein/ml as submitochondrial particles in 50 mM-phosphate buffer, pH 7.4, were added of 5 mM-t-butylhydroperoxide to start the reaction. Concentration of different additions were: 10 μM-antimycin, 3 μM-rotenone, 5.5 mM-succinate, 0.4 mM-NADH, 0.2 mM-cyanide or 5.0 mM-azide. SMP: submitochondrial particles.

Conditions	Maximal Chemiluminescence (counts/sec)
SMP	1750
SMP + antimycin	3300
SMP + rotenone	3050
SMP + NADH	1700
SMP + succinate	1000
SMP + antimycin + NADH	3200
SMP + antimycin + succinate	1700
SMP + cyanide	530
SMP + azide	620

Chemiluminescence of antimycin-inhibited submitochondrial particles was inhibited by relatively high concentration of exogenous superoxide dismutase, 11-30 μg/ml, (Figure 3); these effective concentrations of dismutase are about 10-30 times higher than the dismutase level retained by submitochondrial particles. Inhibition was more effective when antimycin-inhibited submitochondrial particles were supplemented with low amount of t-butylhydroperoxide indicating a competition of superoxide dismutase with the tertiary hydroperoxide; an increase of 3.5-fold in the hydroperoxide concentration to produce half maximal chemiluminescence was observed with the addition of 11.5 μg superoxide dismutase/ml.

The production of O_2^- and H_2O_2 by submitochondrial particles have a similar dependency on pH[16] that was not matched by chemiluminescence assayed under the same conditions (Figure 4). Dependency of chemiluminescence on O_2^- production could be expected to decrease at pH values higher than 7.3. Decrease in light emission seems to indicate the protonated nature of an active group, either as light emitter or as a catalyst, with a pK of about 7.9.

<u>On the mechanism of light emission</u>. The participation of singlet molecular oxygen as the light emitting species is suggested by the effect on chemiluminescence of specific quenchers (β-carotene)[17] or singlet oxygen-dimol emission

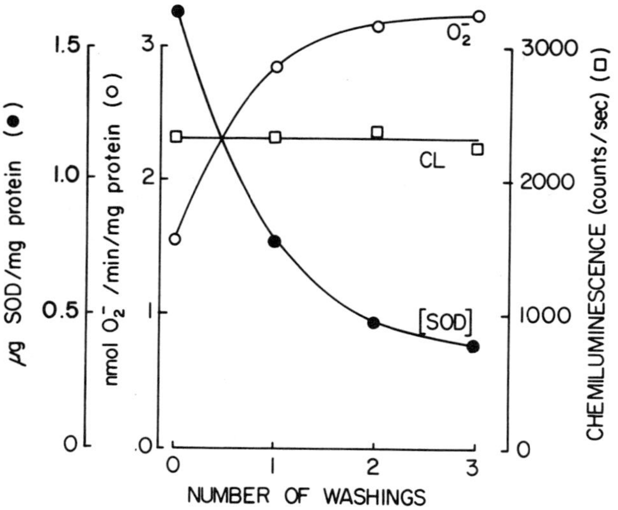

Fig. 2. Hydroperoxide induced chemiluminescence, superoxide anion generation and superoxide dismutase content of submitochondrial particles. Assays were performed in mannitol/sucrose/Tris buffer, pH 7.4. Superoxide dismutase (SOD) activity was measured according to the method described in Ref. 26; superoxide anion was measured by the adrenochrome assay (Ref. 27) and chemiluminescence as described in MATERIALS AND METHODS.

enhancers (DABCO: 1,4-diazabicyclo-(2,2,2)-octane) (Ref. 14). β-Carotene, incorporated into the particles by sonication (12 nmol/mg protein) decreased light emission by 50% (Figure 5, trace B). Superoxide dismutase and catalase addition, scavenging O_2^- and H_2O_2, respectively, accounted for 81% inhibition of the chemiluminescence (Figure 5, trace C), conditions under which the addition of the carotenoid did not produce a further inhibition (Figure 5, trace D).

DABCO enchanced both, the maximal chemiluminescence yield by about twice fold, and the rate of increase of light emission by 3-fold (Figure 6).

The effects of β-carotene and DABCO indicate a likely involvement of singlet oxygen in the hydroperoxide-dependent chemiluminescence of submitochondrial particles. The effect of superoxide dismutase and catalase suggests that O_2^- and H_2O_2 could be triggering a free radical chain reaction with production of light emitting species, and among them, singlet molecular oxygen.

Cytochrome c and chemiluminescence of submitochondrial particles. Because of the enhancing effect of antimycin and rotenone and the inhibitory effect of cyanide on hydroperoxide induced chemiluminescence of submitochondrial particles,[12] the role of an oxidized component of the respiratory chain, located on the oxygen side of the antimycin-sensitive site, as participating in light

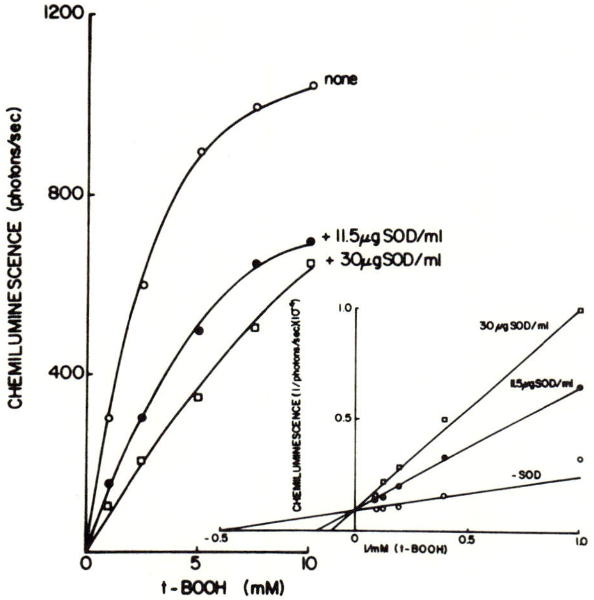

Fig. 3. Effect of superoxide dismutase on hydroperoxide-induced chemiluminescence of submitochondrial particles. Experimental conditions as described in Fig. 1A.

emission, seems quite sensible. Ferricytochrome c increased the hydroperoxide-induced chemiluminescence of submitochondrial particles, and the effect was more marked when antimycin was present (Figure 7), thus implying a minimal possibility of redox reactions between the added cytochrome c and the respiratory chain.

It is known that heavy metal complexes like haem act as catalysts or initiators of lipid peroxidation reactions.[18] Isolated cytochrome c catalyzes the homolytic scission of the hydroperoxide yielding alkoxy and HO· radicals.[19-21] Moreover, ferricytochrome c itself proved to be a source of light emission when supplemented with hydroperoxide.[22]

The light emitting reactions in antimycin-inhibited submitochondrial particles and in ferricytochrome c seemed to be quite similar since both systems became strongly inhibited (75-85%) on addition of β-carotene (Figure 8). β-Carotene proved to be a better inhibitor of the chemiluminescence reactions when added from a methanolic solution than when it was incorporated to the membranes by sonication (Figure 5). Furthermore, the former technique allowed a better and more manageable control of the amount of β-carotene acting in the system.

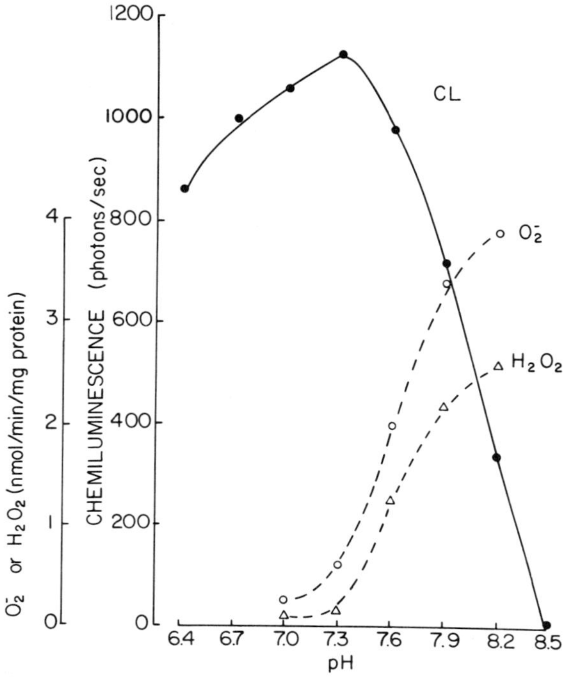

Fig. 4. Effect of pH on chemiluminescence of hydroperoxide-supplemented submitochondrial particles. Assay conditions as in Fig. 1A. The dotted line for O_2^- and H_2O_2 generation were obtained from Boveris and Cadenas (Ref. 16).

DISCUSSION

Components of the respiratory chain participate in reactions that lead to hydroperoxide-dependent chemiluminescence of submitochondrial particles; this view is supported by the effect of respiratory inhibitors on light emission.[13] A main contribution of the oxidized form of cytochrom c to chemiluminescence of submitochondrial particles is suggested by: (a) the enhancing effect of cytochrome c on the hydroperoxide-dependent chemiluminescence of antimycin-inhibited submitochondrial particles, (b) the light emission observed by the system cytochrome c/hydroperoxide,[22] (c) the similar inhibitory effect of β-carotene on chemiluminescence of submitochondrial particles and cytochrome c, and, (d) the already described ability of haemoproteins to catalyze peroxidation reactions.[7,21]

The requirements for chemiluminescence were an organic hydroperoxide, molecular oxygen and either mitochondrial membranes or ferricytochrome c. Light emission that arises from the reaction between cytochrome c and an organic

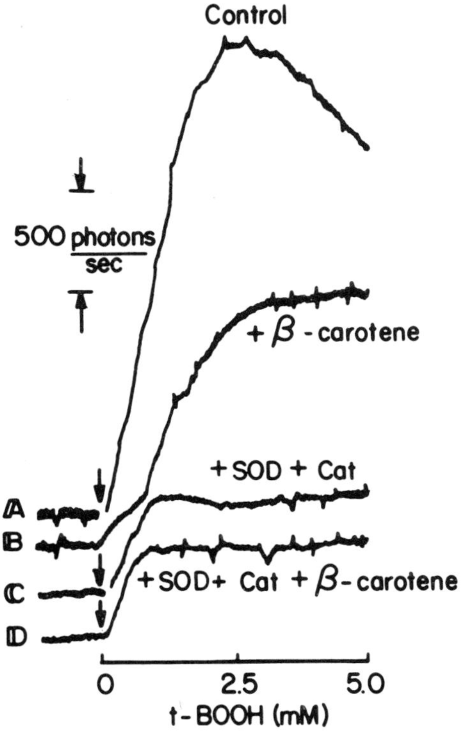

Fig. 5. Effect of β-carotene, superoxide dismutase and catalase on hydroperoxide-induced chemiluminescence of submitochondrial particles. Experimental conditions as described in Fig. 1. (a) Control; (B) submitochondrial particles containing 10 nmol β-carotene/mg protein incorporated by sonication; (C) Control membranes plus 20 μg superoxide dismutase/ml plus 20 μg catalase/ml. (D) The same membranes as (B) plus 20 μg superoxide dismutase/ml plus 20 μg catalase/ml.

hydroperoxide (extensive also to submitochondrial particles and cytochrome c supplemented submitochondrial particles) can be formulated in terms of the following reactions:

$$ROOH + c^{3+} \longrightarrow RO^{\cdot} + c^{3+}\text{-}OH^{\cdot} \qquad (1)$$

$$c^{3+}\text{-}OH^{\cdot} + ROOH \longrightarrow c^{3+} + H_2O + ROO^{\cdot} \qquad (2)$$

$$c^{3+}\text{-}OH^{\cdot} + RH \longrightarrow c^{3+} + H_2O + R^{\cdot} \qquad (3)$$

$$R^{\cdot} + O_2 \longrightarrow ROO^{\cdot} \qquad (4)$$

$$ROO^{\cdot} + ROO^{\cdot} \longrightarrow {}^1O_2 + RO + ROH \qquad (5)$$

$$2\,{}^1O_2 \longrightarrow {}^3O_2 + h\nu \qquad (6)$$

(7)

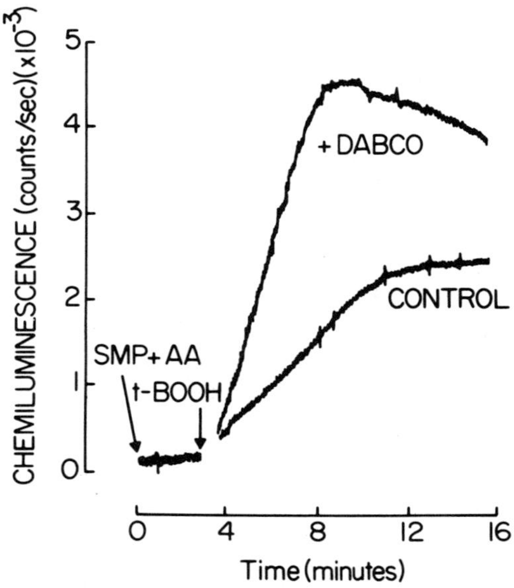

Fig. 6. Effect of DABCO on hydroperoxide-induced chemiluminescence of submitochondrial particles. Experimental conditions as in Fig. 1A; 10 mM-DABCO.

Reaction (1) is the cytochrome c catalyzed homolytic scission of the oxygen-oxygen bond of the hydroperoxide molecule[19] that yields alkoxy radical and cytochrome c bound hydroxyl radical. Iron bound hydroxyl radical may react with either t-butylhydroperoxide (reaction (2)) or RH (reaction (3)); RH meaning the primary carbons of t-butylhydroperoxide molecule or any residue of the cytochrome c molecule abstracting an hydrogen atom[28] and yielding either t-butylperoxy or alkyl radicals, respectively. Peroxy radicals are also formed by incorporation of one molecule of oxygen to the alkyl radical (reaction (4)). Reaction (5) is a radical termination reaction interpreted on the basis of a Russell's mechanism[29] for the self reaction of primary and secondary hydroperoxides. Reactions (6) and (7) give account of the nature of the observed light emission in terms of either a dimol-emission of singlet oxygen[30] or the formation of excited carbonyl groups throughout a dioxetane mechanism.[25]

The previous reactions do not contemplate a participation of O_2^- or H_2O_2 which makes the mechanism only applicable to the hydroperoxide-dependent chemiluminescence insensitive to superoxide dismutase, which is the case for cytochrome c and submitochondrial particles in the absence of substrate. The involvement of O_2^-, H_2O_2 or $OH^\cdot$, for instance in the substrate- and rotenone or

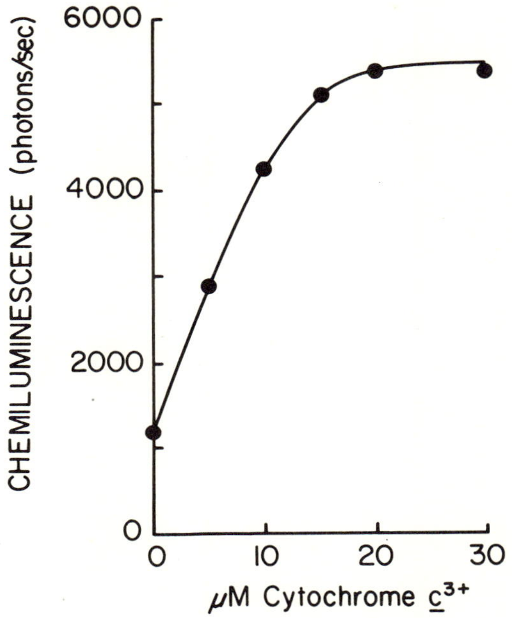

Fig. 7. Effect of externally added ferricytochrome c on chemiluminescence of hydroperoxide-supplemented submitochondrial particles. Assay conditions as described in Fig. 1A.

antimycin-supplemented submitochondrial particles, could be understood in terms of a metal-catalyzed $OH^{\cdot}$ radical formation[15,23-24] able to initiate free radical processes:

$$O_2^- + H_2O_2 \longrightarrow HO^{\cdot} + O_2 + OH^- \qquad (8)$$

Chemiluminescence measures integratively side reactions of the free-radical lipid peroxidation process, which are non-stoichiometric; this fact sets limits for the assessment of the involvement of individual chemical species. The observed light emission, about 10^5 counts/sec/litre corresponds to a rate of 10^{-15} M/sec of the light emitting species (considering a detection efficiency of 0.1%) and to the emission of one photon per about 10^{10} oxygen molecules utilized in the oxidative radical chain.

Although β-carotene effect only indicates that the process is likely to be singlet oxygen-dependent, i.e., that singlet oxygen would either (a) yield light emission by itself on the dimol-emission or (b) lead to the production of another light emitting species (for instance 1-2, dioxetane derivatives), as pointed out by Foote[25]; the DABCO effect suggests that light emission might be due to singlet oxygen dimol-emission.[14]

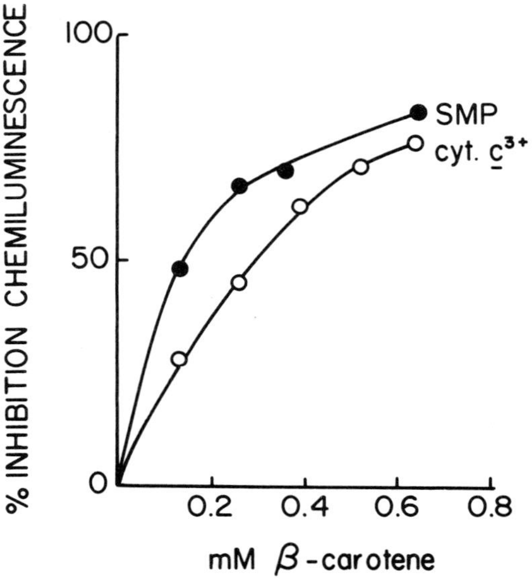

Fig. 8. Effect of β-carotene on hydroperoxide-induced chemiluminescence of cytochrome c and submitochondrial particles. β-Carotene was added as a methanolic solution in the assay mixture for both experiments. Assay conditions as in Fig. 1A.

ACKNOWLEDGEMENTS

This work was supported by USPHS grants TW-02457 and HL-SCOR-15061. EC is a USPHS Fogarty International Research Fellow.

REFERENCES

1. Pederson, T.C. and Aust, S.D. (1973) Biochem.Biophys.Res.Commun. 52, 1071-1078.
2. Fong, K.L., McKay, P.B., Poyer, J.L., Keele, B.B. and Misra, H. (1973) J.Biol.Chem. 248, 7792-7797.
3. Bus, J.S., Aust, S.D. and Gibson, J.E. (1974) Biochem.Biophys.Res.Commun. 58, 749-755.
4. Kellog, E.W. and Fridovich, I. (1975) J.Biol.Chem. 250, 8812-8817.
5. King, M.M., Lai, E.K. and McKay, P.B. (1975) J.Biol.Chem. 250, 6496-6502.
6. Kellog, E.W. and Fridovich, I. (1977), J.Biol.Chem. 252, 6721-6728.
7. Tappel, A.L. (1973) Fed.Proc. 32, 1870-1874.
8. Dillard, C.J., Dumelin, E.E. and Tappel, A.L. (1977) Lipids, 12, 109-114.
9. Nakano, M., Noguchi, T., Sugioka, K., Fukujama, H., Sato, M., Shimizu, Y., Tsuji, Y. and Inaha, H. (1975) J.Biol.Chem. 250, 2404-2406.
10. Sugioka, K. and Nakano, M. (1976) Biochim.Biophys.Acta, 243, 203-216.
11. Boveris, A., Cadenas, E. and Stoppani, A.O.M. (1976) Biochem.J. 156, 435-444.

12. Boveris, A., Chance, B., Filipkowsky, M., Nakase, Y. and Paul, K.G. (1978) in Frontiers of Biological Energetics: Electrons to Tissues, Scarpa, A., Dutton, P.L. and Leigh, J.S.Jr., eds., Vol 2, Academic Press, New York, pp. 975-984.
13. Boveris, A., Reiter, R., Cadenas, E. and Chance, B. (1979) in 23rd Annual Meeting of the Biophysical Society, Atlanta, Georgia: Biophys.J. 25, 48a.
14. Deneke, C.F. and Krinsky, N.I. (1977) Photochem.Photobiol. 25, 299-304.
15. Halliwell, B. (1978) FEBS Lett. 92, 321-326.
16. Boveris, A. and Cadenas, E. (1975) FEBS Lett. 54, 311-314.
17. Foote, C.S. and Denny, R.W. (1968) J.Amer.Chem.Soc. 90, 6233-6235.
18. Demopoulus, H.B. (1973) Fed.Proc. 32, 1959-1961.
19. Desair, I.D. and Tappel, A.L. (1963) Lipid Res. 4, 204-207.
20. Roubal, W.T. and Tappel, A.L. (1966) Arch.Biochem.Biophys. 113, 50-55.
21. Banks, A., Eddice, E. and Smith, J.G.M. (1961) Nature, 190, 908-909.
22. Cadenas, E., Boveris, A., Reiter, R. and Chance, B. (1979) in Abstr. XI° Intern.Cong.Biochem., Toronto, Canada.
23. Cohen, G. (1977) in Superoxide and Superoxide Dismutases, Michelson, A.M., McCord, J.M. and Fridovich, I., eds., Academic Press, London, pp. 317-321.
24. McCord, J.M. and Day, E.D.Jr. (1978) FEBS Lett. 86, 139-142.
25. Foote, C.S. (1976) in Free Radicals in Biology, Pryor, W.A., ed., Vol 2, Academic Press, New York, pp. 85-133.
26. McCord, J.M. and Fridovich, I. (1969) J.Biol.Chem. 244, 6049-6055.
27. Misra, H.P. and Fridovich, I. (1972) J.Biol.Chem. 247, 3170-3175.
28. Pryor, W.A. (1978) Photochem.Photobiol. 28, 787-801.
29. Russell, G.A. (1957) J.Amer.Chem.Soc. 79, 3871-3877.
30. Seliger, H.H. (1964) J.Chem.Phys. 40, 3133-3134.

REGULATION OF MITOCHONDRIAL SUPEROXIDE DISMUTASE BY MITOCHONDRIALLY PRODUCED SUPEROXIDE ANIONS AND HYDROGEN PEROXIDE IN A HIGHER EUKARYOTE

ANNE P. AUTOR AND JEFFREY B. STEVENS

The Toxicology Center, Department of Pharmacology, The University of Iowa, Iowa City, Iowa 52242, USA

INTRODUCTION

In a series of studies from this laboratory, the discovery of a rapid oxygen-dependent induction of neonatal rat lung superoxide dismutase, as well as catalase has been described.[1-4] Several characteristics of this system are now well established. The pulmonary macrophages are the only lung cells with the capacity for rapid synthesis and intracellular accumulation of both enzymes upon exposure to hyperoxia. Furthermore, neither the pulmonary macrophages of adult rats nor any other lung cell of these animals induced superoxide dismutase or catalase under similar experimental conditions. Although the lung content of both CuZn superoxide dismutase and Mn superoxide dismutase is elevated by oxygen exposure, only the latter is enhanced intracellularly.[5] Normobaric oxygen is an effective inducing agent at a concentration of 70% and above, in both *in vivo* and *in vitro* oxygen exposure systems.[6] Under *in vivo* conditions, neonatal rats are maintained in a hyperoxic atmosphere for 24 hours and then the lung cells are isolated for enzymatic assay. Under *in vitro* conditions, isolated pulmonary macrophages are suspended in buffer and incubated under a hyperoxic atmosphere for 5 hours.[3] In addition to these age-dependent differences in inductive response to hyperoxia, the basal content of mitochondrial superoxide dismutase and catalase in pulmonary macrophages changes with age. The content of both enzymes in the macrophages from neonatal animals is four-fold higher than in the cells from adult animals.

Superoxide dismutase and catalase are currently viewed as enzymes which protect intracellular components from damage inflicted by oxygen-centered free radicals generated either chemically or metabolically.[7,8]

The association of superoxide dismutase and catalase with resistance to the toxicity of partially reduced oxygen species is thought to be derived from their respective catalytic activities which together remove the reactants for the non-enzymatic, iron-catalyzed production of the highly reactive hydroxyl radical.[9,10] The following reaction sequence illustrate the interrelationship between these reactions:

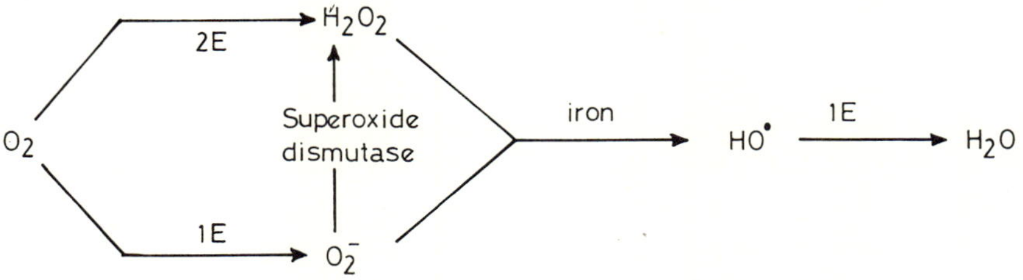

$$2O_2^- + 2H^+ \xrightarrow{\text{Superoxide dismutase}} H_2O_2 + O_2$$

$$2H_2O_2 \xrightarrow{\text{catalase}} 2H_2O + O_2$$

Fridovich and coworkers have clearly shown that superoxide dismutase content in prokaryotic and simple eukaryotic cells is governed by growth conditions which then affect the cellular content of superoxide dismutase.[11-15] Growth under conditions in which oxygen radical generation occurs, i.e., in normobaric 100% oxygen atmosphere, in a hyperbaric oxygen atmosphere or in the presence of autoxidizing chemical agents such as paraquat, yields organisms which contain elevated superoxide dismutase. A direct relationship was observed by these investigators between enhanced superoxide dismutase content and resistance to the toxic consequences of hyperoxia. Microorganisms in which superoxide dismutase content was elevated were relatively unaffected by growth in a hyperbaric atmosphere. Furthermore, a general relationship appears to exist between obligate anaerobic organisms (those which do not metabolize molecular oxygen) and absence of superoxide dismutase.[16-18] A recent report has described a similar direct relationship in plants between intracellular superoxide dismutase and catalase content and resistance to the oxygen radical-producing herbicide, paraquat.[19]

Neonatal rats exhibit a strong resistance to the toxic consequences of hyperoxic exposure when compared with adult rats. Our work has established that the age pattern of elevated superoxide dismutase and catalase content in neonatal rat lung is coincident with the age when these neonates are strongly resistant to oxygen toxicity.[4,5] It has been well documented by the work of Boveris and his colleagues[30] that superoxide radicals are generated from the electron transport chain and hydrogen peroxide is released from intact mitochondria during oxidative metabolism. This information led us to test the possibility that there exists an age-dependent difference in the capacity of the mitochondria of pulmonary macrophages to generate superoxide radicals. If oxygen radicals affect the biosynthesis of superoxide dismutase and catalase in higher eukaryotics, as is true for prokaryotes and simple eukaryotes, a higher steady state concentration of partially reduced oxygen could explain the high basal content of these enzymes in neonatal rat cells compared with the content found in cells from adult animals. To test this possibility, pulmonary macrophages were isolated from neonatal and older rats. Mitochondrial preparations were made from neonatal cells in which a variety of mitochondrial functions were tested and compared with results obtained from the mitochondria of adult rat macrophages.

MATERIALS AND METHODS

Animals, Cells and Tissues Preparations. The Sprague-Dawley rats utilized for these studies were bred and raised at The University of Iowa Animal Care Facility. Mature female rats were exposed to a 12-hour breeding period. The onset of pregnancy was timed from the midpoint of this 12-hour period.

Pulmonary macrophages were isolated by the mincing technique as described earlier.[3] The cells were freed of contaminating blood cells by hypotonic shock. Cell viability was then quantitated by erythrocin B dye exclusion.[20] Cell number was determined with a hemocytometer.

Homogenates of pulmonary macrophages were obtained by disrupting the cells for 20 seconds in a Braun Cell Disruptor. Glass beads (0.45-0.50 mm diameter) were utilized at a concentration of 5 gm/10 ml cell suspension. Cells were suspended in isotonic phosphate-buffered saline at a concentration of 5×10^6 cells/ml. For the mitochondrial experiments, cells were suspended in 0.25 M sucrose containing 1 mM EDTA and 2 mM Tris-Cl, pH 7.4. Mitochondria were then isolated by differential centrifugation in a Sorvall RC2-B refrigerated centrifuge. Protein concentration in all cell samples was assessed by the method of Lowry *et al.*[21]

Enzyme Assays. Superoxide dismutase activity was measured by the method of McCord and Fridovich.[22] The assay mixture contained 10 μM ferricytochrome c, 100 μM xanthine, 50 mM potassium phosphate, pH 7.8, 0.1 mM EDTA, 0.015 to 0.02 units of xanthine oxidase, 5×10^{-5} M NaCN and cell homogenates. Cell homogenates were initially sonicated for 3 minutes at 35 watts at 0-4°C in a Branson Model W185 Sonifier cell disruptor. One unit of superoxide dismutase activity was defined as the amount which inhibited the reference rate of ferricytochrome c reduction by 50%. The reaction rate of a sample containing no cell homogenate (reference rate) was measured at 550 nm in a Varian Spectrophotometer model 635D at 25°C. Increasing volumes of cell preparations were introduced into the reaction mixture until the rate of reduction fell to within 40-60% of the reference rate. The two forms of superoxide dismutase were distinguished by their differential sensitivity to cyanide. Enzyme activity in the cell preparations in the presence of 1 mM CN^- was assumed to reflect the activity of the mitochondrial mangano enzyme.[23]

Catalase activity was measured by the method of Beers and Sizer,[24] glutathione peroxidase activity by the method of Paglia and Valentine[25] and cytochrome oxidase activity by the method of Wharton and Tzagoloff.[26] Reaction rates were measured spectrophotometrically at 550 nm.

Hydrogen Peroxide Generation from Intact Mitochondria. Mitochondrial generation of H_2O_2 was quantitated by the method of Boveris and Chance.[27] Mitochondria isolated from pulmonary macrophages were placed in the following buffer solution: 225 mM mannitol, 75 mM sucrose, 30 mM Tris-MOPS, pH 7.4, 6 mM succinate, 6mM glutamate and 1.8 mg of purified cytochrome c peroxidase. Reactions were run in an Aminco DW-2A spectrophotometer at room temperature with absorbance changes monitored at 424 nm-404 nm. The extinction coefficient used for the cytochrome c peroxidase-H_2O_2 complex was 55 mM^{-1} cm^{-1}. Hydrogen peroxide generation was always measured during State 4 respiration. Cytochrome c peroxidase was purified from yeast by the method of Yonetani and Ray.[28]

ADP:O Ratio Determination. ADP:O ratios of isolated mitochondria were obtained with the use of a Clark electrode fitted into a 3/4 ml oxygraph. Data were obtained in the following solution: 100 mM KCl, 15 mM KPO_4, 50 mM Tris-Cl, pH 7.5, 5% glucose, 10 mM glutamate and 10 mM malate. Increments of 50-100 nmoles ADP were added to obtain State 3 respiration.

Superoxide Anion Generation from Submitochondrial Particles. Submitochondrial particles were prepared by the method of Dionisi *et al*.[29] Upon sonication, the particles were centrifuged at 27,000 rpm for 60 minutes in a Beckmann L3 centrifuge equipped with a 30 rotor, then washed in buffer and centrifuged again. The pelleted material was resuspended in buffer, and superoxide dismutase assays were performed. Only preparations exhibiting little detectable superoxide dismutase activity were used in these studies. Superoxide anion production was measured as the superoxide dismutase-inhibitable reduction of exogenous cytochrome c. The assay system for these studies included: 230 mM mannitol, 70 mM sucrose, 30 mM Tris-MOPS, pH 7.8, 1 mM ferricytochrome c, 1 μM antimycin A, 1 μM rotenone, 6 mM succinate and 50 μM NaCN. Absorbance changes were monitored at 550 nm-540 nm. A blank rate was obtained by adding 200 units/ml purified superoxide dismutase to the above assay mixture.

RESULTS

Homogenates of isolated pulmonary macrophages obtained from minced lung tissue were assayed after disruption for both forms of superoxide dismutase and catalase. The activities of all three enzymes were determined and compared in cells from animals of increasing age from neonates to adults. The activity of both catalase and cyanide-resistant, mitochondrial superoxide dismutase diminished approximately four-fold from a high level at birth (Table 1). CuZn superoxide dismutase, however, did not change with age. For comparative purposes, the activities of two other macrophage enzymes located in different subcellular compartments were also assessed as a function of animal age. Neither cytochrome oxidase nor glutathione peroxidase activity was altered in an age-related manner. Several other enzymes tested, including the cytosolic enzyme lactic dehydrogenase (data not shown), were also unchanged, thus emphasizing the specific character of the age-related decrease of catalase and mitochondrial superoxide dismutase. The inverse relationship with oxygen-mediated induction of catalase and mitochondrial superoxide dismutase in these cells is also shown in Table 1. No change in the other three enzymes was observed under similar experimental conditions.

TABLE 1

ACTIVITY OF REPRESENTATIVE ENZYMES IN RAT PULMONARY MACROPHAGES OBTAINED FROM RATS OF INCREASING AGE

Enzyme Activity (units per 10^6 cells)	Animal Age (days)						
	Newborn	2	5	10	15	30	Adult
CuZn Superoxide Dismutase[a]	0.8	0.7	0.8	0.5	0.6	0.7	0.6
Glutathione peroxidase	0.4	0.5	0.4	0.4	0.5	0.4	0.4
Cytochrome oxidase ($x10^2$)	2.9	2.7	2.9	2.8	2.9	2.8	2.9
Catalase[b]	(22) 22	(20) 23	(22) 15	(23) 10	(5) 4	(4) 3	(3) 3
Mn Superoxide dismutase[b,c]	(4.1) 4.1	(3.9) 3.9	(3.3) 2.5	(3.8) 2.0	(3.5) 1.0	(1.0) 0.9	(1.0) 1.0

[a]Enzyme activity inhibited when assayed in the presence of 1 mM NaCN.
[b]Numbers in parentheses indicate enzymatic activity in cells incubated *in vitro* in 95% oxygen (1 atm.) for 5 hours before assay.
[c]Enzyme activity not inhibited when assayed in the presence of 1 mM NaCN.

To determine whether a relationship might exist between the regulation of the biosynthesis of catalase and mitochondrial superoxide dismutase and intracellularly generated oxygen radicals or partially reduced oxygen species, H_2O_2 generation by intact mitochondria prepared from isolated pulmonary macrophages was determined. With increasing animal age, the H_2O_2 generation from isolated mitochondria diminished four-fold from an initial rate of 1.2 to 1.3 nmol per min per mg of mitochondrial protein in newborns to an average rate of 0.33 nmol per min per mg (Table 2). This latter rate is comparable to reported values for heart mitochondria and is equivalent to approximately 2% of the total mitochondrial oxygen consumption.[31] Typical rates of H_2O_2 production are illustrated in Figure 1. Assays for H_2O_2 generation were conducted in the presence of substrate and antimycin A.

TABLE 2

ADP:O RATIO AND HYDROGEN PEROXIDE GENERATION IN MITOCHONDRIA ISOLATED FROM PULMONARY MACROPHAGES PREPARED FROM RATS OF INCREASING AGE

Animal Age (days)	ADP/O[a]	H_2O_2[b] (nmoles/min/mg of protein)
3	2.3	1.20
5	(not done)	1.30
7	2.3 (Expt.1)	1.25 (Expt.1)
	2.4 (Expt.2)	1.25 (Expt.2)
10	2.5 (Expt.1)	1.25 (Expt.1)
	2.6 (Expt.2)	1.30 (Expt.2)
	2.5 (Expt.3)	1.20 (Expt.3)
15	2.9	0.60 (Expt.1)
		0.70 (Expt.2)
25	2.8	0.50
Adult	2.9 (Expt.1)	0.40 (Expt.1)
	2.8 (Expt.2)	0.25 (Expt.2)
		0.35 (Expt.3)

[a] Determined in the presence of 10 mM glutamate and 10 mM malate.
[b] Determined in the presence of 6 mM succinate and 6 mM glutamate plus 40 μM antimycin A. Each data point represents the results from pooled mitochondria prepared from between 250 to 300 neonatal animals or 30 to 30 adult animals for both sets of data.

The state of coupling of the macrophage mitochondria as a function of animal age was also tested. A small but consistent difference in the ADP/O ratio between mitochondria from neonatal animals and adult rats was apparent. The ratio appeared to increase gradually and to achieve the maximum adult value at approximately 15 days after birth (Table 2). These data can be interpreted to demonstrate partial uncoupling of the mitochondria of neonatal pulmonary macrophages. The significance of the difference between the ADP/O ratios is corroborated by the differences in the respiratory control index (RCI) of mitochondria obtained from cells of neonates and adults. In the former the RCI was 3.6 whereas in the latter the RCI was 6.7. Because of the nature of these determinations it was crucial to demonstrate that the mitochondria used were intact and further, that H_2O_2 was not lost during the assay procedure. Intactness of mitocondria was tested in each preparation by the measurement of NADH impermeability. Mitochondria were only used when it was determined that added NADH would not support H_2O_2 generation. Occasional determination of oxygen consumption were conducted, as well. Hydrogen peroxide was added to mitochondria preparations and approximately 90% was recovered indicating that catalase and/or glutathione peroxidase consumes only 10% or less of the released H_2O_2.

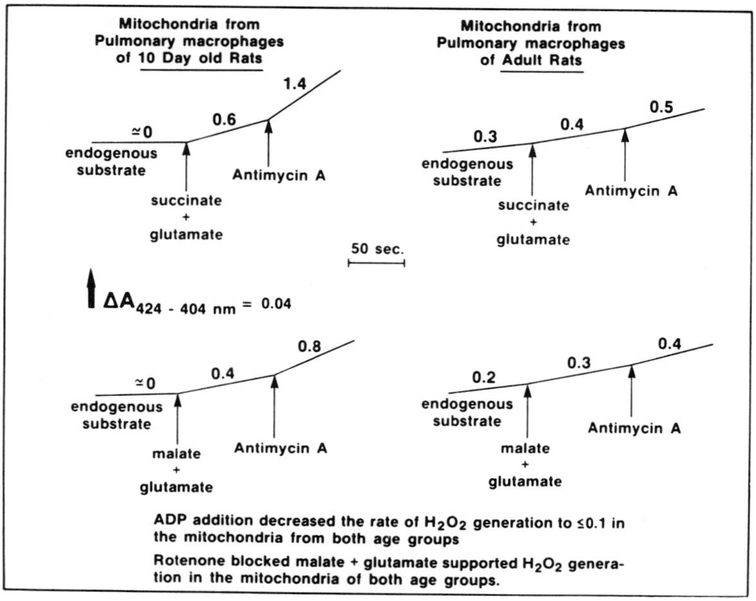

Fig. 1. Generation of H_2O_2 by isolated mitochondria from rat pulmonary macrophages obtained from either 10 day old or adult animals. Hydrogen peroxide was measured as the rate of formation of the H_2O_2-cytochrome c peroxidase complex. The distance marked by the large arrow indicates on the ordinate, $\Delta A_{424-404\ nm}$ = 0.04; and on the abcissa, 50 sec. Assay mixtures contained from 1 to 3 mg of mitochondrial protein per ml. The numbers on the traces indicate nmoles of H_2O_2 per min per mg of protein.

Because of existing reports in the literature strongly suggesting that the source of mitochondrially generated H_2O_2 is O_2^- produced directly from some autoxidizable component or components of the electron transport chain,[30] it was of great interest to determine whether O_2^- release from submitochondrial particles could be detected and whether the same age dependent pattern of generation rate existed as for H_2O_2 generation. Submitochondrial particles were thoroughly washed before assay in order to remove the endogenous superoxide dismutase. The assay used for the detection of O_2^- relied on its reduction of added ferricytochrome c inhibitable by added superoxide dismutase. Intrinsic superoxide dismutase activity in the particles was much less than 1% of the total amount of added superoxide dismutase required to fully inhibit ferricytochrome c reduction. The same inverse relationship between animal age and O_2^- release from submitochondrial particles was observed as with H_2O_2 generation (Table 3). If all H_2O_2 released from mitochondria originates from O_2^- which is then, presumably, enzymatically dismuted to H_2O_2 as has been suggested,[30] then a ratio of O_2^-/H_2O_2 of 2/1 should have been obtained. Much less O_2^- than expected was obtained in these experiments. The reduction of ferricytochrome c is known to be a much less sensitive assay for O_2^- than other established assays, and may not have detected all the O_2^- present. Further studies are planned to test this possibility. The rates of H_2O_2 and O_2^- generation in Tables 2 and 3 are reported on the basis of mitochondrial protein. In order to obviate possible artifacts from variable preparations, rates were also determined on the basis of cytochrome oxidase activity since it was established that this enzyme did not change with increasing animal age (Table 1). The age-dependent differences in rates remained when calculated in this way.

TABLE 3

SUPEROXIDE ANION GENERATION BY MITOCHONDRIA ISOLATED FROM PULMONARY MACROPHAGES PREPARED FROM RATS OF INCREASING AGE

Animal Age (days)	O_2^- Generation (nmoles/min/mg of protein)
3	0.69
5	0.72
7	0.75
10	0.60 (Expt.1)
	0.72 (Expt.2)
15	0.28
25	0.22
Adult	0.17 (Expt.1)
	0.09 (Expt.2)

Interestingly, the same relationship between lowered ADP/O ratios and elevated rates of H_2O_2 generation were observed in another study from this laboratory.[32] Rats were treated in vivo with 2,4-dinitrophenol, and exhibited the classic signs of reduced coupling, i.e. (1) weight loss, (2) elevated body temperature and (3) diminished ADP/O ratio in liver mitochondria (2.0 versus 2.8). When liver superoxide dismutase activity in treated rats was compared with the activity from saline-treated controls, the former activity was increased approximately three-fold in the crude homogenate (Table 4). Subsequent analysis using 1 mM cyanide as an inhibitor of CuZn superoxide dismutase in the assay of isolated mitochondria demonstrated that as with the neonatal rat pulmonary macrophages, only the mitochondrial Mn superoxide dismutase was elevated. Also similar to the previously observed relationship between elevated superoxide dismutase and increased mitochondrial generation of H_2O_2, a dramatically elevated rate of H_2O_2 generation was observed in the liver mitochondria of rats treated _in vivo_ with the uncoupling agent (Table 4). In two different systems in higher eukaryotes, therefore, a direct relationship between elevated Mn superoxide dismutase and enhanced rates of mitochondrial H_2O_2 release has been demonstrated. In addition there is a suggestion that uncoupling of mitochondria results in increased H_2O_2 generation and, further, that the released H_2O_2 originates at least in part from O_2^- release from the electron transport chain.

TABLE 4

SUPEROXIDE DISMUTASE ACTIVITY AND MITOCHONDRIAL HYDROGEN PEROXIDE GENERATION IN LIVER PREPARATIONS FROM CONTROL AND DINITROPHENOL-TREATED RATS[a]

Assay	Control Rats	Dinitrophenol-Treated Rats[b]
Superoxide dismutase activity (units/mg protein):		
Liver homogenates (assayed in 5×10^{-4} M CN^-)	26.0 ± 2.1	70.0 ± 6.9
Liver homogenates (assayed in 1×10^{-3} M CN^-)	7.1 ± 1.2	51.2 ± 1.0
Liver mitochondria (assayed in 5×10^{-4} M CN^-)	8.1 ± 0.3	37.0 ± 4.5
Liver mitochondria (assayed in 1×10^{-3} M CN^-)	7.3 ± 0.2	32.6 ± 4.5
H_2O_2 generation[c] (nmol/min/mg protein):		
Liver mitochondria	1.6 ± 0.6	97.7 ± 7.6

[a]Data for this Table was taken from Ref. 32.
[b]Dinitrophenol was administered intraperitoneally twice per day for 3 days (10 mg/kg of body weight). Control animals received saline (ip).
[c]Determined in the presence of 6 mM succinate and 6 mM glutamate plus 40 μM antimycin A.

DISCUSSION

A hypothetical scheme has been designed to illustrate the observed interactions in the mitochondria of neonatal pulmonary macrophages between partially reduced oxygen species and the enzymes which catalyze their conversion to less reactive products (Figure 2). A sequence of events affecting the regulation of Mn superoxide dismutase and catalase has been proposed in this scheme, as well. In the pulmonary macrophages of neonatal rats, O_2^- generation from mitochondria, monitored via H_2O_2 release from intact mitochondria, occurs through the incomplete reduction of oxygen by the electron transport chain. It is likely that other sources of either O_2^- or H_2O_2 from the mitochondria contribute to only a small extent to the total H_2O_2 produced although superoxide anion production from dihydroorotate dehydrogenase has been reported in another system.[38] Mitochondria of cells from the neonatal animals appear to be partly uncoupled. Although the difference in ADP/O ratio between neonates and adults is small, it is reproducible. Credence is given to this relationship between low ADP/O ratios and elevated H_2O_2 release in mitochondria since it is also seen in the liver mitochondria of dinitrophenol-treated rats. The uncoupled state of the neonatal macrophage mitochondria may indicate that one (or more) component in the electron transport chain is in some way labilized and therefore made more easily autoxidizable. Such a relatively unstable component may reflect the immaturity of the animal. It is well-known that the neonatal rat lung is an immature organ.[32]

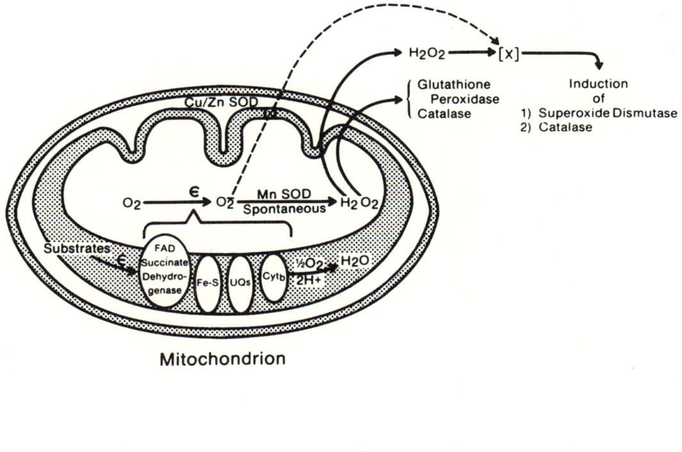

Fig. 2. Proposal for oxygen radical control of superoxide dismutase and catalase.

The question of which component or components of the electron transport chain produces the O_2^- control cannot be answered at this time. There may be one single source in the neonate or there may be many sources. Possible sources are indicated in Figure 2. Superoxide radical production from several isolated components, such as quinones,[34] cytochromes,[35] succinate dehydrogenase,[36] flavins,[34] and iron-sulfur-containing proteins,[37] has been reported. Because of the configuration of these components within the mitochondrial inner membrane, O_2^- is probably released into the matrix space, where under normal circumstances it is immediately converted to H_2O_2 (and O_2) by Mn superoxide dismutase. Hydrogen peroxide freely crosses the two mitochondrial membranes and is released into the cytoplasm when it can be removed by either glutathione peroxidase or peroxisomal catalase. Our data did not fully account for all the O_2^- potentially generated from electron transport particles, using the conversion factor of O_2^-/H_2O_2 of 2/1. As previously mentioned (see RESULTS), this might reflect the relative insensitivity of the assay used. In addition, it is possible that a certain proportion of the O_2^- generated by membrane components is sequestered within the vesicle and not released. This would certainly be true if (1) the component source of the radicals were oriented toward the inner side of the electron transport particle vesicle, and (2) if the O_2^- were not freely permeable across the membrane.

The association between enhanced O_2^-/H_2O_2 generation and elevated Mn superoxide dismutase and catalase content is clearly established in the two systems described in this report. It is possible to speculate that both the observed high basal content and the induced component of Mn superoxide dismutase and catalase result from the action of elevated O_2^-/H_2O_2 or some subsequent product on regulation of the biosynthesis of these enzymes. Under conditions where O_2^- generation is elevated, the intrinsic protective mechanisms could be overwhelmed. In the case of mitochondria, if O_2^- is accumulated in the matrix, both reactants for the production of $HO^\bullet$ would be present. This radical has a high, destructive potential because of its capacity for both hydrogen abstraction and electron withdrawal. If superoxide dismutase content in the mitochondria is inadequate, a proportion of superoxide anion radicals could escape the matrix and perhaps interact with the inner membrane (see Figure 2). Alternatively, formed $HO^\bullet$ could also react with some component of the mitochondrial membrane. Either this component, a metabolic product generated therefrom, or even H_2O_2 (which is also elevated in this system) could act as an agent to stimulate the synthesis of Mn superoxide dismutase and catalase. The diminished basal activity and loss of inductive capacity observed in cells from adult animals could be explained as emanating from either (1) the enchanced respiratory control in the adult cells (2) the inability to generate X (see Figure 2) in adult cells, or (3) the loss of the ability of X to regulate the biosynthesis of catalase and Mn superoxide dismutase. The correlation of radical generation and control of Mn superoxide dismutase is supported by the effect of <u>in vivo</u> dinitrophenol treatment of rats. It is recognized that this process may be different; however, the basic feature of the dependence of enzyme regulation upon enhanced partial reduction of oxygen remains.

ACKNOWLEGEMENTS

This work was supported by USPHS Grant GM-12675 and a Basic Research Grant from The National Foundation - March of Dimes. A.P.A. is the recipient of a Research Career Development Award from the National Institutes of Health (NICHD-00068).

REFERENCES

1. Autor, A.P., Frank, L. and Roberts, R.J. (1976) Pediat. Res. 10, 154-158.
2. Stevens, J.B. and Autor, A.P. (1977) J. Biol. Chem. 252, 3509-3514.
3. Stevens, J.B. and Autor, A.P. (1977) Lab. Invest. 37, 470-478.
4. Autor, A.P. and Stevens, J.B. (1978) Photochem. Photobiol. 28, 775-780.
5. Autor, A.P., Fox, A.W. and Stevens, J.B. in Oxygen: Biochemical and Clinical Aspects, Caughey, W.S., ed., Academic Press, New York, in press.
6. Hoffman, M., Stevens, J.B. and Autor, A.P.: in preparation.
7. Willson, R.L. (1977) Chem. Ind. 183-193.
8. Autor, A.P., McLennan, G. and Fox, A.W. in Molecular Basis of Environmental Toxicity, Bhatnagar, R.J., ed., Academic Press, New York, in press.
9. Haber, F. and Weiss, J. (1934) Proc. Roy. Soc. A, 147, 332-351.
10. Willson, R.L. (1979) in Oxygen Free Radicals and Tissue Damage, Ciba Foundation Symp. 65 (New Series), Excerpta Medica, Amsterdam, pp. 19-42.
11. Gregory, E.M. and Fridovich, I. (1973) J. Bacteriol. 114, 543-548.
12. Gregory, E.M. and Fridovich, I. (1973) J. Bacteriol. 114, 1193-1197.
13. Gregory, E., Goscin, S.A. and Fridovich, I. (1974) J. Bacteriol. 117, 455-460.
14. Yost, R.J., Jr. and Fridovich, I. (1976) Arch. Biochem. Biophys. 175, 514-519.
15. Hassan, H.M. and Fridovich, I. (1978) J. Biol. Chem. 253, 8143-8148.
16. McCord, J.M., Keele, B.B., Jr. and Fridovich, I. (1971) Proc. Nat. Acad. Sci. USA, 68, 1024-1027.
17. Rolfe, R.D., Hentges, D.J., Campbell, B.J. and Barrett, J.T. (1978) Appl. Environ. Microbiol. 36, 306-313.
18. Gregory, E.M., Moore, W.E.C. and Holdeman, L.V. (1978) Appl. Environ. Microbiol. 35, 988-991.
19. Harper, D.B. and Harvey, B.M.R., (1978) Plant Cell Environ. 1, 211-215.
20. Phillips, H.J. (1973) in Tissue Culture: Methods and Applications, Kruse, P.F., Jr. and Patterson, M.K., Jr., eds., pp. 406-408.
21. Lowry, O.H., Rosebrough, N.J., Farr, A.L. and Randall, R.J. (1951) J. Biol. Chem. 193, 265-275.
22. McCord, J.M. and Fridovich, I. (1969) J. Biol. Chem. 244, 6049-6055.
23. Weisiger, R.A. and Fridovich, I. (1973) J. Biol. Chem. 248, 3582-3592.
24. Beers, R.F. Jr. and Sizer, J.W. (1952) J. Biol. Chem. 195, 133-140.
25. Paglia, D.E. and Valentine, W.N. (1967) J. Lab. Clin. Med. 70, 158-169.
26. Wharton, D.C. and Tzagoloff, A. (1967) Methods Enzymol. 10, 245-250.
27. Boveris, A., Oshino, N. and Chance, B. (1972) Biochem. J. 128, 617-629.
28. Yonetani, T. and Ray, G.S. (1965) J. Biol. Chem. 240, 4503-4508.
29. Dionisi, O., Galeotti, T., Terranova, T. and Azzi, A. (1975) Biochim. Biophys. Acta, 403, 292-300.
30. Boveris, A. (1977) Adv. Exp. Med. Biol. 78, 67-82.
31. Boveris, A., Oshino, N. and Chance, B. (1972) Biochem. J. 128, 617-630.
32. Dryer, S.E., Dryer, R.L. and Autor, A.P., J. Biol. Chem., in press.
33. Thurlbeck, W.M. (1975) Amer. Rev. Resp. Dis. 111, 803-844.
34. Misra, H.P. and Fridovich, I. (1972) J. Biol. Chem. 247, 188-192.
35. Loschen, G., Azzi, Y., Richter, C. and Flohé, L. (1974) FEBS Lett. 42, 68-72.
36. Forman, H.J. and Kennedy, J.A. (1974) Biochem. Biophys. Res. Commun. 60, 1044-1050.
37. Misra, H.P. and Fridovich, I. (1971) J. Biol. Chem. 246, 6886-6890.
38. Forman, H.J. and Kennedy, J. (1976) Arch. Biochem. Biophys. 173, 219-224.

BACTERIAL CHEMILUMINESCENCE: OXYGEN DEPENDENCE AND INHIBITION BY SUPEROXIDE DISMUTASE AND CATALASE*

ROBERT C. ALLEN
U.S. Army Institute of Surgical Research, and Clinical Investigation Service, Brook Army Medical Center, Fort Sam Houston, Texas, USA

INTRODUCTION

Chemiluminescence is detected from polymorphonuclear leukocytes and other phagocytes in association with oxidative microbicidal action.[3,5-7,11,12,20,24,28] This luminescence is of biological origin, but the term chemilunescence (CL) is employed to distinguish these ultra weak emissions from classical bioluminescence. In common with bioluminescence, cellular CL is associated with redox activity and requires O_2 (Ref.31). The CL response from activated phagocytes correlates with microbicidal metabolism as measured by hexose monophosphate shunt metabolism of glucose and by O_2 consumption. CL is not detected from resting phagocytes.[4,34]

Recently CL has been detected from anaerobically cultured Streptococcus faecalis following exposure to O_2. This O_2-dependent luminescence was proposed to result from toxic oxidation reactions similar to those effected by activated phagocytes. The inhibition of this CL by addition of superoxide dismutase (SOD) and catalase also suggested the involvement of superoxide anion (O_2^-) and peroxide in bacterial oxygen toxicity.[4] These observations are consistent with the proposal that the bacterial synthesis of SOD and catalase serve to protect the microbe from toxic oxidation reactions.[10,26,27] In an attempt to clarify the mechanism for bacterial generation of CL, a variety of metabolically different bacteria were tested for CL response relative to O_2 tension. The role of O_2^- and H_2O_2 in oxidation reactions resulting in CL was investigated using scavenger enzymes such as catalase and SOD. The generation and accumulation of H_2O_2 was detected by the CL response following addition of myeloperoxidase (MPO) to the bacterial suspension.

MATERIALS AND METHODS

Bacteria. The bacteria used were: Staphylococcus aureus (ATCC 25923);

* The opinions or assertions contained herein are the private views of the author and are not to be construed as representing the views of the Department of the Army or the Department of Defense.

Streptococcus pneumoniae (ATCC 11733); Escherichia coli serotype 014 (clinical isolate); Streptococcus faecalis (clinical isolate 638); Streptococcus faecalis (clinical isolate 654); and a viridans group Streptococcus (clinical isolate). All bacteria were cultured in tryptic soy broth (Difco labs), a soybean-casein digest, under candle jar atmosphere at 37°C. When turbidity was observed, the cultures were transferred to an anaerobic glove box and allowed to incubate for an additional hour. During this period 5.0 ml aliquots of the culture were transferred to sterile scintillation counting vials. Where indicated, bovine erythrocyte SOD and bovine liver catalase were added to the vials. The vials were then sealed gas tight, removed from the anaerobic chamber, and placed on the specimen train of the scintillation spectrometer.

After monitoring the background CL from the anaerobically poised bacterial cultures, the vials were uncapped, exposed to air (approximately 21% O_2), and monitored for CL. The O_2 in the gas above the broth culture was then removed by N_2-gassing, and the effect of O_2 removal on CL response was measured. Finally, the O_2-free atmosphere above the cultures was replaced by gassing with 100% O_2, and the CL response to high P_{O_2} was measured. Where indicated, the vials were again opended to air, MPO was added, and CL was monitored.

The relative population densities of the broth cultures were determined by measuring the absorbance of the cultures at 520 nm using a spectrophotometer blanked against sterile broth. Absorbance determinations were made on cultures at the time of air exposure.

Enzymes. Bovine liver catalase (H_2O_2:H_2O_2 oxidoreductase) was purchased as the thymol-free purified powder from Sigma Chemical Co. One unit of activity decomposes one micromole of H_2O_2 per min. at pH 7.0 at 25°C. (Ref.36). One thousand units of catalase were added to the culture where indicated.

Bovine SOD (superoxide:superoxide oxidoreductase) was purchased as the type 1, lyophilized powder from Sigma Chemical Co. One unit of activity is that amount of enzyme required to inhibit the xanthine oxidase mediated reduction of cytochrome c by 50% as described by McCord and Fridovich.[25] One thousand units of SOD were added to the cultures as indicated. Where indicated, the preparations of catalase and SOD were subjected to heat denaturation by autoclaving at 121°C for 15 minutes.

Myeloperoxidase (Cl^-:H_2O_2 oxidoreductase) was obtained by acid extraction of hypotonically lysed human leukocytes. The unit activity of the extract was determined by o-dianisidine oxidation.[23,36] One unit of peroxidase activity decomposes one micromole of H_2O_2 per minute at 25°C. Five hundred units of MPO were added to the cultures as indicated. Assays for the above enzyme were performed on a Beckman DU or ACTA CIII spectrophotometer.

Chemiluminescence. The single photon counting capability of liquid scintillation counters using the tritium channel and operating in the out-of-coincidence mode was used for detection and quantification of chemiluminescence (CL) as previously described.[2,4] A Beckman Model LS-150 scintillation counter was used for these studies. The counter was equipped with EMI photomultiplier tubes selected for high sensitivity with low dark-current signal. The relative CL intensity is reported as counts per minute.

Where multiple samples were tested, each vial was equally spaced on the same train of the scintillation counter to insure uniform agitation. The CL intensity from each vial was measured for a one minute period. The train continuously cycled the samples for periodic measurement.

RESULTS

Oxygen-dependent chemiluminescence from anaerobically poised Streptococcus faecalis. A variety of bacteria were cultured under microaerophilic and finally under anaerobic conditions. These cultures were then monitored for CL response following exposure to O_2. Of the bacteria tested, only the Strep. faecalis yielded CL of high intensity and long duration.

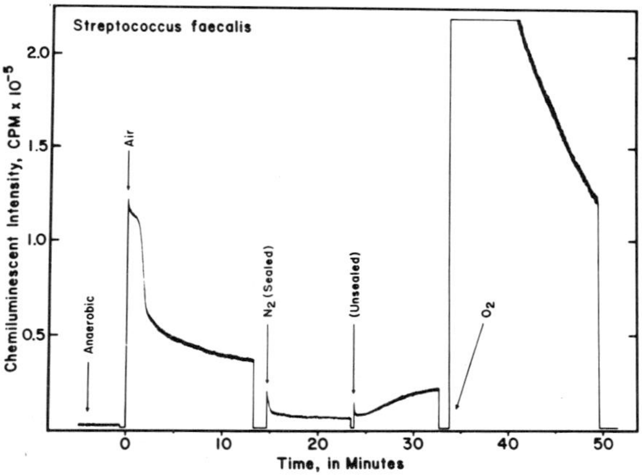

Fig. 1. Temporal traces of CL intensity from Streptococcus faecalis (isolate 638) in response to change in O_2 absorbancy of 0.2 at 520 nm. The anaerobically poised culture was exposed to air (21% O_2) at time zero, then flushed with N_2 to remove O_2. After 10 min., the gas tight seal was released allowing slow re-exposure to air. Finally, the culture was flushed with O_2 (100%).

The continuous plot of CL intensity versus time for Strep.faecalis isolate 638 is presented in Figure 1. No CL above background was detected from the anaerobically poised culture. However, a large CL response followed exposure of the culture to air (21% O_2 by volume). This CL was abolished by N_2-gassing to displace O_2. Gradual re-exposure to O_2, by breaking the gas-tight seal of the vial cap, resulted in a corresponding gradual increase in CL intensity. Finally, exposure of the culture to a high partial pressure of oxygen (P_{O_2}) by gassing with 100% O_2 yielded a CL of great magnitude and was associated with oxidative damage to the bacteria. Strep. faecalis cultured under aerobic conditions did not yield CL even following exposure to 100% O_2. Furthermore, exposure to high P_{O_2} was not associated with bacterial action.

Effect of age and culture density of Streptococcus faecalis chemiluminescence.

The effects of age and population density on the CL response from Strep. faecalis culture was studied by growing the bacteria to two different population densities, and CL was monitored following O_2 exposure. As described in the following section, catalase and SOD were introduced to the anaerobically poised cultures prior to O_2 exposure. The results are depicted in a bar graph presented as Figure 2. The younger culture (A_{520}= 0.81) yielded a larger CL response following expousre to 100% O_2. However, the general pattern of O_2 responsiveness and enzyme inhibition was maintained in both cultures.

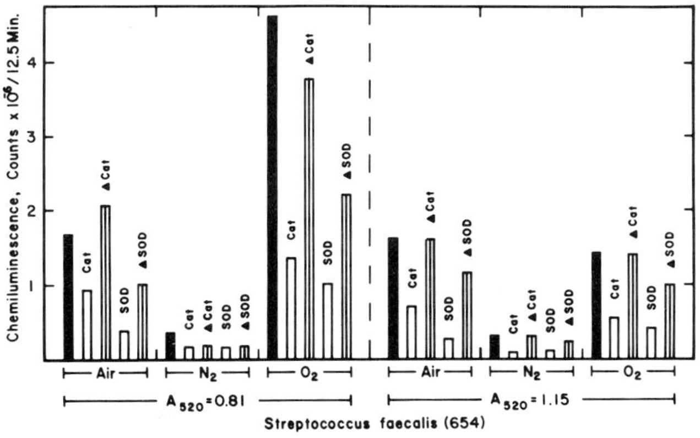

Fig. 2. Comparison of the integral CL responses from culture of Streptococcus faecalis (isolate 654) at two different times during its growth, and under different conditions of O_2 tension. Catalase (Cat) and erythrocyte SOD were added to the culture where indictaed by notation above the bar. The triangle indicates that the enzyme was autoclaved at 121°C for 15 min.

Inhibition of O_2-dependent bacterial chemiluminescence by superoxide dismutase and Catalase. In order to investigate the involvement of O_2^- and H_2O_2 in bacterial CL, scavenger enzymes, such as copper-containing bovine erythrocyte superoxide dismutase (CuSOD) and catalase were added to the anaerobic culture prior to O_2 exposure. The cultures were then exposed to different O_2 tensions with measurement of CL response. The numerical data obtained from these studies of O_2-dependence and enzyme inhibition of CL are presented in Table 1. Inhibition of luminescence by CuSOD, catalase, or a combination of both enzymes was investigated for each bacteria. The effect of heat-denaturation of the enzyme by autoclaving was also studied.

Finally, myeloperoxidase (MPO) extracted from human polymorphonuclear leukocytes was added to the bacterial culture following O_2 exposure. This enzyme is microbicidal when presented with H_2O_2 and with Cl^- as an oxidizable halide cofactor. The oxidative action of MPO is associated with generation of CL, and therefore, the detection of luminescence following addition of MPO to the culture was used as an index of accumulated peroxide. It is noted that a small CL response was detected following addition of MPO to O_2-exposed, sterile broth. This observation is consistent with the non-bacterial generation of peroxides within the medium. No CL was observed following addition of MPO to broth under anaerobic conditions.

Streptococcus faecalis. A large CL response was detected from anaerobically-poised cultures of both isolates of Strep. faecalis (isolate 654) is presented in Figure 3. Gassing with N_2 greatly diminsihed this luminescence, whereas, gassing with O_2 yielded a very large CL response.

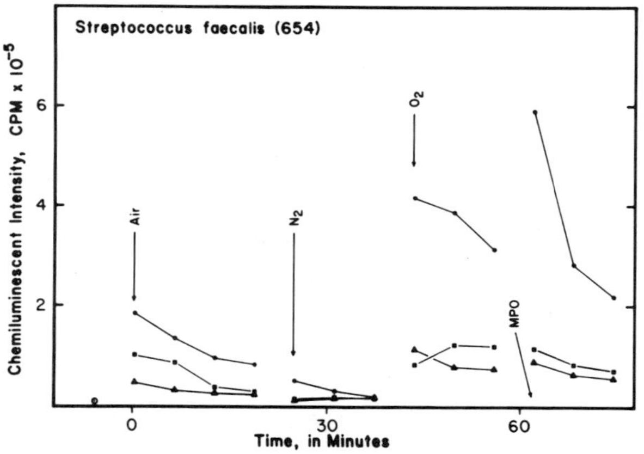

Fig. 3. The O_2-dependent CL response from Streptococcus faecalis (isolate 654). Absorbance of the culture at 520 nm was 0.87 (A_{520}= 0.87). The anaerobically poised culture was exposed to air, flushed with N_2, the exposed to O_2 (100%). Myeloperoxidase (500U) was added following O_2 exposure. The circles (•) depict the CL response from the native system. The square (■) and triangles (▲) depict the effect of added catalase (1000U) and CuSOD (1000U) respectively.

TABLE 1

BACTERIAL CHEMILUMINESCENCE: THE OXYGEN REQUIREMENT, THE INHIBITORY ACTION OF CATALASE AND SUPEROXIDE DISMUTASE AND THE PEROXIDE-DEPENDENT ACTION OF MYELOPEROXIDASE

Organism	Inhibitor Enzyme[a]	Integral Chemiluminescence Total counts ($x 10^{-6}$)/12.5 min. Oxygen Condition[b] Air	N_2	O_2	Air + MPO[c]
Sterile Tryptose Soy Broth A_{520} = 0.00	None	0.20	0.23	0.27	0.80
Streptococcus faecalis (isolate 638) A_{520} = 0.81	None	2.84	0.15	6.76	-
	Catalase	0.62	0.11	1.89	-
	CuSOD	0.23	0.18	1.08	-
	Catalase + CuSOD	0.14	0.13	0.41	-
	Autoclaved catalase	2.33	0.13	5.68	-
	Autoclaved CuSOD	1.59	0.11	3.45	-
Streptococcus faecalis (isolate 654) A_{520} = 0.87	None	1.67	0.36	4.63	4.24
	Catalase	0.94	0.16	1.34	1.06
	CuSOD	0.38	0.15	1.02	0.81
	Autoclaved catalase	2.05	0.18	3.75	3.82
	Autoclaved CuSOD	1.00	0.17	2.22	2.97
Streptococcus pneumoniae (ATCC 11733) A_{520} = 0.81	None	0.18	0.16	0.40	4.97
	Catalase	0.23	0.14	0.66	1.64
	CuSOD	0.09	0.14	0.25	3.93
	Catalase + SOD	0.10	0.11	0.29	1.14
	Autoclaved catalase	0.17	0.15	0.42	5.42
	Autoclaved CuSOD	0.13	0.13	0.29	4.77
Streptococcus (vitidans group) (clinical isolate) A_{520} = 0.88	None	0.14	0.17	0.40	4.22
	Catalase	0.20	0.17	0.23	1.60
	CuSOD	0.11	0.15	0.24	3.93
	Catalase + CuSOD	0.11	0.11	0.34	1.06
	Autoclaved catalase	0.14	0.24	0.35	4.26
	Autoclaved CuSOD	0.11	0.13	0.30	3.63
Escherichia coli (serotype 014) (clinical isolate) A_{520} = 1.04	None	0.14	0.17	0.66	0.41
	Catalase	0.12	0.20	0.46	0.33
	CuSOD	0.10	0.18	0.29	0.16
	Catalase + CuSOD	0.11	0.14	0.21	0.20
	Autoclaved catalase	0.11	0.19	0.60	0.38
	Autoclaved CuSOD	0.13	0.22	0.86	0.29
Staphylococcus aureaus (ATCC 25923) A_{520} = 0.54	None	0.18	0.26	0.46	0.55
	Catalase	0.18	0.14	0.48	0.33
	CuSOD	0.13	0.20	0.24	0.33
	Catalase + CuSOD	0.13	0.14	0.25	0.22
	Autoclaved catalase	0.20	0.13	0.37	0.55
	Autoclaved CuSOD	0.20	0.16	0.29	0.50

[a]Vials containing 1000 units of catalase or CuSOD added under anaerobic conditions prior to aerobic shock. Autoclaved enzymes were exposed to 121°C for 30 min.

[b]Vials were unsealed and exposed to air or to a gentle stream of N_2 or O_2 for 0.25 min. as described in MATERIALS AND METHODS.

[c]Human myeloperoxidase (MPO), 500 o-dianisidine units, was added in 0.25 ml of 0.1 M acetated buffer, pH 5.0.

Luminescence was inhibited by addition of catalase, and to a greater extent, by addition of CuSOD to the culture. Heat denaturation of catalase resulted essentially in a complete loss of its capacity to inhibit CL. To a lesser extent, heat treatment also diminished the inhibitory action of CuSOD. Addition of MPO yielded an increase in CL from the O_2-exposed cultures of Strep. faecalis. This MPO-dependent CL was not detected from cultures containing catalase.

Alpha-Hemolytic Streptococci. Essentially no CL was detected following exposure of anaerobically poised cultures of Strep. pneumoniae or Strep. (viridans group) to air, and only a relatively small CL response followed exposure to 100% O_2. For both bacteria, this small response was inhibited by CuSOD, but not by catalase. The plots of CL intensity versus change in time for Strep. pneumoniae and Strep. (viridans groups) are presented in Figures 4 and 5, respectively.

A very large yield of CL was detected following addition of MPO to the O_2-exposed cultures of either organism. This MPO-dependent CL was greatly inhibited in cultures containing catalase. However, the presence of SOD in the cultures resulted in only slight inhibition of MPO-depedent CL, and the presence of both catalase and SOD was only slightly more inhibitory than catalase alone. Heat-denatured catalase did not inhibit the MPO-dependent CL response from either culture.

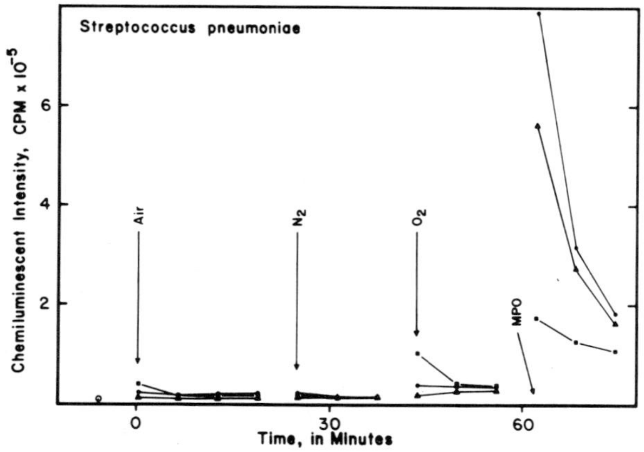

Fig. 4. The O_2-dependent CL response from Streptococcus pneumoniae (ATCC 11733) at a culture density of A_{520} = 0.81. The culture was exposed to different O_2 tensions in the presence and absence of catalase and SOD, as described in Fig.3. Myeloperoxidase (500U) was added following O_2 exposure.

Escherichia coli and Staphylococcus aureus. As depicted in Figure 6, no CL was detected from anaerobically poised cultures of E. coli or S. aureus following exposure to air, and only relatively small yields of CL were detected following exposure to 100% O_2. As was the case for the alpha-hemolytic streptococci, these low yield, O_2-dependent CL responses were inhibited by the presence of CuSOD in the culture. However, unlike the alphahemolytic streptococci, cultures of E. coli and S. aureus did not yield large CL responses following addition of MPO to the O_2-exposed cultures.

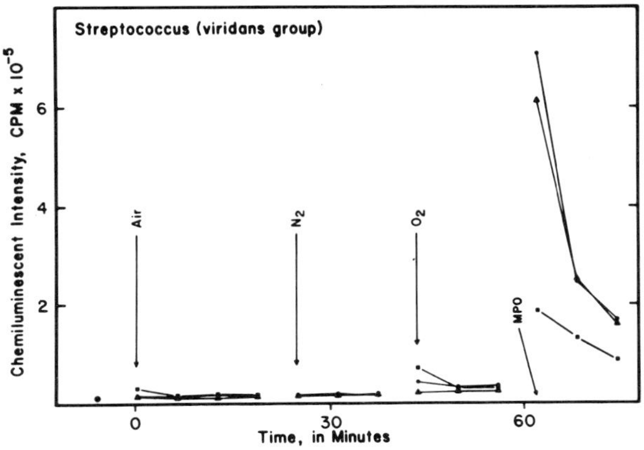

Fig. 5. The O_2-dependent CL response from a viridans group Streptococcus (clinical isolate) at a culture density of A_{520} = 0.88. The conditions of enzyme addition and variation of O_2 tension are the same as those for Fig. 3.

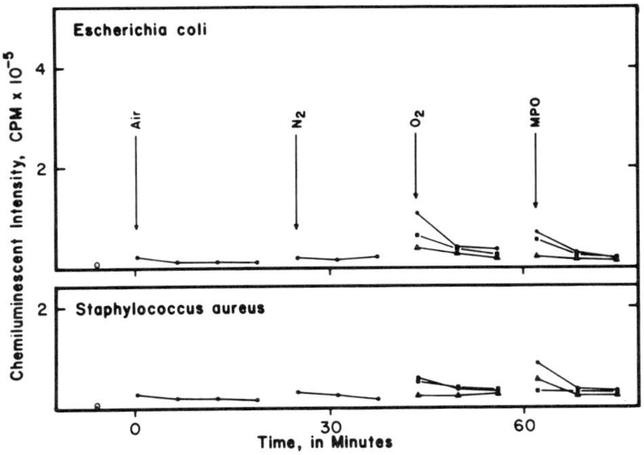

Fig. 6. The O_2-dependent CL response from Escherichia coli (serotype 014) at a culture density of A_{520} = 1.04, and Staphylococcus aureus (ATCC 25923) at a culture density of A_{520} = 0.54. The conditions of enzyme addition and variation of O_2 tension are the same as those for Fig. 3.

DISCUSSION

Streptococci are members of the lactic acid family of bacteria. They are aerotolerant anaerobes lacking the ability to synthesize protoporphyrins, and thus lack catalase and a cytochrome electron transport system. Consequently, their terminal redox reactions are catalyzed by flavoprotein, and their metabolism is essentially fermentive.[13]

The limited respiration observed from certain streptococci appears to result from the action of an NADH:O_2 oxidoreductase (NADH oxidase).[10,16,32,33] This flavoprotein catalyzes the reaction:

$$NADH + H^+ + O_2 \rightarrow O_2^- + H^+ \rightarrow NAD^+ + H_2O_2$$

It is probable that O_2^- is an intermediate product in the generation of the H_2O_2.

Streptococcus faecalis is capable of aerobic growth without accumulation of H_2O_2. However, adaption to high O_2 tension must be gradual to avoid toxicity to the culture. A rapid change from an anaerobic to an aerobic environment is associated with the generation of H_2O_2 (Ref.32). This generation of H_2O_2 is consistent with the action of NADH oxidase. The absence of detectable H_2O_2 from aerobically cultured Strep. faecalis is consequent to the O_2-induced synthesis of an additional flavoprotein having NADH:H_2O_2 oxidoreductase activity (NADH peroxidase). This enzyme catalyzes the reaction:

$$NADH + H^+ + H_2O_2 \rightarrow NAD^+ + 2H_2O$$

In the absence of catalase, NADH perodixase affords a mechanism for the removal of H_2O_2 by expenditure of the reducing equivalents of NADH.

Streptococcus faecalis is also reported to synthesize a manganese-containing SOD (MnSOD).[10,26] Unlike iron-containing bacterial SOD, the MnSOD is resistant to H_2O_2 inhibition.[15] The presence of MnSOD thus affords Strep.faecalis a mechanism for scavenging the O_2^- generated by NADH oxidase. It is probable that the synthesis of MnSOD is controlled by O_2 tension in a manner similar to the O_2-dependent synthesis of MnSOD by other bacteria.[15,16] Therefore, the enzymatic defense mechanisms of Strep. faecalis against oxygen toxicity would include both NADH peroxidase and MnSOD.[10]

Chemiluminescence is detected from anaerobically poised cultures of Strep. faecalis following exposure to O_2. However, no CL is detected from aerobically grown cultures of the same organism. This O_2-dependent CL is proposed to reflect toxic oxidation reactions similar to those involved in phagocyte-mediated oxidative microbicidal action.[4] This thesis is consistent with the inverse relationship between O_2-dependent CL response and the bacterial content of MnSOD and NADH peroxidase. That is, aerobically grown bacteria are protected from toxic oxidations resulting in CL by synthesis of scavenger enzymes. Thus, CL is only observed following exposure of anaerobically grown bacteria lacking enzymatic protection against O_2. Furthermore, the CL response from Strep.

faecalis is inhibited by addition of exogenous CuSOD, and to a lesser extent, by addition of exogenous catalase. The combined addition of both CuSOD and catalase results in greater than 90% inhibition of CL. No inhibition of CL was observed using autoclaved catalase, but a residual inhibitory action was consistently observed from autocalved SOD. These data suggest that the generation and accumulation of O_2^-, and to a lesser extent H_2O_2, are required for CL from Strep. faecalis.

This conclusion is further supported by studies of the CL responses from Strep. pneumoniae and the viridans group streptococcus. These alpha-hemolytic streptococci yielded relatively insignificant CL responses following exposure to O_2. Based on the relatively low SOD content of Strep. pneumoniae, Britton et al.[10] concluded that very little O_2^- is generated by this bacteria. The CL data support this conclusion. However, the generation of H_2O_2 is characteristic of the metabolism of alpha-hemolytic streptococci.[13] When presented with H_2O_2, MPO catalyzes microbicidal oxidations characterized by the generation of CL.[1,3,29] A very large CL response is detected following addition of MPO to either of the alpha-hemolytic streptococci. This MPO-dependent CL is inhibited by addition of catalase, but CuSOD and autoclaved catalase did not effect inhibition of CL. Thus, it is concluded that the generation and accumulation of H_2O_2 does not, in itself, result in bacterial CL.

The last two bacteria tested were aerobes possessing cytochrome systems for electron transport, and catalase. A relatively small CL response was detected following exposure of anaerobically poised, but not aerobically grown, E. coli to high O_2 tension. This CL response was inhibited by CuSOD, and to a lesser extent, by catalase. Hassan and Fridovich[16] reported that the catalase and MnSOD activities of E. coli vary with the degree of oxygen exposure during growth, and that both enzymes are significantly decreased in anaerobic cultures.

The data presented support the thesis that O_2^- and H_2O_2 generation are necessary for bacterial chemiluminescence. Reactions involving O_2^- and H_2O_2 yield potent oxidants such as hydroxyl radical ($^{\bullet}OH$) and either singlet molecular oxygen ($^1\Delta_gO_2$) or an oxidant having the reactive potential of $^1\Delta_gO_2$ (Refs.5,6,8,14,20-22, 35). The detection of bacterial CL is consistent with oxidation reactions yielding intermediate dioxetane products. The breakdown of these dioxetanes yields electronically excited carbonyl chromophores, and the relaxation of these carbonyls to ground state by photon emission yields CL.[3,9,30]

The role of flavin mononucleotide oxidoreductases in bacterial bioluminescence is well established,[17-19] and the possibility remains that riboflavin may also be directly involved in the O_2-dependent CL from non-bioluminescent bacteria. Preliminary filter studies of bacterial-CL indicate that the light is not predominantly red, and thus, it is unlikely that the direct dimolecular emission of singlet oxygen contributes significantly to CL.

It is interesting to consider the mechanism whereby addition of CuSOD and catalase inhibit CL. It seems unlikely that these added enzymes can enter the bacterial cell. Therefore, the redox reactions responsible for radical generation, and ultimately, for CL probably occur on or near the surface of the bacteria. An additional possibility is that the products of O_2 reduction may be released to the suspending medium.

ACKNOWLEDEGEMENTS

I would like to thank Drs. T.R. McNitt and D.L. Stevens for advice, and G.L. Strong and D. Bratten for assistance in preparation of this manuscript. This researcg was supported by the U.S.Army through the Clinical Investigation Service approved protocol C-35-78 and the U.S. Army Institute of Surgical Research approved protocol "Micromethod for assessment of serum opsonic capacity in the burn patient".

REFERENCES

1. Allen, R.C. (1975) Biochem. Biophys. Res. Commun. 63, 675-683.
2. Allen, R.C. (1977) Infect. Immunity. 15, 828-833.
3. Allen, R.C. (1979) in Lysosomes in Biology and Pathology, Vol.6., Dingle, J.T. and Jacques, P.J., eds., North Holland Biomedical Press, Amsterdam, 197-231.
4. Allen, R.C. (1979) Photochem. Photobiol. 30, 157-163.
5. Allen, R.C., Stjernholm, R.L. and Steele, R.H. (1972) Biochem. Biophys. Res. Commun. 47, 679-684.
6. Allen, R.C., Yevich, S.J., Orth, R.W. and Steele, R.H. (1974) Biochem. Biophys. Res. Commun. 60, 909-917.
7. Allen, R.C. and Loose, L.D. (1976) Biochem. Biophys. Res. Commun. 69, 245-252.
8. Babior, B.M., Kipnes, R.S. and Curnutte, J.T. (1973) J. Clin. Invest. 52, 741-744.
9. Bogan, D.J., Surant, J.L., Sheinson, R.S. and Williams, F.W. (1979) Photochem. Photobiol. 30, 3-15.
10. Britton, L., Malinowski, D.P. and Fridovich, I. (1978) J. Bacteriol. 134, 229-236.
11. Cheson, B.D., Christensen, R.L., Sperling, R., Kohler, B.E. and Babior, B.M. (1976) J. Clin. Invest. 58, 789-796.
12. DeChatelet, L.R., Shirley, P.S., McPhail, L.C., Huntley, C.C., Muss, H.B. and Bass, D.A. (1977) Blood, 50, 525-535.
13. Dolin, M.I. (1961) in The Bacteria, Vol. 2, Gunsalus, I.C. and Stanier, R.Y., eds., Academic Press, New York, 425-452.
14. Foote C.S. (1979) in Oxygen: Biochemical and Clinical Aspects, Caughey, W.S. ed., Academic Press, New York, in press.
15. Gregory, E.M. and Fridovich, I. (1973) J. Bacteriol. 114, 543-548.
16. Hassan, H.M. and Fridovich, I. (1977) J. Bacteriol, 129, 1574-1583.
17. Hastings, J.W. (1978) Photochem. Photobiol. 27, 397-404.
18. Hastings, J.W. and Nealson, K.H. (1977) Annu. Rev. Microbiol. 31, 549-571.
19. Jablonski, E. and Deluca, M. (1978) Biochemistry, 17, 672-678.
20. Johnston, R.B., Keele, B.B.,Jr., Misra, H.P., Lehmeyer, J.E., Webb, L.S., Baehner, R.L. and Rajagopalan, K.V. (1975) J. Clin. Invest. 55, 1357-1372.
21. Kellogg, E.W. and Fridovich, I. (1975) J. Biol. Chem. 251, 8812-8817.
22. Khan, A.U. (1970) Science, 168, 476-477.
23. Klebanoff, S.J. (1968) J. Bacteriol. 95, 2131-2138.
24. Klebanoff, S.J., Durack, D.T., Rosen, H. and Clark, R.A. (1977), Infec. Immunity, 17, 167-173.

25. McCord, J.M. and Fridovich, I. (1969) J. Biol. Chem. 244, 6049-6055.
26. McCord, J.M., Keele, B.B.,Jr. and Fridovich, I. (1971) Proc. Nat. Acad. Sci. USA, 68, 1024-1027.
27. Mandell, G.L. (1975) J. Clin. Invest. 55, 561-566.
28. Nelson, R.D., Mills, E.L., Simmons, R.L. and Quie, P.G. (1976) Infec. Immunity, 14, 129-132.
29. Rosen, H. and Klebanoff, S.J. (1977) J. Biol. Chem. 252, 4803-4810.
30. Schuster, G.B., Dizon, B., Koo, J., Schmidt, S.P. and Smith, J.P. (1979) Photochem. Photobiol. 30, 17-26.
31. Seliger, H.H. (1975) Photochem. Photobiol. 21, 335-361.
32. Seeley, H.W. and Vandemark, P.J. (1951) J. Bacteriol. 61, 27-35.
33. Seeley, H.W. and del Rio-Estrada, C. (1951) J. Bacteriol. 62, 649-655.
34. Trush, M.A., Wilson, M.E. and Van Dyke, K. (1978) Methods Enzymol. 57, 462-494.
35. Weiss, S.J., King, G.W. and LoBuglio, A.F. (1977) J. Clin. Invest. 60, 370-373.
36. Worthington Enzyme Manual (1972) Worthington Biochemical Corp., Freehold, N.J.

SUPEROXIDE ANION IN 2-NITROPROPANE DIOXYGENASE REACTION

KENJI SODA,[+] KATSUYUKI TANIZAWA,[+] KATSUMI HASHIZUME,[+] MASAAKI ISHIDA,[+] TOSHIKO KIDO[+] AND KOJI ASADA++
+Laboratory of Microbial Biochemistry, Institute for Chemical Research, Kyoto University, Uji, Kyoto-Fu 611, Japan; ++The Research Institute for Food Science, Kyoto University, Uji, Kyoto-Fu 611, Japan

ABSTRACT

2-Nitropropane dioxygenase (E.C. 1.13.11) purified to homogeneity from a yeast, Hansenula mrakii, is significantly inhibited by thiol compounds, pyrocatechol-3,5-disulfonate disodium salt, thio-bis-tert-butylcresol, nitro blue tetrazolium chloride, and tetranitromethane as well as by superoxide dismutase. The reduction of nitro blue tetrazolium chloride and tetranitromethane occurs concomitantly with the inhibition of enzymatic nitroalkane oxygenation. Superoxide dismutase and scavengers for superoxide anion inhibit the reduction of both compounds. These findings confirm the generation of superoxide anion and its participation in the oxygenation of 2-nitropropane.

Less inhibitory effect of superoxide scavengers on the oxygen consumption than on the nitrite formation in 2-nitropropane oxygenation suggests that superoxide anion plays an important role particularly in the formation of nitrite and acetone.

INTRODUCTION

2-Nitropropane dioxygenase (E.C. 1.13.11) of a yeast, Hansenula mrakii, catalyzes the oxygenation denitrification of 2-nitropropane (Refs. 1,2):

$$2CH_3CH(NO_2)CH_3 + {}^{18}O_2 \rightarrow 2CH_3C^{18}OCH_3 + 2HNO_2$$

The enzyme has been purified to homogeneity from a cell-free extract of H. mrakii grown in a nitroethane medium. It catalyzes incorporation of 2 atoms of molecular oxygen into 2 molecules of acceptor.[2] The absorption spectrum of the enzyme exhibits maxima at 274, 370 and 440 nm and a shoulder at 470 nm. The enzyme contains 1 mole of FAD/mole of enzyme protein as a prosthetic group.[2]

The enzyme is significantly inhibited by superoxide dismutase and various scavengers for superoxide anion such as cytochrome c, epinephrine, thiols and polyhydric phenols.[3,4] The reduction of cytochrome c and the oxidation of epinephrine and NADH are concomitant with the inhibition of enzymatic oxygenation. The enzyme catalyzes also the formation of nitrite from 2-nitropropane by KO_2 added anaerobically. These findings indicate the generation of superoxide anion and its participation in the oxygenation of 2-nitropropane. The formation and participation of superoxide anion have been shown also for indoleamine 2,3-dioxygenase,[5,6] anthranilate hydroxylase,[7] and p-coumarate hydroxylase.[8]

We here report further studies for superoxide anion to elucidate the role of superoxide anion in the enzyme reaction.

MATERIALS AND METHODS

Bovine erythrocyte superoxide dismutase was purchased from Sigma. 2-Nitropropane was obtained from Makarai Chemicals, Kyoto, redistilled and stored at 5°C in the dark. 2-Nitropropane dioxygenase was purified to homogeneity from a cell-free extract of H. mrakii (IFO 0895) as described previously.[1] The other chemicals were analytical grade reagents.

The enzyme activity was assayed by determination of nitrite formed. The standard reaction mixture consisted of 75 µmol of 2-nitropropane, 100 µmol of potassium phosphate buffer (pH 8.0) and enzyme in a final volume of 1.0 ml. In a blank the enzyme was replaced by water. Incubation was carried out at 37°C for 10 min with shaking and the reaction was terminated by addition of 1.0 ml of 99.5% ethanol. Nitrite was determined with sulfanilamide and N-(1-naphthyl)-ethylenediamine according to the method of Ida and Morita.[9] The absorbance was measured at 530 nm with a Beckman SPECTA-20 spectrophotometer with a 1.0-cm light path. Enzyme unit and the specific activity are expressed as described previously.[1]

Oxygen uptake was measured at 25°C in a closed reaction chamber (1.0 ml) with a Hansatech D.W. oxygen electrode. Protein was determined by measuring the absorbance at 280 nm. The absorbance coefficient of the enzyme ($A^{1\%}_{1\ cm}$ - 13.3) was used throughout, obtained by absorbance and dry weight determination. The reduction of nitro blue tetrazolium chloride and tetranitromethane was followed by measuring the absorbance change at 560 and 366 nm, respectively (at 25° C), with a Union Giken SW 401 spectrophotemeter or a Shimadzu MPS-50L spectrophotometer.

RESULTS AND DISCUSSION

Effect of thiol compounds. We have shown that 2-nitropropane dioxygenase is significantly inhibited by several compounds which can undergo oxidation-reduction reaction such as cytochrome c, epinephrine and NADH.[3,4] The inhibition is ascribed to scavenging of superoxide anion generated as an essential intermediate in the enzyme reaction. Various thiol compounds also markedly inhibit the formation of nitrite from 2-nitropropane (Table 1). Half-inhibition was found at 0.027 mM with L- and D-cysteines, 0.05 mM with L- and D-penicillamines, 0.04 mM with glutathione and 0.18 mM with dithiothreitol. The inhibition by thiol

TABLE 1

EFFECT OF THIOL COMPOUNDS ON THE OXYGENATION OF 2-NITROPROPANE[a]

Additions	Final concentration (mM)	Relative activity (Nitrite formation) (%)
None		100
L-Cysteine	1	0
	0.1	18
D-Cysteine	1	0
	0.1	19
L-Penicillamine	1	0
	0.1	24
D-Penicillamine	1	6
	0.1	25
Glutathione	1	0
	0.1	14
Dithiothreitol	1	24
	0.1	58
Dithioerythritol	1	19
	0.1	46
2,3-Dimercapto-1-propanol	1	36
	0.1	73
2-Mercaptoethanol	1	28
	0.1	54

[a]The standard reaction mixture was used except that the inhibitors listed were added. The reaction was started by addition of enzyme.

compounds is reversible; the enzyme recovered about 90% of its original activity by removal of the inhibitors from the reaction mixture by dialysis against 10 mM potassium phosphate buffer (pH 7.0). Thiols are known to protect the cell components from damage by oxygen as a scavenger for activated oxygen. Recently, Asada and Kanematsu[10] studied the reaction of thiols with superoxide anion at weak alkaline pH, and suggested the oxidation of thiols by superoxide and the formation of thiyl radical:

$$RSH + O_2^- + H^+ \rightarrow RS^\cdot + H_2O_2$$

$$2RSH + H_2O_2 \rightarrow 2RS^\cdot + 2H_2O$$

$$2RS^\cdot \rightarrow RSSR$$

Preincubation of the enzyme with thiol compounds had no effect on the enzyme activity. Therefore, the thiol inhibition also results from trapping superoxide formed during the oxygenation of 2-nitropropane.

We have found that the inhibition of 2-nitropropane dioxygenase by NADH is concomitant with the oxidation of NADH to NAD^+ (Refs. 3,4). One mole of NADH is bound per mole of the enzyme and the pro-*R* C-4 hydrogen of the nicotinamide moiety of bound NADH is predominantly transferred to superoxide anion formed enzymatically or infused. Thus, the enzyme can discriminate the diastereotopic hydrogen of NADH. However, there was no stereochemical difference between the enantiomers of cysteine and penicillamine in the inhibition of the enzyme.

Inhibition by pyrocatechol-3,5-disulfonate disodium salt (Tiron) and thio-bis-*tert*-butylcresol. The addition of Tiron (0.1 mM) to the reaction system led to 83% inhibition. Half-inhibition was found at 0.013 mM. We reported previously that 1 g-atom of non-heme iron was determined per mole of enzyme as a prosthetic group.[2] However, we recently demonstrated that the iron is derived from a tightly bound contaminant protein which is separable from the enzyme by repeating a Bio-Gel P-150 column chromatography, and that only FAD is a prosthetic group of the enzyme (K. Tanizawa, K. Hashizume and K. Soda, unpublished data). The inhibition of Tiron was independent of the incubation period (up to 1 hr at 37°C) and the enzyme activity was not affected by previous treatment with 1 mM Tiron at 0°C. The inhibition by Tiron was also reversible; the enzyme regained 95% of its initial activity by dialysis. The inhibition was not counteracted by addition of 1 mM $FeCl_2$, $FeCl_3$, $Fe_2(SO_4)_3$ or $K_3Fe(CN)_6$ into the reaction mixture. These results suggest that the inhibition by Tiron is not due to chelation of iron, but due to scavenging of superoxide anion formed as an intermediate. It has been reported that phenols such as Tiron and oxine (8-hydroxyquinoline) trap superoxide anion effectively.[11,12]

Thio-bis-*tert*-butylcresol, a scavenger for free radicals, also inhibited the enzyme. No nitrite was formed from 2-nitropropane in the presence of 0.1 nM thio-bis-*tert*-butylcresol, and half-inhibition was at 0.011 mM.

Reduction of nitro blue tetrazolium chloride. Beauchamp and Fridovich reported that nitro blue tetrazolium chloride (NBT) is reduced by superoxide anion to form the blue formazan.[13] 2-Nitropropane dioxygenase reaction was carried out in the presence of various concentrations of NBT. This compound strongly inhibited the formation of nitrite from 2-nitropropane. Half-inhibition was found at 0.055 mM. The concomitant reduction of NBT in the reaction mixture was demonstrated by an increase in the absorbance at 560 nm due to formation of the formazan (Figure 1). As shown in Figure 2, the rate of reduction depended on the concentration of 2-nitropropane dioxygenase. On the basis of the molecular absorbance coefficient of the formazan (ε = 7500 at 560 nm), the rate was calculated to be approximately 5800 nmoles per min per mg of enzyme (specific activity, 52,400) with 0.05 mM NBT, whereas the amount of nitrite produced under the same conditions was 28,800 nmoles per min per mg of enzyme (45% inhibition). Thus, 5800 nmoles of NBT was used to intercept superoxide anion which was responsible for the formation of 23,600 nmoles of nitrite. The complete reduction of 1 mole of NBT requires 2 moles of superoxide anion, and consequently the amount of nitrite produced is reduced by 4 moles on the basis of the stoichiometry of 2-nitropropane oxygenation.[3]

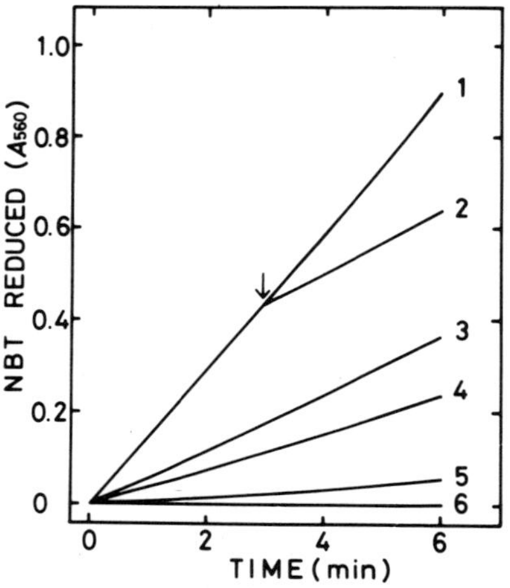

Fig. 1. Reduction of nitro blue tetrazolium chloride and inhibition by superoxide dismutase, Tiron and cysteine. The reaction mixture contained 100 μmoles of potassium phosphate buffer (pH 8.0), 75 μmoles of 2-nitropropane, 015 μmoles of NBT and 3.5 μg of 2-nitropropane dioxygenase in a final volume of 1.0 ml. Curve 1, Complete system; Curves 2 and 3, Superoxide dismutase (8.5 nM) was added at the point indicated by an arrow and before incubation, respectively; Curve 4, L-Cysteine (50 μM) was added before incubation; Curve 5, Tiron (20 μM) was added before incubation; Curve 6, 2-Nitropropane or the enzyme was replaced by water.

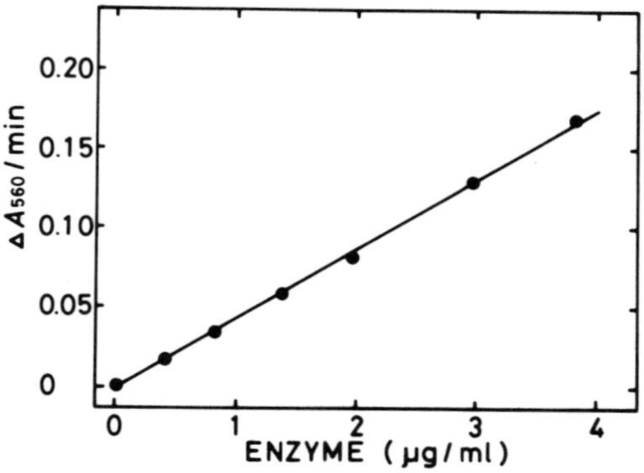

Fig.2. Reduction rate of nitro blue tetrazolium chloride as a function of enzyme concentration. The reaction was carried out under the same conditions as described in Fig. 1 except that various amounts of 2-nitropropane dioxygenase were used.

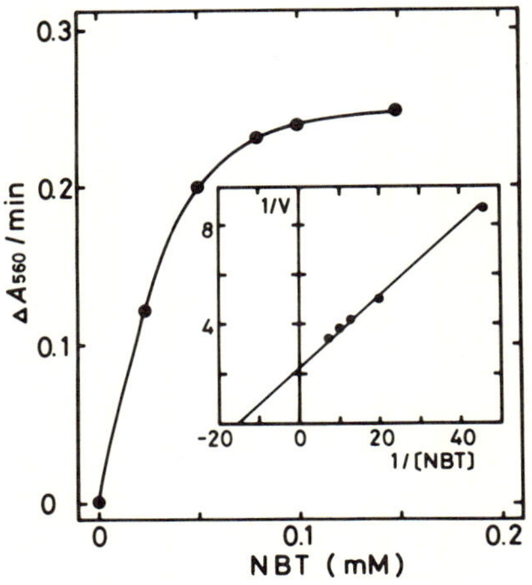

Fig. 3. Effect of nitro blue tetrzolium chloride concentration on reduction rate. The reaction mixtures contained 100 μmoles of potassium phosphate buffer (pH 8.0), 75 μmoles of 2-nitropropane, 5.7 μg of enzyme, and NBT as indicated in a final volume of 1.0 ml. The inset gives the corresponding double-reciprocal plots of the data.

The apparent Km value for the reduction of NBT was estimated from kinetic data (Figure 3) to be 0.067 mM, which is in agreement with the concentration for half-inhibition.

The O_2^--induced reduction of NBT was inhibited by addition of superoxide dismutase, Tiron and cysteine (Figure 1). These findings show that NBT is reduced by superoxide anion generated in the enzyme reaction and that the removal of superoxide anion by scavengers decreases the reduction rate.

Tetranitromethane and its reduction. Further evidence for the participation of superoxide anion in the reaction was obtained by studies on the inhibition of oxygenation by tetranitromethane. The addition of 0.1 mM tetranitromethane to the assay system resulted in 25% inhibition of the oxygenation, and loss of the activity was associated with an increase in the absorbance at 366 nm. This indicates the reduction of tetranitromethane to trinitromethane anion[14,15]:

$$O_2^- + C(NO_2)_4 \rightarrow O_2 + C(NO_2)_3^- + NO_2$$

Tetranitromethane was reduced slightly or not at all in the presence of superoxide dismutase, or in the absence of 2-nitropropane or oxygen (Figure 4). The rate of tetranitromethane reduction depends upon the concentration of 2-nitropropane dioxygenase. This also suggests that the oxidation of 2-nitropropane is mediated by superoxide anion formed through the enzymatic univalent reduction of molecular oxygen with 2-nitropropane.

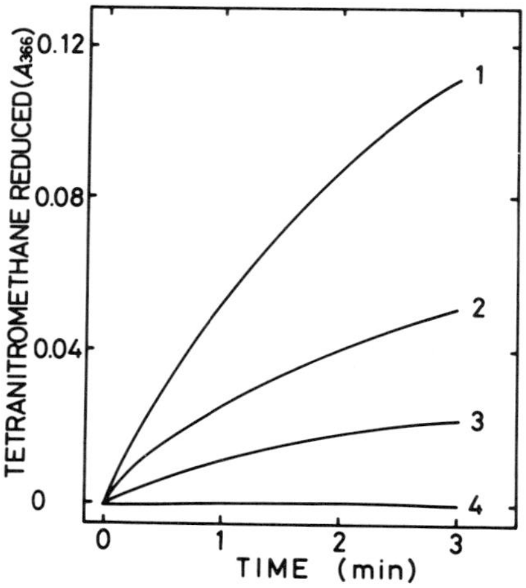

Fig. 4. Reduction of tetranitromethane and inhibition by superoxide dismutase. The reaction mixture contained 100 μmoles of potassium phosphate buffer (pH 7.0), 15 μmoles of 2-nitropropane, 0.1 μmole of tetranitromethane, and enzyme (Curve 1, 57 μg and Curve 2, 11.4 μg) in a final volume of 1.0 ml. Curve 3, Superoxide dismutase (34 nM) was added before incubation in Curve 1; Curve 4, The reaction was carried out in the absence of 2-nitropropane, enzyme or oxygen.

<u>Effect of superoxide anion scavengers on oxygen consumption during 2-nitropropane oxygenation</u>. We have reported that various scavengers for superoxide anion inhibit the enzymatic acetone formation from 2-nitropropane to the same extent as they do the nitrite formation.[3] We investigated the effect of superoxide anion scavengers on the oxygen consumption during the oxygenation of 2-nitropropane (Table 2). All the compounds tested inhibited oxygen consumption less effectively than the formation of nitrite. Since molecular oxygen is produced when superoxide anion is oxidized by cytochrome *c* or disproportionated by superoxide dismutase, we can expect that the scavengers may inhibit more strongly the oxygen consumption than nitrite formation. However, the relative rate of oxygen consumption in the presence of the scavengers was higher than that of nitrite formation. This suggests that superoxide anion is generated by the reaction of the reduced intermediate derived from the substrate and the oxidized enzyme with molecular oxygen and it is essential for the formation of nitrite and acetone:

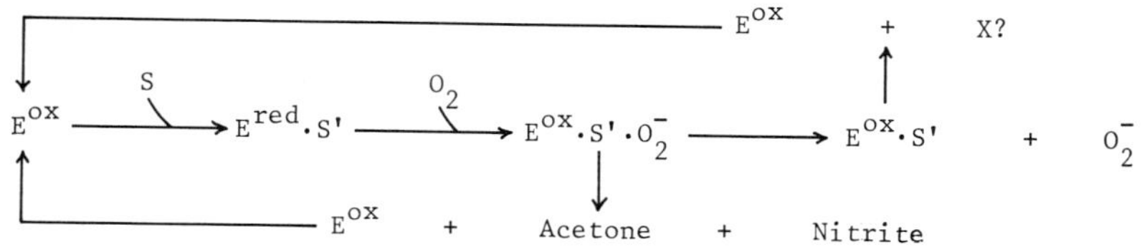

Because superoxide anion scavengers do not affect strongly the consumption of oxygen, the oxidized form of the enzyme is probably regenerated under the conditions without nitrite and acetone formation. The product(s) of 2-nitropropane oxygenation in the presence of superoxide scavengers is not known at present. Further detailed studies are currently under investigation to shed light on the reaction mechanism of 2-nitropropane dioxygenase from a standpoint of the process of electron transfer.

TABLE 2

EFFECT OF VARIOUS SCAVENGERS FOR SUPEROXIDE ANION ON OXYGEN CONSUMPTION DURING OXYGENATION OF 2-NITROPROPANE[a]

Additions	Final concentration (μM)	Relative activity (Oxygen consumption) (%)
None		100
Superoxide dismutase	1	49
	5	43
NADH	100	62
Cytochrome _c_	17	66
α-Naphthol	100	29
8-Hydroxyquinoline	1000	34
L-Cysteine	1000	37
Tiron	100	42

[a]The standard reaction mixture was used except that inhibitors listed were added. The reaction was carried out in a closed reaction chamber (1.0 ml) and initiated by addition of enzyme. The reaction was followed by determination of the rate of oxygen consumption.

REFERENCES

1. Kido, T., Yamamoto, T. and Soda, K. (1976) J. Bacteriol. 126, 1261-1265.
2. Kido, T., Soda, K., Suzuki, T. and Asada, K. (1976) J. Biol. Chem. 251, 6994-7000.
3. Kido, T., Soda, K. and Asada, K. (1978) J. Biol. Chem. 253, 226-232.
4. Kido, T., Hashizume, K., Soda, K. and Asada, K. (1978) Photochem. Photobiol. 28, 729-732.
5. Hirata, F. and Hayaishi, O. (1971) J. Biol. Chem. 246, 7825-7826.
6. Hirata, F. and Hayaishi, O. (1975) J. Biol. Chem. 250, 5960-5966.
7. Kumar, R.P., Ravindranath, S.D., Vaidynathan, C.S. and Rao, N.A. (1972) Biochem. Biophys. Res. Commun. 49, 1422-1426.
8. Halliwell, B. (1975) Eur. J. Biochem. 55, 355-360.
9. Ida, S. and Morita, Y. (1973) Plant Cell Physiol. 14, 661-664.
10. Asada, K. and Kanematsu, S. (1976) Agr. Biol. Chem. Tokyo, 40, 1891-1892.
11. Miller, R.W. and Rapp, U. (1973) J. Biol. Chem. 248, 6084-6090.
12. Greenstock, C.L. and Miller, R.W. (1975) Biochim. Biophys. Acta, 396, 11-16.
13. Beauchamp, C. and Fridovich, I. (1971) Anal. Biochem. 44, 276-287.
14. Rabani, J., Mulac, W.A. and Matheson, M.S. (1965) J. Phys. Chem. 69, 53-70.
15. McCord, J.M. and Fridovich, I. (1969) J. Biol. Chem. 244, 6049-6055.

PHYLOGENIC DISTRIBUTION OF THREE TYPES OF SUPEROXIDE DISMUTASE IN ORGANISMS AND IN CELL ORGANELLES

KOZI ASADA, SUMIO KANEMATSU, SHUNYA OKAKA AND TAKAHIKO HAYAKAWA
The Research Institute for Food Science, Kyoto University, Uji, Kyoto 611, Japan

Three distinct types of superoxide dismutase have been isolated from various kinds of organisms. Each type contains different prosthetic metals, Cu,Zn-, Fe- and Mn-enzymes. In addition to the prosthetic metals, these three types are distinguished by their amino acid sequences; the sequence of the Cu,Zn-enzyme has no homology with sequences of the Fe- and Mn-enzymes.[1-3] Although a high degree of homology between the Fe- and Mn-enzymes has been shown, the micro-environments of the ligands for metal binding should be specific for each type, because the reconstitution of the active Fe- and Mn-enzymes from their respective apoenzymes is successful only with the original metal ions.[4,5] Further, the Fe- and Mn-enzymes from the same organisms have been immunologically distinguished (Figure 1).

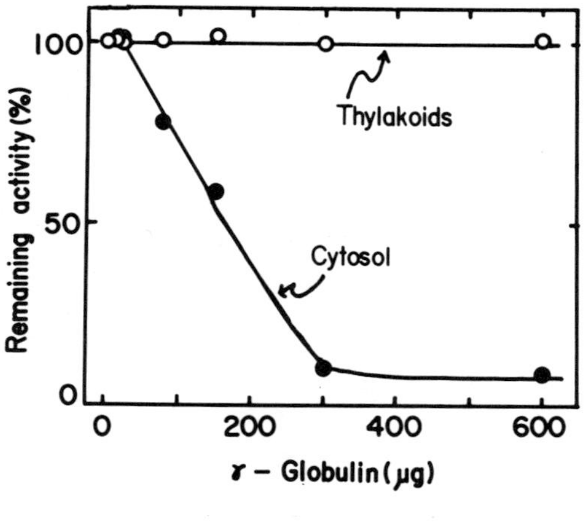

Fig. 1. Effect of the anti-Fe-superoxide dismutase antibody on Plectonema boryanum cytosol and thylakoid superoxide dismutase (Okada, Kanematsu and Asada, (Ref. 86). The cytosol and thylakoid fractions containing 1-3 units of superoxide dismutase were incubated with the indicated amounts of immunoglobulin in 0.15 M NaCl containing 20 mM potassium phosphate, pH 7.8, at 25°C for one hour. Enzymatic activity was determined by the xanthine-xanthine oxidase cytochrome c system.

In respect to the catalysis of the disproportionation of O_2^-, however, little difference has been found among the three types. Their reaction rate constant between O_2^- and the enzyme is about $10^9\ M^{-1}\ s^{-1}$ at a neutral pH. The sole difference is that the catalytic rate of the Cu,Zn-enzyme is constant between pH 5.5 and 10 and that of the Fe- and Mn-enzymes becomes progessively lower as the pH is raised. Nevertheless, distribution of the three types of superoxide dismutase in the organisms is characteristic of the stage of evolution and of the cell organelles. This characteristic distribution must reflect the evolutionary history of the acquisition of the defenses against oxygen toxicity due to the increase in atmospheric oxygen by photosynethetic organisms, and should provide insight into the physiological functions of superoxide dismutase and its molecular evolution.

The distribution of the three types of enzyme in organisms has been determined by its isolation and characterization. Further, the insensitivity of the Mn-enzyme to both cyanide and H_2O_2, and the sensitivities of the Cu,Zn-enzyme to cyanide and the Cu,Zn- and Fe-enzymes to H_2O_2[6] allow us to distinguish the three types of enzyme in crude extracts. Different ratios of inhibition by azide with the three types of enzyme also can be used to distinguish them.[7] The present review summarizes the distribution of the three types of superoxide dismutase in organisms that range from anaerobic prokaryotes to higher organisms. In addition, the sub-organelle distribution of the enzyme in chloroplasts and mitochondria is described. Based on the distribution of the three types of enzyme, their physiological functions and evolutionary aspects are discussed.

Superoxide dismutase in anaerobic bacteria; the Fe-enzyme. The presence of superoxide dismutase in photosynthetic anaerobes, sulfate reducing bacteria, and the fermentative, anaerobic *Clostridium* was first reported by Hewitt and Morris.[8] Since then the enzyme has been found in obligate anaerobes including bacteroids and in other fermentative bacteria.[9-14] The enzyme also occurs in anaerobically grown protozoa.[53] We isolated the enzyme from photosynthetic anaerobes; the purple sulfur bacterium, *Chromatium vinosum*,[10,15] and the green sulfur bacterium, *Chlorobium thiosulfatophilum*.[16] Both bacteria contain only the Fe-enzyme. Further, Hatchikian and Henry[17] have characterized the superoxide dismutase from a sulfate reducing bacterium, *Desulfovibrio desulfuricans*, as the Fe-enzyme.

The molecular properties of the Fe-superoxide dismutase from these anaerobes are similar to those from aerobic prokaryotes,[15-17] although there is a slight difference between the Fe-enzymes from prokaryotes and from the eukaryotic alga, *Euglena*.[18] The amino acid sequence of the Fe-enzyme from *Euglena* differs from sequences of the prokaryotic enzymes according to the statistical analysis of the amino acid composition.[18] The amino acid sequence of the NH_2-terminal region of the Fe-superoxide dismutase from the sulfate reducing bacterium has a high degree of homology with that of the Fe- and Mn-enzymes from other organisms, including vertebrates.[19] The sequence analysis in the NH_2-terminal region of our preparations from both anaerobic sulfur bacteria by the late Dr. J.I. Harris has indicated a similar high degree of homology (unpublished). Thus, anaerobes contain only the Fe-enzyme that has properties similar to those of the Fe-enzyme from other organisms.

It admits no doubt that anaerobic prokaryotes are the most primitive organisms on the evolutionary tree; therefore, the Fe-enzyme of anaerobes may be the ancestor of the Fe- and Mn-enzymes of aerobes. On the other hand, to what extent the environmental concentration of oxygen was raised by abiotic reactions such as the photodissociation of water vapor when anaerobes appeared on the earth before the blue-green algae, is an open question.[20,21] The presence of superoxide dismutase in anaerobes suggests that there was an accumulation of oxygen in the early Precambrian period even if at an extremely low concentration, if the superoxide dismutase in anaerobes functions to protect the cells from the active oxygen produced by temporary exposure to oxidized environments.[21] Recently some correlation between the superoxide dismutase content and the air-tolerance of anaerobic bacteria has been shown, which supports the above role for the enzyme.[12] However, circumstantial evidence that the Fe-enzyme is biosynthesized under the strict anaerobic conditions such as in a medium of sulfur bacteria containing sulfide and thiosulfate has cast a doubt on the above role for the enzyme. The contents of superoxide dismutase in anaerobic *Chromatium* and *Chlorobium* are 7 and 13 units per mg protein,[15,16] and the content in aerobically-grown *Escherichia coli* is about 20 units. Catalysis of oxidation-reductions other than the disporportionation of O_2^- by the enzyme cannot be excluded, at least, in anaerobes, although no such reaction has yet been found.

Superoxide dismutase in facultative and aerobic bacteria; the Fe- and Mn-enzymes. In contrast to anaerobic bacteria, aerobic and facultative anaerobic bacteria contain the Fe- or Mn-enzyme or both. The only exception to this is *Photobacterium leiognathi*, a luminescent bacterium, that contains the Cu,Zn-enzyme in addition to the Fe-enzyme.[22] However, this bacterium is a symbiont with fish and gene-transfer of the Cu,Zn-enzyme from the host fish has been speculated.[23]

The Mn-enzyme has been isolated from *E. coli*,[24] *Mycobacterium phlei*,[25] *M.* sp. Strain Takeo,[26] *M. lepraemurium*,[27] *Streptococcus mutans*,[28] *S. faecalis*,[29] *Bacillus stearothermophilus*,[30] *Thermus aquaticus*,[31] *T. thermophilus*[32] and *Rhodo pseudomonas spheroides*.[33] In contrast, the Fe-enzyme has been purified from *Mycobacterium tuberculosis*,[34] *Photobacterium sepiae*,[23] *P. leiognathi*,[23] *E. coli*,[35,36] *Pseudomonas ovalis*,[37] *Bacillus megaterium*,[96] and cyanobacteria (see below).

Some comments on the above list are in order: a) Britton et al.[29] carried out the H_2O_2-test on twenty four species of aerobic bacteria, and showed that most Mn-enzyme-containing bacteria are gram-positive, and that the Fe-enzyme- or both-enzyme-containing bacteria are generally gram-negative. The above list shows the same trends, but there are several exceptions as in the results of Britton et al. b) In contrast to anaerobic photosynthetic bacteria, aerobic photosynthetic bacteria contain the Mn-enzyme. c) Three thermophilic bacteria have the Mn-enzyme and the luminescent bacteria the Fe-enzyme. d) The enzyme content in M. lepraemurium is the highest in any of the bacteria surveyed so far; the Mn-enzyme constitutes 6% of its soluble protein.[27]

Because of the broad divergence of prokaryotes and the difficulty in correlating their evolutionary relationships, more extensive surveys of the distribution of Fe- and Mn-superoxide dismutases in prokaryotes are needed and the type of enzyme would provide one taxonomic criterion for prokaryotes. We have determined the type of enzyme in aerobic diazotrophs in terms of its relationship to the oxygen concentration in the cells.

Superoxide dismutase in aerobic diazotrophs; the Fe-enzyme. Most diazotrophs (Clostridium, Chromatium, Chlorobium and Desulfovibrio) are anaerobes, but, bacteria such as Azotobacter, Rhizobium and the blue-green algae are aerobes. Nitrogenase is a highly oxygen labile enzyme; its half life in air is only 4.5 min (Ref. 38); therefore, for the fixation of nitrogen, the concentration of oxygen inside the cells must be extremely low or near zero, even in the aerobic diazotrophs. In fact, a sharp gradient for oxygen concentration has been shown in the "unstirred layer" surrounding of Rhizobium cells, and the oxygen concentration on the surface of the cells have been estimated to be less than 3×10^{-8} to 3×10^{-9} M (Ref. 39). As described above the superoxide dismutase in anaerobes is the Fe-enzyme. Facultative anaerobic bacteria such as E. coli contain only the Fe-enzyme when cells have been cultered under anaerobic states. Biosynthesis of the Mn-enzyme and of "hybrid" Fe-superoxide dismutase[40] is induced by O_2 (Ref. 41) or O_2^- (Ref. 42). Thus, the Fe-enzyme appears to be biosynthesized when the oxygen concentration inside the cells is extremely low. Aerobic diazotrophs should contain only the Fe-enzyme because of the very low oxygen concentration inside the cells; we found that this is the case.[43]

Table 1 shows the high content of superoxide dismutase in Azotobacter, compared with other aerobes. A similar high content also has been reported by Nishie and Yano.[44] Exogeneous O_2^- inhibits the nitrogen-fixation of Azotobacter, which implies a physiological role for superoxide dismutase.[45] Isoelectric focusing on polyacrylamide gel of the extracts of Azotobacter shows a single band of superoxide dismutase that corresponds to a pI value of 4.1. Nitrogen in the form of N_2, NH_3, NO_3^- or in organic form did not affect the content of the enzyme appreciably and the same single activity band was observed. Incubation of the cells in 100% O_2 for 24 hours increased the enzyme by 20-50%, but, again, no new activity band appeared. This single Azotobacter superoxide dismutase is the Fe-enzyme according to the H_2O_2-test. We further confirmed this by the

TABLE 1

SUPEROXIDE DISMUTASE ACTIVITY IN AZOTOBACTER IN COMPARISON WITH THAT OF SEVERAL AEROBIC BACTERIAL[a]

Species	Medium[b]	Culture time[b] (hours)	Superoxide dismutase activity (units/mg protein)
Azotobacter vinelandii	N-free	18	48
		24	42
		38	40
	NO_3^--N	24	43
	NH_3-N	24	43
	TSY	24	39
	N-free	38[c]	63
Azotobacter chroococcum	NO_3^--N	18[c]	61
Acinetobacter aerogenes	TSY	10	20
Bacillus sphericus	TSY	10	5
Pseudomonas aeruginosa	TSY	10	15

[a]Kanematsu and Asada (Ref. 43).
[b]The cells were cultured with shaking, in 500 ml-medium, with 50 ml of the inoculum which had been cultured for 18-24 hours. N-free; Modified Burk's N-free medium. NO_3^--N and NH_3-N; To the Burk's N-free medium 8 nM KNO_3 or NH_4Cl was added. TSY; 3% Trypticase soy broth and 0.5% yeast extract.
[c]The cells were cultured in a jar-fermenter under vigorous agitation with air.

isolation of several grams of the enzyme. Table 1 shows that about 2% of the soluble protein is the Fe-enzyme; thus, Azotobacter is a good source for the isolation of the Fe-superoxide dismutase. The isolation and properties of the Azotobacter enzyme and the Mössbauer spectrum of the ^{57}Fe-reconstituted enzyme will be published elsewhere.

The bacteroids from soybean root-nodules, Rhizobium japonicum, also contain only the Fe-enzyme. Although the superoxide dismutase in host cells of the nodules is mainly the Cu,Zn-enzyme, enzymatic activity in the isolated bacteroids was insensitive to cyanide and to the antibody against the spinach Cu,Zn-enzyme (Figure 2), but it was inactivated by H_2O_2. Further, the bacteroid enzyme was inhibited by 70% with 10 mM azide. All this evidence shows that bacteroids contain only the Fe-enzyme, and there is no gene-transfer for the Cu,Zn-enzyme from the host cells, as postulated in the case of the photobacterium.[23]

In the blue-green alga, Anabaena cylindrica, superoxide dismutase is found in the heterocysts where nitrogenase is localized, although the type of enzyme has not been determined.[46] Because the Mn-enzyme binds to the thylakoids in blue-green algae, the major superoxide dismutase in the heterocysts is considered to be the Fe-enzyme.

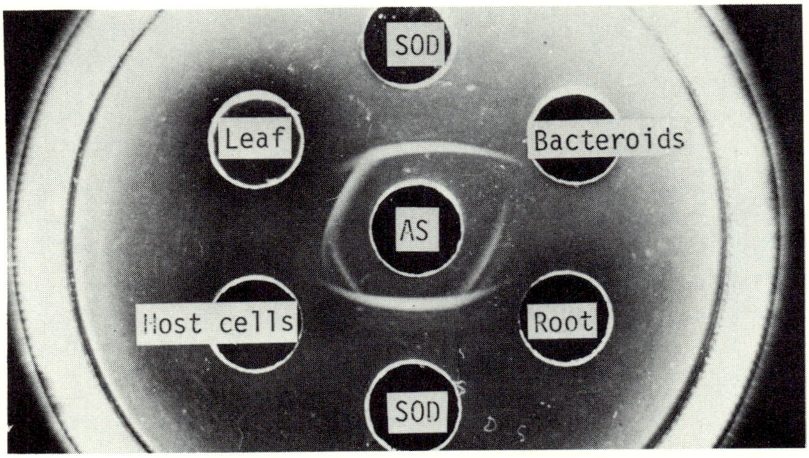

Fig. 2. Ouchterlony double immunodiffusion plates of the spinash Cu,Zn-superoxide dismutase antiserum and superoxide dismutase in cell-free extracts from soybean tissues and bacteroids in soybean root-nodules (Kanematsu and Asada, Ref. 52). Center well (AS) contained 260 μg of the antiserum to spinach, Cu,Zn-enzyme, Outer wells; cell-free extracts containing 3 to 7 units of enzyme from leaves and roots of the soybean and from host cells and bacteroids of the root-nodules, SOD; spinach Cu,Zn-superoxide dismutase (1.1 μg, 3 units).

Thus, anaerobes, anaerobically grown-facultative anaerobes, and aerobic diazotrophs contain only the Fe-superoxide dismutase and lack the Mn-enzyme. The property common to the three bacterial groups is an extremely low or nearly zero concentration of O_2 or O_2^- inside the cells. We propose that the biosynthesis of the Fe-enzyme occurs only when the concentration of O_2 or O_2^- is nearly zero inside the cells, and that the Mn-enzyme is formed only in cells that contain O_2 or O_2^- above a definite concentration. If further evidence supports this proposal, the type of superoxide dismutase should be useful in predicting the concentration of O_2 or O_2^- inside the cells.

Superoxide dismutases in algae; the Fe- and Mn-enzymes. Prokaryotic blue-green algae contain only the Fe- and Mn-enzymes as do other aerobic prokaryotes.[11,47] The Fe-enzymes from *Plectonema boryanum*[6,48] *Spirulina platensis*[33] and *Anacystis nidulans*[49] have been characterized. As described below, the Mn-enzyme localizes in the thylakoids.

Eukaryotic algae, except the phragmoplast algae,[50] lack the cyanide-sensitive enzyme and contain the Fe- and/or Mn-enzymes.[6,9-11,47,50] The Fe-enzyme from *Euglena*[18] and the Mn-enzyme from red algae (*Porphyridium cruentum*[51] and *Porphyra tenera*[52]) have been isolated. In *Euglena*, the Mn-enzyme binds to the thylakoids.[18]

Superoxide dismutases in protozoa; the Fe- and Mn-enzymes. No Cu,Zn-superoxide dismutase is detectable in Euglenophyta,[11] Flagellata (Tritrichomonas foetus, Monocercomonas sp.[53] and Crithidia fasciculata[11]), and Ciliata (Tetrahymena pyriformis[53]). These organisms contain only the cyanide-insensitive enzyme; the Fe-enzyme has been isolated from Euglena.[18] Thus, like most eukaryotic algae, protozoa are a group of eukaryotes that lack the Cu,Zn-enzyme and contain only the Fe- and Mn-enzymes. It should be noted here that anaerobically grown T. foetus contains superoxide dismutase,[53] but, which type has not been determined. This indicates that the presence of the enzyme in anaerobically grown cells is not limited to prokaryotes. On the other hand, the content of the enzyme in Euglena cultured under photoautotrophic conditions is higher than that in cells cultured under heterotrophic conditions,[11] which reflects the photoproduction of O_2^- in chloroplasts.[54-56]

Superoxide dismutases in animals; the Cu,Zn- and Mn-enzymes. In contract to protozoa, the animals, including invertebrates and vertebrates, tested so far contain the Cu,Zn-enzyme in addition to the Mn-enzyme. Although the mammalian enzymes have been extensively studied, only a few surveys have been done on invertebrates. The enzyme has been detected only in the hemerythrocytes of sipunculids that contain hemerythrin in place of hemoglobin,[57] and in the sea anemone.[58] However, no determination of the type of enzyme has been made. We surveyed the type of superoxide dismutase in twenty five species of primitive marine invertebrates with the cyanide-test; the results are summarized in Table 2.[52] Even the most primitive animals such as Porifera and Coelenterata have the Cu,Zn-superoxide dismutase and, as expected, more evolved invertebrates, Mollusca, Echinodermata and Arthropoda, also contain the Cu,Zn-enzyme. Thus, animals

TABLE 2

DISTRIBUTION OF Cu,Zn-SUPEROXIDE DISMUTASE AND CYANIDE-INSENSITIVE Fe- AND Mn-SUPEROXIDE DISMUTASES IN PRIMITIVE INVERTEBRATES[a]

Family	Number of species tested	Cu,Zn-SOD	Cyanide-insensitive SOD
Porifera	2	+	+
Coelenterata			
Hydrozoa	1	+	+
Anthozoa	9	+	+
Mollusca	4	+	+
Echinodermata	6	+	+
Arthropoda	3	+	+

[a]Kanematsu and Asada (Ref. 52).

acquired the Cu,Zn-enzyme at the early step of evolution from unicellular protozoa to multicellular primitive invertebrates. The fossil record of Porifera and Coelenterata shows that they appeared on the earth in the Cambrian period.

Superoxide dismutase has been purified from many mammalian species and tissues. A complete list is beyond the scope of this review; we cite only the classic work on the Cu,Zn-enzyme from bovine erythrocytes[59] and the Mn-enzyme from human liver.[60] A survey of the enzyme in vertebrates, including fish has been conducted[61,62] and the Cu,Zn-enzyme has been isolated from swordfish.[63] In liver mitochondria the Mn-enzyme localizes in the matrix including the inner membrane, whereas, the Cu,Zn-enzyme is in the intermembrane space.[60,64,65] The Mn-enzyme also has been found in the cytosol of the liver.[60] To our knowledge, however, no Fe-superoxide dismutase has been detected in animal tissues.

Superoxide dismutase in fungi and slime molds; the Cu,Zn- and Mn-enzyme. The Cu,Zn-enzyme has been isolated from *Saccharomyces*,[66,67] *Fusarium*[68] and *Neurospora*,[69] and the Mn-enzyme has been isolated from the luminescent fungus, *Pleurotus olearius*,[23] and from the mitochondria of yeast.[70] In addition, we have tested two species of slime molds (Myxomycota) and forty two species of fungi (Eumycota) covering Mastigomycotina, Zygomycotina, Ascomycotina, Basidiomycotina, and Deuteromycotina for their sensitivity to cyanide in the extracts.[47] Without exception, enzyme activity was inhibited by cyanide which indicates the presence of the Cu,Zn-enzyme in Myxomycota and Eumycota. Thus, among the microorganisms, only those of this group contain the Cu,Zn-enzyme. No one has reported the Fe-enzyme in fungi.

A classification of organisms based on recent biological data has been proposed by Whittaker, who divides the organisms into the five kingdoms; Monera, Protista, Plantae, Fungi and Animalia.[71] In this classification the fungi are regarded as an independent kingdom characterized by its evolutionary development in the absorption of nutrients, and one of the three kingdoms of higher organisms. Yamanaka has proposed that eukaryotic algae and protozoa appeared earlier than fungi from his analysis of the reactivity of the cytochrome *c* of algae and yeast with cytochrome oxidase.[72] Thus, several lines of evidence indicate that the fungi are developed organisms; therefore, the absence of the Cu,Zn-enzyme in algae and protozoa and its presence in fungi is not surprising,

Superoxide dismutases in land plants; the Cu,Zn- and Mn-enzymes. All the land plants tested so far including the mosses, ferns and seed plants (gymnosperms and angiosperms) contain the Cu,Zn-enzyme in addition to the cyanide-insensitive enzyme.[10,11] The enzyme has been demonstrated both in leaves and in other tissues such as seeds, fruits, shoots, and roots.[73-78] The Cu,Zn-enzyme has been purified from green pea,[73,74] wheat germ,[75] spinach leaves[79] and kidney bean;[80] its properties are very similar to those of the enzyme from animals and fungi. Cyanide-insensitive superoxide dismutase has been detected in wheat,[75]

Jerusalem artichoke,[81] corn,[74] and peas.[83] The H_2O_2-test shows the presence of the Mn-enzyme and the absence of the Fe-enzyme in kidney bean, spinach and tea.[80,82] We have purified the Mn-enzyme from spinach thylakoids.[84]

Thus, the Fe-enzyme is absent in plants (but see Ref. 97) as well as in animals and fungi, and localization of the enzyme in the mitochrondria of plants is also the same as in animals and fungi; the Mn-enzyme is in the matrix and the Cu,Zn-enzyme is in the intermembrane space.[81,85] And, as described below, in chloroplasts the Mn-enzyme binds to the thylakoids and the Cu,Zn-enzyme is localized in the stroma.

Binding of Mn-superoxide dismutase to thylakoids. As described above, prokaryotic and eukaryotic algae contain the Fe- and Mn-enzymes. In contrast, land plants have the Cu,Zn- and Mn-enzymes. Previously we showed the localization of superoxide dismutase in spinach chloroplasts,[79] where O_2^- is produced by autooxidation of the primary electron acceptor in photosystem I,[55] especially when the physiological light energy acceptor, CO_2, is deficient.[56] On the other hand, manganese is an essential component for the photosynthetic evolution of oxygen, therefore, it is interesting to determine whether Mn-superoxide dismutase occurs in chloroplasts as in mitochondria.

Blue-green algae are prokaryotes that can use water as the electron donor for the electron transport system in the thylakoids that produce molecular oxygen. First, we determined the subcellular distribution of the Mn- and Fe-enzymes in three species of blue-green algae. The algal spheroplasts were prepared by incubation of the cells with lysozyme, and were disrupted by a brief sonication, after which, thylakoids and cytosol were separated by centrifugation. Immunological results (Figure 1) and the H_2O_2-test (Figure 3) have shown that the Fe-enzyme is localized in the cytosol, but the Mn-enzyme exclusively binds to the thylakoids.[86] The binding of Mn-superoxide dismutase to thylakoids has also been observed in Euglena chloroplasts. In this green algal protozoa the Fe-enzyme is believed to be the superoxide dismutase of the stroma.[18]

Localization of Cu,Zn-superoxide dismutase in chloroplasts from higher plants has been shown,[79,85,87-90] but conflicting results have been reported as to whether the chloroplasts contain Mn-superoxide dismutase; both its presence[10,90] and absence [85,89,90] have been claimed. We show here that the Mn-enzyme is bound to the thylakoids of spinach chloroplasts.[84] Intact spinach chloroplasts were isolated by brief homogenization of the leaves, followed by differential centrifugation, then silica-sol density centrifugation (Figure 4). The fractionation allows the separation of intact chloroplasts with an envelope and separation of mitochondria. Because of the intact state of the chloroplasts, the superoxide dismutase content is about three times higher than that in class II chloroplasts.[79] Our present results confirm the presence of both cyanide-sensitive and -insensitive superoxide dismutases in chloroplasts.

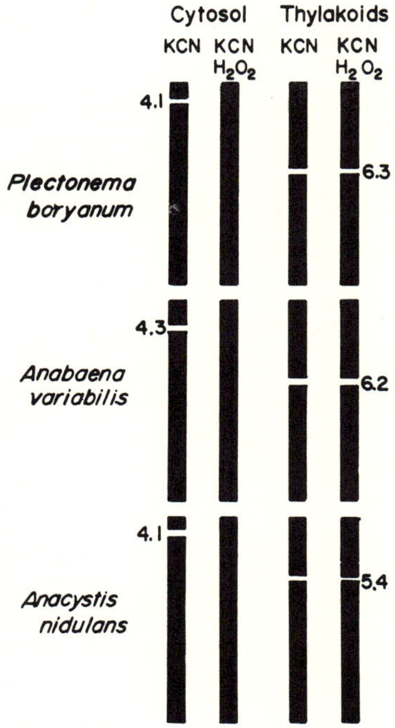

Fig. 3. Isoelectric focussing pattern of superoxide dismutase in the cytosol and thylakoids of blue-green algae (Okada, Kanematsu and Asada, Ref. 86.) Each fraction containing 3-4 units of superoxide dismutase was electrofocused with Ampholine (pH 3.5-10.0) on a polyacrylamide gel disc. Thylakoids were dissolved in 3% Triton X-100 before application. Where indicated, the sample had been incubated in 5mM H_2O_2, containing 5 mM KCN for 12 hours at $0^{o}C$. The acromatic zone indicates the enzyme; the numbers show its pI value.

Further fractionation of chloroplasts into stroma and thylakoids indicates that the Cu,Zn and Mn-enzyme are bound to the thylakoids, and that the Cu,Zn-enzyme is localized in the stroma. The thylakoid enzyme was inhibited by only half to one third with cyanide (Figure 5), but the stroma enzyme was completely inhibited. Further cyanide-insensitive activity in the thylakoids was not inactivated by H_2O_2 (Figure 6).

The thylakoid-enzyme is not released by a hypotonic medium or salt. Full activity was observed only when the assay mixture contained a detergent such as Triton X-100 (the stroma enzyme was not affected by Triton X-100). A similar membrane-bound latency has been observed for the hydrogenosome-bound superoxide dismutase of T. foetus.[53] The effect of Triton X-100 on spinach thylakoid

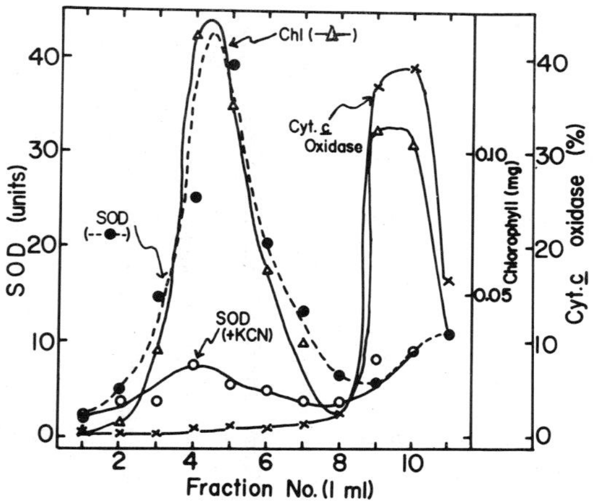

Fig. 4. Subcellular distribution of superoxide dismutase in spinach leaves (Kanematsu, Hayakawa and Asada, Ref. 84). Spinach leaves were homogenized in a Waring bendor for 3 s in three volumes of a grinding medium consisting of 0.33 M sorbitol, 1 mM EDTA, 1mM $MgCl_2$, 50 mM Tricine-KOH, pH 7.8, and 0.5% bovine serum albumin. The crude, intact chloroplast fraction was obtained by centrifugation at 1700 g for 1 min after filtration through eight layers of gauze, then it was suspended in a small volume of the grinding medium. This suspension (1 ml) was immediately layered onto 10 ml of a linear Percoll (Pharmacia Fine Chemicals) gradient (10-90%, v/v) containing 0.33 M sorbitol, 1mM $MgCl_2$, 1 mM EDTA, 50 mM Tricine-KOH, pH 7.8, and 1% bovine serum albumin, then centrifuged at 5000 g for 15 min. Superoxide dismutase activity was assayed by the xanthine-xanthine oxidase cytochrome c system containing 1% Triton X-100 in the presence or absence of 1 mM KCN.

enzyme shows that the Mn-enzyme is bound more tightly to the membranes than is the thylakoid-Cu,Zn-enzyme (Figure 5). A similar latency also was found for the thylakoid-enzyme of blue-green algae and Euglena.[18,86]

Recently superoxide dismutase activity has been found in thylakoid subfragments and chlorophyll-protein complexes,[91,92] which support the binding of superoxide dismutase to the thylakoids, but the chlorophyll protein itself is not superoxide dismutase. Purification of the Mn-enzyme from blue-green alga and spinach thylakoids[84,85] separated superoxide dismutase from the chlorophyll-containing fractions.

Thus, the thylakoid-bound superoxide dismutase of prokaryotic and eukaryotic algae, and of land plants is the Mn-enzyme. The cytosol of blue-green algae and the stroma of green algal chloroplasts contain the Fe-enzyme, providing evidence for the symbiotic origin of the chloroplasts of green algae from prokaryotic cyanobacteria. The stroma enzyme in Euglena, however, is replaced by the Cu,Zn-enzyme in land plants. Our preliminary results show that Euglena mitochondria contain the Fe-enzyme, which suggests that the intermembrane space

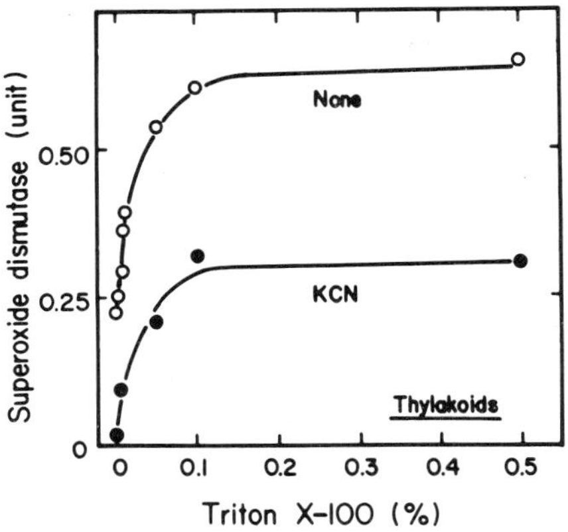

Fig. 5. Effect of Triton X-100 on spinach thylakoid-bound superoxide dismutases (Kanematsu, Hayakawa and Asada, Ref. 84). Thylakoids were obtained from osmotically-ruptured spinach chloroplasts by centrifugation. The thylakoids (28 μg of chlorophyll) were washed three times with 50 mM potassium phosphate, pH 7.8, 2 mM $MgCl_2$, 1 mM EDTA and 0.1 M NaCl. Enzymatic activity in the washed thylakoids was assayed in the presence of the indicated amounts of Triton X-100 by the xanthine-xanthine oxidase cytochrome *c* system. Where indicated 1mM KCN was added.

Fe-enzyme of algal and protozoal mitochondria is replaced by the Cu,Zn-enzyme in plants, fungi, and animals. In mitochondria from animals, fungi, and plants the Mn-enzyme is localized in the matrix including the inner membrane.[60,64,65,70,81,85] The thylakoids of chloroplasts correspond to the inner membrane and matrix of mitochrondria in regard to their functions of electron transport; therefore, the sub-chloroplast and -mitochondrial distribution of Mn-superoxide dismutase suggests that both organelles originate from a common prokaryote. Sub-chloroplast and -mitrochondrial distributions of the three types of enzyme in organisms at different stages of evolution are schematically summarized in Figure 7.

We found intact spinach chloroplasts to contain about 100 units of superoxide dismutase per mg chlorophyll, of which about 10% was Mn-enzyme (Figure 4). This corresponds to 0.38 molecule of the Cu,Zn-enzyme and 0.036 molecule of the Mn-enzyme per 400 molecules of chlorophyll (photosynthetic unit), assuming specific activities of 3000 and 2500 units per mg protein and molecular weights of 32,000 and 40,000 for the Cu,Zn- and Mn-enzymes, respectively. Thylakoids of blue-green algae contain 0.1 molecule of the Mn-enzyme per photosynthetic unit.[86] Thus, the concentration of superoxide dismutase in intact spinach chloroplasts is 2.1×10^{-5} M for the Cu,Zn-enzyme and 2.3×10^{-6} M for the Mn-enzyme, if we assume that the chlorophyll concentration in intact chloroplasts is 2.5×10^{-2} M.[93] In the binding of the Mn-enzyme to thylakoids, the local concentration of the Mn-enzyme is similar to that of the Cu,Zn-enzyme in the stroma if the

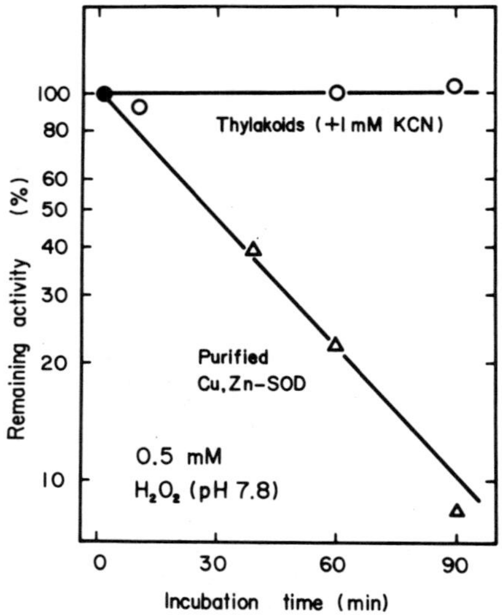

Fig. 6. Effect of hydrogen peroxide on spinach thylakoid-bound Mn- and purified Cu,Zn-superoxide dismutases (Kanematsu, Hayakaya and Asada, Ref. 84). Spinach thylakoids were incubated with 0.5 mM H_2O_2, 10 mM potassium phosphate, pH 7.8, 0.1 mM EDTA, 1mM KCN and 0.1% Triton X-100, at 25°C. Samples (about 1 unit) were withdrawn at intervals and the enzymatic activity was assayed by the xanthine-xanthine oxidase cytochrome c system containing 1% Triton X-100 and 1 mM KCN. The thylakoid-bound Cu,Zn-enzyme was suppressed by cyanide. The effect of H_2O_2 on purified Cu,Zn-superoxide dismutase was measured under the same conditions except that KCN and Triton X-100 were omitted.

volume of the thylakoids is 10% of the intact chloroplasts. Superoxide dismutase at this concentration lowers the steady state concentration of O_2^- in chloroplasts by 10^{-4}.[56]

The scavenging of O_2^- in the thylakoid membranes is one probable function of the thylakoid-bound Mn-enzyme, but the possibility that the Mn-enzyme is a component that catalyzes the oxidation of water cannot be excluded, even though the thylakoid-bound Mn-superoxide dismutase constitutes only several percent of the thylakoid-bound manganese.[84,86] We have purified the spinach thylakoid-enzyme and have been studying whether this protein participates in the dehydrogenation of water in the thylakoids.

Concluding remarks. The distribution of three types of superoxide dismutase in organisms from anaerobic prokaryotes to higher organisms is summarized in Figure 8. In addition to anaerobes, the aerobes such as aerobic diazotrophs whose concentration of oxygen inside the cells is supposed to be near zero contain only the Fe-enzyme. Aerobic and facultative bacteria, prokaryotic algae,

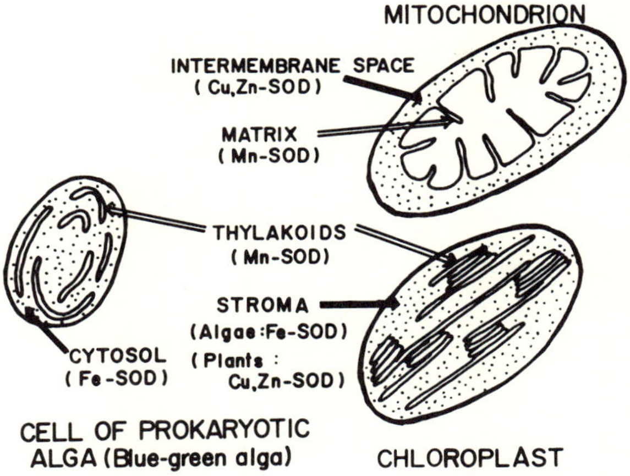

Fig. 7. Sub-chloroplast and sub-mitochondrial distribution of three types of superoxide dismutase in organisms at various stages of evolution.

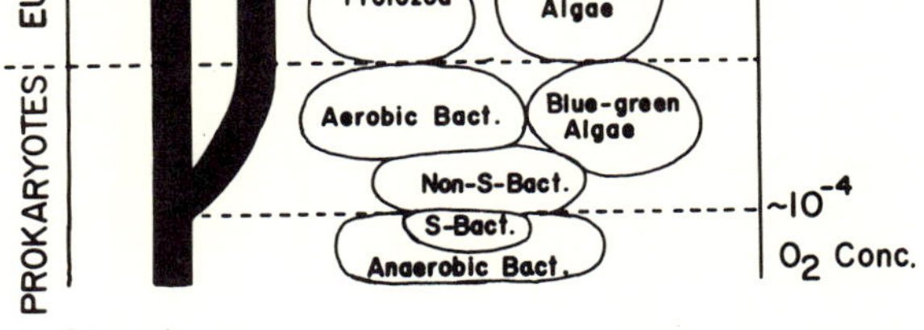

Fig. 8. Phylogenic distributions of three types of superoxide dismutase in organisms at various stages of evolution. O_2 Conc. shows the concentrations of O_2 in the atmosphere, relative to the present concentration, when the indicated organisms appeared on the earth.

protozoa and most eukaryotic algae contain the Fe- or Mn-enzyme or both; the Cu,Zn-superoxide dismutase is absent in these organisms. Anaerobically grown-facultative prokaryotes contain only the Fe-enzyme, and biosynthesis of the Mn-enzyme is induced by oxygen. The Cu,Zn-enzyme has been found in animals, fungi and plants; even in such primitive organisms, as Porifera, Coelenterata, slime molds, ferns, and mosses. These organisms lack the Fe-enzyme, and the Mn-enzyme occurs mainly in the mitochondrial matrix and chloroplast thylakoids. Superoxide dismutase in the intermembrane space of the mitochrondria and in the stroma of chloroplasts (cytosol in the blue-green algae) is the Fe-enzyme in algae and protozoa and the Cu,Zn-enzyme in animals, fungi and plants (Figure 7).

The Fe- and Mn-enzymes in aerobes may have evolved from a common ancestor, the Fe-enzyme of anaerobes, because of the high degree of homology of the amino acid sequences of the two types of enzyme. Divergence of the Mn-enzyme from the Fe-enzyme should reflect an increase in the oxygen concentration of the biosphere. When aerobic prokaryotes appeared on the earth and to what extent the free oxygen was produced by abiotic reactions are points of controversy,[20,21] but, we may assume that divergence of the Mn-enzyme from the Fe-enzyme came in the early Precambrian period, 30 to 20 x 10^8 years ago when the oxygen concentration was about 10^{-4} that of the present atmospheric level.[94] The acquistion of the Cu,Zn-enzyme is believed to have occurred in the Cambrian period, 6 to 4.5 x 10^8 years ago, because of the appearance of primitive animals such as Porifera and Coelenterata in fossils. According to Berkner and Marshall, the oxygen concentration in the atmosphere at that time was 10^{-2} of the present concentration.[94]

This distribution of three types of superoxide dismutase in organisms and in cell organelles and their molecular evolution raises basic questions; why did organisms evolve three types of enzyme that have similar catalytic properties in respect of the disproportionation of O_2^-? And why did the Fe-enzyme disappear and a Cu,Zn-enzyme was acquired in higher organisms? The most basic and simplest answer to these questions is that each type of enzyme has its own biological function that includes not only the scavenging of O_2^- but some unknown reaction. The determination of the biological role of the Fe-enzyme in anaerobes and of the thylakoid-bound Mn-enzyme may provide clues toward the answer of these questions. Further, the insensitivity of the Mn-enzyme to hydrogen peroxide,[6] and the reported peroxidase[56,95] and superoxidase activity[56] of superoxide dismutase also deserve further investigation.

SUMMARY

The distribution of cuprozinc, ferric and mangano superoxide dismutases in organisms from anaerobic prokaryotes to higher organisms, and sub-organelle distribution of the enzymes are summarized. Anaerobes contain only the Fe-enzyme and aerobic diazotrophs, whose oxygen concentration inside the cells is extremely low, also contain only the Fe-enzyme. Facultative and aerobic bacteria, prokaryotic algae, most eukaryotic algae, and protozoa have the Fe- or Mn-enzymes or both and lack the Cu,Zn-enzyme. The Cu,Zn-enzyme is contained in animals, plants and fungi, even in primitive forms. These higher organisms have also the Mn-enzyme, mainly in their mitchondrial matrix and chloroplast thylakoids, but they lack the Fe-enzyme (but see Ref. 97). The Mn-enzyme is bound to the thylakoids from blue-green algae, Euglena and plants, but, the stroma Fe-enzyme in algae is replaced by the Cu,Zn-enzyme in the stroma of plant chloroplasts. Based on the distribution of superoxide dismutase, its physiological function and molecular evolution are discussed.

REFERENCES

1. Steinman, H.M. and Hill, R.L. (1973) Proc. Natl. Acad. Sci. USA, 70, 3725-3729.
2. Bridgen, J., Harris, J.I. and Northrop, F. (1975) FEBS Lett.49, 392-395.
3. Steinman, H.M. (1978) J. Biol. Chem. 253, 8708-8720.
4. Ose, D.E. and Fridovich, I. (1976) J. Biol. Chem. 251, 1217-1218.
5. Yamakura, F., Suzuki, K. and Omori, O. (1978) J. Japan Biochem. Soc. 50, 971.
6. Asada, K., Yoshikawa, K., Takahashi, M., Maeda, Y. and Enmanji, K. (1975) J. Biol. Chem. 250, 2801-2807.
7. Misra, H.P. and Fridovich, I. (1978) Arch. Biochem. Biophys. 189, 317-322.
8. Hewitt, J. and Morris, J.G. (1975) FEBS Lett. 50, 315-317.
9. Lumsden, J. and Hall, D.O. (1975) Nature, 257, 670-672.
10. Asada, K., Kanematsu, S., Takahashi, M. and Kono, Y. (1976) in Iron and Copper Proteins, Yasunobu, K.T., Mower, H.F. and Hayaishi, O., eds., Plenum Press, New York, pp.551-564.
11. Asada, K., Kanematsu, S. and Uchida, K. (1977) Arch. Biochem. Biophys. 179, 243-256.
12. Tally, F.P., Golden, B.R., Jacobus, N.V. and Gorbach, S.L. (1977) Infec. Immunity, 16, 20-25.
13. Carlsson, J., Werthau, C. and Beckman, G. (1977) J. Clin. Microbiol. 6, 280-284.
14. Gregory, E.M., Moore, W.E.C. and Holdeman, L.V. (1978) Appl. Environ. Microbiol. 35, 988-991.
15. Kanematsu, S. and Asada, K. (1978) Arch. Biochem. Biophys. 185, 473-482.
16. Kanematsu, S. and Asada, K. (1978) FEBS Lett. 91, 94-98.
17. Hatchikian, E.C. and Henry, Y.A. (1977) Biochimie, 59, 153-161.
18. Kanematsu, S. and Asada, K. (1979) Arch. Biochem. Biophys. 195, 535-545.
19. Bruschi, M., Hatchikian, E.C. Bonicel. J., Bovier-Lapierre, G. and Couchoud, P. (1977) FEBS Lett. 76, 121-124.
20. Schwarz, R.M. and Dayhoff, M.O. (1978) Science, 199, 395-403.
21. Towe, K.M. (1978) Nature, 274, 657-661.
22. Puget, K. and Michelson, A.M. (1974) Biochem. Biophys. Res. Commun. 58, 830-838.
23. Puget, K., Lavelle, F. and Michelson, A.M. (1977) in Superoxide and Superoxide Dismutases, Michelson, A.M., McCord, J.M. and Fridovich, I., eds., Academic Press, New York, pp. 139-150.
24. Keele, B.B., McCord, J.M. and Fridovich, I. (1970) J. Biol. Chem. 245, 6176-6181.
25. Chikata, Y., Kusunose, E., Ichihara, K. and Kusunose, M. (1975) Osaka City Med. J. 21, 127-136.
26. Kusunose, M., Noda, Y., Ichihara, K. and Kusunose, E. (1976) Arch. Microbiol. 108, 65-73.
27. Ichihara, K., Kusunose, E., Kusonose, M. and Mori, T. (1977) J. Biochem. Tokyo, 81, 1427-1433.
28. Vance, P.G., Keele, B.B. and Rajagopalan, K.V. (1972) J. Biol. Chem. 247, 4782-4786.
29. Britton, L., Malinowski, D.P. and Fridovich, I. (1978) J. Bacteriol. 134, 229-236.

30. Harris, J.I. (1977) in Superoxide and Superoxide Dismutase, Michelson, A.M., McCord, J.M. and Fridovich, I., eds. Academic Press, New York, pp. 151-157.
31. Sato, S. and Harris, J.I. (1977) Eur. J. Biochem. 73, 373-381.
32. Sato, S. and Nakazawa, K. (1978) J. Biochem. Tokyo, 83, 1165-1171.
33. Lumsden, J., Cammack, R. and Hall, D.O. (1976) Biochim. Biophys. Acta, 438, 380-392.
34. Kusunose, E., Ichihara, K., Noda, Y. and Kusunose, M. (1976) J. Biochem. Tokyo, 80, 1343-1352.
35. Yost, F.J. and Fridovich, I. (1973) J. Biol. Chem. 248, 4905-4908.
36. Slykhouse, T.O. and Fee, J.A. (1976) J. Biol. Chem. 251, 5472-5477.
37. Yamakura, F. (1976) Biochim. Biophys. Acta, 422, 280-294.
38. Israel, D.W., Howard, R.L. Evans, H.J. and Russel, S.A. (1974) J. Biol. Chem. 249, 500-508.
39. Wittenberg, J.B., Bergersen, F.J., Appleby, C.A. and Turner, C.L. (1974) J. Biol. Chem. 249, 4057-4066.
40. Dougherty, H.W., Sadowski, S.J. and Baker, E.E. (1978) J. Biol. Chem. 253, 5220-5223.
41. Hassan, H.M. and Fridovich, I. (1977) J. Bacteriol. 129, 1574-1583.
42. Hassan, H.M. and Fridovich, I. (1977) J. Biol. Chem. 252, 7667-7672.
43. Kanematsu, S. and Asada, K. (1977) Proc. Agr. Chem. Soc. Japan, p.266.
44. Nishie, H. and Yano, K. (1977) Proc. Agr. Chem. Soc. Japan, p.266.
45. Buchanan, A.G. (1977) Can. J. Microbiol. 23, 1548-1553.
46. Henry, L.E.A., Gogotov, I.V. and Hall, D.O. (1978) Biochem. J. 174, 373-377.
47. Asada, K. and Kanematsu, S. (1978) in Evolution of Protein Molecules, Matsubara, H. and Yamanaka, T., eds., Japan Scientific Societies Press, Tokyo, pp.361-3720.
48. Misra, H.P. and Keele, B.B. (1975) Biochim. Biophys. Acta, 379, 418-425.
49. Cséke, C., Horvath, L.I., Simon, P., Borbély, G., Keszthelyi, L. and Farkas, G. (1979) J. Biochem. Tokyo, 85, 1397-1404.
50. Henry, L.E.A. and Hall, D.O. (1977) in Photosynthetic Organelles, Miyachi, S., Fujita, Y. and Shibata, K., eds., Japanese Soc. Plant Physiol., Kyoto, pp.377-382.
51. Misra, H.P. and Fridovich, I. (1977) J. Biol. Chem. 252, 6421-6423.
52. Kanematsu, S. and Asada, K. unpublished.
53. Lindmark, D.G. and Müller, M. (1974) J. Biol. Chem. 249, 4634-4637.
54. Asada, K. and Kiso, K. (1973) Eur. J. Biochem. 33, 253-257.
55. Asada, K., Kiso, K. and Yoshikawa, K. (1974) J. Biol. Chem. 249, 2175-2181.
56. Asada, K., Takahashi, M., Tanaka, K. and Nakano, Y. (1977) in Biochemical and Medical Aspects of Active Oxygen, Hayaishi, O. and Asada, K., eds., Japan Scientific Societies Press, Tokyo, pp.45-63.
57. Manwell, C. (1977) Comp. Biochem. Physiol. 58B, 331-338.
58. Lin, A.L. (1979) Comp. Biochem. Physiol. 62B, 425-431.
59. McCord, J.M. and Fridovich, I. (1969) J. Biol. Chem. 244, 6049-6055.
60. McCord, J.M., Boyle, J.A., Day, E.D., Rizzolo, L.J. and Salin, M.L. (1977) in Superoxide and Superoxide Dismutases, Michelson, A.M., McCord, J.M. and Fridovich, I., eds., Academic Press, New York, pp. 129-138.
61. Maral, J., Puget, K. and Michelson, A.M. (1977) Biochem. Biophys. Res. Commun. 77, 1525-1535.
62. Mazeaud, F., Maral, J. and Michelson, A.M. (1979) Biochem. Biophys. Res. Commun. 86, 1161-1168.
63. Bannister, J.V., Anastasi, A. and Bannister, W.H. (1977) Comp. Biochem. Physiol. 56B, 235-238.
64. Weisiger, R.A. and Fridovich, I. (1973) J. Biol. Chem. 248, 3582-3592.
65. Peeters-Joris, C., Vandevoorde, A. and Baudhuin, P. (1975) Biochem. J. Tokyo, 150, 31-39.
66. Weser, U., Fretzdorff, A. and Prinz, R. (1972) FEBS Lett. 27, 267-269.
67. Goscin, S.A. and Fridovich, I. (1972) Biochim. Biophys. Acta, 289, 276-283.

68. Rapp, U., Adams, W.C. and Miller, R.W. (1973) Can. J. Biochem. 51, 158-171.
69. Misra, H.P. and Fridovich, I. (1972) J. Biol. Chem. 247, 3410-3414.
70. Ravindranath, S.D. and Fridovich, I. (1975), J. Biol. Chem. 250, 6107-6112.
71. Whittaker, R.H. (1969) Science, 163, 150-163.
72. Yamanaka, T. and Fukomori, Y. (1978) in Evolution of Protein Molecules, Matsubara, H. and Yamanaka, T., eds., Japan Scientific Societies Press, Tokyo, pp.387-399.
73. Sawada, Y., Ohyama, T. and Yamazaki, I. (1972) Biochim. Biophys. Acta, 268, 305-312.
74. Giannopolitis, C.N. and Ries, S.K. (1977) Plant Physiol. 59, 315-318.
75. Beauchamp, C.O. and Fridovich, I. (1973) Biochim. Biophys. Acta, 317, 50-64.
76. Baker, J.E. (1976) Plant Physiol. 58, 644-647.
77. Hai, D.Q., Kovacs, K. Matkovics, I. and Matkovics, B. (1975) Biochem. Physiol. Pflanz. 167, 357-359.
78. Giannopolitis, C.N. and Ries, S.K. (1977) Plant Physiol. 59, 315-318.
79. Asada, K., Urano, M. and Takahashi, M. (1973) Eur. J. Biochem. 36, 257-266.
80. Kono, Y., Takahashi, M. and Asada, K. (1979) Plant Cell Physiol., in press.
81. Arron, G.P., Henry, J.M. Palmer, J.M. and Hall, D.o. (1976) Biochim. Soc. Trans. 4, 618-620.
82. Hakamada, K., Kanematsu, S. and Asada, K. (1978) Study of Tea, No. 54, 6-10.
83. Del Rio, L.A., Sevilla, F., Gomez, M., Yanez, J. and Lopex, J. (1978) Planta, 140, 221-225.
84. Kanematsu, S., Hayakawa, S. and Asada, K. unpublished.
85. Jackson, C., Dench, J., Moore, A.L., Halliwell, B., Foyer, C.H. and Hall, D.O. (1978) Eur. J. Biochem. 91, 339-344.
86. Okada, S., Kanematsu, S. and Asada, K. (1979) FEBS Lett. 103, 106-110.
87. Halliwell, B. (1975) Eur. J. Biochem. 55, 355-360.
88. Doll, S., Lutz, C. and Ruppel, H.G. (1976) Z. Pflanzenphysiol. 80, 166-176.
89. Elstner, E.F. and Heupel, A. (1975) Planta, 123, 145-154.
90. Lumsden, J. and Hall, D.O. (1975) Biochem. Biophys. Res. Commun. 58, 35-41.
91. Van Ginkel, G. and Brown, J.S. (1978) FEBS Lett. 94, 284-286.
92. Foyer, C.H. and Hall, D.O. (1979) FEBS Lett. 101, 324-328.
93. Nobel, P.S. (1974) in Introduction to Biophysical Plant Physiology, Freeman C., San Francisco, p.201.
94. Berkner, L.C. and Marshall, L.C. (1965) J. Atmos. Sci. 22, 225-261.
95. Hodgson, E.K. and Fridovich, I. (1975) Biochemistry, 14, 5299-5303.
96. Anastasi, A., Bannister, J.V. and Bannister, W.H. (1976) Int. J. Biochem. 7, 541-546.
97. Salin, M.L. and Bridges, S.M. this volume.

PRODUCTION OF SUPEROXIDE ANION AND SUPEROXIDE DISMUTASES IN PARACOCCUS DENTITRIFICANS

PAULETTE M. VIGNAIS, MICHELE-FRANCE HENRY, ARIELLE TERECH AND JACQUELINE CHABERT
Biochimie (CNRS/ERA n° 36, INSERM U.191), Département de Recherche Fondamentale, Centre d'Etudes Nucleaires, 85X, 38041 Grenoble-Cedex, France

Production of superoxide anion. Mitochondria are known to produce superoxide anions during the oxidation of biomolecules by molecular oxygen. Several components of the respiratory chain located before the antimycin-sensitive site[1,2] are able to reduce O_2 monoelectronically and therefore to produce O_2^- (Ref. 3). There is no evidence for O_2^- production by electron transfer from cytochrome c to oxygen.[4]

In similar fashion, *Paracoccus denitrificans* which contains a mitochondrion-like respiratory chain[5] can generate O_2^- (Refs. 6, 7). To measure the production of O_2^- by *P. denitrificans* cells were grown on succinate with either O_2 or nitrate as terminal electron acceptors. When grown in presence of nitrate, the cells (S-N) had no spectrophotometrically detectable cytochrome oxidase; this lack of cytochrome aa_3 was the main difference between the cells grown in aerobiosis (S-O) and those grown on nitrate (S-N) (Figure 1). Cytochrome aa_3 with a high affinity for O_2 and its ability to catalyze a rapid four electron reduction of O_2 to H_2O probably contributes to a lowering of O_2^- production during aerobic metabolism. Indeed the rate of production of O_2^- by S-O membranes oxidizing NADH, nearly doubled when antimycin A was added to interrupt electron transfer between cytochrome b and c (not shown). The rate of O_2^- production by S-O and S-N membranes is reported in Table 1. In presence of antimycin A which increases the concentration of reduced low potential components and hence of O_2^- production, the rate of generated O_2^- was practically the same with membranes from cells grown aerobically or anaerobically (Table 1).

Occurrence of two proteins having superoxide dismutase activity. The ability of membranes obtained after anaerobic growth to produce O_2^- in the presence of oxygen, may explain why superoxide dismutase (SOD) activity was found in anaerobically-grown cells.[6,9]

In *P. denitrificans*, activity was associated with the soluble fraction of a cell-free homogenate prepared by disruption of the cells by sonication. When electrophoretograms were stained for SOD activity, two bands were revealed on 12% polyacrylamide gels (Figure 2). The fast band migrating with an Rf of 0.77, did not appear when the gels were stained in the presence of 2 mM KCN (Figure 2), suggesting that this latter activity was caused by a cyanide-sensitive Cu,Zn-SOD.

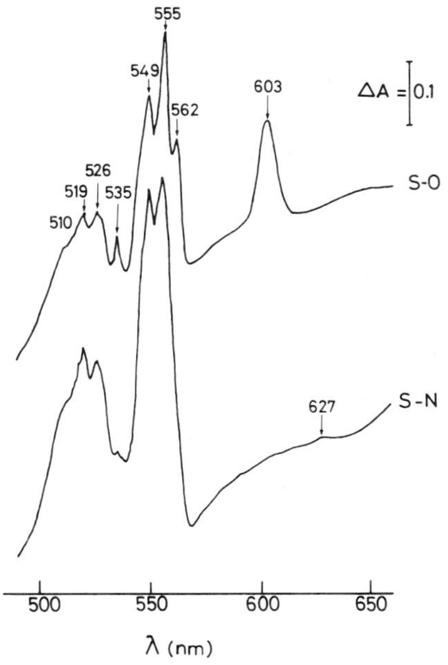

Fig. 1. Low temperature absorption difference spectra of membrane particles from P. denitrificans. The reduced (dithionite) minus oxidized (H_2O_2 + catalase) difference spectra were recorded at 77K. The membranes were obtained after disruption of cells by sonication and collected by centrifugation at 105,000 x g for 3 hr after unbroken cells and cell debris had been discarded by low speed centrifugation. S-O and S-N refer to membranes isolated from cells grown on succinate-oxygen and succinate-nitrate respectively. Protein concentrations were: S-O membranes : 12.6 mg protein.ml^{-1}; S-N membranes : 13.3 mg protein.ml^{-1}.

TABLE 1

SUPEROXIDE PRODUCTION BY ISOLATED MEMBRANE PARTICLES FROM P. DENITRIFICANS

Membranes from cells grown on	Superoxide production[a] (nmol O_2^- .min^{-1}.mg $protein^{-1}$)
Succinate-oxygen	2.3
Succinate-nitrate	4.7

[a]The rates of O_2^- production were measured spectrophotometrically from the rate of adrenochrome formation (Ref. 8). The medium consisted of 10 mM HEPES, 2 mM $MgCl_2$ buffer, pH 7.75, supplemented with 1 mM epinephrine, 0.2 mM NADH, 13 µg antimycin A and 0.3 mg membrane protein/ml, in a total volume of 3 ml.

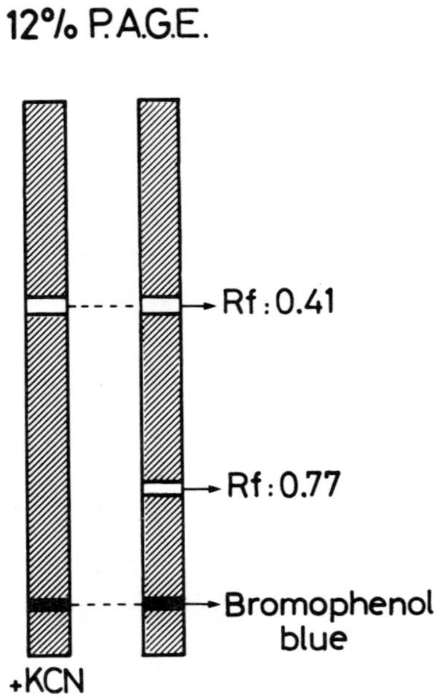

Fig. 2. Separation on 12% polyacrylamide gels of soluble extracts from P. denitrificans. Cell free extracts (100 µg protein) were applied onto the gels together with one crystal of sucrose and bromophenol blue as tracking dye. The running buffer was made of 3g Tris; HCl; 14.4g glycine/l. Proteins were stained for SOD activity by the method of Beauchamp and Fridovich (Ref. 10).

Separation of the two SOD proteins. Ammonium sulfate precipitation of the crude cell-free extract, followed by chromatography on DEAE-cellulose, allowed the separation of the two fractions with SOD activity (Figure 3). The active fraction n° 1 contained the cyanide-insensitive SOD and was eluted first, while the cyanide-sensitive enzyme, present in active fraction n° 2 appeared afterwards at a higher NaCl concentration.

The SOD-containing fractions were collected and pooled as shown on Figure 3 and further purified to give only a single protein band on gel electrophoretograms. Analysis of their metal content indicated that the cyanide-insensitive SOD contained iron at the active site, while the cyanide-sensitive enzyme contained both copper and zinc at the catalytic site.

The purification and analysis of the Fe-containing SOD of P. denitrificans will be described elsewhere (Térech and Vignais, in preparation). Data concerning some physical properties of this enzyme are summarized in Table 2.

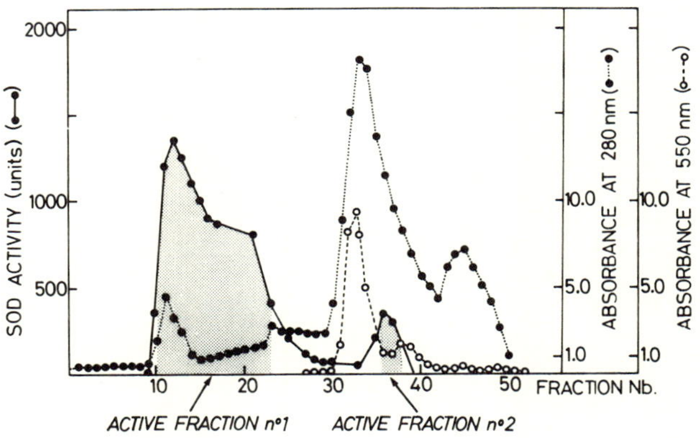

Fig. 3. Chromatography of the soluble SOD from P. denitrificans on DEAE-cellulose. The DE-52 column (3 cm x 8 cm) equilibrated in 20 mM phosphate buffer, pH 7.8, was eluted with a linear gradient of NaCl (50 to 500 mM); 10 ml fractions were collected.

TABLE 2

PHYSICAL PROPERTIES OF THE Fe-SOD FROM P. DENITRIFICANS

M.W. 41,000
2 Fe/molecule
2 subunits of 20,500
$pH_i = 4.5$

Assay of the cyanide-sensitive SOD by NMR. The cyanide-sensitive SOD has been the object of a separate study. To characterize a Cu-SOD, the NMR method recently published[11,12] has been used. This novel assay considered by the authors to be quite specific for Cu-SOD, is based on the increased nuclear magnetic relaxation of ^{19}F by SOD and the competitive reversal effect of CN^- in the case of Cu-SOD. We observed that the addition of the cyanide-sensitive SOD (active fraction n° 2 in Figure 3) to a NaF solution led to a broadening of the signal due to ^{19}F from 2 Hz up to 13.5 Hz, which was narrowed to a width of 4 Hz upon addition of KCN (Figure 4).

The NMR method was also used to determine the localization in the cell of the Cu-SOD. For this purpose the cells were fractionated by treatment with lysozyme which releases the proteins from the periplasmic space and yields spheroplasts. The proteins from the cytoplasmic compartment were obtained after osmotic lysis of the spheroplasts. The NMR assay indicated the occurrence of Cu-SOD in the

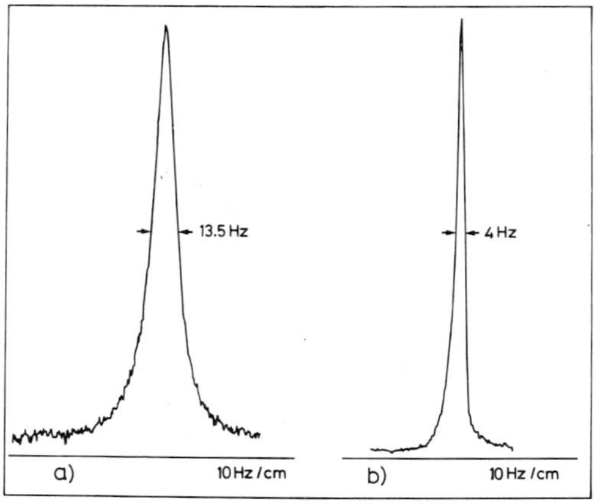

Fig. 4. Broadening of the ^{19}F signal by active fraction n° 2. Antagonistic effect of KCN. The nuclear magnetic relaxation of ^{19}F was measured using a Brüker WP-60 high resolution nuclear magnetic resonance spectrometer operating at 46.4 MHz. Trace a) represents the line due to 0.5 M NaF in D_2O + the SOD-containing fraction n° 2. Trace b) was obtained after further addition of cyanide to the SOD-containing sample.

cytoplasm only. However, by gel electrophoresis, the cyanide-sensitive and the cyanide-insensitive SOD were found in both the periplasmic and the cytoplasmic space. No clear conclusion can be drawn as yet regarding the cellular localization of these enzymes.

Concluding remarks. The isolation of the Cu,Zn-superoxide dismutase from the free-living bacterium, P. denitrificans constitutes the second example of the occurrence of such an enzyme in a prokaryote. The first case found so far, was described by Puget and Michelson[13] in a marine bacterium, Photobacterium leiognathi living in symbiosis with Pony fish.

The presence of two superoxide dismutases in P. denitrificans constitutes an intriguing clue for the establishment of the evolutionary or phylogenetic position of the organism. As emphasized by John and Whatley,[5] P. denitrificans offers many features in common with mitochondria, so that referring to the endosymbiotic theory, these authors proposed that the present-day P. denitrificans and the mitochondrion might have originated from a common ancestral aerobic bacterium. However, our studies demonstrated the occurrence of a Fe-SOD in P. denitrificans (Térech and Vignais, in preparation) while mitochondria have been shown to contain a Mn-SOD.[14]

The second intriguing problem is the presence, in a prokaryote, of a Cu,Zn-SOD considered, up to now, to be typically an eukaryotic enzyme. As mentioned above, the occurrence of a Cu,Zn-SOD had already been observed in a symbiotic prokaryote but the question was raised whether such an enzyme was not the result of gene transfer by a plasmid from the eukaryotic host cell to the symbiont. The occurrence of a Cu,Zn-SOD in a free-living prokaryote, as reported here, is the first case known so far. Unless this SOD activity represents a side by-reaction of an enzyme having other functions, the presence of a Cu,Zn-SOD in P. denitrificans instead of a Mn-SOD as in mitochondria requires a reconsideration of evolutionary trees proposed so far for bacterial evolution.[15,16]

ACKNOWLEDGEMENTS

The assistance of Miss Marie-Louise DHEU (CERMAV, Grenoble) for NMR experiments is gratefully acknowledged.

REFERENCES

1. Loschen, G., Azzi, A., Richter, C. and Flohe, L. (1974) FEBS Lett. 42, 68-72.
2. Boveris, A. and Cadenas, E. (1975) FEBS Lett. 54, 311-314.
3. Misra, H.P. and Fridovich, I. (1972) J. Biol. Chem. 247, 188-192.
4. Chance, B., Saronio, C. and Leigh, J.S., Jr. (1975) Proc. Nat. Acad. Sci. USA, 72, 1635-1640.
5. John, P. and Whatley, F.R. (1975) Nature, 254, 495-498.
6. Henry, M.-F., Térech, A., Chabert, J. and Vignais, P.M. (1979) XIth Int. Congress Biochemistry, Toronto, 8-13 July 1979, Abstr. 04-2-S85.
7. Henry, M.-F. and Vignais, P.M. (1980): Submitted for publication.
8. Misra, H.P. and Fridovich, I. (1972) J. Biol. Chem. 247, 3170-3175.
9. Vignais, P.M., Térech, A., Chabert, J. and Henry, M.-F. (1980): Submitted for publication.
10. Beauchamp, C. and Fridovich, I. (1971) Anal. Biochem. 44, 276-287.
11. Rigo, A., Viglino, P., Argese, E., Terenzi, M. and Rotilio, G. (1979) J. Biol. Chem. 254, 1759-1760.
12. Viglino, P., Rigo, A., Stevanato, R. and Ranieri, G.A. (1979) J. Magnetic Res. 34, 265-274.
13. Puget, K. and Michelson, A.M. (1974) Biochem. Biophys. Res. Commun. 58, 830-838.
14. Weisiger, R.A. and Fridovich, I. (1973) J. Biol. Chem. 248, 3582-3592.
15. Dickerson, R.E., Timkovich, R. and Almassy, R.J. (1976) J. Mol. Biol. 100, 473-491.
16. Brock, C.J. and Harris, J.I. (1977) The Evolution of Metalloenzymes, Metalloproteins and Related Materials, Leight, G.J., ed., Symposium Press, London, pp. 85-99.

PROPERTIES OF THE IRON-CONTAINING SUPEROXIDE DISMUTASE FROM PROPIONIBACTERIUM SHERMANII

BEATE MEIER AND ARNOLD C. SCHWARTZ
Botanisches Institut der Rheinischen Friedrich-Wilhemls-Universität, D-5300 Bonn 1, W. Germany.

INTRODUCTION

Susceptibility and tolerance of anaerobic bacteria towards oxygen are determined by the balance between oxygen-sensitive processes and defence mechanisms in their cells. The defences include low activities of superoxide dismutase (E.C.1.15.1.1) in various species and genera of anaerobic bacteria.[1-4] Elevated specific activities of the enzyme in cell-free extracts were correlated to greater oxygen tolerance of the respective anaerobes.[5] The induction of the enzyme decreased the rate of killing caused by oxygen in *Bacteroides fragilis*.[6] The low activities in anaerobes suggest the presence of interesting variants of the enzyme, which may possess impaired catalytic properties. This possibility was investigated in *Propionibacterium shermanii*, as the oxygen relations of this organism are already known to some extent. It is a relatively aerotolerant anaerobe with a very low aerobic respiration.[7] Its anaerobic nature was ascertained by estimating the inhibition zones caused by the diffusion of oxygen into deep agar cultures.[8] Detrimental effects of oxygen observed in this genus include the inhibition of growth and cytochrome biosynthesis,[9] and the inactivation of glycerol, lactate and propionate oxidation.[8,10] Pritchard et al.[11] showed that the membrane-bound dehydrogenases and cytochromes, which were already present under anaerobic conditions, went through a complex pattern of inactivation and reactivation, when chemostat cultures of *P. shermanii* were exposed to increasing partial pressures of oxygen. Under the same conditions superoxide dismutase exhibited an optimum at 85 mm Hg p_{O_2} (11.3 kPa). The purification and some properties of the superoxide dismutase of *P. shermanii* are reported in this communication.

MATERIALS AND METHODS

Growth of *Propionibacterium freudenreichii* sp. *shermanii* PZ3 and preparation of cell-free extracts have been described.[7,8] The enzyme was estimated with the cytochrome c competition assay,[12] and the polarographic assay with the dropping mercury electrode at 25°C (Ref.13) using 50 mM sodium phosphate buffer, pH 9.5 with 5% methanol and 1 mM triphenylphosphine oxide. Other methods included standard procedures for protein determination with the biuret reagent, polyacrylamide-gel electrophoresis with and without sodium dodecyl sulfate, amino acid analysis with the Beckman Multichrom 4235, the estimation of respiration with the oxygen electrode, and further, the activity staining of the enzyme on gels,[14] and the determination of catalase activity.[15] Cytochrome c and xanthine oxidase were obtained from Sigma.

RESULTS

Purification of the enzyme. Cell-free extracts of *P. shermanii*, prepared with 50 mM potassium phosphate buffer pH 7.8 (used throughout) contained 3-5 units per mg protein of superoxide dismutase activity (cytochrome c competition assay). After the removal of DNA with DNAase the enzyme was purified with a heat step (3 min at 70°C in buffer plus 0.1 M KCl), ammonium sulfate fractionation between 70 and 95% saturation, gel filtration in Sephadex G-100 and chromatography on DEAE-cellulose. The enzyme was eluted from the DEAE-cellulose with a gradient of 0-0.8 M KCl in the buffer mentioned above. After concentration by filtration through an Amicon PM 30 membrane, and a repetition of the gel filtration step, the enzyme was enriched 100-fold, and considered to be homogeneous as judged by polyacrylamide-gel electrophoresis with SDS. The yield was 80%. The enzyme, dissolved in the buffer mentioned above, was crystallized by dialysis against distilled water at 4°C.

Properties of the enzyme. The molecular weight of the holoenzyme was determined by polyacrylamide-gel electrophoresis in the presence of SDS, and found to be 73,000 (Figure 1). Boiling for 10 min in the presence of SDS split the enzyme into subunits of 32,000 molecular weight, irrespective of the presence of 1% mercapto-ethanol. It was concluded that the holoenzyme consisted of two identical subunits which were not linked by disulfide bonds.

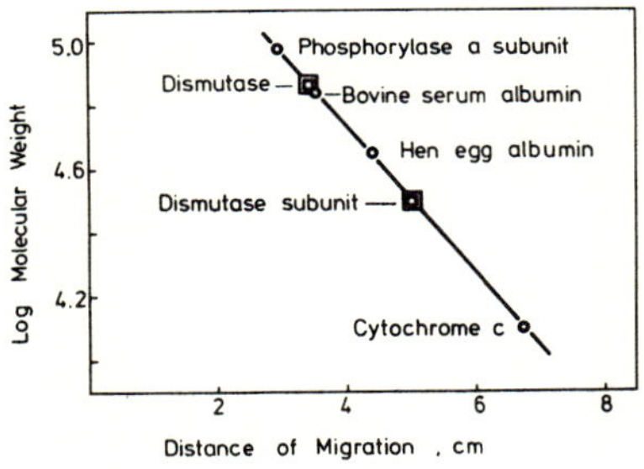

Fig. 1. Determination of the molecular weights of the superoxide dismutase from *Propionibacterium shermanii* and its subunit by SDS polyacrylamide-gel-electrophoresis.

Metal determination by atomic absorption spectroscopy indicated the presence of 0.13% iron in the dry matter of the dialyzed and lyophilized enzyme. The highest impurity was manganese (0.018%). The estimations correspond to 1.7 gram atoms of iron per mol of the holoenzyme (manganese 0.23 gram atom mol^{-1}). It was assumed that the enzyme may contain one iron atom per subunit.

The amino acid composition of the enzyme (Table 1) is difficult to compare to that of other superoxide dismutases because of the difference in the molecular weights. The enzyme contains neither cysteine, which is in accordance with the iron superoxide dismutases from *Photobacterium sepia*[16] and *Thiobacillus denitrificans*,[17] nor methionine, in accordance with the iron-enzyme from *Escherichia coli*.[18]

TABLE 1

AMINO ACID COMPOSITION OF THE IRON-CONTAINING SUPEROXIDE DISMUTASE FROM *PROPIONIBACTERIUM SHERMANII*

Amino acid	Residues per subunit[a]	Variance[b]
Lysine	21	2.1
Histidine	12	0.2
Arginine	9	0.29
Tyrptophan[b]	15	5.8
Aspartic acid	48	6.8
Threonine	11	0.59
Serine	18	3.92
Glutamic acid	32	1.13
Proline	15	0.54
Glycine	26	0.14
Alanine	40	3.73
Half-cystine[c]	0	0
Valine	17	2.07
Methinone	0	0
Isoleucine	10	1.5
Leucine	31	3.27
Tyrosine[b]	8	0.17
Phenylalanine	15	1.33

[a]Calculated from 6 estimations from 2 runs.

[b]Determined from the UV absorption (Ref.27).

[c]Cysteine assayed with Ellman's reagent (Reg.28), cystine according to Cavallini et al. (Ref.29).

Michaelis constant. The Michaelis constant of the enzyme was determined with the polarographic assay.[13] Different O_2^- concentrations were obtained by bubbling the reaction vessel with mixtures of O_2 and N_2, until constant O_2 concentrations were established. The actual O_2 concentrations were determined from the mean limiting currents observed with the polarograph, and were checked with a calibration curve obtained from the polarographic determination of Cd^{2+} in 0.1 M

KCl (diffusion coefficient of Cd^{2+} 7.15 x 10^{-4} mm^2 s^{-1},(Ref.19)). As O_2 is rapidly reduced to O_2^- at the electrode, the effective O_2^- concentration is considered to be equal to the O_2 concentration.[20] The straight line in the Lineweaver-Burk diagram (Figure 2) was obtained from the experimental data with the method of the least squares. The Michaelis constant for O_2^- was K_m = 9.8 x 10^{-4}M at pH 9.5.

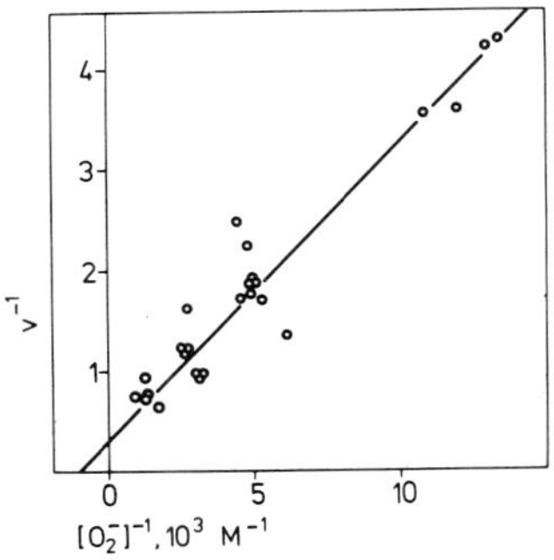

Fig. 2. Lineweaver-Burk diagram for the determination of the Michaelis constant for O_2^- at pH 9.5 of the superoxide dismutase from *Propionibacterium shermanii*.

Rate constants. For the determinations with the cytochrome c competition assay at pH 8.5, the rate constant for the reduction of cytochrome c by O_2^- in the absence of superoxide dismutase, k = 1.1 x 10^5 $M^{-1}s^{-1}$ (Ref.21) was used in spite of the objections of Koppenol et al.,[22] as the purpose of this determination was the comparison with rate constants reported in the literature.[23] The corrections which are necessary according to Koppenol et al.[22] would result in even lower values for the rate constants of superoxide dismutases determined with the cytochrome c competition assay.

The steady state concentration of O_2^- in our estimations was calculated to be 1.3 x 10^{-8} M in the absence of superoxide dismutase. Fifty percent inhibition in the assay was obtained with 1.4 x 10^{-8} M enzyme. The rate constant at pH 8.5 for the reaction of the enzyme with its substrate was calculated to be k = 7.1 x 10^7 M^{-1} s^{-1}. The same result was obtained when the O_2^- was supplied by adding a saturated solution of KO_2 in dry dimethylsulfoxide with 10^{-3} M crown ether 18-crown-6.

The rate constant was also determined with the polarographic assay at pH 9.5. The reaction showed linear dependence on the enzyme concentration in the range of 2-10 nM, and was independent of the drop time of the mercury electrode in the same range. The rate constant with this method was 4.3 x 10^8 M^{-1} s^{-1}. In order to check the accuracy of our determinations, the rate constant of a commercial copper-zinc superoxide dismutase (bovine erythrocyte from Sigma) was determined. The value obtained, 2.3 x 10^9 M^{-1} s^{-1} coincided with that reported by Rigo et al. (Ref.13).

The rate constants of the iron superoxide dismutase from E. coli decrease at higher pH values.[23] This was also observed for the enzyme from P. shermanii (Figure 3). For the calculations with the data from the cytochrome c competition assay at pH values other than 8.5, the rate constants of the bovine copper-zinc enzyme obtained by pulse radiolysis (as cited by Forman and Fridovich[23]) were used as standard values. The results in Figure 3 further indicate that the rate

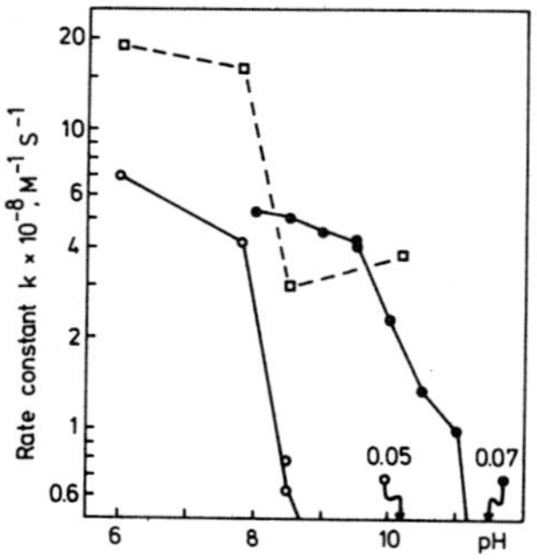

Fig. 3. Dependence on the pH of the rate constants of the superoxide dismutase from Propionibacterium shermanii. Cytochrome c competition assay (O), polarographic assay (●), iron-containing superoxide dismutase from Escherichia coli (□, cytochrome c assay, Ref.23).

constants obtained with the polarographic assay are considerably higher than those estimated with the cytochrome c competition assay. The rate constants of the iron enzyme from E. coli[23] are also included in Figure 3. In the physiological pH range this enzyme is more active, by a factor of about 3, than the superoxide dismutase from P. shermanii.

Inhibitors. The inhibitory effect of H_2O_2 on iron superoxide dismutases was first reported by Asada et al.[24] The enzyme from P. shermanii was inhibited by H_2O_2 in the activity stain on polyacrylamide gels. Effects of H_2O_2 on the enzyme revealed by electron spin resonance spectroscopy are reported in another contribution (H. Twilfer et al. 1979, this volume).

The effect of EDTA on the enzyme was studied in some detail. This chelator is routinely added in the cytochrome c competition assay in order to complex cation impurities which might simulate superoxide dismutase activity.[12] It is known now, however, that these complexes themselves display superoxide dismutase activity.[25] At pH 6.0 and 7.8 the rate constants of the enzyme were not affected by 10^{-4} M EDTA in this test, whereas the rate constants were about one half in the presence of the chelator at pH 8.5 and 10.2. When the EDTA concentration was raised from 10^{-9} to 10^{-3} M at pH 10.2, the rate constants decreased proportionately from 8.8 to 1.9×10^6 M^{-1} s^{-1}. Effects of the chelator on the xanthine oxidase were excluded by repeating some estimations with O_2^- added in dimethylsulfoxide, and effects on cytochrome c by using the photometric nitroblue tetrazolium assay.[14] It is not yet known whether EDTA removes the iron from the enzyme, or influences the reaction by binding to the active site.

Influence of iron nutrition. Gregory et al.[26] reported that cells of E. coli grown in an iron-deficient medium contained only about 25% of the amount of iron superoxide dismutase found in iron-saturated cells. The culture of P. shermanii in a defined medium with various concentrations of $FeSO_4.7H_2O$ was without effect on the superoxide dismutase activity in the cell-free extracts prepared from the respective cell batches (Figure 4). On the other hand, growth, lactate-stimulated respiration of cells, and catalase activities of the cell-free extracts strongly depended on the iron concentration of the growth medium. The effect of the iron variation in the medium on the iron contents of the superoxide dismutase has not yet been investigated.

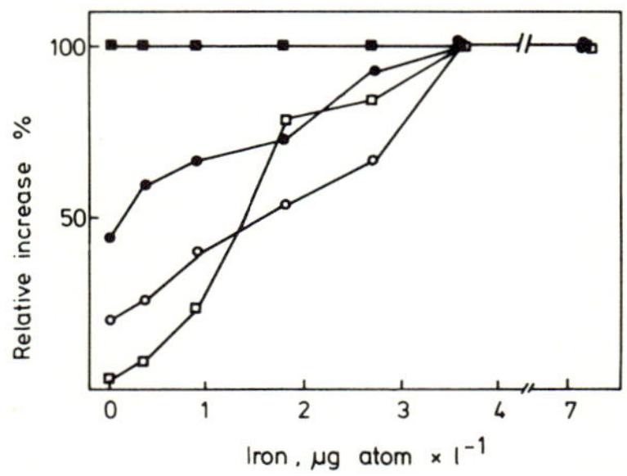

Fig. 4. Effect of iron variation in a defined growth medium on superoxide dismutase and catalase activity, growth, and lactate-stimulated respiration of Propionibacterium shermanii. Superoxide dismutase (■), 100% = 3 U·mg^{-1} protein in the cell-free extract, cytochrome c competition assay; growth (●), 100% = 6 g.l^{-1} fresh weight of cells; lactate respiration (○), 100% = 5.2 μmol·min^{-1}·g^{-1} O_2 consumed per fresh weight of cells; catalase activity (□), 100% = 24 μmol·s^{-1}·mg^{-1} H_2O_2 split per protein of the cell-free extract.

DISCUSSION

The iron-containing superoxide dismutase of Propionibacterium shermanii is notable because of the following unusual properties. It has the highest molecular weight reported so far among iron-containing representatives of the enzyme with two subunits, exceeding their values by about 30,000 for the holoenzyme. In view of its function in protecting the cells against the harmful effects of O_2^- and the subsequent reaction products of this radical, the catalytic properties of the enzyme are of particular interest. The comparison with other iron superoxide dismutases is useful only to a limited extent, as only few of these enzymes have been purified so far (Table 2). The total activity in the cell-free extract of P. shermanii is about similar to that of the anaerobic Desulfovibrio desulfuricans. With regard to the specific activities of the homogeneous

TABLE 2

ACTIVITIES OF IRON-CONTAINING SUPEROXIDE DISMUTASES

Organism	Amount of enzyme (total activity[a] 10^5 units kg^{-1} wet cells)	Specific activity of the pure enzyme ($U\ mg^{-1}$)	Rate constant at pH 7.8[a] ($k \cdot 10^{-8} M^{-1} s^{-1}$)	Refs
Propionibacterium shermanii	1.9	400	4.2	
Desulfovibrio desulfuricans	2.4	2,100	-	(30)
Chlorobium thiosulfatophilum	-	4,100	-	(31)
Chromatium vinosum	6.0	4,500	24	(32)
Escherichia coli B	22.0	5,200	16	(23, 33)
Bacillus megaterium	0.77	1,100	-	(34)
Thiobacillus denitrificans	5.4	1,900	-	(17)
Plectonema boryanum	18	7,700	-	(24)

[a]Based on the cytochrome c competition assay (Ref.12)

enzymes, and the rate constants of the enzyme-catalysed reaction, the enzyme from P. shermanii exhibits by far the lowest values. They are exceeded by those of the other enzymes by factors of 4 to 20, except for the enzyme of the aerobic Bacillus subtilis. The enzyme from P. shermanii thus possesses a very low catalytic efficiency, indeed. It is not surprising to find such an enzyme in an anaerobic organism, to which it is useful, of course, in spite of any catalytic shortcomings. The known defences of P. shermanii against oxygen include now the superoxide dismutase identified in this communication, besides the low respiration and the catalase activity observed previously. The results obtained with this organism emphasize the fact that the relation of an anaerobe to oxygen can only be understood, when all relevant detrimental and protective mechanisms in its metabolism are taken into consideration.

ACKNOWLEDGEMENT

This investigation is supported by the Deutsche Forschungsgemeinschaft.

REFERENCES

1. McCord, J.M., Keele, B.B..Jr. and Fridovich, I. (1971) Proc. Nat. Acad. Sci. USA, 68, 1024-1027.
2. Hewitt, J. and Morris, J.G. (1975) FEBS Lett. 50, 315-318.
3. Gregory, E.M., Moore, W.E.C. and Holdeman, L.V. (1978) Appl. Environ. Microbiol. 35, 988-991.
4. Uesugi, I. and Yajima, M. (1978) Z. Allg. Mikrobiol. 18, 593-601.
5. Tally, F.P., Goldin, B.R., Jacobus, N.V. and Gorbach, S.L. (1977) Infect. Immunity. 16, 20-25.
6. Privalle, C.T. and Gregory, E.M. (1979) J. Bacteriol. 138, 139-145.
7. Schwartz, A.C. and Sporkenbach, J. (1975) Arch. Microbiol. 102, 261-273.
8. Schwartz, A.C. (1973) Z. Allg. Mikrobiol. 13, 681-691.
9. De Vries, W., van Wijck-Kapteijn, W.M.C. and Stouthamer, A.H. (1972) J. Gen. Microbiol. 71, 515-524.
10. Schwartz, A.C., Mertens, B., Voss, K.W., Hahn, H. (1976) Z. Allg. Mikrobiol. 16, 123-131.
11. Pritchard, G.G., Wimpenny, J.W.T., Morris, H.A., Lewis, M.W.A. and Hughes, D.E. (1977) J. Gen. Microbiol. 102, 223-233.
12. McCord, J.M. and Fridovich, I. (1969) J. Biol. Chem. 244, 6049-6055.
13. Rigo, A., Viglino, P. and Rotilio, G. (1975) Anal. Biochem. 68, 1-8.
14. Beauchamp, C. and Fridovich, I. (1971) Anal. Biochem. 44, 276-287.
15. Aebi, K. (1974) in Methoden der enzymatischen Analyse, Vol.1, Bergmeyer, H.U., ed., Verlag Chemie, Weinheim, pp.713-724.
16. Puget, K. and Michelson, A.M. (1974) Biochimie, 56, 1255-1267.
17. Baldensperger, J.B. (1978) Arch. Microbiol. 119, 237-244.
18. Yost, F.J., Jr. and Fridovich, I. (1973) J. Biol. Chem. 248, 4905-4908.
19. von Stackelberg, M., Pilgram, M. and Toome, V. (1953) Ber. Bunsenges. physikal. Chemie, 57, 342-350.
20. Heyrovsky, J. and Kuta, J. (1965) Grundlagen der Polarographie (Translation from Czech, Schwabe, K., ed.), Akademie-Verlag, Berlin.
21. Land, E.J. and Swallow, A.J. (1971) Arch. Biochem. Biophys. 145, 365-372.
22. Koppenol, W.H., van Buuren, K.J.H., Butler, J. and Braams, R. (1976) Biochim. Biophys. Acta, 449, 157-168.
23. Forman, H.J. and Fridovich, I. (1973) Arch. Biochem. Biophys. 158, 396-400.
24. Asada, K., Yoshikawa, K., Takahashi, M., Maeda, Y. and Emanji, K. (1975) J. Biol. Chem. 250, 2801-2807.
25. Halliwell, B. (1975) FEBS Lett. 56, 34-38.
26. Gregory, E.M., Yost, F.J., Jr. and Fridovich, I. (1973) J. Bacteriol. 115, 987-991.
27. Witkop, B. (1961) Adv. Protein Chem. 16, 221-260.
28. Criddle, R.S., Edwards, D.L. and Petersen, T.G. (1966) Biochemistry, 5, 578-582.
29. Cavallini, D., Graziani, M.T. and Dupre, S. (1966) Nature, 212, 294.
30. Hatchikian, E.C. and Henry, Y.A. (1977) Biochimie, 59, 153-161.
31. Kanematsu, S. and Asada, K. (1978) FEBS Lett. 91, 94-98.
32. Kanematsu, S. and Asada, K. (1978) Arch. Biochem. Biophys. 185, 473-482.
33. Slykhouse, T.O. and Fee, J.A. (1976) J. Biol. Chem. 251, 5472-5477.
34. Anastasi, A., Bannister, J.V. and Bannister, W.H. (1976) Int. J. Biochem. 7, 541-546.

INVESTIGATION OF THE ACTIVE CENTRE OF THE IRON-CONTAINING SUPEROXIDE DISMUTASE FROM PROPIONIBACTERIUM SHERMANII BY ELECTRON SPIN RESONANCE SPECTROSCOPY

HANS TWILFER,[+] KLAUS GERSONDE,[+] BEATE MEIER[++] AND ARNOLD C. SCHWARTZ[++]
+ Abteilung Physiologische Chemie, Sektion Physikalische Chemie der Proteine, Rheinisch-Westfälische Technische Hochschule Aachen, D-5100 Aachen, W. Germany
++ Botansiches Institut der Rheinischen Friedrich-Wilhelms-Universität, D-5300 Bonn 1, W. Germany

INTRODUCTION

The superoxide dismutase (E.C. 1.15.1.1.) from the anaerobic, yet aerotolerant Propionibacterium freudenreichii sp. shermanii has a molecular weight of 73,000 and contains two iron atoms per mol. The enzyme consists of two probably homologous subunits of a molecular weight of 32,000 (B. Meier and A.C. Schwartz, this volume). Sulfur atoms as ligands of the iron are excluded, because the protein contains neither cysteine nor methionine. Among other amino acids the protein contains 12 histidines, 8 tyrosines, and 15 tryptophans per subunit. The enzyme displays a remarkably low specific activity (400 units per mg protein of the homogeneous enzyme) in the xanthine oxidase-cytochrome c assay of McCord and Fridovich.[1] This low specific activity is in accordance with the fact that the organism is an anaerobe. In the present contribution the active centre of the enzyme is investigated with the aid of electron spin resonance spectroscopy.

MATERIALS AND METHODS

The organism was grown and the enzyme was prepared as described elsewhere (B. Meier and A.C. Schwartz, this volume). The purification included a heating step, an ammonium sulfate fractionation, gel filtration on Sephadex G-100 and chromatography on DEAE-cellulose. After a 100-fold enrichment the enzyme was homogeneous, as established by polyacrylamidegel-electrophoresis with and without sodium dodecylsulfate. The yield was 80%.

For ESR measurements the superoxide dismutase was dissolved in 0.05 M phosphate buffer pH 7.8. The sample tubes were filled with 0.3 ml of the enzyme solution at room temperature and were frozen in liquid nitrogen. In the case of the spectra presented, the samples were permitted to react for 10 min at room temperature with the anions or with H_2O_2, before they were quickly frozen in liquid nitrogen. The reaction time was decreased to 5 min for the titrations with anions or H_2O_2.

ESR spectra were recorded at 77 K on an X-band spectrometer (Type B-ER 420, Bruker-Analytik, Karlsruhe). Quartz tubes with an inner diameter of 3.3 mm were used. The amplitude of the 100 kHz field modulation was 0.8 mT. The microwave power was attenuated to 25 mW; under the conditions used saturation phenomena did not occur. The microwave frequency was measured with a frequency counter, the magnetic field strength with a nuclear magnetic resonance oscillator.

RESULTS AND DISCUSSION

The ESR spectrum of the native enzyme. The oxidized state of the protein is ESR-active and shows stronger signals in the region og g = 4 (see Figure 1) and a weak signal at g = 9.2. The titrations of the enzyme with the anions CN^-, F^- and N_3^-, and with H_2O_2, lead to the conclusion that the spectrum of the oxidized form consists of two overlapping spectral species, one with rhombic symmetry ($g_1' = 4.95$, $g_2' = 3.85$, $g_3' = 3.50$) and the other one with axial symmetry ($g_1 = 4.32$, $g_2 = 4.18$). Ligation with these anions leads to a loss of the rhombic spectral species. In the presence of CN^- the axial species remains unchanged; however, binding of F^- and N_3^- causes small rhombic distortions of the axial species. The ratio of the amplitudes of the rhombic and the axial species in the spectra of the native enzyme was constant for all preparations.

The oxidized enzyme is reduced by dithionite or ascorbate, and reoxidized by $K_3Fe(CN)_6$. The reduced state is ESR-silent. Adding ferricyanide to the native protein does not increase the amplitude of the ESR-signals, i.e. both iron atoms are in the oxidized Fe(III) state.

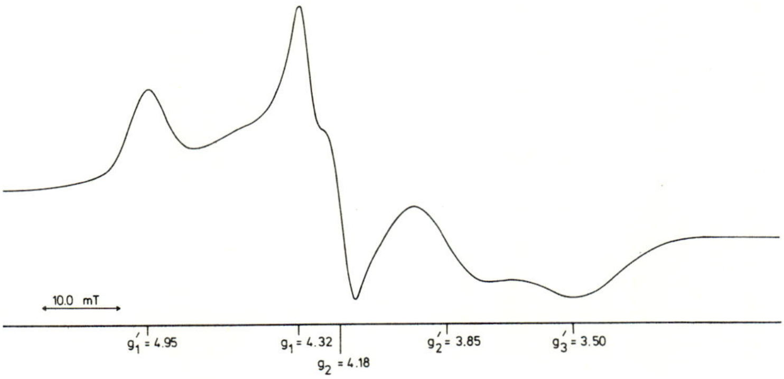

Fig. 1. ESR spectrum of the oxidized superoxide dismutase in 50 mM phosphate buffer pH 7.8; temperature 77 K.

Effects of cyanide. A titration of the enzyme with CN^- is shown in Figure 2. The comparison of the ESR intensities of the axial and the rhombic species, respectively, indicates the decrease of the intensity of the rhombic species with increasing amounts of CN^-, and an increase of the intensity of the axial species. At a molar ratio of $Fe:CN^- = 1 : 100$, based on the molar concentration of the enzyme and assuming 2 iron atoms per molecule of the holoenzyme, the signals with rhombic symmetry are completely lost, whereas the amplitude of the axial species has increased to twice its original intensity. It may be concluded from this finding that the rhombic species is transformed into the axial one.

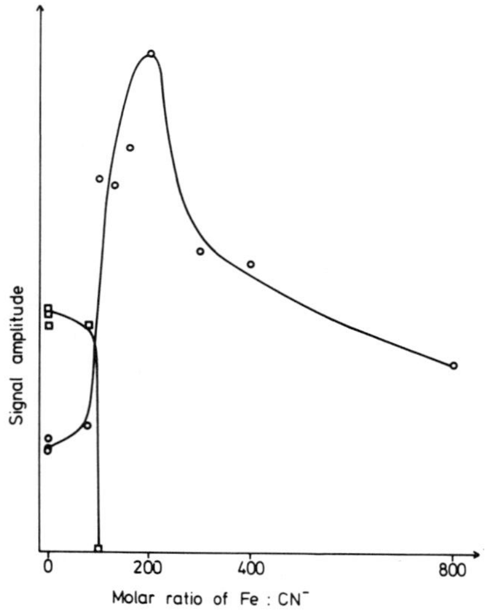

Fig. 2. Titration of the superoxide dismutase with cyanide in 50 mM phosphate buffer 7.8. Relative change of the signal amplitudes of the species with rhombic symmetry (□), and with axial symmetry (o). Iron concentration 5.4×10^{-4}M; temperature 77 K.

A further increase of the CN^- concentration to molar ratios greater than 1 : 100 produces a further increase of the amplitude of the axial spectrum. A maximum is reached at a molar ratio of $Fe : CN^- = 1 : 200$. The ESR spectrum of the species at this molar ratio is shown in Figure 3. At present, the ESR spectrum changes produced by CN^- at molar ratios above 1 : 100 cannot be explained.

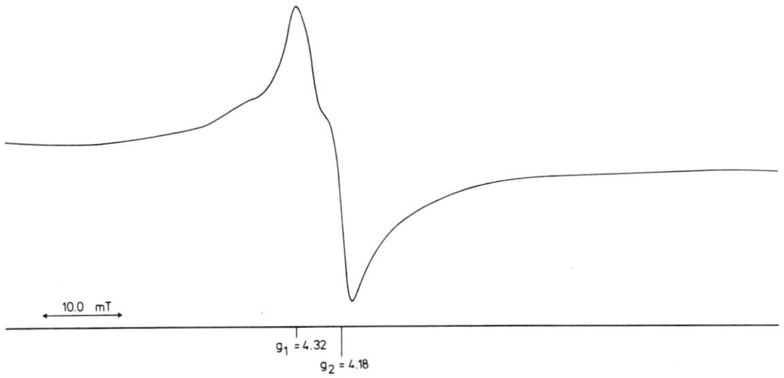

Fig. 3. ESR spectrum of the CN^--ligated superoxide dismutase in 50 mM phosphate buffer pH 7.8 at 77 K.

At molar ratios of Fe : CN^- exceeding 1 : 200, the amplitude of the axial spectrum decreases, i.e. the axial species is transformed into an ESR-inactive form. Possible explanations for the decrease of the amplitude of the axial species include the displacement of an amino acid residue by CN^- from its binding-site on the iron, or changes in the relaxation time or the spin state of the iron. Even a magnetic coupling between the iron atoms might occur. In contrast to these results, CN^- did not alter the ESR spectral properties of the iron-containing superoxide dismutase from Echerischia coli.[2]

Effects of azide and fluoride. On binding azide ions the spectrum with rhombic symmetry is lost and the axial one is rhombic distorted with $g_1 = 4.36$, $g_2 = 4.18$ and $g_3 = 3.98$ (see Figure 4). Fluoride ions,too, produce a spectrum with rhombic symmetry ($g_1 = 4.36$, $g_2 = 4.25$, $g_3 = 4.08$) and a hyperfine (see Figure 5). At present it is not possible to determine the number of fluoride ions bound to the iron atoms by analyzing the hyperfine structure. Tentatively assuming only one fluoride, we calculate hyperfine constants of $a_1 = 2.2$ mT (in g_1) and $a_3 = 3.0$ mT (in g_3). A final decision will be possible after Q-band measurements, which are in progress.

The binding curves of azide (see Figure 6) and fluoride (see Figure 7) are almost linear with increasing concentration of the anions and reach saturation. The affinities of azide and fluoride to the enzyme-bound iron are much higher than that of cyanide, as evident from the low concentrations needed for satura tion with these anions. This accords with the fact that F^- and N_3^- are stronger ligands than CN^- for free Fe^{3+} in water.

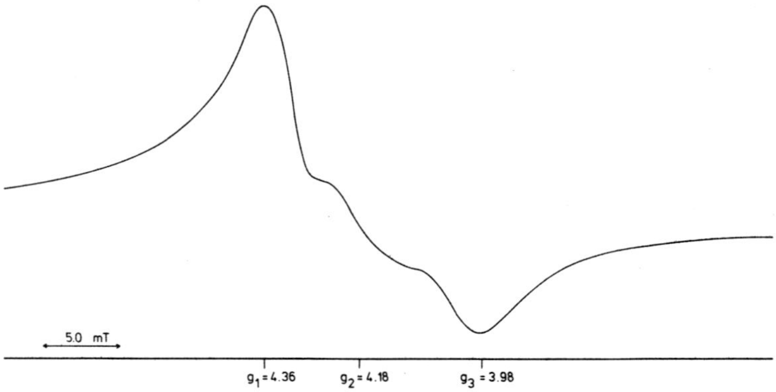

Fig. 4. ESR spectrum of the N_3^--ligated superoxide dismutase in 50 mM phosphate buffer pH 7.8 at 77 K.

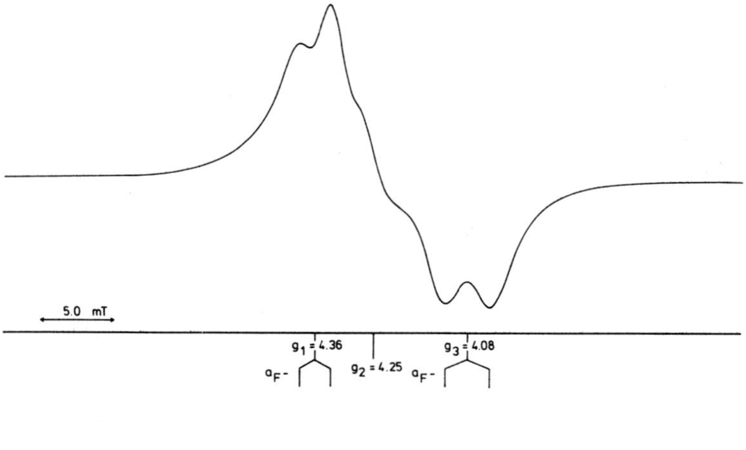

Fig. 5. ESR spectrum of the F^--ligated superoxide dismutase in 50 mM phosphate buffer pH 7.8 at 77K.

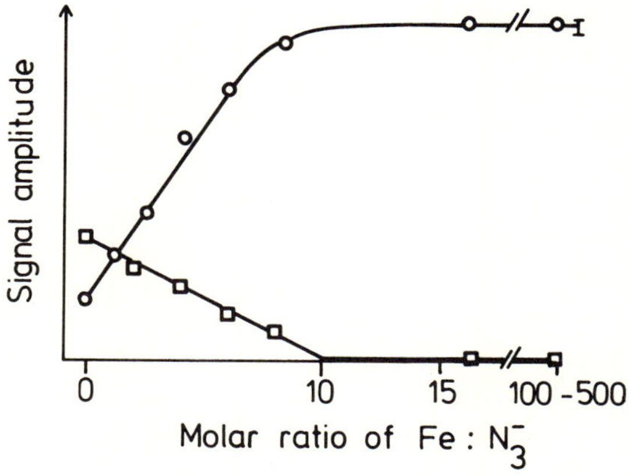

Fig. 6. Titration of the superoxide dismutase with azide in 50 mM phosphate buffer pH 7.8. Relative change of the signal amplitudes of the species with rhombic symmetry (□), and with axial symmetry (o). Iron concentration 4.2 x 10^{-4} M; temperature 77 K.

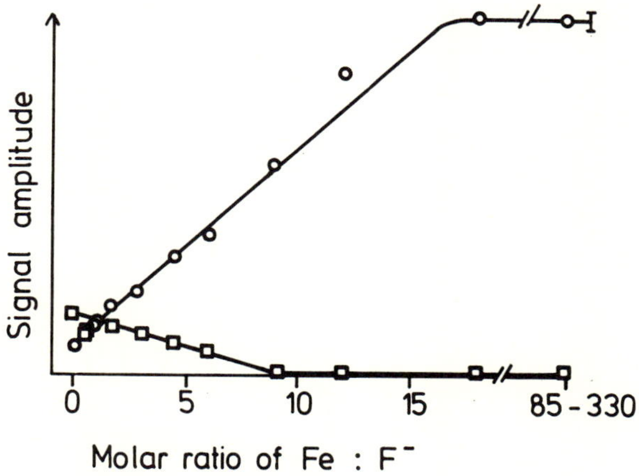

Fig. 7. Titration of the superoxide dismutase with fluoride in 50 mM phosphate buffer pH 7.8. Relative change of the signal amplitudes of the species with rhombic symmetry (□), and with axial symmetry (o). Iron concentration 4.2 x 10^{-4} M; temperature 77 K.

The ESR spectral changes brought about by azide and fluoride differ from those observed for the iron-containing superoxide dismutase from *Echerschia coli*[2]. Hyperfine splitting with F^- has not been observed with other iron-containing superoxide dismutase.

Fluoride, azide and cyanide can be removed by dialysis against buffer, and the signals of the oxidized protein are regained. The binding of the anions does not inactivate the enzyme, as evident from the activity of undialyzed samples tested with the cytochrome c competition assay.

Effects of hydrogen peroxide. As H_2O_2 is a product of the reaction cycle, the influence on the ESR spectrum is of interest. In the presence of H_2O_2 an ESR spectrum with axial symmetry (g_1 = 4.32, g_2 = 4.20) is observed (Figure 8). Titration with H_2O_2 results in the loss of the rhombic species, as well, and in the twofold increase of the amplitude of the axial species exactly at saturation (see Figure 9). It is again concluded that the rhombic species is transformed into the axial one, as in the case of CN^-, but without the additional complication introduced by that anion. The signals of the spectrum are not changed after removal of free H_2O_2 by dialysis. The spectral change effected by H_2O_2 is not reverted upon the addition of ferricyanide as oxidant, catalase, or $FeSO_4$. Likewise, this spectrum cannot be quenched any more by the addition of dithionite or ascorbate. Samples of the enzyme ligated with H_2O_2 are fully active in the cytochrome c competition assay.

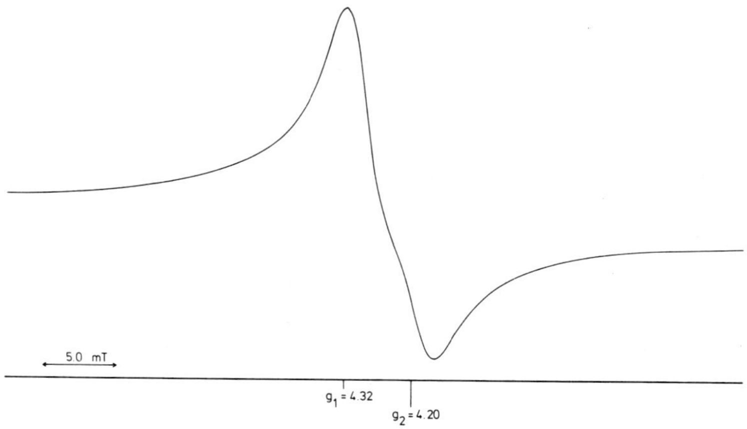

Fig. 8. ESR spectrum of the superoxide dismutase in the presence of H_2O_2 in 50 mM phosphate buffer pH 7.8 at 77 K.

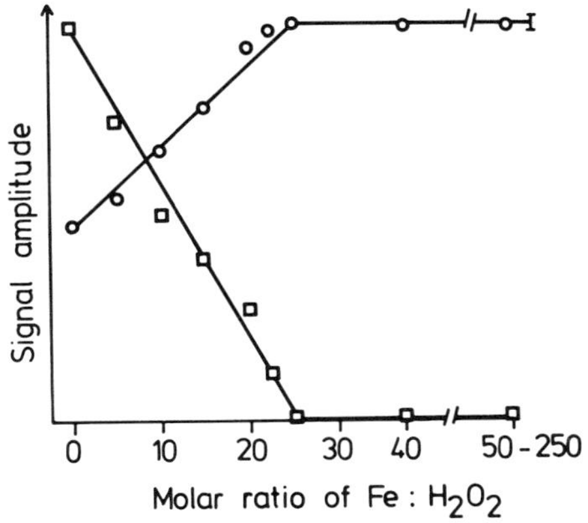

Fig. 9. Titration of the superoxide dismutase with H_2O_2 in 50 mM phosphate buffer pH 7.8. Relative change of the signal amplitudes of the species with rhombic symmetry (□), and with axial symmetry (o). Iron concentration 4.7 x 10^{-4} M; temperature 77 K.

The transformation of the original ESR spectrum to a spectrum with a single species at g = 4.3 was also observed for the iron-containing superoxide dismutase from Photobacterium leiognathi[3] and attributed to the reduction by H_2O_2 of that part of the iron which was shown to be active in the catalysis. The iron represented by g = 4.3 in the presence of H_2O_2 was regarded as an impurity by these authors. It is obvious that H_2O_2 reacts in a different manner with the iron of the enzyme from P. shermanii, i.e. by binding to it instead of reducing it. In our view the iron of the axial species in this enzyme represents one of two possible states of the iron atom within the protein moiety, and, in the order of magnitude observed, not an impurity in the enzyme.

ACKNOWLEDGEMENT

This investigation was supported by grants from the Deutsche Forschungsgemeinschaft.

REFERENCES

1. McCord, J.M. and Fridovich, I. (1969) J. Biol. Chem. 244, 6049-6055.
2. Slykhouse, T.O. and Fee, J.A. (1976) J. Biol. Chem. 251, 5472-5477.
3. Lavelle, F., McAdam, M.E., Fielden, E.M., Roberts, P.B., Puget, K. and Michelson, A.M. (1977) Biochem. J. 161, 3-11.

ISOLATION AND CHARACTERIZATION OF AN IRON-CONTAINING SUPEROXIDE DISMUTASE FROM BRASSICA CAMPESTRIS

M.L. SALIN AND S.M. BRIDGES
Department of Biochemistry, Mississippi State University, Mississippi State, Mississippi 39762, USA.

INTRODUCTION

Since its initial isolation and characterization in 1969 by McCord and Fridovich,[1] superoxide dismutase has been the subject of increasing study. The enzyme has been isolated and characterized from many sources, both eukaryotic and prokaryotic. Several excellent reviews on the subject of superoxide dismutases have been written. Some chronicle the discovery of the various isozymes[2,3] some document evolutionary trends,[4] while others examine the current literature on the toxicity of the superoxide radical anion.[5,6] At least one point which emerges from ten years of study is the necessity of superoxide dismutase for all oxygen metabolizing organisms. This should hold especially true for plants.

Photosynthesizing plants, in producing oxygen, generate a strong oxidant capable of removing four electrons from water and at the same time generate a reductant with the capacity of reducing viologen dyes. On either side of the photosynthetic electron transport chain, there is the capacity to generate O_2^-. On the oxidation side, superoxide, hydrogen peroxide and hydroxyl radical all represent intermediates in electron abstraction from water. On the reducing side, once a compound such as a viologen dye is reduced, it can autoxidize in one electron increments, yielding O_2^- (see Ref. 7). Photosynthesizing plants would therefore have a vital need for superoxide dismutase. Asada et al. were first to purify and characterize the enzyme from spinach.[8] It was a cuprozinc enzyme with many characteristics of the bovine cuprozinc protein. Lumsden and Hall[9] found evidence for cyanide sensitive (cuprozinc) as well as cyanide insensitive isozymes of superoxide dismutase in spinach chloroplasts. The latter was found tightly bound to the lamella membrane, and although not purified, beleived to be either the mangano or iron-containing superoxide dismutase. The mangano enzyme appeared to be a strong candidate for several reasons. First, this metal is intimately involved in photosynthetic oxygen evolution. Secondly, there is no evidence to date of other than the mangano or cuprozinc superoxide dismutases being found in eukaryotic organisms.[6,10-12] It is widely believed that the iron-containing superoxide dismutases are restricted to prokaryotes. Upon examining extracts of mustard leaves for superoxide dismutase, we noticed several cyanide insensitive superoxide dismutase bands on polyacrylamide gels. We therefore undertook a purification and characterization of one of the cyanide insensitive superoxide dismutases.

MATERIALS AND METHODS

Enzyme Purification. About 1500g of fresh, locally obtained, demidribbed Brassica campestris (mustard) leaves were washed and blended thoroughly in a Waring blender. The extraction medium consisted of: 50mM potassium phosphate pH 7.8, 1mM dithiothreitol, and 0.3% Triton X-100. The homogenate was squeezed through several layers of cheesecloth, then spun at 10,000 x g for ten minutes to remove debris. To the supernatant, 2.5% streptomycin sulfate was added and the suspension stirred for 20 min at room temperature. Solid ammonium sulfate was then added to bring the extract to 35% saturation. The suspension was stirred at room temperature for 1 hr after which time it was centrifuged at 10,000 x g for 10 min. To the supernatant, solid $(NH_4)_2SO_4$ was added to bring the final saturation to 65%. After stirring for 45 min at room temperature, the suspension was once more centrifuged at 10,000 x g for 10 min. The tan pellet was resuspended in ice cold 10mM potassium phosphate, pH 7.8 and dialyzed for 48 hrs into 10mM potassium phosphate, pH 7.8. The dialysis buffer was changed frequently.

The sample was clarified by centrifugation and applied at $4^{o}C$ to a DEAE-cellulose column equilibrated with 10mM potassium phosphate pH 7.8. After loading on the column a gradient of 10 to 100mM potassium phosphate was run over the column. Fractions were collected, assayed for cyanide insensitive superoxide dismutase activity, pooled, and dialyzed for 18 hrs in 10mM potassium phosphate pH 7.8. The sample was once more applied to a DEAE-cellulose column. A gradient of 10 to 100mM potassium phosphate pH 7.8 was applied. Fractions were collected, assayed for activity, pooled and dialyzed into 10mM potassium phosphate. The samples was applied to a third DEAE-cellulose column under the same conditions as the first two column steps. Fractions with cyanide inhibitable SOD activity were pooled and concentrated by ultrafiltration. The protein was next applied to a 1.6 x 84 cm Sephadex G-150 column. Fractions with activity were pooled and concentrated. All column steps were performed at $4^{o}C$. Less than 0.5 mg of purified protein (see Figure 1) was obtained from 1500g of demidribbed leaves. Table 1 summarizes the purification of the protein. Yields were good until the last step, gel exclusion, where both yield and activity were drastically diminished. A small amount of an isozyme of the cuprozinc containing superoxide dismutase was carried through all ion-exchange chromatography steps. Several methods were tried to rid our enzyme of the contaminant. We found the best procedure to be a Sephadex G-150 column.

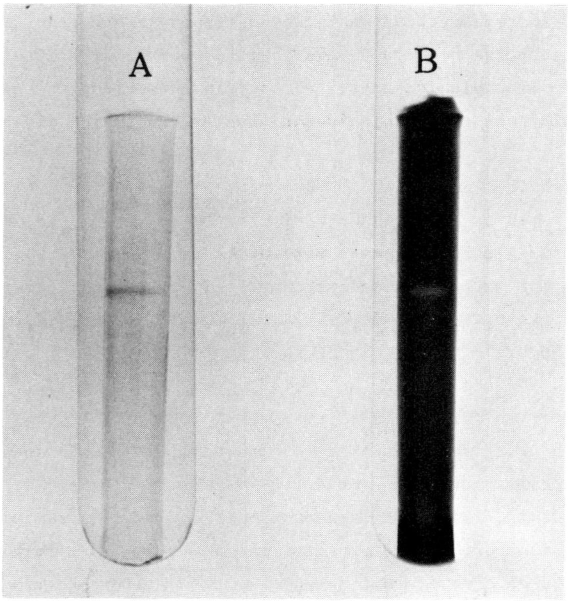

Fig. 1. Polyacrylamide gel of iron-containing superoxide dismutase isolated *Brassica campestris*. (A) 5 μg protein stained with Coomassie blue, (B) 150 μg protein stained for activity with nitroblue tetrozolium.

TABLE 1

PURIFICATION OF IRON-CONTAINING SUPEROXIDE DISMUTASE FROM *BRASSICA CAMPESTRIS*

Purification step	Volume (ml)	Protein Concentration (mg/ml)	Total Protein (mg)	Specific Activity (unit/mg)	Total Activity (unit)	Yield %	Purification Fold
Crude extract after centrifugation	2,320	1.55[a]	3,596	7.74	27,840	100	1
Streptomycin supernatant	2,325	1.50[a]	3,487	7.97	28,800	99.8	1.03
35-65% $(NH_4)_2SO_4$ dialyzed	335	4.36[a]	1,460	12.9	18,859	67.7	1.67
DEAE cellulose	385	.560[a]	215.6	82.0	17,681	63.5	10.6
DEAE cellulose	240	.235[a]	56.4	221.2	12,480	44.8	28.6
DEAE cellulose and ultrafiltration	1.2	7.84[a]	9.4	1311.0	12,320	44.2	169.4
Sephadex G-150	9	.048[b]	0.43	726	312	1.1	94

[a] Lowry *et al.*[13]

[b] Murphy and Keis[14]

Fractions containing small volumes were collected and those with any cyanide sensitive activity were discarded. Due to the similarity of molecular weights between the iron and cuprozinc enzymes considerable overlap occurred. Consequently, a substantial portion of the iron enzyme was discarded due to contamination. Moreover, there was a loss of activity which occurred over the course of the purification, even though all steps were performed at 4°C. We feel that both factors combined to cause the drastic reductions in yield and activity. The highest specific activity we observed after the Sephadex G-150 step was 1600. A unit of activity was defined as the amount of enzyme needed to give 50% inhibition on the standard cytochrome c assay.[1] After one week the specific activity was reduced by half even though the sample was refrigerated.

RESULTS

Molecular weight. The molecular weight of the purified enzyme was determined by gel filtration on a Sephadex G-150 column. By comparison with markers of known molecular weight, the data obtained yielded a value of 41,000 for the plant enzyme. Subunit size was determined by sodium dodecyl sulfate gel electrophoresis in the presence and absence of 2-mercaptoethanol.[15] Under both conditions a single band was obtained which, in comparison to standards, yielded a value of 20,000. It would therefore appear that the protein is composed of two identical subunits that are not linked by disulphide bonds.

Absorption spectra. The visible and ultraviolet absorption spectra are seen in Figure 2. The enzyme exhibited a peak at 278nm, typical of most proteins. A shoulder was observed at 290nm and 260nm. The molar extinction coefficient was $7.4 \times 10^4\ M^{-1}\ cm^{-1}$ at 278nm. A weak broad featureless absorption in the range of 500nm to 300nm was observed. There was no evidence of a shoulder at 350nm as seen in visible spectra of iron superoxide dismutases published from other sources.[16,18]

Metal analysis. The protein was dialyzed extensively against 5mM potassium phosphate pH 7.8 containing 0.5mM EDTA followed by buffer lacking EDTA and then subjected to atomic absorption analysis. The value obtained for iron was 1.6 ± .3 atoms per dimer. Manganese and copper values were less than 0.3 and 0.1 atoms per dimer, respectively. Our value for iron content is in agreement for that found for enzymes isolated from *Photobacterium leiognathi*[9] as well as *Desulfovibrio desulfuricans*.[20]

Amino acid analysis. The protein was hydrolyzed in vacuo in 6M HCl at 110°C for 24 hours, dissolved in 0.02N sodium citrate and subjected to amino acid analysis. Cysteine was determined as S-sulfocysteine after hydrolysis with constant boiling HCl followed by reduction with dithiothreitol and sodium tetrathionate.[21] Tryptophan was measured spectrophotometrically in 6M guanidine hydrochloride.[22] Table 2 shows the amino acid composition of the enzyme isolated from *Brassica*

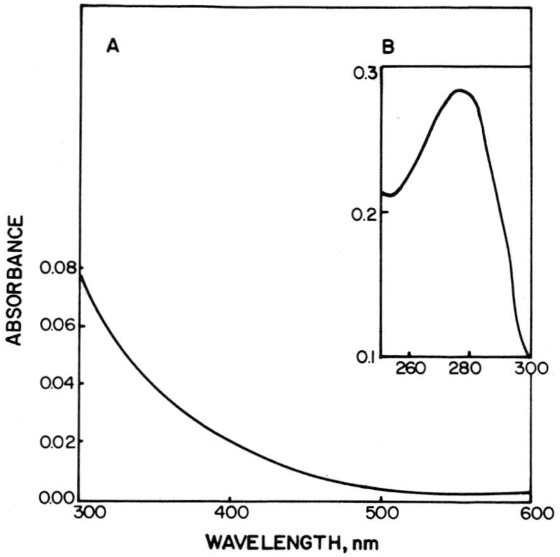

Fig. 2. Absorption spectra of iron-containing superoxide dismutase from Brassica campestris. Protein concentrations were 150 μg/ml.

campestris. For comparison, the amino acid compositions of Plectonema boryanum, Escherichia coli, Spirulina platensis and Rhodopseudomonas spheroides iron SOD are included. There is a great deal of similarity between the amino acid composition of the iron enzyme from the eukaryotic and prokaryotic sources listed. The arginine content is higher than that reported for the blue green algae, S. platensis and P. boryanum.

Isoelectric point. The enzyme was applied to a thin glass plate covered with ampholytes and a matrix of Sephadex G-75. After electrophoresis at 900 volts for three hours, a paper impression of the plate was made and stained for protein. The corresponding pH values were determined directly off the glass plate by insertion of a glass electrode into the matrix. A comparison of protein stain with pH showed one spot with a pI of 4.5.

Localization of iron SOD in higher plants. Mustard leaves were ground in a Waring blender containing 50mM phosphate buffer pH 7.8 and 0.3% Triton X-100. The extract was placed on 7.5% polyacrylamide gels and electrophoresed.[24] The gels were then stained for SOD activity.[25] A banding pattern as seen in Figure 3 was obtained. The three lowest bands were inhibited by cyanide and presumed to be the cuprozinc superoxide dismutase. The upper two bands were cyanide

TABLE 2

COMPARATIVE AMINO ACID COMPOSITION OF IRON-CONTAINING SUPEROXIDE DISMUTASE

Amino Acid	Residues per mole subunit[a]				
	Brassica campestris	Plectonema boryanum[b]	Escherichia coli[c]	Spirulina platensis[d]	Rhodopseudomonas spheroides[d]
Lysine	11	12	10	10	15
Histidine	4	5	6	6	8
Arginine	5	2	4	0	3
Aspartic acid	18	23	22	26	19
Threonine	9	10	13	10	8
Serine	11	9	10	12	11
Glutamic acid	16	16	16	17	22
Proline	10	10	9	8	7
Glycine	16	15	16	13	19
Alanine	17	24	26	25	18
Cysteine[e]	2	1	1	1	2
Valine	11	11	11	9	12
Methionine	1	2	0	3	4
Isoleucine	6	3	8	7	5
Leucine	14	17	15	16	15
Tyrosine	6	5	6	5	7
Phenylalanine	7	13	10	14	10
Tryptophan[f]	4	6	3	3	4

[a]Values given to the nearest integer

[b]From Misra and Keel[23]

[c]From Yost and Fridovich[16]

[d]From Lunsden and Hall[9]

[e]Determined as S-Sulfocysteine after reduction with DTT and oxidation with sodium tetrathionate (ref. 21)

[f]Determined spectrophotometrically in 6M guanidine hydrochloride (Ref. 22)

resistant. The second band from the top had the same R_f value as the purified iron-containing enzyme and was inhibited by 2mM H_2O_2. The uppermost band has not been purified but wasnot only resistant to cyanide but to hydrogen peroxide as well. Asada et al. have shown this to be a characteristic of manganese containing superoxide dismutases from algae.[17] It is possible therefore that the uppermost diffuse achromatic band may represent a mangano superoxide dismutase.

Fractions enriched in various plant organelles were prepared and electrophoresed in an identical fashion. Chloroplasts, mitochondria and cytoplasmic fractions each showed a banding pattern similar to that of the crude extract. The iron-containing superoxide dismutase along with the cuprozinc and presumed mangano enzyme were present in all fractions. There was considerable breakage of chloroplasts during preparation (at least 50%) as measured by marker enzymes. Determination of the amount of superoxide dismutase in the cytosolic fraction is therefore difficult to assess. However, it is apparent that the iron enzyme was present in all fractions. It was also found in the mitochondrial fraction from etiolated mustard seedlings.

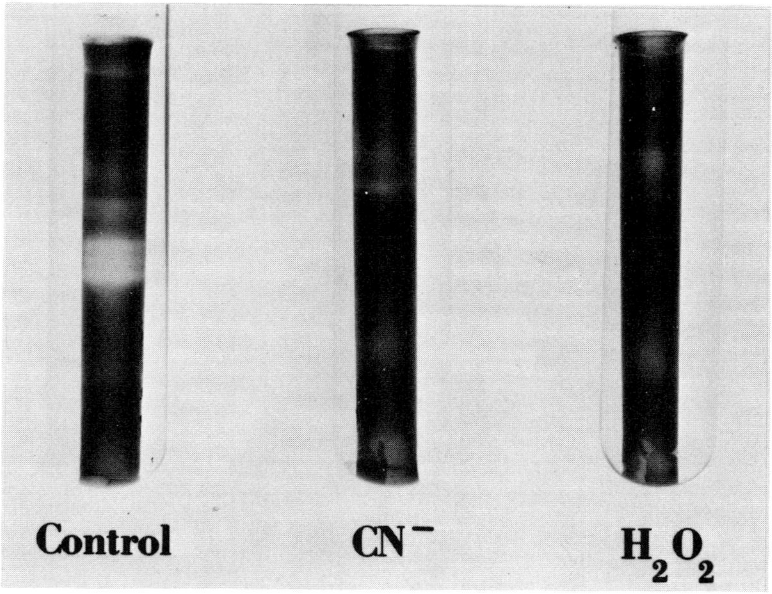

Fig. 3. Polyacrylamide gel electrophoresis of extract from Brassica campestris. Extract was stained for activity with nitroblue tetrozolium. Control, no additions to staining mixture; CN^-, 2mM KCN present in staining mixture; H_2O_2; 2mM H_2O_2 added.

DISCUSSION

The isolation of an iron containing superoxide dismutase from mustard leaves is the first finding of the enzyme in eukaryotes. All previous sources were prokaryotes.[6,10-12] The iron-containing superoxide dismutase from Brassica campestris leaves is similar in physical composition and spectral properties to previously purified iron-containing superoxide dismutases. All enzymes are composed of two equally sized subunits with a dimeric weight of about 40,000. The iron content is 1.6 iron atoms per molecule. Amino acid compositions as well, are comparable. Similarities between the enzymes are most intriguing in light of the diverse sources. Preparation of an antibody to either the prokaryotic or eukaryotic iron-containing superoxide dismutase would be desirable in order to determine if the enzymes from the various sources are antigenically similar. In view of the known sequence homologies of several iron and mangano superoxide dismutases,[4] the addition of the amino acid sequence of the eukaryotic, iron-containing superoxide dismutase now purified, would be most valuable. Based upon similarities to the known prokaryotic enzymes a close homology indicative of little evolutionary divergence may be anticipated.

Preliminary evidence indicates that the iron containing superoxide dismutase is loosely bound to the chloroplast. It was also found in the mitochondrial fraction obtained from green leaves as well as from etiolated seedlings where contamination by chloroplasts would be minimal. The cytosolic fraction contained the iron enzyme, but whether this represents protein which leaked out of the chloroplast or was originally there is difficult to determine due to the significant chloroplast breakage. It should be noted that the iron-containing superoxide dismutase was leached easily from chloroplasts. This is in agreement with a report of a readily solubilized cyanide insensitive superoxide dismutase associated with the chloroplast.[26]

Finally, it has been presumed that the cyanide insensitive superoxide dismutase associated with plants was the mangano enzyme. This was due in part to the manganese requirement for photosystem II as well as to the previous belief that the iron enzyme was restricted only to prokaryotes. The finding of an iron-containing superoxide dismutase in plants would necessiate a reevaluation in the assumption that cyanide insensitive activity represents only the mangano superoxide dismutase.

ACKNOWLEDGEMENTS

We would like to acknowledge the help of Mr. William E. Poe for carrying out the amino acid analysis, and Mr. Michael Aide for help with the atomic absorption.

REFERENCES

1. McCord, J.M. and Fridovich, I. (1969) J. Biol. Chem. 244, 6049-6055.
2. Fridovich, I. (1974) Adv. Enzymol. 41, 35-97.
3. McCord, J.M. and Fridovich, I. (1977) in Superoxide and Superoxide Dismutases, Michelson, A.M., McCord, J.M. and Fridovich, I, eds., Academic Press, New York, pp. 1-10.
4. Harris, J.I. and Steinman, H.M. (1977) in Superoxide and Superoxide Dismutases, Michelson, A.M., McCord, J.M. and Fridovich, I., eds., Academic Press, New York, pp. 225-230.
5. Fridovich, I. (1975) Annu. Rep. Med. Chem. 10, 257-264.
6. McCord, J.M. (1979) in Current Topics in Biological and Medical Research, Rattazzi, M., Scandalious, J. and Whitt, G.S., eds., Alan R. Liss, New York: in press.
7. Allen, J.F. (1977) in Superoxide and Superoxide Dismutases, Michelson, A.M. McCord, J.M. and Fridovich, I., eds., Academic Press, New York, pp.417-436.
8. Asada, K., Urano, M. and Takahashi, M. (1973) Eur. J. Biochem. 36, 257-266.
9. Lumsden, J. and Hall, D.O. (1974) Biochem. Biophys. Res. Commun. 58, 35-41.
10. Fridovich, I. (1975a) Annu. Rev. Biochem. 44, 147-159.
11. Asada, K., Takahashi, M., Tanaka, K. and Nakano, Y. (1977) in Biochemical and Medical Aspects of Active Oxygen, Hayaishi, O. and Asada, K., eds., Japan Scientific Societies Press, Tokyo, pp. 45-63.

12. Asada, K. and Kanematsu, S. (1978) in Evolution of Protein Molecules, Mutsubara, H. and Yamanaka, T., eds., Japan Scientific Societies Press, Tokyo, 361-372.
13. Lowry, O.H., Rosebrough, N.J., Farr, A.L. and Randall, R.J. (1951), J. Biol. Chem. 193, 265-275.
14. Murphy, J.B. and Kies, M.W. (1960) Biochim. Biophys. Acta, 45, 382-384.
15. Weber, K. and Osborn, M. (1969) J. Biol. Chem. 244, 4406-4412.
16. Yost, F.J. and Fridovich, I. (1973) J. Biol. Chem. 248, 4905-4908.
17. Asada, K., Yoshikawa, K., Takahashi, M., Maeda, Y. and Enmanji, K. (1975) J. Biol. Chem. 250, 2801-2807.
18. Baldensperger, J.B. (1978) Arch. Microbiol. 119, 237-244.
19. Puget, K. and Micelson, A.M. (1974) Biochemie, 56, 1255-1267.
20. Hatchikian, C.E., Legall, J. and Bell, G.R. (1977) in Superoxide and Superoxide Dismutases, Michelson, A.M., McCord, J.M. and Fridovich, I., eds., Academic Press, New York, pp. 159-172.
21. Inglis, A.S. and Liu, T.Y. (1970) J. Biol. Chem. 245, 112-116.
22. Edelhoch, H. (1967) Biochemistry, 6, 1948-1954.
23. Misra, H.P. and Keele, B.B., Jr. (1975) Biochim. Biophys. Acta, 379. 418-425.
24. Davis. B.J. (1964) Ann. N.Y. Acad. Sci. 121, 404-427.
25. Beauchamp, C. and Fridovich, I. (1971) Anal. Biochem. 44, 276-287.
26. Jackson, C., Dench, J., Moore, A.L., Halliwell, B., Foyer, C.H. and Hall, D.O. (1978), Eur. J. Biochem. 91, 339-344.

PRELIMINARY CHARACTERIZATION OF A Mn-CONTAINING SUPEROXIDE DISMUTASE FROM A HIGHER PLANT (PISUM SATIVUM L.).

F. SEVILLA, J. LOPEZ GORGE, AND L.A. DEL RIO.
Unidad de Bioquimica Vegetal, Estación Experimental del Zaidin,
CSIC, Professor Albareda 1, Granada, Spain

INTRODUCTION

Superoxide dismutases isolated from different organisms have been classified into three types containing either iron, manganese or copper plus zinc as the prosthetic group.[1] The most generally accepted function for these enzymes is the protection of aerobic cells against superoxide toxicity though specific reactions involved in the postulated deleterious effect of this radical have not been unequivocally documented.[2] Some authors have also suggested that superoxide dismutase might have a function other than the disproportionation of superoxide.[3,4]

We have previously shown that leaf extracts from Pisum sativum contain three electrophoretically distinct superoxide dismutases (Del Rio et al.)[5], two of which are inhibited by 1 mM cyanide, and are probably Cu, Zn-SOD, while the third one was CN-insensitive and, therefore, could be either a Mn- or an Fe-containing SOD. The same isozyme patterns was found in germinating seeds and roots of this plant (unpublished results). The CN-resistant SOD activity, as judged by its dependency on Mn nutrient levels supplied to the plants, was considered to be a Mn-SOD dismutase (Del Rio et al).[5]

In higher plants little information is available on the CN-insensitive Mn- and Fe-containing SODs in comparison with the Cu, Zn-SODs that have been characterized from spinach leaves[6], wheat germ[7] and pea seeds.[8] Cyanide-insenstive SODs have been detected in crude extracts from wheat germ[7], kidney bean[9], corn[10], and in several mosses and ferns.[11] A spinach leaf CN-resistant SOD has been partially purified and its cellular locus associated to chloroplast lamelar membranes,[12] but this has been refuted by other workers.[13,14] More recently, the CN-insensitive SOD cellular location has been revised using three C_3 plants, including spinach, with the conclusion that the enzyme was located outside the chloroplasts and might have been adsorbed nonspecifically to the envelop during isolation.[15] Furthermore, CN-insensitive SOD activity has also been located in mitochondria from corn[10], spinach[15] and Jerusalem artichoke tubers.[16]

However, in higher plants no Mn-SOD has as yet been fully purified, characterized and definitely differentiated from the also CN-resistant enzyme Fe-SOD, whose presence appears to be limited to prokaryotic organisms. In our laboratory, we have recently carried out, for the first time, the purification to homogeneity of a Mn-SOD from Pisum sativum leaves.[17] This report describes the preliminary characterization of this enzyme.

MATERIALS AND METHODS

Superoxide dismutase activity was assayed in terms of its ability to inhibit the reduction of nitro blue tetrazolium (NBT) by the O_2^- -generating system, xanthine oxidase plus xanthine.[18] One unit of SOD activity, defined as that amount of enzyme which caused 50% inhibition of the initial rate of NBT reduction, was found to be equivalent to 0.33 μg of purified enzyme.

Polyacrylamide-gel electrophoresis was carried out at pH 8.9 on 10% gels, polymerized with riboflavin, according to Davies.[19] Superoxide dismutase was visualised on the gels by a photochemical method.[20] The isoelectric point was measured by polyacrylamide-gel isoelectric focussing[21] with photochemical visualization of SOD activity.[20] Determination of molecular weight by gel filtration on Sephadex G-100 was performed as described by Andrews.[22] Estimation of molecular weight by disc gel electrophoresis was made by the method of Hedrick and Smith.[23] The subunit molecular weight was determined by electrophoresis on 10% polyacrylamide gels in the presence of sodium dodecyl sulphate,[24] using the following subunit molecular weight standards for calibration of gels: bovine serum albumin (67000), ovoalbumin (45000), chymotrypsinogen A (25000), myoglobin (17800), and cytochrome c (12400).

Protein concentrations in the purified fractions were determined by the method of Lowry et al.,[25] and in crude extracts a modification of this assay in the presence of phenols and pectins was followed.[26] Metal concentrations were determined by atomic absorption spectrophotometry using a Perkin-Elmer 503 apparatus equipped with a heated graphite atomizer model HGA-74.

Purification of enzyme. Pea plants (Pisum sativum L., var. Lincoln) were grown in the greenhouse in aerated full nutrient solutions as previously described[27], and the manganese nutrient level was 0.5 μg/ml which is the optimum for this plant. Pea leaves (3 kg) were cold homogenized in 50 mM Tris-HCl buffer, pH 7.5, and 0.1 mM EDTA (1:2; w/v), and the homogenate was filtered through nylon cloth and centrifuged at 1500 g for 15 min. The supernatant was heated at 60°C for 5 min, then cooled in an ice bath, and the precipitated proteins removed by centrifugation at 10,500 g for 15 min. The resulting solution was subjected to fractional precipitation by ammonia sulphate at 40°C, and the enzyme, that precipitated in the range 40-70% saturation, was recovered by centrifugation at 10,500 g for 20 min. The pellet was dissolved in 50 mM potassium phosphate buffer, pH 7.8, and dialyzed against this buffer. The enzyme solution was applied to a 2.5 x 24 cm column of DEAE-cellulose (DE-52) equilibrated in 50 mM potassium phosphate buffer, pH 7.8. By washing the column with buffer, the Cu, Zn-SODs were eluted and completely separated from the Mn-SOD, which was eluted with a linear gradient of NaCl at a concentration range of 0.14-0.28 M NaCl. Occasionally, a batch step with DE-23 cellulose was introduced before the DE-52 column chromatography.

Active fractions from the ion-exchange chromatography were dialyzed against 50 mM potassium phosphate buffer, pH 7.8, and concentrated by ultrafiltration on an Amicon PM-10 membrane. This sample was passed through a Sephadex G-100 column (2.5 x 74 cm) equilibrated in 50 mM phosphate buffer, pH 7.8. The peak showing SOD activity was dialyzed against 50 mM Tris-HCl buffer, pH 7.5, and concentrated by ultrafiltration to a volume of about 1 ml. This sample was subjected to preparative polyacrylamide gel electrophoresis at pH 8.9 in a Canalco Prep-Disc apparatus using 10% gels. Fractions were eluted with 50 mM Tris-HCl buffer, pH 7.5, and those containing SOD activity were poooled, and concentrated by ultrafiltration. The yield was 1.2 mg pure enzyme per kg leaf.

RESULTS

Purity. The different enzyme preparations used for the characterization studies were purified about 250-times. Figure 1 shows the results obtained when crude extracts of pea leaves and the purified Mn-SOD were subjected to polyacrylamide disc gel electrophoresis. The purified enzyme revealed a single band of protein which coincided with the band of activity, and had a similar mobility to the enzyme in crude extracts, indicating the absence of any alteration in the enzyme molecule as a result of the purification process. The isoelectric point of the enzyme was at pH 4.5.

The slower moving activity band in the pure enzyme, shown in Figure 1, appeared only after several days storage at $4^{o}C$, and may possibly be due to enzyme polymerization as has been described for SODs from other sources.[20,28,29]

Spectral Properties. The absorption spectrum of the Mn-SOD is shown in Figure 2. The absorption in the near UV is that of a typical protein, with a maximum at about 275 nm, whose molar extinction coefficient was approximately 120,300 $M^{-1}cm^{-1}$, There were also shoulders at 225, 280, and 290 nm. The visible spectrum showed a weak and broad absorption band in the range 350-700 nm, with two shoulders at about 370-400 nm and 570-600 nm, respectively.

Molecular weight. According to the data obtained from gel filtration on Sephadex G-100, Mn SOD appeared at a position corresponding to a molecular weight of 84600 daltons (Figure 3). By polyacrylamide-gel electrophoresis, on the basis of the molecular weight-slope relation of different standard

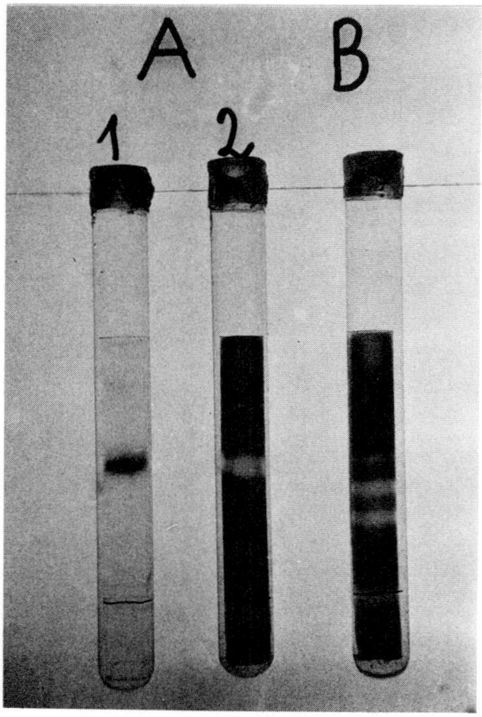

Fig. 1. Polyacrylamide disc gel electrophoresis of pea leaf Mn-superoxide dismutase. Electrophoresis was carried out on 10% polyacrylamide gels at pH 8.9. A: pure enzyme stained for protein with Amino Black (1), and for superoxide dismutase activity (2). B: crude extracts from *Pisum sativum* leaves stained for superoxide dismutase activity.

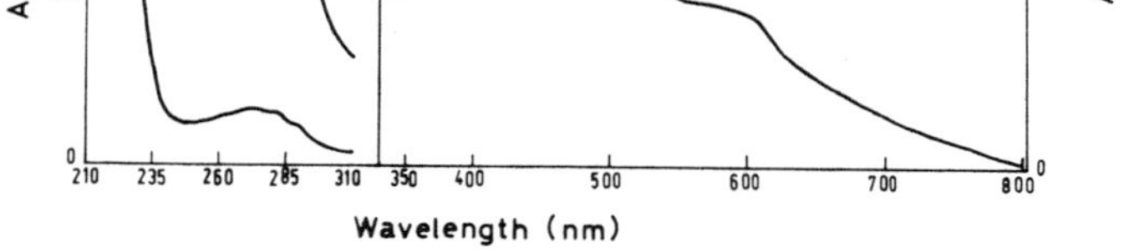

Fig. 2. Absorption spectrum of Mn-superoxide dismutase from pea leaves. Enzyme concentrations were as follows: spectrum A, 0.145 mg/ml; spectrum B, 1 mg/ml; spectrum C, 0.180 mg/ml.

proteins at varying gel concentrations (6-12%), the molecular weight calculated for the enzyme was 94500 daltons (Figure 4).

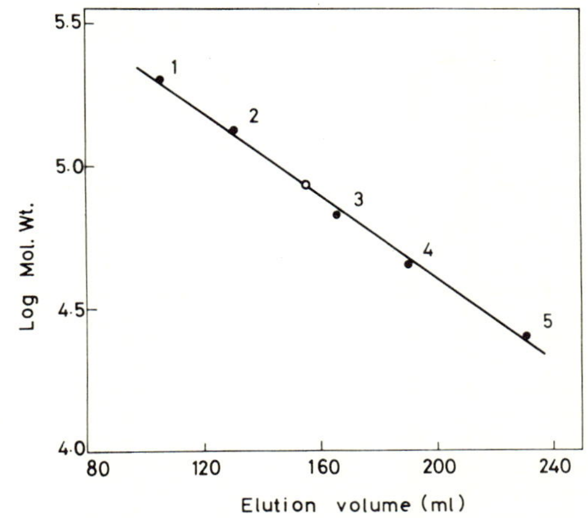

Fig. 3. Determination of the molecular weight of pea Mn-superoxide dismutase by gel filtration. The enzyme after the DEAE-celluose step was dialyzed against 50 mM potassium phosphate buffer, pH 7.8, and passed through a Sephadex G-100 column (2.5 x 74 cm) equilibrated in the same buffer. SOD activity was determined according to Ref.20. The column was calibrated with proteins of known molecular weights: 1, bovine serum albumin (trimer); 2, bovine serum albumin (dimer); 3, bovine serum albumin (monomer); 4, ovalbumin; 5, chymotrypsinogen A.

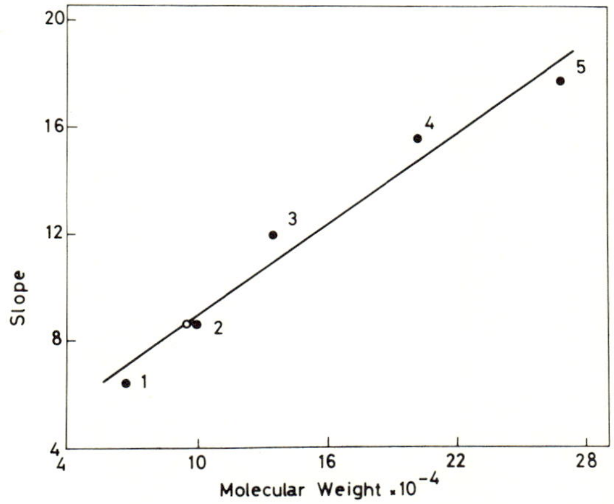

Fig. 4. Determination of the molecular weight of pea Mn-superoxide dismutase by disc gel electrophoresis at different polyacrylamide concentrations. The purified enzyme was subjected to electrophoresis in 6-12% gels. Molecular weight standards were also run in separate gels: bovine albumin (monomer, dimer, trimer, and tetramer) (1)(3)(4)(5), and hexokinase (2). For the enzyme and each molecular weight standard, the log of mobility relative to the dye front was plotted versus the gel concentration, and the corresponding slopes calculated. The enzyme molecular weight was finally determined by plotting the different slopes obtained against the molecular weight of the standards.

Subunit molecular weight. As shown in Table 1, after incubation with 1% SDS at 37°C, 50°C and 100°C for different times, with and without 2-mercaptoethanol, SDS-polyacrylamide gel electrophoresis of the enzyme revealed a major band and a minor band. The comparison of their relative mobilities with those of molecular weight standards indicated the molecular weight of the two protein bands to be 23,500 and 46,300, respectively. However, when the enzyme was

TABLE 1. EFFECT OF DIFFERENT INCUBATION CONDITIONS ON THE DISSOCIATION OF Mn-SUPEROXIDE DISMUTASE INTO SUBUNITS.[a]

Incubation mixture	Heating conditions	Monomer (23500)	Dimer (46300)	Ratio monomer/ dimer
		Percent[b]		
1% SDS + 8% mercaptoethanol	37°C 2h	64.4	35.6	1.8
1% SDS + 8% mercaptoethanol	50°C 2 min	72.6	27.4	2.6
1% SDS + 8% mercaptoethanol	100°C 2 min	68.4	31.6	2.2
1% SDS + 1% mercaptoethanol	100°C 5 min	64.6	35.4	1.8
1% SDS - mercaptoethanol	100°C 5 min	69.3	30.7	2.3
1% SDS + 8% mercaptoethanol	100°C 15 min	65.2	34.8	1.9
1% SDS + 8M urea + mercaptoethanol[c]	100°C 5 min	66.8	33.2	2.0
1% SDS + 8M urea - mercaptoethanol[c]	37°C 2h	90.2	9.8	9.2
1% SDS + 8M urea - mercaptoethanol[c]	100°C 5 min	100	-	-

[a]SDS-polyacrylamide gel electrophoresis was performed on 10% gels in 0.1M sodium phosphate buffer, pH 7.0.

[b]Band percentages were determined by densiometry after staining with Coomassie blue.

[c]Gels contained the same concentrations of the incubation mixture except for SOD which was 0.1%.

heated at 100°C for 5 min in incubation mixtures containing 1% SDS plus 8M urea only a single band was obtained which corresponds to the lower molecular weight (23,500) subunit. The incubation with 8M urea at 37°C for 2 hours produced the nearly complete conversion of the enzyme into the elementary subunit. These results imply that pea leaf Mn-SOD is composed of four subunits of equal size. The addition of 2-mercaptoethanol to the urea incubation medium resulted in the partial conversion of the monomeric into the dimeric subunit (ca. 33%), suggesting an abnormal aggregation in the presence of the reagent. In most cases where the monomeric and dimeric subunits were produced the ratio of percentages of the former to the latter was about 2.0.

The activity of subunits from the enzyme after incubation with 8M urea at 100°C for 5 min or 37°C for 2 hours, in the presence and absence of 2-mercaptoethanol, respectively, was assayed by photochemical staining of the gels after electrophoresis.[20] In these conditions no SOD was detected in the monomeric and dimeric subunits. However, in those gels containing urea plus 2-mercaptoethanol a dark blue band appeared on a bluish background which had an identical mobility to the enzyme monomeric form. Apparently, in the presence of 2-mercaptoethanol the elementary subunit of the manganese superoxide dismutase brings about the reduction of NBT, the dye used in the enzyme stain.

Metal content. The pure superoxide dismutase was dialyzed exhaustively first against 50 mM Tris-HCl buffer plus 0.1 mM EDTA, pH 7.5, and then against this buffer lacking EDTA. The enzyme was assayed for manganese, iron and copper by atomic absorption spectroscopy, both before and after dialysis. This was repeated with two different preparations of the enzyme, and twelve determinations were made on each sample. On a molecular weight basis of 94,500, the undialyzed SOD contained 0.82±0.04 atoms of manganese per molecule, and traces of iron and copper. These two metals were removed by the dialysis, whereas the manganese content was 0.99±0.07 atoms per molecule, i.e. about 0.06% of manganese per molecule. It thus appears that the pea leaf superoxide dismutase contains 1 atom of manganese per molecule of native tetramer.

DISCUSSION

Manganese-containing superoxide dismutase have been found in prokaryotes, lower eukaryotes, and in the matrix space of mitochondria from higher eukaryotes. The Mn-SOD from *Pisum sativum* leaves, reported here, is the first one characterized so far from a higher plant.

The Mn-SOD comprises about 20% of the total SOD activity present in leaves from the flowering *Pisum sativum* plants used for the enzyme purification. On the basis of its specific activity, Mn-SOD accounts for approximately 0.2-0.4% of the total soluble proteins of *Pisum sativum* leaves, which is in good agreement with the values described for total SOD activity in corn, peas and oats.[29]

The enzyme is composed of four subunits of equal size apparently bound by noncovalent forces, and its tetrameric structure is similar to the Mn-SOD from chicken liver,[20] bovine heart mitochondria,[30] rat liver,[31] human liver,[32] yeast,[33] the fungus *Pleurotus olearis*,[34,35] and the bacteria *Thermus aquaticus*,[36] *Thermus thermophilus*,[37] and *Myobacterium phlei*.[38] In Fe-SODs, the only tetrameric enzyme described is that from *Mycobacterium tuberculosis*.[39]

The remaining Mn-SODs so far characterized are built up from two subunits, and occur in bacteria,[40-43] including a photosynthetic organism[44] and a red alga.[45] The only exception is the enzyme from *Mycobacterium* species, strain Takeo, that appears to be a trimer.

Pisum sativum Mn-SOD shows a remarkable resistance to dissociation by SDS. Only after heating at 100°C for 5 min in the presence of 8M urea can the enzyme be completely converted into the monomeric form. Other SODs that have been described to be resistant to SDS denaturation are the Mn-SOD from Mycobacterium species, strain Takeo,[28] the Fe-SOD from Mycobacterium tuberculosis,[39] and the cuprozinc enzymes from bovine erythrocytes,[46] chicken liver.[20]

The complete conversion of pea leaf Mn-SOD into monomers of 23,500 molecular weight by treatment with 8M urea and SDS is in agreement with the demonstration by Bannister et al.[47] and by Marmocchi et al.[48], by gel filtration experiments, that Cu,Zn-SODs from different sources are dissociated into monomers in the presence of 8M urea. However, while the mentioned authors found the monomers from those enzymes catalytically active, the elementary subunit of pea leaf Mn-SOD was revealed to be inactive after staining the SDS-polyacrylamide gels for SOD activity. This situation contrasts with the monomeric and dimeric forms of a Cu,Zn-SOD from wheat germ (isoenzyme I) that under the same circumstances showed SOD activity.[49]

One point that remains to be explained is the systematic appearance of the dimeric form of pea Mn-SOD (ca.33%) when the enzyme was incubated in 8M urea plus 2-mercaptoethanol. Leaving aside the possibility of an artifact derived from the electrophoretic method and/or reagents used, a somewhat speculative explanation of this finding could be based on the existence of intrachain disulfide bridges in the enzyme protomer produced by the action of 8M urea, and the cleavage of these bridges by 2-mercaptoethanol, as has been suggested by Weisiger & Fridovich[20] for the Mn-SOD from chicken liver mitochondria. In these conditions, the subunits could be less compact and correspondingly less mobile, which would be seen as an apparent increase in molecular weight. Another explanation for the dimer formation could be the aggregation of monomers by interactions between intact intrachain disulfide bridges of a monomer molecule and the mercaptoethanol-produced sulphydryl residues of another molecule, resulting in the binding of two protomers by a disulfide bridge to form a dimer. In any case, further experimental evidence is necessary before this explanation could be accepted.

The Mn-SOD from Pisum sativum leaves contains one atom of manganese per tetramer, and considering that the electrophoretic mobility of the pure enzyme is identical to that of the crude extract enzyme, it does not seem very probable that some manganese had been lost during the purification process, although this possibility cannot be ruled out. It must be added that Pisum sativum plants were grown under optimum manganese nutrient levels which is against a lower manganese content in the enzyme as a response to manganese nutritional limitations.

One atom of manganese per a tetramer of 94,000 molecular weight is an unusual stoicheiometry for a Mn-SOD. The enzymes of this type so far characterized contain either 1 atom Mn per dimer, 2 atoms per tetramer or 4 atoms per tetramer.

In the first category are classified Mn-SODs from a number of bacteria[40-43], including a photosynthetic organism,[44] and a red alga.[45] Manganese superoxide dismutases containing 2 atoms Mn per tetramer are those of bacterial and fungal origin[35-37] chicken liver[20] and bovine heart mitochondria,[30] whereas enzymes -with 4 atoms Mn per tetramer are restricted to yeast[33] and mammals.[31,32]

As to the prokaryotic Fe-SODs, though results described by some authors are conflicting,[50-51] also show three different metal-protein stoicheiometries: 1 atom/Fe dimer,[44,51-55] 2 atoms Fe per dimer,[4,50,56-58] and 4 atoms Fe per tetramer.[39]

In the light of the different metal-protein stoichiometries of the Mn-SODs presently known, it may be of significance to note that those enzymes character-ized from photosynthetic organisms, the bacterium Rhodopseudomonas spheroides[44] and the red alga Porphyridium cruentum,[45] both have 1 atom Mn per dimer. Con-sidering the increase in the molecular weight of certain proteins that takes place during the transition from lower to higher stages of the phylogenetic scale, as in the case of ferrodoxins, it does not appear surprising that the Mn-SOD from ahigher plant, reported here, has a stoicheiometry of 1 atom Mn per tetramer.

The ultraviolet absorption spectrum of pea leaf Mn-SOD is similar to that of many proteins containing tyrosine and tryptophan, and shows a shoulder at 225 nm that has also been noted in mitochondrial manganese superoxide dismutases from yeast[33], chicken liver,[20] bovine heart[30] and human liver.[32] This shoulder is apparently due to the manganese chromophore.[32] However, the absorption in the visible did not show the typical peak at 470-480 nm, characteristic of most Mn-SODs, but a weak and broad absorption band with a shoulder at about 600 nm that has also been reported for Mn-SODs from Pleurotus olearius,[34] Thermus thermophilus[37], Thermus aquaticus[36], Mycobacterium species, strain Takeo,[28] and chicken[20] and human liver.[32] We must emphasize that the low content of Mn-SOD in Pisum sativum leaves joined to the poor enzyme recoveries obtained did not allow the use of higher protein concentrations, which very probably would have evidenced sharper absorption differences in the visible region. This point remains to be clarified.

Further characterization of this Mn-SOD, including the study of additional physico-chemical properties, amino acid composition, enzyme activity reconstitu-tion, and subcellular localization now in progress in our laboratory, will provide more information on the relation between this higher plant enzyme and other SODs from different organisms. These results could also contribute to the ellucidation of a possible function for Mn-SOD in the photosynthetic process, as well as unravelling the complex evolutionary history of these enzymes.

ACKNOWLEDGEMENTS

The authors are indebted to Dr.M.Gómez Ortega (Plant Physiology Unit) for the growth of pea plants under greenhouse conditions, to Dr.J.Yañez Fernández (Unit of Applied Analytical Chemistry) for performing metal analysis by atomic absorption spectrophotometry, and to Miss M.Garrido Ortiz and Mr.A. Melgar Porcel for skilful technical assistance. This work was supported by a research grant from the Comisión Asesora de Investigación Cientifica y Técnica.

REFERENCES

1. Fridovich, I. (1975). Annu. Rev. Biochem. 44, 147-159.
2. Fee, J.A. and McClune, G.J. (1978) in Mechanisms of Oxidizing Enzymes, Singer, T.P. and Ondarza, R.N., eds., Elsevier/North Holland, Amsterdam, pp.273-284.
3. Tyler, D.D. (1975) Biochem. J. 147, 493-504.
4. Kanematsu, S. and Asada, K. (1978) Arch. Biochem. Biophys. 185, 473-482.
5. Del Rio, L.A., Sevilla, F., Gómez, M., Yañez, J. and López, J. (1978) Planta, 140, 221-225.
6. Asada, K., Urano, M. and Takahashi, M. (1973) Eur. J. Biochem. 36, 257-266.
7. Beauchamp, C.O. and Fridovich, I. (1973) Biochim. Biophys. Acta, 317, 50-64.
8. Sawada, Y., Ohyama, T. and Yamazaki, I. (1972) Biochim. Biophys. Acta, 268. 305-312.
9. Asada, K., Kanematsu, S., Takahashi, M., and Kono, Y. (1976) in Iron and Copper Proteins, Yasunobu, K.T., Mower, H.F. and Hayaishi, O., eds., Plenum Publishing Corporation, New York, pp.551-564.
10. Giannopolitis, C.N. and Ries, S.K. (1977) Plant Physiol. 59, 309-314.
11. Asada, K., Kanematsu, S. and Uchida, K. (1977) Arch. Biochem. Biophys. 179, 243-256.
12. Lumsden, J. and Hall, D.O. (1974) Biochem. Biophys. Res. Commun. 58, 35-41.
13. Elstner, E.F. and Heupel, A. (1975) Planta, 123, 145-154.
14. Doll, S., Lutz, C. and Ruppel, H.G. (1976) Z. Pflanzenphysiol. 80, 166-176.
15. Jackson, C., Dench, J., Moore, A.L., Halliwell, B., Foyer, C.H. and Hall, D.O. (1978) Eur. J. Biochem. 91, 339-344.
16. Arron, G., Henry, L., Palmer, J.M. and Hall, D.O. (1976) Biochem. Soc. Trans. 4, 618-620.
17. Sevilla, F., López Gorgé, J. and Del Rio, L.A. (1978) XVII Meeting of the Spanish Biochemical Society (SEB), Madrid, Abstract No.59.
18. Beauchamp, C.O. and Fridovich, I. (1971) Anal. Biochem. 44, 276-287.
19. Davis, B.K. (1964) Ann. N.Y. Acad. Sci. 121, 404-427.
20. Weisiger, R,A, and Fridovich, I. (1973) J. Biol. Chem. 248, 3582-3592.
21. Wrigley, C. (1968) Science Tools (LKB Instrument Journal), 15, 17-23.
22. Andrews, P. (1965) Biochem. J. 96, 595-606.
23. Hedrick, J.L. and Smith, A.J. (1968) Arch. Biochem. Biophys. 126, 155-164.
24. Weber, K. and Osborn, M. (1969) J. Biol. Chem. 244, 4406-4412.
25. Lowry, O.H., Rosebrough, N.J., Farr, A.L. and Randall, R.J. (1951) J. Biol. Chem. 193, 265-275.
26. Potty, V.H. (1969) Anal. Biochem. 29, 5351539.
27. Del Rio, L.A., Gómez, M., Yañez, J., Leal, A. and López Gorgé, J. (1978), Plant & Soil, 49, 343-353.
28. Kusunose, M., Noda, Y., Ichihara, K. and Kusunose, E. (1976) Arch. Microbiol. 108, 65-73.
29. Giannopolitis, C.N. and Ries, S.K. (1977) Plant Physiol. 59, 315-318.
30. Marklund, S. (1978) Int. J. Biochem. 9, 299-306.
31. Salin, M.L., Day, E.D.,Jr. and Crapo, J.D. (1978) Arch. Biochem. Biophys. 187, 223-228.

32. McCord, J.M., Boyle, J.A., Day, E.D., Jr., Rizzolo, L.J. and Salin, M.L. (1977) in Superoxide and Superoxide Dismutases, Michelson, A.M., McCord, J.M. and Fridovich, I., Eds., Academic Press, New York, pp.129-138.
33. Ravindranath, S.D. and Fridovich, I. (1975) J. Biol. Chem. 250, 6107-6112.
34. Lavelle, F., Durosay, P. and Michelson, A.M. (1974) Biochimie, 56, 451-458.
35. Lavelle, F. and Michelson, A.M. (1975) Biochimie, 57, 375-381.
36. Sato, S. and Harris, J.I. (1977) Eur. J. Biochem. 73, 373-381.
37. Sato, S., and Nakazawa, K. (1978) J. Biochem. Tokyo, 83, 1165-1171.
38. Chikata, Y., Kusunose, E., Ichihara, K. and Kusunose, M. (1975) Osaka City Med. J. 21, 127-136.
39. Kusunose, E., Ichihara, K., Noda, Y. and Kusunose, M. (1976) J. Biochem. Tokyo, 80, 1343-1352.
40. Keele, B.B., McCord, J.M. and Fridovich, I. (1970) J. Biol. Chem. 245, 6176-6181.
41. Vance, P.T., Keele, B.B., Jr. and Rajagopalan, K.V. (1972) J. Biol. Chem. 247, 4782-4786.
42. Harris, J.I. (1977) in Superoxide and Superoxide Dismutases, Michelson, A.M., McCord, J.M. and Fridovich, I., eds., Academic Press, New York, pp.151-157.
43. Ichihara, K., Kusunose, E., Kusunose, M. and Mori, T. (1977) J. Biochem. Tokyo, 81, 1427-1433.
44. Lumsden, J., Cammack, R. and Hall, D.O. (1976) Biochim. Biophys. Acta, 438, 380-392.
45. Misra, H.P. and Fridovich, I. (1977) J. Biol. Chem. 252, 6421-6423.
46. Keele, B.B., McCord, J.M. and Fridovich, I. (1971) J. Biol. Chem. 246, 2875-2877.
47. Bannister, J.V., Anastasi, A. and Bannister, W.H. (1978) Biochem. Biophys. Res. Commun. 81, 469-472.
48. Marmocchi, F., Venardi, G., Bossa, F., Rigo, A. and Rotilio, G. (1978) FEBS Lett. 94, 109-111.
49. Marmocchi, F., Venardi, G., Caulini, G. and Rotilio, G. (1974) FEBS Lett. 44, 337-339.
50. Asada, K., Yoshikawa, K., Takahashi, M., Maeda, Y. and Enmanji, K. (1975) J. Biol. Chem. 250, 2801-2807.
51. Misra, H.P. and Keele, B.B. Jr. (1975) Biochim. Biophys. Acta, 379, 418-425.
52. Yamakura, F. (1976) Biochim. Biophys. Acta, 422, 280-294.
53. Dougherty, H.W., Sadowski, S.J. and Baker, E.E. (1978) J. Biol. Chem. 253, 5220-5223.
54. Anastasi, A., Bannister, J.V. and Bannister, W.H. (1976) Int. J. Biochem. 7, 541-546.
55. Yost, F.J. Jr. and Fridovich, I. (1973) J. Biol. Chem. 248, 4905-4908.
56. Puget, K. and Michelson, A.M. (1974) Biochimie, 56, 1255-1267.
57. Kanematsu, S. and Asada, K. (1978) FEBS Lett. 91, 94-98.
58. Hatchikian, E.C. and Henry, Y.A. (1977) Biochimie, 59, 153-161.

SUPEROXIDE DISMUTASE FROM CHROMAFFIN GRANULES

N.A. GRIGORYAN AND R.M. NALBANDYAN
Institute of Biochemistry, Academy of Sciences of the Armenian SSR,
Yerevan, 375044, USSR

INTRODUCTION

Chromaffin granules are specific catecholamine-containing subcellular particles located in adrenal medulla and extramedullar chromaffin tissues. The role of these particles consists in the accumulation, storage and release of catecholamines.[1-3] The concentration of catecholamines within the granules is unusually high, about 20% of the dry weight.

It is known that the autoxidation of catecholamines is accompanied with the formation of superoxide radicals.[4] One may therefore suggest the presence of superoxide dismutase in chromaffin granules. Earlier data indicated that adrenal medulla contains high level of superoxide dismutase activity[5] and that the properties of the cytosolic enzyme are similar to those of the cytosolic enzymes from other tissues.[6] Here we report the purification and properties of superoxide dismutase from chromaffin granules, isolated from adrenal medulla.

MATERIALS AND METHODS

Chromaffin granules were prepared from bovine adrenal medulla essentially according to Hoffmann et al.[7] Electron microscopic assay of the preparation obtained indicated that admixture of mitochondria and lysosomes was less than 5%. The preparation of the granules had no cytochrome oxidase activity.

To prepare the lysate from 60 g (wet weight) of granules obtained from 600 g of bovine adrenal medulla, the granules were suspended in 1 L of 0.01 M acetate buffer, pH 6.1 containing 2 mM β-mercaptoethanol. The suspension was sonicated on an instrument "UZDN-1" (USSR) at 22 kHz for 5 min, and the sonicate was dialyzed against 15 L of 0.1 M NaCl for 48 h with changes of the NaCl solution every 12 h. The suspension was further dialyzed against 0.01 M acetate buffer, pH 6.1, for 24 h. The dialyzate was centrifuged at 37000 x g for 30 min. The sediment was removed and the supernatant was then used for the isolation of SOD (see below).

The assay of SOD activity in fractions was performed according to Nishikimi et al.[8] The same method was also used for the determination of SOD zones in polyacrylamide gels after electrophoresis according to Davis.[9] Molecular weight of the protein was determined electrophoretically.[10] Determination of iron was carried out according to Cameron.[11] Protein was estimated by the method of Lowry et al.[12] Absorption spectra were recorded at 22°C in 10 mm cells.

RESULTS AND DISCUSSION

Figure 1 shows the main stages of the purification procedure used. The preparation had two SOD-active zones with two corresponding protein zones (Figure 2). The absorption spectrum of SOD from chromaffin granules is shown in Figure 3. The protein contained 1.3 atom of iron per molecular weight of 40,000 daltons. However, the determination of molecular weight by the method of Weber and Osborn[10] indicates a minimal molecular weight of 18,000 daltons. Because the protein contains iron it was expected that the enzyme should be cyanide insensitive. However, incubation of the SOD with 10 mM CN^- for 30 min. led to complete inhibition (Figure 4).

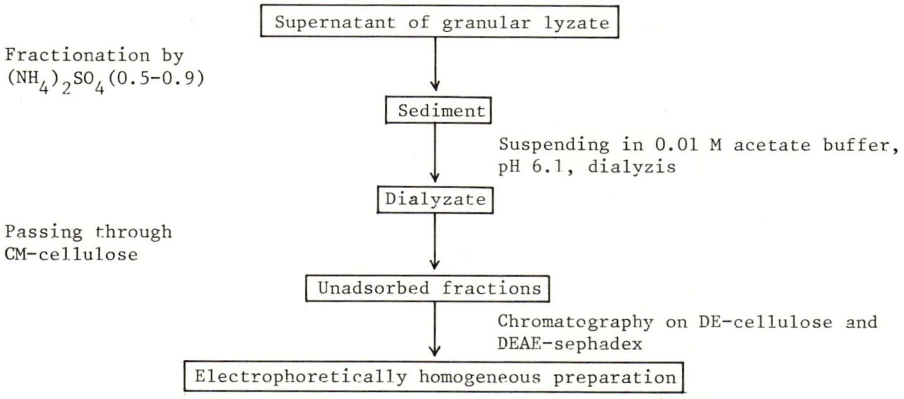

Fig. 1. Purification of SOD from lysate of chromaffin granules.

In many respects SOD isolated from chromaffin granules resembles iron-containing SOD from microorganisms.[13] The resemblance concerns absorption spectra, subunit composition and sensitivity to thiol reagents. There are, however, certain differences between the microbial and granular enzymes. Thus, SOD from chromaffin granules is present as at least two isoenzymes, whereas microbial iron SOD, probably consists of a single enzyme. Besides, microbial iron SOD is insensitive to cyanide, whereas granular SOD is sensitive to cyanide. Sensitivity to cyanide is more characteristic of cuprozinc SOD.

Although iron SOD was isolated from the water-soluble fraction of the granular lysate, it is difficult at present to state that the enzyme is located in the inner soluble compartment of chromaffin granules rather than in the granular membrane. The formation of superoxide radicals proceeds in the course of

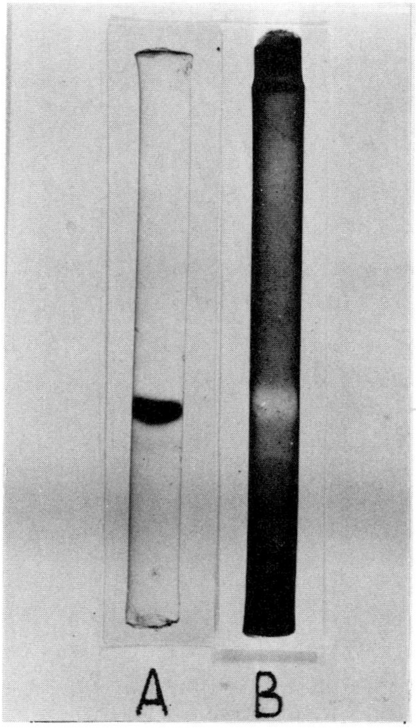

Fig. 2. Electrophoretical patterns of granular SOD. A - stained as proteins; B - stained as SOD active zones.

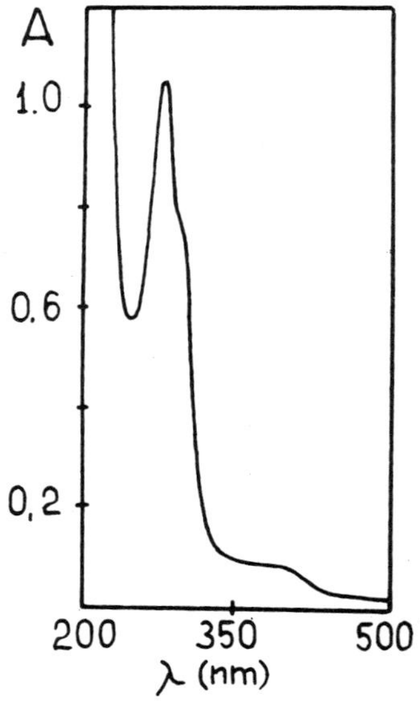

Fig. 3. Absorption spectrum of SOD from chromaffin granules.

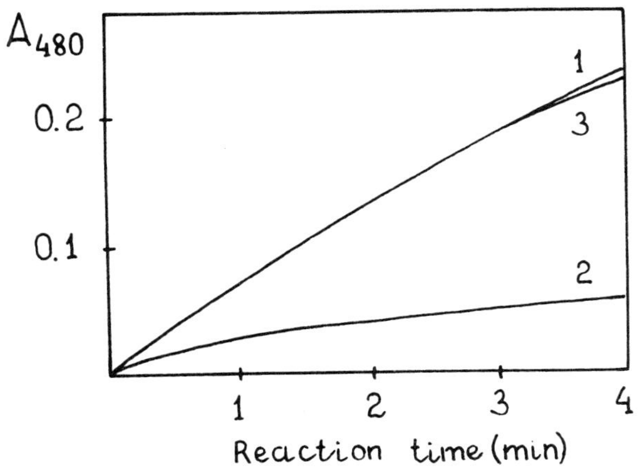

Fig. 4. Inhibition of SOD by 10 mM CN^-. The protein was pre-incubated for 30 min with cyanide and then the tetrazolium system used for assay of SOD activity was added. 1 - formation of formazan in the absence of SOD; 2 - formation of formazan in the present of 5 x 10^{-9} M SOD; 3 - formation of formazan in the presence of 5 x 10^{-9} M SOD and 10 mM CN^-.

oxidation of catecholamines, which is known to be located in the inner soluble compartment of the granules. Therefore, it may be suggested that iron SOD is also located in this compartment. However, it should be borne in mind that superoxide radicals can be also formed in the course of other terminal oxidative processes. In particular, they are formed in the course of the reaction catalyzed by dopamine-β-hydroxylase.[14] This enzyme is present in chromaffin granules in two forms: soluble and membrane-bound. Hence, the question of the localization of iron SOD in chromaffin granules needs further investigation.

This communication is the first indication of the presence of SOD in chromaffin granules. The data obtained indicate certainly that this SOD is remarkably different from the cuprozinc enzyme localized in the cytosol of adrenal medulla.

REFERENCES

1. Blaschko, H. and Muscholl, E. (1972) Catecholamines, Springer-Verlag, New York.
2. Njus, D. and Radda, G.K. (1978) Biochim. Biophys. Acta, 463, 219-244.
3. Johnson, R.G., Carlson, N.J. and Scarpa, A. (1978) J.Biol. Chem. 253, 1512-1521.
4. Bors, W., Michel, C., Saran, M. and Lengfelder, E. (1978) Biochim. Biophys. Acta, 540, 162-172.
5. Hartz, J.W., Funakoshi, D. and Deutsch, H.F. (1973) Clin. Chim. Acta, 46, 125-132.
6. Grigoryan, N.A., Gjulchandanian, G.V., Symonyan, M.A. and Nalbandyan, R.M. (1977) Biokhimiya (Russ.) 42, 1499-1504.

7. Hoffman, P.G., Linder, O., Bonner, W.M. and Pollard, H.B. (1976) Arch. Biochem. Biophys. 176, 375-388.
8. Nishikimi, M., Rao, N.A. and Jagi, K. (1972) Biochem. Biophys. Res. Commun. 46, 849-856.
9. Davis, B.J. (1964) Ann. N.Y. Acad. Sci. 121, 404-427.
10. Weber, K. and Osborn, M. (1969) J.Biol. Chem. 244, 4406-4412.
11. Cameron, B.F. (1965) Anal. Biochem. 11, 164-169.
12. Lowry, O.H., Rosebrough, N.J., Farr, A.J. and Randall, R.J. (1951) J.Biol. Chem. 193, 265-275.
13. Jamakura, F. (1976) Biochim. Biophys. Acta, 422, 280-294.
14. Henry, J-P., Hirata, F. and Hayaishi, O. (1978) Biochem. Biophys. Res. Commun. 81, 1091-1099.

BOVINE SUPEROXIDE DISMUTASE PREPARATIONS: COMPARISON OF THEIR BIOCHEMICAL AND BIOLOGICAL CHARACTERISTICS.

R.A. ZEPP, W.S. CHELACK AND A. PETKAU
Medical Biophysics Branch, Whiteshell Nuclear Research Establishment, Atomic Energy of Canada Limited, Pinawa, Manitoba, Canada ROE 1LO

ABSTRACT

Five preparations of the bovine cupro-zinc superoxide dismutase were compared in terms of their enzyme activity, electrophoretic banding patterns and capacity to radioprotect mouse bone marrow stem cells in vitro. The range in enzyme activity was 208 ± 46 to 360 ± 34 ng/unit. Differences in their electrophorectic banding patterns were accentuated in diluted Tris glycine buffer pH 8.2-8.6 and minimized in full strength Tris borate EDTA buffer at pH 9.1. Their capacity to protect X-irradiated stem cells was linearly related to the enzyme activity.

INTRODUCTION

Superoxide dismutase (SOD, E.C.1.15.1.1) which catalyzes the dismutation of superoxide radicals, has found increasing usage in biological chemistry,[1-17] radiobiology[18-24] and medicine.[25-26] In most of the applications,the cupro-zinc enzyme from bovine erythrocytes was used. The standard extraction procedure[27] utilizes a chloroform-ethanol precipitation step which has been shown to introduce conformational changes in the enzyme and ostensibly induces some loss (∿8%) of copper in vitro.[28] Moreover, these conformational changes could vary in severity from one preparation to another and, in theory, affect the enzyme's biological activity. As well, it appears that chromatographic purification of the bovine enzyme differs between preparations, which also may influence its specific activity and effectiveness within biological systems undergoing oxidative stress by exposure to superoxide radicals. The possible influence of these factors was assessed by comparing different bovine SOD preprations in terms of their biochemical characteristics and biological effectiveness in irradiated bone marrow cells. The results are described in this report.

MATERIALS

Bovine erythrocyte SOD was obtained from Truett Laboratories, Dallas, Texas;

Sigma Chemical Co., St.Louis, Missouri; Miles Laboratories, Elkhart, Indiana; and Dr.J.V.Bannister, University of Malta, Msida, Malta (present address: Inorganic Chemistry Department, South Parks Road, Oxford OX1 3QR, England). Cupro-zinc SOD from bovine liver was also obtained from Diagnostic Data Inc., Mountain View, California. Glycine, monobasic and dibasic potassium phosphate, riboflavin and Tris hydroxymethyl aminoethane (Tris) were obtained from Sigma Chemical Co.,; ammonium persulphate, glycerol, Cyanogum 41 and sodium hydroxide from Fisher Chemicals; EDTA and borate from J.T. Baker Chemical Co.; N,N,N'N'-tetramethylethylene-diamine (TEMED) and nitroblue tetrazolium (NBT) from Eastman Chemicals; and bromophenol blue-zinc sulfate from Hartman-Leddon Co. All other chemicals were reagent grade.

METHODS

Gel Electrophoresis. Polyacrylamide gel electrophoresis (PAGE) was performed according to Davis[29] and the gels stained for activity and for protein as previously described.[30,31] Tris (0.05 M)-glycine (0.38 M) was diluted 1:5 and 1:10 as indicated in Figure 1, the pH adjusted with 2N NaOH or 2N HCl and used as one buffer system. Another buffer used was Tris (0.21 M)-borate (0.20 M)-EDTA (0.0068) diluted 1:5 and 1:10 as required and the pH adjusted as above. Gels were freshly prepared with 2.0 g Cyanogum 41 dissolved in 20 ml of the apprpriate pH-adjusted buffer, followed by the addition of 20 μl of TEMED and 0.01 g ammonium persulfate. The gels were pre-run for 1 h at a current of 2mA/tube. Samples were applied to the gel in 20% glycerol. Bromophenol blue was used as an indicator dye to estimate the length of electrophoretic runs. The first 10 min of each electrophoresis was done at 0.5 mA/tube and the balance at 2.0 mA/tube.

Enzyme and Biological Activity. The *in vitro* enzyme activity of the different SOD preparations was measured with the photochemical NBT assay as previously described.[30] Their biological effectiveness in protecting irradiated mammalian cells *in vitro* was assessed in murine bone marrow stem cells as previously described[23,24] but with the following modification. Fetal calf serum contains variable amounts of SOD ($\leq$ to 30 μg/ml), so it was avoided except in the cell plating procedure. Bone marrow cells were eluted from femurs of female Swiss white mice in phosphate buffered Hanks Balanced Salts (HBS) at pH 6.4. The erythrocytes were removed by differential lysis in distilled water for 20 sec. The nucleated cells were washed in normal saline and suspended in phosphate buffered HBS (pH 7.4) at a concentration of 2×10^7 cells/ml. The cell suspension was divided into two lots, Millipore^R-filtered SOD was added to one lot to a final concentration of 105 μg/ml while the other lot served as an untreated control. Both lots were incubated for 1 h at 37°C. Superoxide dismutase in inhibited by HBS. With Malta SOD, this inhibition in the absence of phenol red increases the amount of protein per unit of activity by 69%. Therefore, the 105 μg/ml of enzyme was functionally equal to $\approx$ 62 μg/ml of activity, a level roughly equivalent to that present previously[24] if the endogenous fraction in fetal calf serum ($\leqslant$ 30 μg/ml) is added to the 35 μg/ml of exogenous SOD actually used for treatment. Each lot of cells was maintained at 37°C, bubbled with air containing 4% CO_2 and X-irradiated to 50, 100, 150, 200, 250 and 300 rad at a dose rate of 100 rad/min with a 250 kV, 15mA X-ray beam. Survival of stem cells was determined by their in vitro ability to develop colony forming units (CFU_c) as previously described.[23] Colonies with $\geqslant$ 25 cells were counted after 10 and 14 days of incubation at 37° with 100% relative humidity and aeration with purified air and 7.5% CO_2. The procedures of washing and incubating the bone marrow cells in HBS in the absence of fetal calf serum before X-irradiation may remove subsets of cells which stimulate or inhibit colony formation.[32] This loss of cells may affect the plating efficiency and possibly slow down the development of colonies; hence the colony counts after 10 and 14 days, respectively. An additional reason for scoring the number of colonies on the two days was that the 14-day CFU_c progenitor cells are the putative precursors of the earlier appearing CFU_c (Ref.32,33). Thus, the biological effectiveness of the enzyme preparations was assessed sequentially on the basis of the early and late appearing CFU_c. The colonies were counted at 25x magnification after staining with methylene blue.

Chelation. To assess the effect of loss of metal atoms on the electrophoretic pattern of the cupro-zinc enzyme, a solution of the Malta SOD (30 μg/ml) was dialysed against acetate buffer, pH 3.8 containing 1 mM EDTA as previously described.[27] Samples were withdrawn over a period of 24 h, assayed by the NBT assay, subjected to PAGE and stained for protein and enzyme activity as before.

RESULTS

The activities of the five SOD preparations are given in Table 1. Relative to the Truett enzyme, our reference preparation on the basis of previous experience[18,21,23,24] only the Malta SOD from Dr.J.V.Bannister possessed an activity that was statistically higher while one preparation, that from Sigma, was significantly less active.

The PAGE banding patterns in Tris-glycine buffer of the five preparations are shown in Figure 1. In general and irrespective of pH, the number of bands is increased when the buffer is used at 1:5 or 1:10 dilution, rather than at full ionic strength. At both 1:5 and 1:10 dilution of the buffer, the Malta SOD is characterized by two bands at pH 8.2, 8.6 and 9.1, whereas the other four preparations usually show one or more extra bands with greater mobility towards the anode. This suggests greater purity and/or less degradation of the Malta preparation, consistent with its higher activity (Table 1).

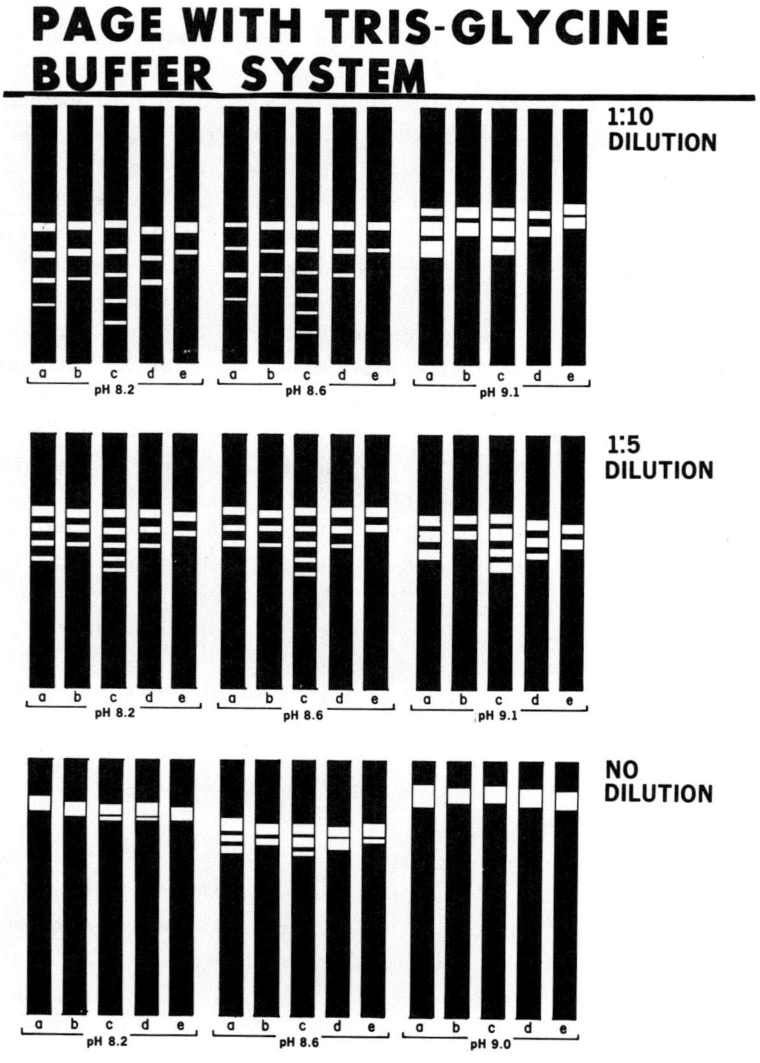

Fig. 1. Electrophoretic banding patterns of different superoxide dismutase (SOD) preparations on PAGE gels. Upper panel: buffer diluted 1:10; middle panel: buffer diluted 1:5; lower panel: buffer undiluted. 1 µg of enzyme applied to each gel. SOD from the following sources was utilized: (a) Truett, (b) Sigma, (c) Miles, (d) Diagnostic Data Inc., (e) Malta.

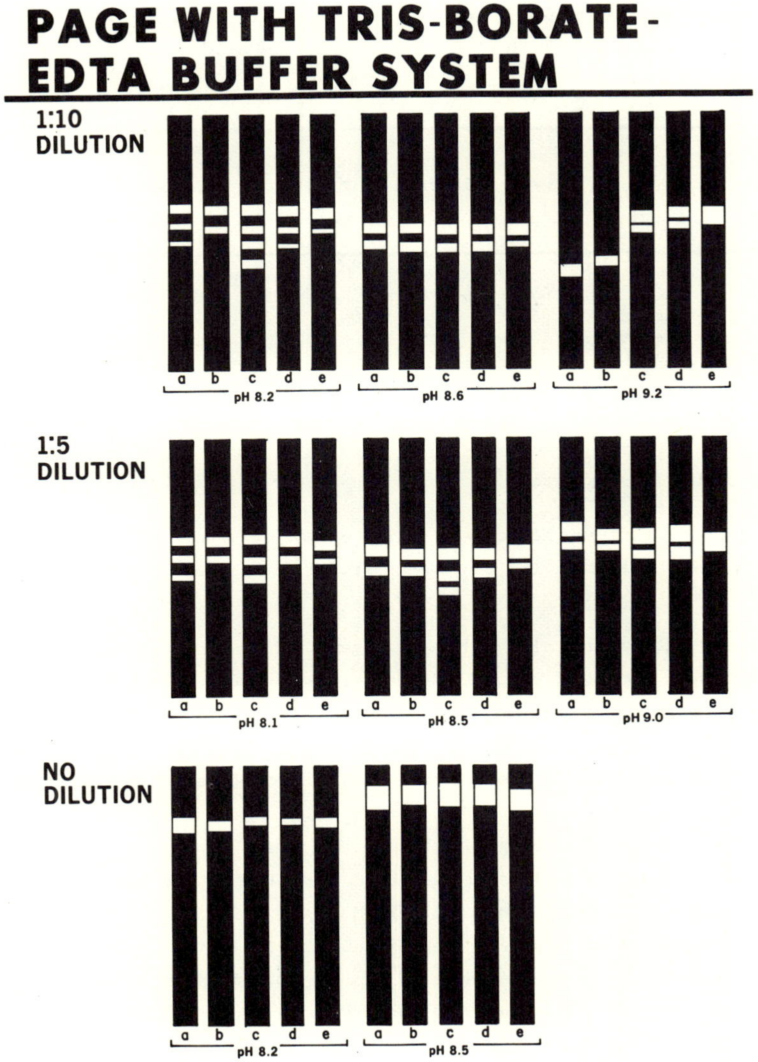

Fig. 2. Electrophoretic banding patterns of different superoxide dismutase (SOD) preparations on PAGE gels. Upper panel: buffer diluted 1:10; middle panel: buffer diluted 1:5; lower panel: buffer undiluted. 1 μg of enzyme applied to each gel. Gels identified in caption to Fig.1.

TABLE 1

COMPARATIVE ENZYME ACTIVITY AND BIOLOGICAL EFFECTIVENESS OF DIFFERENT SUPEROXIDE DISMUTASE PREPARATIONS. BIOLOGICAL EFFICACY DETERMINED WITH MOUSE BONE MARROW STEM CELLS X-IRRADIATED IN VITRO AT 37° IN PHOSPHATE BUFFERED HANKS BALANCED SALTS pH 7.4. CONCENTRATION OF ENZYME IN CELL SAMPLES DURING IRRADIATION WAS 105 μG/ML.

Enzyme		D_o (mean ± σ in rad)[a]					
		After 10 days of growth			After 14 days of growth		
Source	Activity ng/unit ± σ	Control	With enzyme	DMF[b]	Control	With enzyme	DMF[b]
Malta	208 ± 46[c]	67 ± 3	87 ± 4	1.30	68 ± 5	86 ± 2	1.29
Miles	262 ± 32	68 ± 2	84 ± 10	1.23	69 ± 2	92 ± 4	1.33
Diagnostic Data Inc	266 ± 80	60 ± 3	71 ± 1	1.19	71 ± 2	82 ± 2	1.16
Truett	280 ± 63	58 ± 2	73 ± 2	1.20	68 ± 2	89 ± 3	1.31
Sigma	360 ± 34[c]	61 ± 3	68 ± 1	1.11	62 ± 2	78 ± 4	1.26

[a]D_o = radiation dose which decreases the colony forming units by 63%. Rad = 0.01 Gray.

[b]DMF = dose magnifying factor = D_o (with enzyme)/D_o (Control)

[c]Significantly different (5% confidence limit) from Truett SOD, our reference preparation.

The results in Figure 1 also indicate, in general, fewer bands for the other preparations when the pH of the diluted buffer was increased from 8.2 to 9.1. Therefore, both the pH and ionic strength of the buffer are factors which affect the PAGE characteristics of the cupro-zinc enzyme. Polarography and potentiometric titrations have previously been used to demonstrate ionic and pH effects on SOD.[34,35]

The differences in banding patterns of the five SOD preparations are minimized in diluted (1:5 or 1:10) Tris-glycine buffer at pH 9.1. On the contrary, by decreasing the pH of the diluted buffer to 8.6 or 8.2, these differences are accentuated and under this test the Malta SOD is still resolved into only two bands, a consistency characteristic of a highly purified protein.[34] Similar results are observed when the SOD preparations were electrophoresed in Tris-borate-EDTA buffer (Figure 2), with the exception that the resolution of the preparations into more than two bands is more noticeable at pH 8.2 than at 8.6. The extra, faster moving bands are restricted to the preparations supplied by Truett, Miles and Diagnostic Data Inc.

A recent study[36] has suggested that the extra, faster moving bands are due to loss of metal from the enzyme. Since the Malta SOD had the most consistent banding pattern, it was selected for this test and chelated with EDTA for 24 h.

Its activity was determined and the enzyme electrophoresed in 1:10 Tris-glycine buffer, pH 8.6 as before and the gels stained for both protein and activity. The results are summarized in Figure 3. The enzyme activity drops

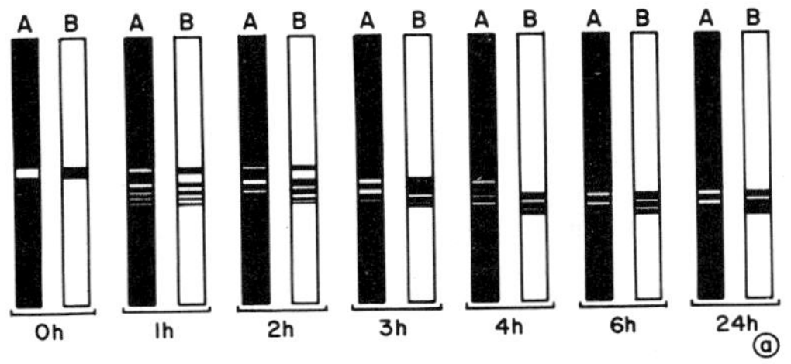

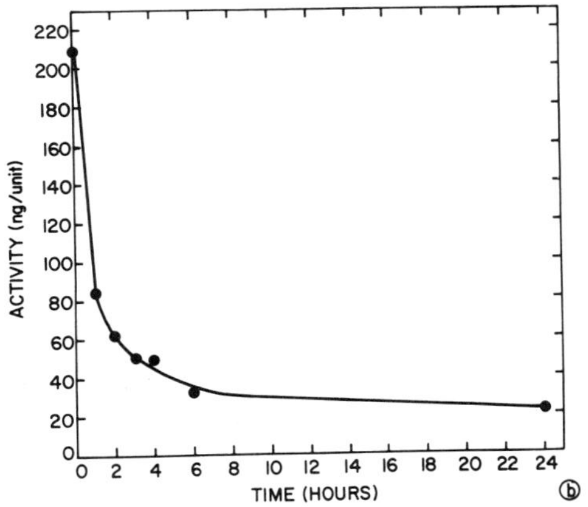

Fig. 3. Electrophoretic patterns on PAGE gels (upper panel) and enzyme activity by NBT assay (lower panel) versus chelation time. PAGE was in Tris-glycine buffer 1:10 dilution at pH 8.6. A: gels stained for enzyme activity; B: gels stained for protein. Amount of enzyme per gel = 1.0 ± 0.1 μg.

rapidly and at 6 and 24 h is down to about 15 and 5%, respectively. The position of the protein and activity bands are co-terminus thus indicating that the two banding patterns are qualitatively equivalent. Their number increases to a maximum of five after 2 h of chelation (corresponding to approximately 73% inhibition of enzyme activity) and then decreases thereafter. There is therefore no quantitative correlation between the PAGE pattern and the residual level of enzyme activity. Nonetheless, the initial loss of enzyme activity is accompanied by an increase in activity and protein bands, consistent with the earlier work.[36] It is noted that the control (0 h) gel (Figure 3) shows three bands whereas, in Figure 1, the Malta SOD exhibited only two. The extra, faster moving band is attributable, by the fore-going argument, to loss of activity through storage (in lyophilized form and with refrigeration) as four months elapsed between the chelation experiment and the earlier ones. This loss of coordinated metal atoms and activity could involve conformational changes and processes related to aging of the enzyme[28,36] and likely accounts for the extra bands shown by the Truett SOD (Figures 1 and 2) which several years ago characteristically gave only two. The Truett preparation is no longer available.

In an earlier publication,[24] it was shown that SOD (Truett) gave good in vitro protection to mouse bone marrow stem cells against X-ray induced loss of colony forming units (CFU_c). This model cell system was therefore used to compare the various SOD preparations as radioprotective agents. Their biological effectiveness, expressed by the dose modifying factor (DMF):D_o(with enzyme)/ D_o (control), is compared in Table 1 on the basis of both the 10- and 14-day CFU_c data (fifth and last columns, respectively). The former comparison is graphically illustrated in Figure 4 which suggests that the biological effectiveness of the different enzyme preparations is linearly related to their enzyme activity and that the Malta SOD is the best, in keeping with its high purity.[34] The latter comparison, however, indicates no correlation as all but one of the five preparations are about equally effective (Table 1, last column), the exception being the enzyme from Diagnostic Data Inc. This preparation is no more effective on the 14 day CFU_c than on those scored on the tenth day. The similarity in effectiveness of the four best preparations is likely the result of their action on the 14 day CFU_c progenitor cells and/or their progeny. The latter alternative is consistent with the enzyme's capacity for post-irradiation protection.[24] Inasmuch as the 10-day CFU_c are derived from the 14-day CFU_c (Ref.33), equality of protection of the latter might be extended to these faster proliferating cells as well. This is apparently not the case (Figure 4) and suggests that protection of the 10-day CFU_c by SOD is more critically dependent on the quality of the enzyme. By this criterion, the Malta SOD appears the best. Thus, this enzyme preparation,

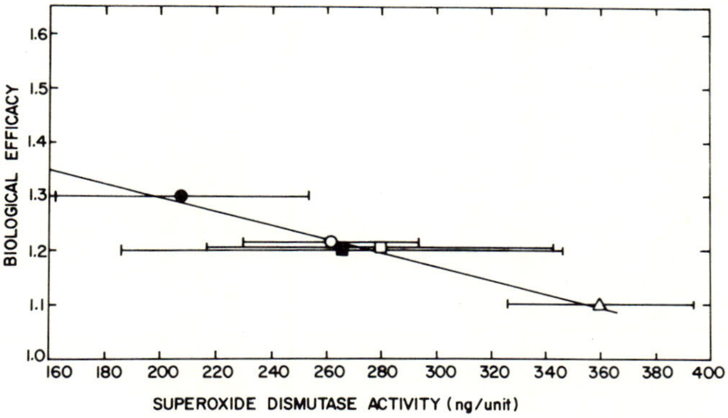

Fig. 4. Biological efficacy of various superoxide dismutase (SOD) preparations against their in vitro activity (Table 1). Biological efficacy expressed as dose modifying factors: D_o (with enzyme/D_o) (control) for X-irradiated mouse bone marrow stem cells. Sources of SOD: Malta (●) Miles (O), Diagnostic Data Inc. (■), Truett (□) and Sigma (Δ). Equation of best-fitted line: $y = 1.56 - 0.0013x$; correlation coefficient -0.98

which has the highest activity and electrophoretically is the most invariant, is also the most effective as a radioprotectant of bone marrow stem cells *in vitro*.

DISCUSSION

Radioprotection of the stem cells, by SOD in the present instance is small relative to that previously found when a DMF of 2.8 was reported.[24] Several reasons are suggested which could account for this difference. The enzyme was several years older and some degradation had occurred as evidenced by the extra bands on PAGE (Figures 1 and 2). As well, the stem cells were irradiated in HBS which, as already noted, inhibits the enzyme consistent with previous observations.[36] The standard HBS recipe contains phenol red which also inhibits the enzyme's activity as measured by the photochemical assay. Elimination of phenol red from HBS did not change the radiation sensitivity of our stem cells in the absence of superoxide dismutase but led to improved dose modifying factors for both the Malta and Truett preparations (Table 2). With the former, a DMF of 1.52 was obtained in respect to the 14-day CFU_c. Further, the changes in medium and cell preparation techniques may have selected a different subset of cells from the composite stem cell population. In this context, it is noted (Table 1) that the mean control D_o values (of 63-67 rad) are substantially

TABLE 2.

EFFECT OF PHENOL RED IN HANKS BALANCED SALTS (HBS) ON THE DOSE MODIFYING FACTOR (DMF) OF SUPEPOXIDE DISMUTASE PREPARATIONS[a] ON MOUSE BONE MARROW STEM CELLS

Enzyme Source	DMF: After 10 days of growth was phenol red in HBS?			DMF: After 14 days of growth was phenol red in HBS?		
	Yes	No	Change	Yes	No	Change
Malta	1.30	1.37	+0.07	1.29	1.52	+0.23
Truett	1.20	1.32	+0.12	1.31	1.34	+0.03
Sigma	1.11	1.18	+0.07	1.26	1.23	-0.03

[a]Concentration 105 μg/ml

lower than 105 ± 6 rad, the previously published value.[24] On the other hand, these lower values may be due to the absence during irradiation of fetal calf serum and its variable content of endogenous SOD. The lower control D_o values mean that the dose modifying factors were calculated at or near the shoulder region of the cell survival curves and this will reduce the DMF attainable. Therefore, the failure to achieve the DMF of 2.83 previously reported[24] was not unexpected. Nonetheless, the difference between it and our present maximum value of 1.52 is being investigated in greater detail.

Aging of SOD is facilitated by chloroform-ethanol and results in the loss of copper and activity.[28,36] All the preparations used in the present comparison were subjected to this apolar solvent system and, therefore, the aging process was most likely a factor in contributing to the differences in their electrophoretic mobility, enzyme activity and biological effectiveness. Residual impurities might also influence the enzyme's radiobiological effectiveness and it is known that even a crystallizable preparation of SOD contains some impurities.[28] Nonetheless, it seems significant that the Malta SOD, which by physicochemical criteria is of recognized purity,[34] is the most effective as a radioprotective agent of stem cells. This radioprotective property is lost when the activity of superoxide dismutase is destroyed.[24,37]

REFERENCES

1. McCord, J.M. and Fridovich, I. (1969) J. Biol. Chem. 244, 6056-6063.
2. Goscin, S.A. and Fridovich, I. (1972) Arch. Biochem. Biophys. 153, 778-783.
3. Rotilio, G., Morpurgo, L., Calabrese, L. and Mondovi, B. (1972) Biochim. Biophys. Acta, 302, 229-235.
4. Rotilio, G., Calabrese, L., Finnazzi Agro, A., Argento-Ceru. M.P., Autuori, F. and Mondovi, B. (1973) Biochim. Biophys. Acta, 321, 98-102.
5. Gregory, E.M. and Fridovich, I. (1973) J. Bacteriol. 114, 1193-1197.
6. Zimmerman, R., Flohe, L., Weser, U. and Hartmann, H.J. (1973) FEBS Lett. 29, 117-120.

7. Webb, L.S., Keele, B.B. and Johnston, R.B. (1974) Infection and Immunity, 9, 1051-1056.
8. Halliwell, B. (1974) New Phytol. 73, 1075-1086.
9. Sawada, Y. and Yamazaki, I. (1973) Biochim. Biophys. Acta, 327, 257-265.
10. Fridovich, I. (1975) Adv. Enzymol. 41, 35-97.
11. Tjioe, S.A., Manian, A.A. and O'Neill, J.J. (1975) Psychopharmacol. Commun. 1, 373-381.
12. Nishikimi, M. (1975) Arch. Biochem. Biophys. 166, 273-279.
13. Halliwell, B. (1975) FEBS Lett. 56, 34-42.
14. Hodgson, E.K. and Fridovich, I. (1975) Biochemistry, 14, 5299-5303.
15. Markossian, K.A. and Nalbandyan, R.M. (1975) Biochem. Biophys. Res. Commun. 67, 870-876.
16. Winterhalter, K.H. (1976) Chimia, 30, 9-12.
17. Masugi, F. and Nakamura, T. (1976) Int. J. Vit. Nutr. Res. 46, 187-191.
18. Petkau, A. and Chelack, W.S. (1974) Int. J. Radiat. Biol. 26, 421-426.
19. Lavelle, F., Michelson, A.M. and Dimitrijevic, L. (1973) Biochem. Biophys. Res. Commun. 55, 350-357.
20. Goscin, S.A. and Fridovich, I. (1973) Radiat. Res. 56, 565-569.
21. Petkau, A., Kelly, K., Chelack, W.S. and Barefoot, C. (1976) Biochem. Biophys. Res. Commun. 70, 452-458.
22. Bartosz, G., Leyko, W., Kedziora, J. and Jeske, J. (1977) Int. J. Radiat. Biol. 31, 197-200.
23. Petkau, A., Chelack, W.S. and Pleskach, S.D. (1978) Life Sic. 22, 867-882.
24. Petkau, A. (1978) Photochem. Photobiol. 28, 765-774.
25. McCord, J.M. (1974) Science, 185, 529-531
26. Proceedings of Active Oxygen and Medicine Conference, A. Autor, ed. January 1979, Raven Press, Honolulu: in press.
27. McCord, J.M. and Fridovich, I. (1969) J. Biol. Chem. 244, 6049-6055.
28. Stansell, M.J. and Deutsch, H.F. (1965) J. Biol. Chem. 240, 4306-4311.
29. Davis, B.J. (1964) Ann. N.Y. Acad. Sci. 121, 404-427.
30. Beauchamp, C. and Fridovich, I. (1971) Anal. Biochem. 44, 276-287.
31. Diezel, W., Kopperschläger, G. and Hofmann, E. (1972) Anal. Biochem. 48, 617-620.
32. Broxymeyer, H.E. and Moore, M.A.S. (1978) Biochim. Biophys. Acta, 516, 129-166.
33. Jacobsen, N., Broxymeyer, H.E., Grossbard, E. and Moore, M.A.S. (1979) Cell Tissue Kinet. 12, 213-226.
34. Lawrence, G.D. and Sawyer, D.T. (1979) Biochemistry, 18, 3045-3050.
35. Rigo, A., Viglino, P., Rotilio, G. and Tomat, R. (1975) FEBS Lett. 50, 86-88.
36. Henry, L.E.A., Palmer, J.M. and Hall, D.O. (1978) FEBS Lett. 93, 327-330.
37. Petkau, A. and Chelack, W.S. (1974) Int. J. Radiat. Biol. 26, 421-426.

STRUCTURAL COMPARISONS OF SUPEROXIDE DISMUTASES*

JOHN E. WALKER+, ANTHONY D. AUFFRET+, CHRIS J. BROCK+** AND HOWARD M. STEINMAN++
+ MRC Laboratory of Molecular Biology, Hills Road, Cambridge CB2 2QH, England;
++ Albert Einstein College of Medicine of Yeshiva University, 1300 Morris Park Avenue, Bronz, N.Y. 10461, USA

Superoxide dismutases have been identified in a wide range of organisms. The enzymes fall into three classes characterised by their metal contents. The Cu/Zn type with one excpetion (see Ref.1 and P.M. Vignais et al., this volume) is restricted to eukaryotes and appears to be unrelated in amino acid sequence to the Mn and Fe types. The latter two classes have been identified in eukaryotes, prokaryotes, mitochondria and chloroplasts and have homologous amino acid sequences (for an earlier review see Harris and Steinman[2]). As part of a detailed structural analysis of prokaryote Mn-superoxide dismutases we have recently determined the complete amino acid sequences of two such enzymes isolated from *Escherichia coli*[3] and from the moderate thermophile *Bacillus stearothermophilus* (this volume). Both enzymes are dimers of identical subunits. This enables us to compare the primary and secondary structures of these Mn-containing enzymes and the Cu/Zn enzyme from bovine erythrocytes.[4] We have also extended earlier partial sequence studies of the Mn/Fe groups[2] to include enzymes isolated from anaerobic organisms.

The manganese enzymes from B. Stearothermophilus and E. coli. (a) Primary structure. The amino acid sequences of these dimeric enzymes are compared in Figure 1. They are very similar, 119 residues (59%) being identical. Conserved amino acids are to be found throughout the polypeptide and especially in the regions encompassing residues 4-43 and 158-196 in the *B. stearothermophilus* sequence. In order to maximise homologies a two-amino acid deletion has been placed at residues 90-91 in the *E. coli* enzyme, a six-amino acid insertion at residues 155-160 and a single insertion at residue 165. These regions cannot be important for catalysis and may well be external regions in the three-dimensional structures of the respective enzymes. No significant resemblance can be found between these sequences and that of the Cu/Zn enzyme from bovine erythrocytes. Can a similar conclusion be made about their secondary structure?

---

*Thispaper is dedicated to the memory of the late J. Ieuan Harris who initiated and directed earlier studies of superoxide dismutase in Cambridge.

**Present address: Department of Biochemistry, University of Bristol, Bristol BS8 1TD, England.

```
                              10                  20                  30                40
B. stearothermophilus   P F E L P A L P Y P Y D A L E P H I D K E T M N I H H T K H H N T Y V T N L N A
E. coli                 S Y T L P S L P Y A Y D A L E P H F D K Q T M E I H H T K H H Q T Y V N N A N A

                                  50                  60                  70                80
                        A L E G H P D L Q N K S L E E L L S N L E A L P E S I R T A V R N N G G G H A N
                        A L E S L P E F A N L P V E E L I T K L D Q L P A D K K T V L R N N A G G H A H

                                  90                 100                 110               120
                        H S L F W T I L S P N G G G E P T G E L A D A I N K K F G S F T A F K D E F S K
                        N S L F W K G L K K G T T L Q - - G D L K A A I E R D F G S V D N F K A E F E K

                                 130                 140                 150               160
                        A A A G R F G S G W A W L V V N N G E L E I T S T P N Q D S P I M E G K - - - -
                        A A A S R F G S G W A W L V L K G D K L A V V S T A N Q D S P L M G E A I S G A

                                 170                 180                 190               200
                        - - - T P I L G L D V W E H A Y Y L K Y Q N R R P E Y I A A F W N V V N W D E V A
                        S G F P I L G L D V W E H A Y Y L K F Q N R R P D Y I K E F W N V V N W D E A A

                        K R Y S E A K A K
                        A R F A A K K
```

'Fig. 1. The amino acid sequence of Mn-superoxide dismutase from *B. stearothermophilus* and *E. coli* B.

(b) Secondary structure. We have attempted to predict the secondary structures of the Mn enzymes from their primary structures by the method of McLachlan.[5] This method applies the laws of statistical inference to amino acid frequency data from proteins of known structure. It calculates the probability of α-helix, β-sheet, bends and irregular structure along the length of a given sequence. We first applied the method to the sequence of the Cu/Zn enzyme from bovine erythrocytes. Crystallographic analysis of this protein has demonstrated that it contains eight antiparallel β-strands folded in a barrel structure.[4] As shown in Figure 2 (part a) seven of these strands and intervening β-bends were correctly predicted by McLachlan's method.

The secondary structures of the Mn enzymes predicted by the same procedure (Figure 2, parts b and c) not surprisingly are very similar to each other (Figure 3) but differ substantially from that of the bovine erythrocyte enzyme. The percentages of α-helix, β-sheet and β-turn predicted in the two Mn enzymes are determined contents in the Cu/Zn enzyme. It appears unlikely that the Mn enzymes contain the eight-stranded β-barrel structure characteristic of the Cu/Zn enzyme. In general concordant secondary structures are obtained for the Mn-containing proteins. The only significant difference is in the region of residues 80-90: an α-helix is weakly predicted (40% probability) in the *E. coli* enzyme and a β-sheet (70% probability) in the *B. stearothermophilus* protein, although the sequences in this region are similar. The insertions in the *E. coli* sequence are to be found in β structures and in the *B. stearothermophilus* enzyme in a β-bend preceding an α-helix (Figure 2).

On closer inspection of the α-helical regions a number of interesting features can be discerned. Thus the first helix (residues 13-31) is almost identical in

TABLE 1

SUMMARY OF PREDICTED CONTENTS OF α-HELIX AND β-STRUCTURES IN MANGANESE SUPEROXIDE DISMUTASE FROM *B. STEAROTHERMOPHILUS* AND *E. COLI*

% Content	*B.stearothermophilus* SOD	*E. coli* SOD	Cu/Zn SOD[a] predicted	Cu/Zn SOD[a] observed[b]
α-helix	45	61	11	3-8
β-sheet	34	27	61	45
β-turn	17	9	26	-
Indeterminate	2	2	1	-

[a] Bovine erythrocyte superoxide dismutase

[b] From the structure of Richardson *et al.* (Ref.4)

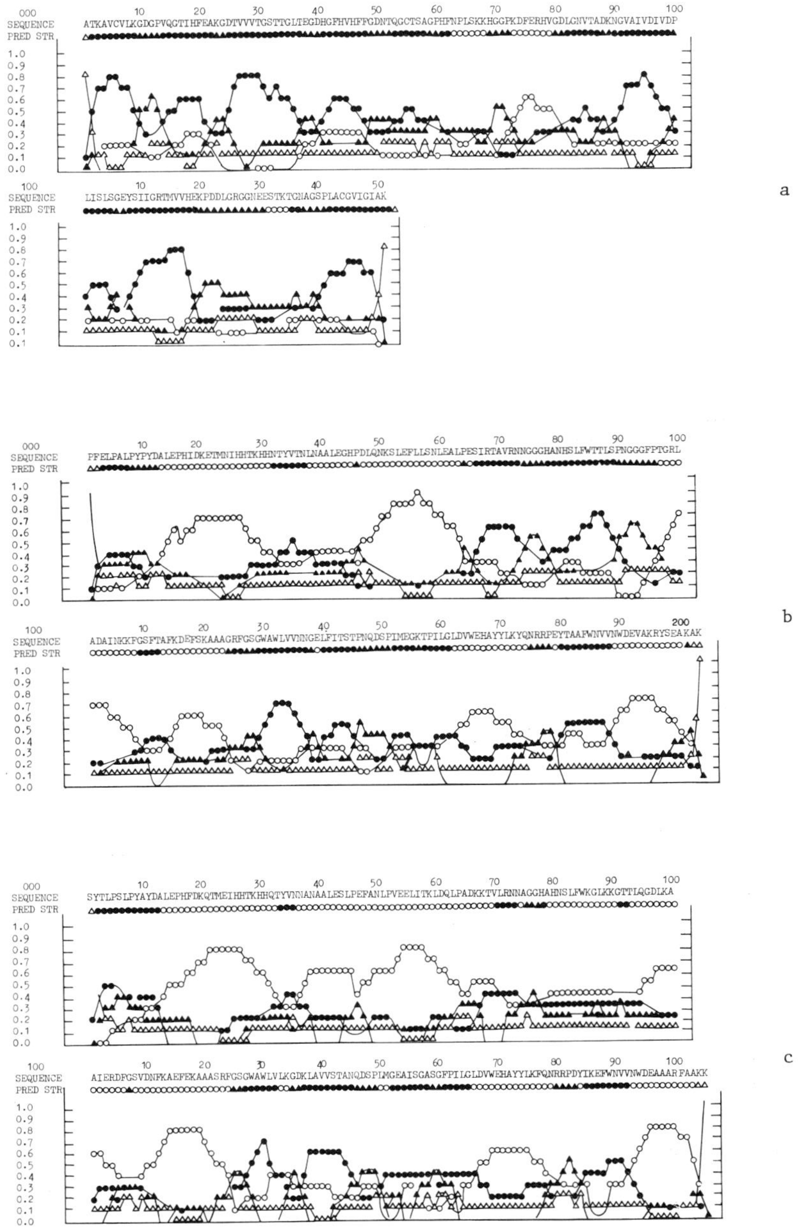

Fig. 2. The predicted secondary structures of (a) the Cu/Zn superoxide dismutase from bovine erythrocytes and the Mn enzymes from (b) B. stearothermophilus and (c) E. coli B. —○— , α-helix; —●— , β-sheet; —▲— , β-turn; —△— indeterminate.

both proteins. It contains four histidine residues and a lysine residue which are in close proximity to each other in the α-helix (Figure 4, part a). Three of the four histidine residues are conserved in all Mn/Fe enzymes sequenced hitherto (Table 2) and nuclear magnetic resonance experiments (A. Cass and H.A.O. Hill, unpublished work) implicate histidine in Mn-binidng. Thus it may be that this cluster of histidine residues provides some of the ligands required to bind the single Mn(III) present in the dimeric B. stearothermophilus enzyme in an octahedral complex

TABLE 2

COMPARISON OF THE AMINO TERMINAL SEQUENCES OF Mn AND Fe SUPEROXIDE DISMUTASES

Organism	Classification[a]		SOD type	Amino acid sequence	Foot-notes
Rhodopseudomonas sphaeroides	Purple nonsulphur	Anaerobic phototrophe	Mn	A P T L P D L P Y A H D A L A A L G M M K E T M E Y H H D I H H K A Y V D N G N K L I A G T	b
Chromatium vinosum	Purple sulphur	"	Fe	M H E L P A L P Y E K N A L F P V I - S A E T I E Y H Y G X X H Q T Y V T N L G	b
Chlorobium thiosulphatophilum	Green sulphur	"	Fe	A Y Z Z P A L P Y A B B A L Z P H I - X A Z T I G F H Y G K H H A A Y V K T Y X G L V	b
Pseudomonas ovalis	Gram -ve rod	Obligate aerobe	Fe	A F E L P P L P Y A H D A L Q P H I - S K E T L E Y H H B K H H N T Y V V N L N N L V P G X T F	b
Thermus aquaticus	"	"	Mn	P Y P F K L P E L G Y P Y E A L E P H I - D A H T M E I H H Q K H H G A Y V T N L N A A L E K Y P Y L Q G A E V Z T L	b,c
Escherichia coli	Entero-bacteria Gram -ve rod	Faculative aerobe	Mn	S Y T L P S L P Y A Y D A L E P H F - D K Q T M E I H H T K H H Q T Y V N N A N A A L E S L P E F A N L P V E E L	d
"	"	"	Fe	S F E L P A L P Y A K D A L A P H I - S A E X I E Y H Y G K	e
Photobacterium leiognathi	Gram -ve rod	"	Fe	A F E L P A L P F A K N A L E P H I - S Q E T L Q Q X X X K H H	b,f
Desulphovibrio desulphuricans	Gram -ve	Strict anaerobe	Fe	S I F V L P D L P Y A K D A L X P K I - S A K T P D P X X G K	g
Bacillus stearothermophilus	Endospore forming Gram +ve rod	Aerobic (sometimes faculative)	Mn	P F E L P A L P Y P Y D A L E P H I - D K E T M N I H H T K H H N T Y V T N L N A A L E G H P D L Q N K S L E E L	h,i
Plectonema boryanum	Cyano-bacteria	Photo-synthetic	Fe	A Y T Q P P L P F P K D A L E P Y X M X A E X F	j
Spirulina platensis	"	"	Fe	A F E L P S L P F D Q D A L E S S K M S A N T L S Y H H G K H H A A Y V K N L N A A I Z G T B M A B	b
Saccharomyces cerevisiae	Fungi	Faculative anaerobe mitochondrial enzyme	Mn	K V T L P D L K W D F G A L E P Y I - S G Q I N E L H Y T	j
Chicken liver mitochondria	-	Obligate aerobe	Mn	K H T L P D L P Y D Y G A L E P H I - S A E I M Q L H X X K	e
Human liver mitochondria	-	"	Mn	K H S L P D L P Y D Y G A L E P H I - N A Q I M Q L H X S K	j

Footnotes

a Bergey's Manual of Determinative Bacteriology (1974).

b Walker, Auffret, Northrop and Harris (1979)

c Sato and Harris (1977)

d Steinman (1978)

e Steinman and Hill (1973)

f The sequence reported previously by Harris and Steinman (1977) was incorrect, because of copying errors

g Bruschi, Hatchikian, Bonicel, Bovier-Lapierre and Couchoud (1977)

h Bridgen, Harris and Northrop (1975)

i Brock and Walker (1979)

j Steinman, unpublished data

OF Mn AND Fe SUPEROXIDE DISMUTASES

TABLE 2. COMPARISON OF THE AMINO TERMINAL SEQUENCES OF MN AND FE SUPEROXIDE DISMUT

[a]Ref.24 [b]Ref.25 [c]Ref.26 [d]Ref.3 [e]Ref.7

[f]The sequence reported previously by Harris and Steinman (Ref.2) was incorrect, because of copying errors.

[g]Ref.27 [h]Ref.28

[i]C.J. Brock and J.E. Walker, this volume

[j]H.M. Steinman, unpublished data.

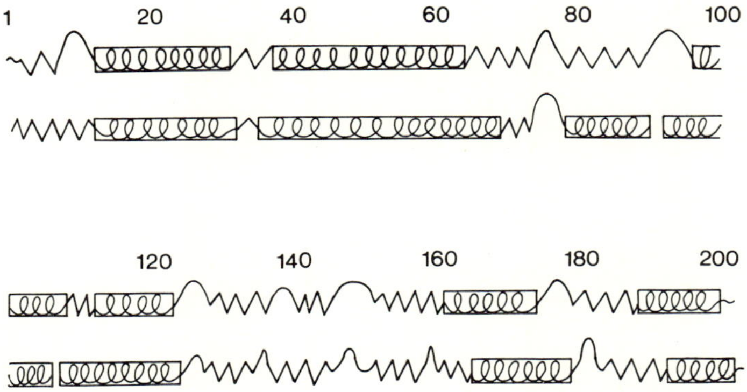

Fig. 3. Comparison of the secondary structures of superoxide dismutase from B. stearothermophilus (upper) and E. coli B (Lower) β-sheet; α-helix; β-bend.

Another histidine residue conserved in both enzymes is in the penultimate helix (residues 162-174, B. stearothermophilus; 166-179, E. coli). This helix is also highly conserved. In addition to the histidine residue, which is next to two acidic residues, the helix contains a patch of four aromatic amino acids (Figure 4, part c).

Another hydrophobic patch is to be found in helix 2 (Figure 4, part b). This extends down one vertical face of the helix, and is adjacent to a face containing a number of charged residues. These features can be taken as additional support for the notion that this region is α-helical and are likely to be important in making contacts between structural elements in the folded protein.

Sequence homology between Mn and Fe enzymes. Amino acid sequence homology in the amino terminal of a wide range of Fe and Mn enzymes has been demonstrated earlier.[2] We have extended these studies to encompass photosynthetic anaerobic microorganisms (Table 2) (oxygen has a bacteriostatic rather than a bactericidal effect on these organisms and the level of superoxide dismutase activity correlates with their tolerance to oxygen[6]). From these sequences a percentage homology matrix has been calculated (Figure 5). From this matrix a number of points can be made. Firstly, there is no clear distinction between the Mn and Fe proteins from prokaryotes. Thus it would not be possible to decide to which group an enzyme belonged on the basis of its amino terminal sequence. Secondly, there is wide variation between bacterial enzymes and the enzymes cannot be grouped

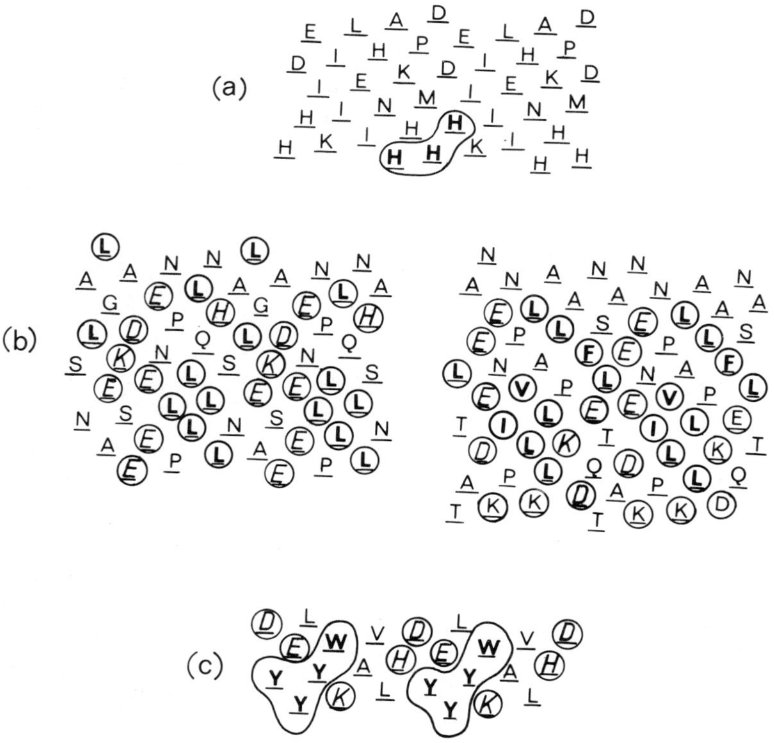

Fig. 4. The sequence of three of the predicted helices of B. stearothermophilus superoxide dismutase plotted on a helix surface with 3.5 residues per turn. Sequence read downwards right to left as if projected onto a sheet of paper wrapped twice round the helix axis and then unrolled. Each residue appears twice. (a) Helix 1 of the B. stearothermophilus enzyme. The encircled area contains three histidine residues in close proximity in the helix. These residues are conserved in all sequenced Mn and Fe enzymes. (b) Helix 2 of B. stearothermophilus (left) and E. coli enzyme (right). Leucine residues are in heavy type and encircled, charged residues in italics and encircled. (c) Penultimate helix of B. stearothermophilus enzyme residues (162-174) showing a cluster of aromatic residues in heavy type and charged residues in italics and encircled.

according to bacterial taxonomic groups. This great variation in bacterial proteins has also been noted in the case of cytochrome c_2 from the Rhodospirillacea. The sequence of this protein varies more in this family alone than in the whole range of eukaryotic mitochondrial c-type cytochromes[29]. Thirdly, the yeast mitochondrial enzyme is more similar to the other mitochondrial enzymes than to the bacterial enzymes.

			1	2	3	4	5	6	7	8	9	10	11	12	13	14	15
1.	Rhodopseudomonas sphaeroides	Mn															
2.	Chromatium vinosum	Fe	42														
3.	Chlorobium thiosulphatophilum	Fe	44	64													
4.	Pseudomonas ovalis	Fe	62	61	61												
5.	Thermus aquaticus	Mn	40	50	46	47											
6.	Escherichia coli	Mn	48	50	51	53	54										
7.	Escherichia coli	Fe	52	78	75	72	45	55									
8.	Photobacterium leiognathi	Fe	45	63	71	72	48	52	59								
9.	Desulphovibrio desulphuricans	Fe	43	55	57	57	41	46	67	52							
10.	Bacillus stearothermophilus	Mn	48	60	54	64	65	67	62	65	50						
11.	Plectonema boryanum	Fe	48	48	62	48	35	48	48	42	45	48					
12.	Spirulina platensis	Fe	52	50	51	51	38	44	52	59	46	50	46				
13.	Saccharomyces cerevisiae (mitochondria)	Mn	31	38	39	38	35	45	39	38	37	38	38	31			
14.	Chicken liver mitochondria	Mn	43	59	54	54	57	57	59	56	52	61	48	43	67		
15.	Human liver	Mn	35	50	54	45	52	52	50	45	46	55	38	34	61	89	

Fig. 5. Percentage similarity matrix for superoxide dismutases. The values are calculated over the sequence lengths shown in Table 2. For the comparison Asx was taken as identical to Asp and Asn, Glx to both Glu and Gln. Deletions were scored as a 21st amino acid and unidentified residues were not scored at all.

Sequence homologies and the endosymbiotic hypothesis. Structural comparisons among the superoxide dismutases have been presented as evidence in favour of the endosymbiotic hypothesis of mitochondrial origin.[2,7,8] Amino terminal sequences of the mitochondrial dismutases are highly homologous to both Mn- and Fe-superoxide dismutases of prokaryotes, yet non-homologous to the amino terminal sequence of the Cu/Zn eukaryote cytosolic dismutase (Table 2). If these amino terminal homologies extend throughout the entire sequence, then one may predict that the mitochondrial and prokaryote forms are similar in their secondary structures as well, with neither bearing any resemblance to the β-barrel, distinctive of the bovine erythrocyte superoxide dismutase. Opponents of endosymbiosis have refuted such arguments, attributing mitochondrial-prokaryote similarities to a common ancestor of the prokaryote and eukaryote cell lines, rather than to an endosymbiotic event.[9,10] In the case of the superoxide dismutases, this refutation may not be valid, because these enzymes are likely to have evolved after the eukaryote/prokaryote divergence, i.e. only when levels of atmospheric oxygen became high enough to necessitate an enzymatic defence against oxygen toxicity.[3,8] The endosymbiotic hypothesis has other implications. Hence in humans the gene for the Mn (mitochondrial) superoxide dismutase is found on chromosome 6 (Ref.11). Similarly the yeast mitochondrial enzyme is encoded in the nucleus[12]. (It appears that mitochondrial DNA encodes only about twelve proteins, some of which are involved in respiration[13,14]). Hence if the endosymbiotic origin of mitochondria is correct then in at least two cases the superoxide dismutase gene must have been transposed from the mitochondrion to the nucleus at some stage. (One mitochondrial ATPase subunit, the DCC binding protein, is encluded in the mitochondrion in yeast and in the nucleus in Neurospora; the amino acid sequence of the proteins is highly conserved[15-18]). The organism must also evolved a mechanism of specifically returning the Mn enzyme (but not the Cu/Zn enzyme) once synthesised to the mitochondrion. It is unlikely that this requirement for specific transport explains the Cu/Zn and the Mn enzymes since selectivity of transport also operates on isozymes of similar structure. Thus porcine heart contains two isozymes of aspartate aminotransferase. One is located in the cytosol, the other in the mitochondrion. The amino acid sequences of the isozymes are highly homologous.[19-22] Yet intact mitochondria are permeable to the mitochondrial but not the cytosol enzyme.[23]

Thus selective transport operates for pairs of enzymes from common ancestors and for proteins exemplified by Cu/Zn and Mn superoxide dismutase which have evolved convergently.

SUMMARY

1. The sequences of the Mn superoxide dismutases from Bacillus stearothermophilus and Escherichia coli B are highly homologous, 59% of the amino acid residues being identical.

2. No sequence homology between these enzymes and the Cu/Zn enzyme from bovine erythrocytes is apparent.

3. Secondary structures of the manganese proteins suggest that the eight-stranded β-barrel of the Cu/Zn enzyme is not present in the Mn enzymes.

4. The latter are predicted to contain a number of α-helices. In the amino terminal helix three highly conserved histidine residues may be clustered together.

5. Amino terminal sequences of a wide variety of Mn/Fe superoxide dismutases do not permit the Mn and Fe types to be distinguished.

6. Bacterial enzymes in the Mn/Fe class show wide sequence variability which cannot be related to taxonomic groups.

ACKNOWLEDGEMENTS

We thank the following colleagues for generous gifts of superoxde dismutases: K. Asada, Chlorobium thiosulphatophilum and Chromatium vinosum; J. Lumsden, Rhodopseudomonas sphaeroides and Spirulina platensis; A.M. Michelson, Photobacterium leiognathi; F. Yamakura, Pseudomonas ovalis.

REFERENCES

1. Puget, K. and Michelson, A.M. (1974) Biochem. Biophys. Res. Commun. 58, 830-834.
2. Harris, J.I. and Steinman, H.M. (1977) in Superoxide and Superoxide Dismutases, Michelson, A.M., McCord, J.M. and Fridovich, I., eds., Academic Press, New York, pp.225-230.
3. Steinman, H.M. (1978) J. Biol. Chem. 253, 8708.8720.
4. Richardson, J.S., Thomas, K.A., Byron, H.R. and Richardson, D.C. (1975) Proc. Nat. Acad. Sci. USA, 72, 1349-1353.
5. McLachlan, A.D. (1977) Inter. J. Quantum Chem. 12, Suppl. 1, 371-385.
6. Hewitt, J. and Morris, J.G. (1975) FEBS Lett. 55, 282-285.
7. Steinman, H.M. and Hill, R.L. (1973) Proc. Nat. Acad. Sci. USA, 70, 3725-3829.
8. Fridovich, I. (1974) Life Sci. 14, 819-826.
9. Uzzell, T. and Spolsky, C. (1974) Amer. Scientist 62, 334-343.
10. Mahler, H.R. and Raff, R.A. (1975) Int. Rev. Cytol. 43, 1-124.
11. Creagan, R., Tischfield, J., Ricciuti, F. and Ruddle, F.H. (1973) Humangenetik, 20, 203-209.
12. Weisiger, R.A. and Fridovich, I. (1973) J. Biol. Chem. 248, 4793-4796.
13. Attardi, G. (1979) Proc. ICN/UCLA Meeting, in press.
14. Schatz, G. and Mason, T.L. (1974) Annu. Rev. Biochem. 43, 51-87.
15. Hensgens, L.A.M., Grivell, L.A., Borst, P. and Bos, J.L. (1979) Proc. Nat. Acad. Sci. USA, 76, 1663-1667.
16. Macino, G. and Tzagoloff, A. (1979) J. Biol. Chem. 254, 4617-4623.
17. Sebald, W., Graf, Th. and Wild, G. (1976) in Genetics and Biogenesis of Chloroplasts and Mitochindria, Bücher, Th. et al., eds., Elsevier/North-Holland, Amsterdam, pp.167-174.
18. Sebald, W. and Wachter, E. (1978) in 29th Mosbacher Colloquium on Energy Conservation in Biological Membranes, Schäfer, G. and Klingenberg, M., eds., Springer-Verlag, Berlin, pp.228-236.
19. Ovchinnikov, Yu A., Egorov, C.A., Aldanova, N.A., Feigina, M. Yu, Lipkin, V.M., Abdulaev, N.G., Grishin, E.V., Kiselev, A.P., Modyanov, N.N., Braunstein, A.E., Polyanovsky, D.L. and Nosikov, V.V. (1973) FEBS Lett. 29, 31-33.
20. Doonan, S., Doonan, H.J., Hanford, R., Vernon, C.A., Walker, J.M., Airaldi, L.P. da S., Bossa, F., Barra, D., Carloni, M., Fasella, P. and Riva, F. (1975) Biochem. J. 149, 497-506.
21. Kagamiyama, H., Sakakibara, R., Wada, H., Tanese, S. and Morino, Y. (1977) J. Biochem. Tokyo, 82, 291-294.

22. Barra, D., Bossa, F., Doonan, S., Fahmy, H.M.A., Hughes, G.J., Kakoz, K.Y., Martini, F. and Petruzzelli, R. (1977) FEBS Lett. 83, 241-244.
23. Marra, E., Doonan, S., Sacoone, C. and Quagliariello, E. (1978) Eur. J. Biochem. 83, 427-435.
24. Buchanan, R.E. and Gibbons, N.E. (1974) Bergey's Manual of Determinative Bacteriology, Williams and Wilkins, Baltimore.
25. Harris, J.I., Auffret, A.D., Northrop, F.D. and Walker, J.E. (1979), submitted for publication.
26. Sato, S. and Harris, J.I. (1977) Eur. J. Biochem. 73, 373-381.
27. Bruschi, M., Hatchikian, E.C., Bonicel, J., Bovier-Lapierre, G. and Couchoud, P. (1977) FEBS Lett. 76, 121-123.
28. Bridgen, J., Harris, J.I. and Northrop, F.D. (1975) FEBS Lett. 49, 392-395.
29. Ambler, R.P., Daniel, M., Hermoso, J., Meyer, T.E., Bartsch, R.G. and Kamen, M.D. (1979) Nature, 278, 659-660.

STUDIES ON THE PROTEIN STRUCTURE OF COPPER-ZINC SUPEROXIDE DISMUTASES

F. BOSSA,[+] D. BARRA,[+] F. MARMOCCHI,[+] F. MARTINI,[+] G. ROTILIO,[+] J.V. BANNISTER[++] AND W.H. BANNISTER[+++]
+ Institutes of Biological Chemistry, Universities of Rome and Camerino and C.N.R. Centre for Molecular Biology, Rome, Italy; ++ Inorganic Chemistry Department, University of Oxford, Oxford OX1 3QR, England; +++ Nuffield Department of Clinical Biochemistry, Radcliffe Infirmary, Oxford OX2 6HE, England.

Do active monomers of copper-zinc superoxide dismutase actually exist? The superoxide dismutases that are primarily localized in the cytoplasm of eukaryotic organisms are dimers composed of two identical subunits of approximately 16,000 molecular weight, each containing one copper and one zinc atom.[1] All the proteins of this class have very similar catalytic and spectroscopic properties related to the substantial identity of the structure of their bimetal clusters, which include the active copper[2] and the subsidiary zinc.[3] The protein moiety has been relatively neglected, even though it is clear that its properties are essential to the enzymic function. The enzyme from bovine erythrocytes is the most studied one in this respect, since its amino acid sequence[4] and high-resolution X-ray structure[5] have been determined. Moreover a great deal of evidence has been accumulated for this enzyme, indicating that the subunits interact between each other in a number of experimental conditions.[6] Concerning the influence of the dimeric assembly on the property of the enzyme there are in the literature apparently conflicting reports on the possible formation of active monomers by the effect of various dissociation agents.[7-9] Divergent interpretatations of the pertinent experimental data have led to controversial ideas on the relationships between "native" dimeric arrangement and catalytic activity in this class of proteins and on the possibility that such a relationship could be regulated by mechanisms to some extent common to copper-zinc superoxide dismutases from widely distant organisms. For example, SDS has been used to attempt dissociation of these enzymes with different results; the bovine enzyme has been reported to be resistant to dissociation, while formation of practically inactive monomers has been described following treatment with SDS of isoenzyme I from wheat germ.[7] These monomers revert to fully active dimers on removal of SDS, contrary to those obtained with all copper-zinc superoxide dismutases by treatment with SDS and mercaptoethanol. However, it is not ruled out that the absence of activity of the monomer is due to specific binding of SDS. This hypothesis is implicitly supported by work on swordfish superoxide dismutase, where evidence for dissociation in 8 M urea, with formation of active protomers, has been presented on the basis of gel filtration experiments.[8]

As the latter work was not itended to define whether this behaviour was unique to the swordfish enzyme or represented a more general property of copper-zinc superoxide dismutases, a more detailed study was undertaken of the effects of urea on gel-filtration behaviour and on some active site-linked parameters of two different superoxide dismutases, widely distant in the phylogenetic tree, namely, the yeast and the bovine enzymes. Preliminary results on the gel-filtration experiments have already been published;[9] the elution volumes of the two proteins from an Ultrogel AcA 54 column equilibrated in 0.1 M phosphate buffer, pH 7.4, and calibrated with the usual set of standard proteins, are commensurate with a molecular weight corresponding to the dimer molecule; when the column is equilibrated with 8 M urea, either with or without β-mercaptoethanol, both proteins elute at a position corresponding to the monomer. Dependence of the elution volumes on the urea concentration shows that this parameter changes abruptly between 6 and 7 M urea for both proteins (Figure 1).

Decreasing the urea concentration by either dilution or dialysis results in complete reconversion to the elution volumes corresponding to the dimer structure. These experiments demonstrate that the gel-filtration behaviour in urea of different superoxide dismutases is identical also in cases representative of those enzymes that have been reported to be resistant (the bovine enzyme, Ref.10) or sensitive (the yeast enzyme, Ref.11) to the dissociative action of SDS alone. However, as it has been already pointed out,[9] any conclusion on the existence of monomers of these enzymes in urea is hampered by consideration of the reliability of the gel-filtration technique in assessing molecular weights of proteins. In fact, special caution is indicated in the case of copper-zinc superoxide dismutase, a protein which has been shown to be particularly resistant to denaturing agents[12] in comparison with other proteins routinely used as molecular weight standards. This caution applies to the interpretation of the significant amount of enzyme activity measured in the presence of 8 M urea[8] and which has been attributed to substantial identity of catalytic power for monomers and dimers.

With this limitation in mind a study of the effect of urea on some active site-linked parameters of bovine and yeast enzymes was undertaken. The results obtained, already published elsewhere[13] will be briefly summarised here. Activity measurements indicate that addition of 8 M urea has differential effects on the catalytic efficiency of the two proteins: while the bovine enzyme is practically unaffected (even after long incubation in the presence of urea), the yeast enzyme undergoes a slow inactivation (a decrease to 0.1% activity

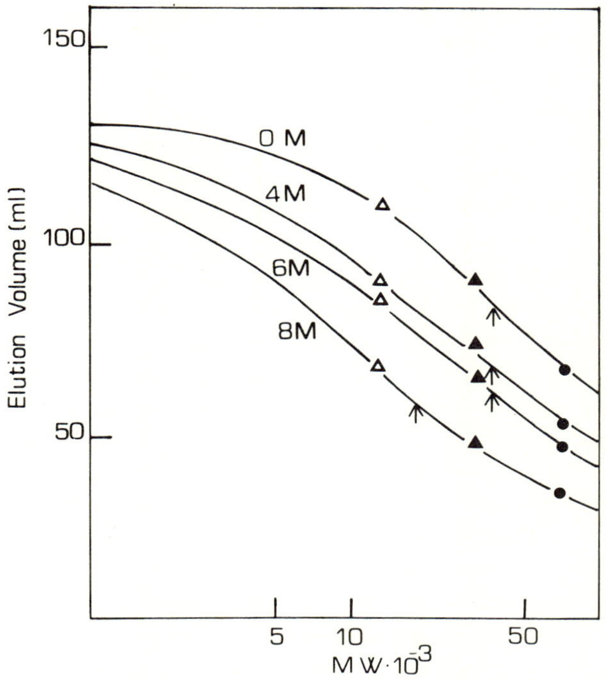

Fig. 1. Gel-filtration chromatography of bovine and yeast copper-zinc superoxide dismutase (↑) in the presence of various concentrations of urea (indicated in the Figure). The elution volumes of only three protein standards - cytochrome c (Δ), carbonic anhydrase (▲) and bovine serum albumin (●) - are reported for simplicity. Other experimental conditions as in Ref.9.

after 4 days) during incubation in urea. Enzymatic activity is fully recovered after extensive dialysis. Parallel changes of EPR spectra of the two enzymes are observed: addition of urea brings about an immediate, slight change in the copper signal of the bovine enzyme, whereas a more pronounced immediate change followed by a slow modification towards a drastically altered spectrum takes place in the yeast enzyme. Also these spectral modifications are fully reversible. These results indicate that the yeast enzyme in 8 M urea undergoes a more substantial unfolding which would seem incompatible with maintenance of the dimeric structure. However, the identical behaviour of the two enzymes in gel filtration experiments makes this argument not decisive.

As an independent approach, production of molecular hybrids has been attempted. Incubation in 8 M urea of either bovine or yeast enzyme with another superoxide dismutase characterized by a suitably different elctrophoretic mobility (such as porcine or wheat germ enzyme) followed by slow removal of urea by dialysis, leads to the formation of active hybrid dimers, as unequivocally demonstrated by activity-stained polyacrylamide gel electrophoretograms. This is in line with dissociation of dimers in urea. However, it is not a demonstration that dissociation actually occurs, as these results can also be explained by facilitated subunit exchange between dimers in urea.

In conclusion, unequivocal evidence for integrity of the catalytic mechanism of these enzymes in the isolated subunits and even for the actual production of such subunits by urea is still lacking. The only unambiguously demonstrated monomeric form of copper-zinc superoxide dismutase, capable of recovering the native dimeric structure, is inactive.[7]

Primary structure of yeast and human superoxide dismutases. A preliminary study of the chemical structure of the yeast enzyme was undertaken to provide a firmer ground for the comparative analysis of the similarities and differences observed in the response of yeast and bovine superoxide dismutases to urea treatment. The presence of one intrachain disulfide bond per subunit has been demonstrated also in the yeast enzyme, thus confirming the general importance of this feature for the structure and function of copper-zinc superoxide dismutases. Our sequence data included the N-terminal (residues 1-89) and C-terminal (last 20 residues) regions, thus accounting for 72% of the molecule.[13] While this work was in progress, we became aware of the engagement of a Danish group in the determination of the primary structure of this enzyme- in fact the complete structure of the yeast enzyme has now been determined[14] and detailed discussion on the structural comparison with the bovine enzyme need not be given here. It should, however, be pointed out in this context that there is a full conservation of the primary structure of the regions that are expected to be critical for the integrity of the metal-binding sites: in particular the peptides including amino acid side chains that provide the ligand groups to the copper and the zinc and those containing the residues adjacent to the disulfide bridge in the bovine enzyme[5] are practically invariant. Therefore the structural basis for the differential sensitivity to urea of the two enzymes has to be found in more subtle differences and demands a more expanded comparative analysis of variable regions of copper-zinc superoxide dismutases. This project is currently being pursued in our laboratories, and the first step has been the sequence of the human enzyme.

The determination of the primary structure of copper-zinc superoxide dismutase from human erythrocytes has first been undertaken because this protein has very distinctive properties from the structural standpoint. For example, at variance with the bovine enzyme, it contains one tryptophan residue per subunit, no tyrosine, and one group reacting like a persulfide, responsible for an absorption band centered around 330 nm apparently unique to the human enzyme.[15]

A sample of apoenzyme was reduced and carboxymethylated with radioactive iodoacetate and the digested with trypsin. The peptide mixture was fractionated by column chromatography (gel-filtration and ion-exchange); final purification was achieved by paper techniques (chromatography and/or high-voltage electrophoresis). Pure peptides were sequenced mainly by the manual dansyl-Edman technique[16] complemented by auxiliary procedures, such as subdigestion with a variety of proteolytic enzymes (exo- or endopeptidases), determination of the electrophoretic mobility at pH 6.5, etc. Two additional samples of human apo-superoxide dismutase were digested respectively with chymotrypsin and with the protease from S. aureus and the peptide mixtures were processed through a similar purification and analysis procedure. Unfortunately, cleavage with CNBr, successful in the case of the bovine enzyme,[14] could not be attempted for the human protein which does not contain methionine residues.

The structural information obtained from the whole of the various peptides is shown in Figure 2; the peptides could unambiguously be aligned on the basis of homology considerations along the sequence of the bovine enzyme, which is also reported in the Figure for purposes of comparison. Information on about 91% of the sequence is presently available; we have at the moment only preliminary data on a stretch 13 residues long (assuming no insertions or deletions in comparison with the bovine protein) between positions 86-98. However, a number of observations can already be made at this stage of the work. The overall homology with the corresponding portions of the bovine enzyme is very high: only 19% of the residues are substituted.

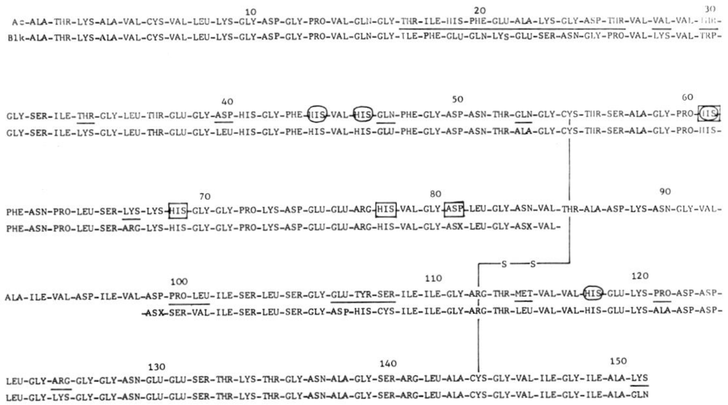

Fig. 2. Comparison of the primary structure of bovine (upper sequence; Ref.4) and human (lower sequence) copper-zinc superoxide dismutases. For the bovine enzyme residues that are ligands of the copper (circles) and the zinc (squares) atoms and the position of the disulfide bond are also indicated. Lines between the two sequence mark residues not homologous.

Again, the more vital residues, that is those liganding metals and those adjacent to the cysteines, which presumably form the structurally essential disulfide bond also in the human enzyme, are conserved. As far as the position of aromatic residues is concerned, the single tryptophan residue of the human enzyme does not substitute for the single tyrosine residue in the bovine enzyme. In fact these two residues occupy positions very distant in terms of primary structure in the two proteins - Trp 30 and Tyr 108 in the human and in the bovine enzyme, respectively - although the positions are not that distant in the folded molecule.[5]

The human enzyme contains 4 half-cysteine residues per subunit instead of 3 in the bovine and the extra cysteine occupies position 109, within a sequence of three residues which is unique to the human enzyme. It is very likely that Cys 109 is the residue responsible for the formation of the persulfide group present in preparations of the human enzyme[15] and that its peculiar location may in some way be related to the modification of its sulfur moiety.[18]

Finally, the distribution of substitutions along the polypeptide chain deserves some comment, since it appears clearly not in uniform: 12 out of the 26 residues found up to now to be substituted are clustered in the tract of sequence from residue 17 to 30. Looking at the model of the three-dimensional folding of the polypeptide backbone of the bovine enzyme,[5] it appears that this "hypervariable" tract is located on the external side of the barrel, near to another loop of the backbone, where the second most significant tract of substituted residues, the above-metnioned sequence of three residues including the cysteine which presumably forms the persulfide group, is located. The substitutions observed in the tract 17-30 seem compatible, according to the predictive rules of Chou and Fasman,[19] with the maintenance of the barrel structure characteristic of the bovine enzyme; if this is the case, the bulky tryptophanyl residue in position 30 should point towards the solvent, well in agreement with chemical modification experiments and fluorescence data.[20]

The presence of such a specific "hypervariable" region is scarcely perceivable when one looks at the sequence of the yeast superoxide dismutase[14] where a number of tracts characterized by a great extent of substitutions can be found, the yeast enzyme being much less homologous to the bovine. The uniqueness of this structural feature is striking when sequences of relatively highly homologous proteins, such as the bovine, the human and the equine enzymes[21] are compared. The finding of this species-specific cluster of residues located on the surface of the protein molecule, especially if confirmed by similar observations on other copper-zinc superoxide dismutases,should be taken into account when considering the antigenic properties of this class of enzymes, which show no cross-reaction between each other (in our preliminary experiments), and the differential behaviour of some of these enzymes towards a number of agents. The case of urea sensitivity has been mentioned here, and a more detailed investigation of various chemical and enzymic effects on a wide comparative scale is in progress in our laboratories.

ACKNOWLEDGEMENTS

W.H.B. and J.V.B. acknowledge the support received from the Wellcome Trust.

REFERENCES

1. Fridovich, I. (1974) Adv. Enzymol. 41, 33-97.
2. Fielden, E.M., Roberts, P.B., Bray, R.C., Lowe, D.J., Mautner, G.N., Rotilio, G. and Calabrese, L. (1974) Biochem. J. 139, 49-60.
3. McAdam, M.E., Fielden, E.M., Lavelle, F., Calabrese, L., Cocco, D. and Rotilio, G. (1977) Biochem. J. 167, 271-274.
4. Steinman, H.M., Naik, V.R., Abernethy, J.L. and Hill, R.L. (1974) J. Biol. Chem. 249, 7326-7338.
5. Richardson, J.S., Thomas, K.E., Rubin, B.M. and Richardson, D.C. (1975) Proc. Nat. Acad. Sci. USA, 72, 1349-1353.
6. Rotilio, G., Rigo, A., Viglino, P. and Calabrese, L. (1977) in Superoxide and Superoxide Dismutases, Michelson, A.M., McCord, J.M. and Fridovich, I., eds., Academic Press, New York, pp.207-214.
7. Rigo, A., Marmocchi, F., Cocco, D., Viglino, P. and Rotilio, G. (1978) Biochemistry, 17, 534-537.
8. Bannister, J.V., Anastasi, A. and Bannister, W.H. (1978) Biochem. Biophys. Res. Commun. 81, 469-472.
9. Marmocchi, F., Venardi, G., Bossa, F., Rigo, A. and Rotilio, G. (1978) FEBS Lett. 94, 109-111.
10. Keele, B.B. Jr., McCord, J.M. and Fridovich, I. (1971) J. Biol. Chem. 246, 2875-2880.
11. Goscin, S.A. and Fridovich, I. (1972) Biochim. Biophys. Acta, 289, 276-283.
12. Forman, H.J. and Fridovich, I. (1973) J. Biol. Chem. 248, 2645-2649.
13. Barra, D., Bossa, F., Marmocchi, F., Martini, F., Rigo, A. and Rotilio, G. (1979) Biochem. Biophys. Res. Commun. 86, 1199-1205.
14. Haseman, V., Owe-Petersen, K., and Johansen, J.T., Carlsberg Res. Commun. in press.
15. Calabrese, L., Federici, G., Bannister, W.H., Bannister, J.V., Rotilio, G. and Finazzi-Agro, A. (1975) Eur. J. Biochem. 56, 305-309.
16. Hartley, B.S. (1970) Biochem. J. 119, 805-822.
17. Evans, H.J., Steinman, H.M. and Hill, R.L. (1974) J. Biol. Chem. 249, 7315-7325.
18. Barra, D., Martini, F., Bossa, F., Rotilio, G., Bannister, J.V. and Bannister, W.H. (1978) Biochem. Biophys. Res. Commun. 81, 1195-1200.
19. Chou, P.Y. and Fasman, G.D. (1974) Biochemistry, 13, 222-245.
20. Finazzi-Agro, A., Cocco, D., Calabrese, L., Bannister, W.H. and Bossa, F. (1977) Int. J. Biochem. 8, 389-393.
21. Ammer, D. and Lerch, K., this volume.

PRIMARY STRUCTURE OF HORSE LIVER SUPEROXIDE DISMUTASE

D. AMMER AND K. LERCH
Biochemisches Institut der Universität Zürich, Zürichbergstrasse 4,
CH-8021 Zürich, Switzerland.

INTRODUCTION

Superoxide dismutase catalyzes the dismutation of the O_2^- radical to molecular oxygen and hydrogen peroxide.[1] The enzyme has been found in all aerobic organisms where it is believed to cope with the toxic effects of oxygen by scavenging O_2^- radicals.[2,3] Almost all eukaryotic organisms contain a cytoplasmatic superoxide dismutase which is a dimer with a subunit molecular weight of approximately 16,000 containing one copper and one zinc atom per subunit. The amino acid sequence and the X-ray structure of the bovine erythrocyte superoxide dismutase have been determined.[4,5] The three-dimensional structure of this molecule shows the active site metal center to be located on the outside of a "barrel" of eight antiparallel strands of β-pleated sheet. The two metal ions are held 6 Å apart and are ligated by 6 histidines and 1 aspartate in a peculiar fashion sharing histidine 61 as a bridging ligand.[5]

The physical chemical properties of the copper and zinc-containing superoxide dismutases from a wide variety of sources are known to be rather similar.[2] Since the similarity of these properties is expected to be reflected in the primary structure especially in the protein regions which are critically involved in maintaining the integrity of the active site and the dimeric structure, we have started to investigate the structural and functional relationship within this class of superoxide dismutase by comparative amino acid sequence analysis. In this report we present the complete primary structure of copper-zinc superoxide dismutase from horse liver, whose enzymatic properties have been described earlier.[6] The comparison of the sequence of this enzyme with that of bovine erythrocyte and the recently reported primary structure of yeast superoxide dismutase[7] reveals a high degree of sequence homology among the three enzymes indicating that they arose from a common evolutionary ancestor.

MATERIALS AND METHODS

Horse liver superoxide dismutase was obtained as a side-fraction of metallothionein preparations according to Bühler and Kägi.[8] The enzyme was further purified by gel filtration on Sephadex G-75 and DEAE-cellulose chromatography.

The molecular and enzymatic properties of the purified horse liver superoxide dismutase were found to be identical to those reported earlier[6] using a different purification procedure. Aposuperoxide dismutase was obtained from the native enzyme by incubation in 50% formic acid and 10 mM EDTA and subsequent gel filtration on Sephadex G-25 in 7% formic acid. The metal-free protein was carboxymethylated as described for bovine superoxide disumtase[4] and cleaved with cyanogen bromide in 70% formic acid for 24 hr at 25°C. The resulting fragments were separated by gel filtration on Sephadex G-50 superfine in 7% formic acid. Enzymatic digestions with trypsin, pepsin, thermolysin and subtilisin were performed by established procedures. The resulting peptides were fractionated on a column (0.9 x 20 cm) of Beckman M-72 resin at 55°C using pyridine/acetate buffer. The peptides were further purified by high-voltage electrophoresis on paper at pH 1.9 or 6.5 or by gel filtration on Sephadex G-25 fine in 50% acetic acid. Depending on the size and composition of the pure fragments, sequence analyses were carried out by either automated or manual Edman degradation techniques. Automated sequence analysis was performed on a Beckman sequencer model 890B (updated) using the Edman and Begg procedure[9] as modified by Hermoson et al.[10] The phenylthiohydantoin derivatives were identified by high pressure liquid chromatography[11] and by amino acid analysis after reconversion to the free amino acids by hydrolysis with 5.7 M HCl containing 0.1% $SnCl_2$ (Ref.12). Amino acid analyses were performed on a Durrum (model D-500) amino acid analyzer. The identification of the blocked amino terminus was carried out by hydrazinolysis and thin-layer chromatography.[13] ^{14}C-iodoacetic acid was obtained from Amersham, England.

RESULTS AND DISCUSSION

The strategy of the amino acid sequence determination of horse liver superoxide dismutase is schematically outlined in Figure 1. Cleavage of the protein with cyanogen bromide yielded three major fragments and one minor overlap peptide. Automatic sequence analysis of the cyanogen bromide fragments CB 2 and CB 3 allowed the unambiguous sequence determination of 14 and 24 residues, respectively. CB 1 was not susceptible to Edman degradation due to the presence of a blocking group at the N-terminus. The amino composition of the overlap fragment CB 2-3 was identical with the sum of the corresponding peptides CB 2 and CB 3 and because homoserine was absent in CB 2 the three cyanogen bromide fragments were aligned in the order CB 1 - CB 3 - CB 2 (Figure 1). Subdigestion of the three cyanogen bromide fragments with trypsin followed by chromatographic separation allowed the isolation in pure form of all expected tryptic peptides with the exception of a fragment of CB 2 (residues 123-128) containing glutamine as the amino terminal amino acid residue. Since this residue is know to undergo cyclisation to pyrrolidone carboxylic acid, the ensuing pyroglutamyl peptide

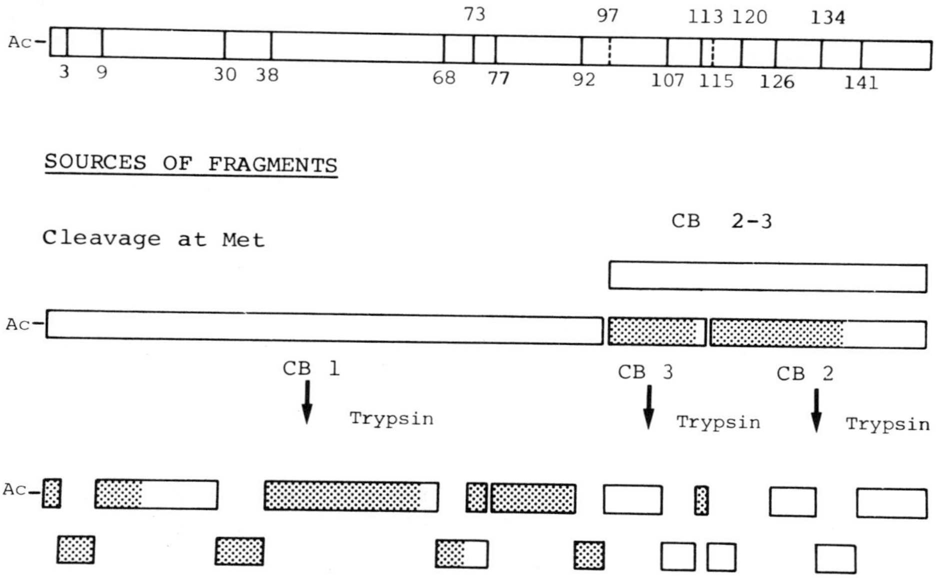

Fig. 1. Strategy of the amino acid sequence determination of horse liver superoxide dismutase. The top bar represents the entire polypeptide chain. Specific cleavage points are indicated by vertical lines drawn inside the bars. Residue numbers below and above the bars refer to the sequence in Fig. 2. The top bar shows the points of cyanogen bromide cleavage at met (broken lines) and of tryptic cleavage (solid lines), The stippled area of each bar represents the portion of the sequence determined by automated or manual Edman degradation of the intact fragment. CB, fragments obtained by cyanogen bromide cleavage.

would be expected to have two negative charges at pH 6.5 and was probably lost on high-voltage electrophoresis. Automated and manual Edman degradation of the intact tryptic peptides furnished more than 50% of the sequence information as shown by the stippled areas in Figure 1. The tryptic fragments which could only be partially sequenced were subdigested further by thermolysin and pepsin (not shown in Figure 1) and the purified peptides subjected to complete sequence analysis.

The tryptic tripeptide (Ala_1 Leu_1 Lys_1) of CB 1, found to be resistant to Edman degradation, was treated for 5 min with 6 N HCl at 110°C. Separation of the peptide mixture by high-voltage electrophoresis at pH 1.9 resulted in free lysine, alanine, leucine and the two peptides (Ala_1 Leu_1) and (Leu_1 Lys_1). These data unequivocally establish the sequence of the tripeptide as Ala-Leu-Lys. The zero net charge of the tryptic tripeptide at pH 6.5 indicates an acid-labile blocking group on the terminal alanine residue. The blocking group of this tripeptide was identified by hydrazinolysis and subsequent chromatographic separation of the hydrazides by thin-layer chromatography. A comparison of the R_f-value of the unknown acylhydrazide with those of synthetic formyl and acetyl hydrazide unequivocally established the sequence of the blocked tripeptide as N-Acetyl-Ala-Leu-Lys.

The complete amino acid sequence of horse liver superoxide dismutase is shown in Figure 2. The amino acid composition calculated from the sequence data is in good agreement with the one reported by Albergoni and Cassini (Ref. 6, Table 1). Figure 2 contains also the primary structures of bovine erythrocyte[4] and yeast superoxide dismutase[7] for comparatibe purposes. A comparison of the three sequences (Figure 2) reveals a number of interesting structural features which deserve further comments.

	Position	Sequence
Horse liver	1	AC-A L K A V C V L K G D G P V H G V I H F E Q Q Q E G G P V V
Bovine erythrocyte		AC-A T K A V C V L K G D G P V Q G T I H F E A K G D T V V
Yeast		V Q A V A V L K G D A G V S G V V K F E Q A S E S E P T T
	31	L K G F I E G L T K G D H G F H V H E F G D N T Q G C T T
		V T G S I T G L T E G D H G F H V H Q F G D N T Q G C T S
		V S Y E I A G N S P N A E R G F H I H E F G D A T N G C V S
	60	A G Q H F N P L S K K H G G P K D E E R H V G D L G N V T A
		A G P H F N P L S K K H G G P K D E E R H V G D L G N V T A
		A G P H F N P F K K T H G A P T D E V R H V G D M G N V K T
	90	D E N G K A D V D M K D S V I S L S G K H S I I G R T M V V
		D K N G V A I V D I V D P L I S L S G E Y S I I G R T M V V
		D E N G V A K G S F K D S L I K L I G P T S V V G R S V V I
	120	H E K Q D D L G K G G N E E S T K T G N A G S R L A C G V I
		H E K P D D L G R G G N E E S T K T G N A G S R L A C G V I
		H A G Q D D L G K G D T E E S L K T G N A G P R P A C G V I
	150	G I A P
		G I A K
		G L T N

Fig. 2. Comparison of the amino acid sequence of the copper and zinc-containing superoxide dismutases from horse liver (this study), bovine erythrocytes (Ref. 4) and yeast (Ref. 7). Identical residues are indicated by boxes. The different sequences were aligned such as to obtain maximal homology. One - letter notation for amino acid sequences is used.

TABLE 1

AMINO ACID COMPOSITION OF S-CARBOXYMETHYL HORSE LIVER SUPEROXIDE DISMUTASE[a]

Amino Acid	Sequence	Acid hydrolysis
Lysine	13	13.3
Histidine	10	9.0
Arginine	3	3.3
S-CM Cysteine	3	2.9
Aspartic acid	17	17.7
Threonine	8	8.2
Serine	7	7.7
Glutamic acid	16	17.0
Proline	5	5.6
Glycine	26	24.3
Alanine	8	9.5
Valine	14	13.4
Methionine	2	2.0
Isoleucine	7	6.4
Luecine	9	9.7
Tyrosine	–	0.2
Phenylalanine	5	5.2
Tryptophan	–	0.1

[a] Residues per molecule of subunit calculated from the sequence or from amino acid analysis based on a subunit molecular weight of 16,500 (Ref. 6).

Contrary to yeast superoxide dismutase which has a free amino terminal valine residue, the horse liver and the bovine erythrocyte enzymes are both blocked by an acetyl group. Considering the difference of species and cell origin of these two enzymes it is quite likely that the amino termini of superoxide dismutases from higher eukaryotic organisms are also blocked. The subunit molecular weights of horse liver and yeast superoxide dismutases are slightly higher (16,000 and 15,950, Ref. 7) than the one from the bovine enzyme (15,600), which is reflected by the difference in the total number of amino acid residues. Thus both horse liver and yeast[7] superoxide dismutase are composed of 153 amino acid residues in contrast to 151 residues found for bovine erythrocyte superoxide dismutase.[4]

To obtain maximum sequence homology between all three enzyme forms, the alignment shown in Figure 2 was necessary. The two additional residues found in horse liver and yeast superoxide dismutase had to be inserted between positions 27 and 28 of bovine erythrocyte dismutase. In the case of the yeast enzyme a further insertion (between residues 37 and 38 of horse liver superoxide dismutase) was required because the enzyme lacks the first residue of bovine erythrocyte and horse liver superoxide dismutase. The over-all sequence identity of the three copper-zinc dismutases (as indicated by boxes in Figure 2) was found to be 50.3%. A comparison of the sequence of horse liver superoxide dismutase with that of the bovine erythrocyte enzyme indicated 79.8% sequence identity, whereas 56.2% and 53.6% identity was found for the horse/yeast and the bovine/yeast pairs respectively.

Further it is apparent from Figure 2 that certain regions of the sequence are highly conserved whereas others are variable. The sequence stretch which involves the short insertions mentioned above is especially variable. Small insertions are usually expected to occur on the "loops" connecting adjacent elements of secondary structure[14]. The results of the X-ray structure of bovine erythrocyte superoxide dismutase are in complete agreement with this conjecture, demonstrating the insertions to be found at the loops between β-strands A and B and B and C.[5]

A rather high degree of sequence homology is found in the regions containing the two long loops.[5] The first loop which contributes three ligands to the zinc and one to the copper and involves the disulfide bridge shows more than 65% sequence identity. The same number is calculated for the second exposed loop. As expected, the positions of all six histidine residues and the single aspartic acid residue known to constitute the ligands of the active site metal ions are identical in the three enzymes.[5]

A prominent feature of bovine erythrocyte superoxide dismutase is the high content of glycine and the uniform distribution of this amino acid throughout the sequence. From the X-ray work on this enzyme this residue was shown to be of major importance for the pronounced protein folding. The general validity of the critical involvement of glycine in the architecture of copper/zinc containing superoxide dismutases is substantiated from a sequence comparison of the three representatives in Figure 2. Thus, the positions of 19 glycine residues out of a total of 25 (bovine erythrocyte), 26 (horse liver) and 21 (yeast) were found to be identical in all three enzymes.

In contrast to bovine erythrocyte superoxide dismutase which contains a single tyrosine residue (residue 110, Figure 2) the enzyme from horse liver is devoid of this amino acid. From the X-ray data of the bovine erythrocyte enzyme the tyrosine residue was shown to be exposed to solvent. The non-conservative substitution of the tyrosine by histidine (horse liver) and threonine (yeast) suggests that this residue is not of any functional importance.

Besides the disulfide loop already mentioned above, the C-terminal region has been demonstrated to be also involved in the subunit contact of bovine erythrocyte superoxide dismutase.[5] The high sequence homology of the C-terminal amino acid residues between the three dismutases (Figure 2) indicates similar subunit interactions in the horse liver and yeast enzymes.

In conclusion, the sequence comparison of the three copper-zinc containing superoxide dismutases from horse liver, bovine erythrocytes and yeast reveals an outstanding sequence homology. The conservation of the amino acid residues involved in the metal binding, the maintenance of the intramolecular disulfide bridge and the preponderance of glycine residues in horse liver and yeast superoxide dimutase strongly suggests that these two enzymes possess three dimensional structures similar to the one reported for the bovine erythrocyte enzyme.

ACKNOWLEDGMENT

This work has been supported by the Swiss National Science Foundation (grant no. 3.420.78). We would like to thank Marietta Sutter, Christine Kaspar and Hanspeter Lutz for technical assistance.

REFERENCES

1. McCord, J.M. and Fridovich, I. (1969) J. Biol. Chem. 244, 6049-6055.
2. Fridovich, I (1975) Annu. Rev. Biochem. 44, 147-159.
3. Fridovich, I. (1978) in Frontiers in Physicochemical Biology, Pullman, B. ed. Academic Press, New York, pp.269-281.
4. Steinman, H.M., Naik, V.R., Abernethy, J.L. and Hill, R.L. (1974) J. Biol. Chem. 249, 7326-7338.
5. Richardson, J.S., Thomas, K.A., Rubin, B.H., and Richardson, D.C. (1975) Proc. Natl. Acad. Sci. USA, 72, 1349-1353.
6. Albergoni, V. and Cassini, A. (1974) Comp. Biochem. Physiol. 47 B, 767-777.
7. Johansen, J.T., Petersen, C.O., Martini, B., Haseman, V., and Svendsen, I. (1979) XIth International Congress of Biochemistry, Toronto, Canada (Abstract).
8. Bühler, R.H.O. and Kägi, J.H.R. (1974) FEBS Lett. 39, 229-234.
9. Edman, P. and Begg, G. (1976) Eur. J. Biochem. 1, 80-91.
10. Hermodson, M.A., Ericsson, L.H., Titani, K., Neurath, H., and Walsh, K.A. (1972) Biochemistry 11, 4493-4502.
11. Frank, G. and Strubert, W. (1973) Chromatographia 6, 522-524.
12. Mendez, E. and Lai, C.K. (1975) Anal. Biochem. 68, 47-53.
13. Narita, K. (1958) Biochim. Biophys. Acta, 28, 184-191.
14. Stroud, R.M., Kay, L.M., and Dickerson, R.E. (1971) Cold Spring Harbor Symp. Quant. Biol. 36, 125-140.

SUPEROXIDE DISMUTASE FROM BACILLUS STEAROTHERMOPHILUS: METAL BINDING AND COMPLETE AMINO ACID SEQUENCE*

CHRIS J. BROCK** AND JOHN E. WALKER
MRC Laboratory of Molecular Biology, Hills Road, Cambridge CB2 2QH, England

The thermophilic bacterium Bacillus stearothermophilus contains a typical manganese superoxide dismutase. The enzyme is a dimer of molecular weight 45,000. The subunits are identical and the enzyme contains one atom of manganese (III) per dimeric molecule.[1] The enhanced stability of enzymes from thermophilic microorganisms makes them particularly amenable to structural analysis,[2] and this property has again proved of advantage in the study of this enzyme. Hence, it can be readily crystallised[3,4] and denatured and reconstituted with a range of metal ions. Here we describe the primary structure of the enzyme, an essential adjunct of crystallographic studies and reconstitution experiments which demonstrate a remarkable specificity for manganese.

Metal binding sites. A stable dimeric apo-superoxide dismutase can be prepared by treatment of the B. stearothermophilus enzyme with a chelating agent (EDTA) in acidic 8 M urea (pH 3.7) followed by a multiple step dialysis procedure to return the polypeptide chain to its native conformation.[1] Reconstitution of the apo-protein with Mn, Fe, Co, Ni and Cu was achieved by addition of an excess of a divalent salt of the appropriate transition metal (e.g. $MnCl_2$, $FeCl_2$, $CoCl_2$, $NiSO_4$, $CuCl_2$) under these dissociating conditions followed by renaturation.[5]

The apo-protein and all reconstituted products were indistinguishable from the native enzyme by gel filtration and polyacrylamide-gel electrophoresis under native conditions. The apo-enzyme was devoid of metal and the reconstituted proteins contained one metal atom per dimer in all cases (Table 1). However, only the native enzyme and enzyme reconstituted with manganese had the same specific activity as the native enzyme. The valence states of the metal in some of the reconstituted samples were also determined. Those with manganese had absorption spectra identical to the native enzyme, indicating that oxidation to the tervalent

*This paper is dedicated to the memory of the late J. Ieuan Harris who initiated and directed earlier studies of superoxide dismutase in Cambridge.

**Present address: Department of Biochemistry, University of Bristol, Bristol BS8, England.

form had taken place upon binding. The absorption spectrum of the enzyme reconstituted with iron was similar to that of an iron containing superoxide dismutase and treatment with o-phenanthroline[6] confirmed that the iron was present in the ferric state. Cobalt may also have been oxidised since its absorption spectrum had a broad maximum in the range 540-570 nm indicative of a tervalent complex. It is unlikely that nickel and copper had changed their valence states upon binding however, since their tervalent ions are less stable.[7]

Since treatment with acidic 8 M urea was required to release the manganese from its binding site, the unfolding of the protein in urea with decreasing pH was examined by optical rotatory dispersion and tryptophan fluorescence techniques, as illustrated in Figure 1. These measurements demonstrate that the ordered structure in the molecule broke down gradually in 8 M urea as the pH was lowered from 6 to 3. The increasing exposure of the tryptophanyl residues as the protein denatured was more concerted, however, taking place between pH 6 and pH 4 in 8 M urea. The conditions for maximum tryptophan fluorescence were similar to those which allowed release of the metal from its binding site.

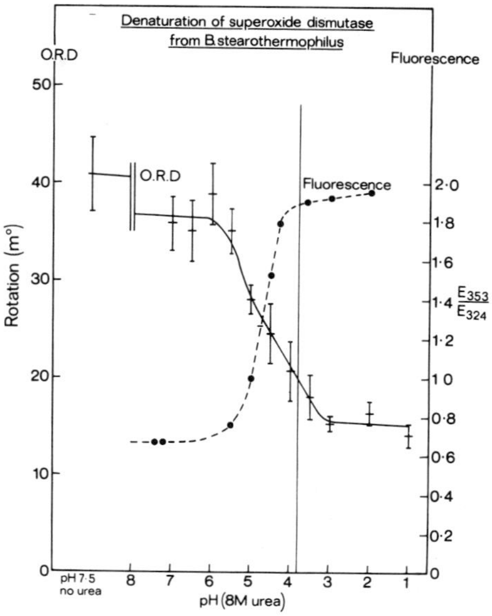

Fig. 1. Denaturation of manganese superoxide dismutase from *B. stearothermophilus*. Optical rotatory dispersion was measured in m° on a Bendix Polarmatic 62. Tryptophan fluorescence was measured as the ratio of emmission strengths at 353 nm and 324 nm on a Perkin-Elmer MPF-3 fluorescence spectrophotometer.

The recconstitution studies demonstrated that the role of the manganese is principally catalytic rather than structural. The apo-protein and all reconstituted samples containing metals other than manganese were indistinguishable from the native enzyme by the criteria of electrophoresis and gel filtration under native conditions, but were nevertheless completely inactive. The similarity of the gross structures of the native and apoenzymes has been confirmed by comparison of their proton nuclear magnetic resonance spectra (A. Cass and H.A.O. Hill, unpublished work). The two spectra were identical except for shifts in a few resonance peaks, most of which were due to histidine residues. This implies that the overall structures of the apo- and holo-enzymes are similar.

These metal binding experiments are indicative of the extent to which the manganese superoxide dismutase has diverged from its likely ancestor, an early form of the iron enzyme. Dougherty _et al._[8] have reported that _E. coli_ grown aerobically elaborates not only iron and manganese containing superoxide dismutases but also a small amount of an active hybrid dimer constituted from an iron type subunit, a manganese type subunit and with iron as the active site metal. Thus it appears that when the two types of enzyme have been subject to similar evolutionary pressures, such as would be exerted by coexistence in the same organism, the two forms remain functionally compatible. In _B. stearothermophilus_, in which only the manganese enzyme has been detected, presumably the enzyme has been subjected to a unique set of evolutionary pressures, and has evolved so that it can bind and oxidise iron to the ferric state but can no longer function catalytically.

The observed stoichiometry of one manganese atom per dimeric molecule coupled with an apparent inaccessibility of the metal demonstrated by the requirement for unfolding of the protein for its removal, suggests that the metal may well be buried between the two subunits. This is required if the dimer is symmetric.

Amino acid sequence. The complete amino acid sequence of _B. stearothermophilus_ superoxide dismutase, consisting of 203 amino acids, has now been determined and is shown in Figure 2. The primary structure was elucidated by sequence analysis of peptides derived from cleavage of the protein with cyanogen bromide, hydroxylamine, trypsin, chymotrypsin and the protease from _Staphylococcus aureus_. Some 98% of the sequence was derived by automated sequence analysis, the remainder being derived by the dansyl-Edman procedure and C-terminal analysis of the protein with carboxypeptidase Y (C.J. Brock and J.E. Walker, in preparation).

This sequence is homologous with the recently and independently determined sequence of the manganese superoxide dismutase from _E. coli_ B. Fifty nine percent (119/203) of the residues are identical. Homology is maximised in the regions corresponding to residues 4-34 and 158-196. A deletion of two

TABLE 1

METAL BINDING STOICHIOMETRIES AND ACTIVITIES OF B. STEAROTHERMOPHILUS MANGANESE SUPEROXIDE DISMUTASE AND THE PRODUCTS OF RECONSTITUTION EXPERIMENTS

Material	Binding stoichiometry[a]	Activity[e]
Superoxide dismutase	0.9 - 1.3[b,c]	$3.9 \times 10^8\ M^{-1}s^{-1}$
Apo-superoxide dismutase	0.0 - 0.02[b,c]	$2 \times 10^6\ M^{-1}s^{-1}$
Superoxide dismutase reconstituted with Mn	0.8 - 1.1[b,c]	$3.5 - 3.7 \times 10^8 M^{-1}s^{-1}$
Superoxide dismutase reconstituted with Fe	0.7[b,d]	Absent
Superoxide dismutase reconstituted with Co	0.9 - 1.3[b,c]	Absent
Superoxide dismutase reconstituted with Ni	0.5 - 1.4[b]	Absent
Superoxide dismutase reconstituted with Cu	0.8[b]	Absent
Superoxide dismutase reconstituted with Mn from Co-reconstituted enzyme	0.9 - 1.3[b,c]	Present

[a]All binding stoichiometries are based upon a molecular weight of 40,000 for the enzyme. Range of values is that observed in a number of experiments.

[b]Determined by atomic absorption analysis.

[c]Determined by neutron activation analysis.

[d]Determined by the method of Massey (1957).

[e]Specific activities were determined by pulse radiolysis (Ref.11) and are expressed as second order rate constants for the turnover of O_2^- under conditions of exponential decay. In other cases activity was determined by a specific stain after polyacrylamide-gel electrophoresis (Ref.12).

amino acids in the E. coli enzyme occurs in the region of residue 90, a six amino acid insertion in the region of residues 155-160 and a single insertion at residue 165. (All residue numbers refer to the B. stearothermophilus sequence). Thus these particular regions may well be external in the three-dimensional structure of the enzyme and it is likely that they play no important role in catalysis. In contrast there is no detectable homology between the primary structure of B. stearothermophilus superoxide dismutase and that of the copper-zinc enzyme from bovine erythrocytes[10] indicating that the enzymes have evolved in a convergent manner.

This conclusion is borne out by a comparison of the secondary structures of the two enzymes described by Walker et al. (this volume). Its confirmation rests upon the determination of the three-dimensional structure which is currently in progress elsewhere.

```
                                                   10
H2N-Pro-Phe-Glu-Leu-Pro-Ala-Leu-Pro-Tyr-Pro-Tyr-Asp-Ala-Leu-Glu
                           20                                   30
  -Pro-His-Ile-Asp-Lys-Glu-Thr-Met-Asn-Ile-His-His-Thr-Lys-His
                                                   40
  -His-Asn-Thr-Tyr-Val-Thr-Asn-Leu-Asn-Ala-Ala-Leu-Glu-Gly-His
                           50                                   60
  -Pro-Asp-Leu-Gln-Asn-Lys-Ser-Leu-Glu-Glu-Leu-Leu-Ser-Asn-Leu
                                                   70
  -Glu-Ala-Leu-Pro-Glu-Ser-Ile-Arg-Thr-Ala-Val-Arg-Asn-Asn-Gly
                           80                                   90
  -Gly-Gly-His-Ala-Asn-His-Ser-Leu-Phe-Trp-Thr-Ile-Leu-Ser-Pro
                                                   100
  -Asn-Gly-Gly-Gly-Glu-Pro-Thr-Gly-Glu-Leu-Ala-Asp-Ala-Ile-Asn
                           100                                  120
  -Lys-Lys-Phe-Gly-Ser-Phe-Thr-Ala-Phe-Lys-Asp-Glu-Phe-Ser-Lys
                                                   130
  -Ala-Ala-Ala-Gly-Arg-Phe-Gly-Ser-Gly-Trp-Ala-Trp-Leu-Val-Val
                           140                                  150
  -Asn-Asn-Gly-Glu-Leu-Glu-Ile-Thr-Ser-Thr-Pro-Asn-Gln-Asp-Ser
                                                   160
  -Pro-Ile-Met-Glu-Gly-Lys-Thr-Pro-Ile-Leu-Gly-Leu-Asp-Val-Trp
                           170                                  180
  -Glu-His-Ala-Tyr-Tyr-Leu-Lys-Tyr-Gln-Asn-Arg-Arg-Pro-Glu-Tyr
                                                   190
  -Ile-Ala-Ala-Phe-Trp-Asn-Val-Val-Asn-Trp-Asp-Glu-Val-Ala-Lys
                           200         203
  -Arg-Tyr-Ser-Glu-Ala-Lys-Ala-Lys-COOH
```

Fig. 2. The complete amino acid sequence of manganese dismutase from _B. stearothermophilus_.

ACKNOWLEDGEMENT

C.J.B. thanks the Wellcome Trust for support.

REFERENCES

1. Brock, C.J., Harris, J.I. and Sato, S. (1976) J. Mol. Biol. 107, 175-178.
2. Walker, J.E. (1979) Proc. 12th FEBS Meeting, Vol.52, Hofmann, E., ed., Pergamon Press, Oxford, pp.211-225.
3. Bridgen, J., Harris, J.I. and Kolb, E. (1976) J. Mol. Biol. 105, 333-335.
4. Smit, J.D.G., Pulver-Sladek, J. and Jansonius, J.N. (1977) J. Mol. Biol. 112, 491-494.
5. Brock, C.J. and Harris, J.I. (1977) Biochem. Soc. Trans. 5, 1537-1539.
6. Massey, V. (1957) J. Biol. Chem. 229, 763-770.
7. Cotton, F.A. and Wilkinson, G. (1972) Advanced Inorganic Chemistry, J. Wiley and Sons Inc., New York.
8. Dougherty, H.W., Sadowski, S.J. and Baker, E.E. (1978) J. Biol. Chem. 253, 5220-5223.
9. Steinman, H.M. (1978) J. Biol. Chem. 253, 8708-8720.
10. Steinman, H.M., Naik, V.R., Abernethy, J.L. and Hill, R.L. (1974) J. Biol. Chem. 249, 7326-7338.
11. Rotilio, G., Bray, R.C. and Fielden, E.M. (1972) Biochim. Biophys. Acta, 268, 605-609.
12. Beauchamp, C.O. and Fridovich, I. (1971) Anal. Biochem. 44, 276-287.

METAL REPLACEMENT STUDIES AND CRYSTALLOGRAPHIC DATA ON IRON-SUPEROXIDE DISMUTASE FROM *PSEUDOMONAS OVALIS*

FUMIYUKI YAMAKURA,[+*] KOJI SUZUKI,[+] GREGORY A. PETSKO[++] AND DEMETRIUS TSERNOGLOU[+++]
[+] Department of Chemistry, School of Medicine, Juntendo University, Narashino, Chiba 275, Japan; [++] Department of Chemistry, Massachusetts Institute of Technology, Cambridge, Massachusetts 02139, USA; [+++] Department of Biochemistry, Wayne State University, School of Medicine, Detroit, Michigan 48201, USA.

ABSTRACT

The Cd-, Cr- or Mn-substituted derivatives of iron-superoxide dismutase from *Pseudomonas ovalis* was prepared from its apoenzyme by using the alkaline treatment method described before (see Ref.5) with a few modifications. The Cd-substituted enzyme had 1.16 gram atoms of Cd per mole of enzyme and no visible absorption. The Cr-substituted enzyme had 1.27 gram atoms of Cr per mole of enzyme and had optical absorption maxima at 530 nm and 670 nm with a shoulder around 370 nm. The Mn-substituted enzyme had 1.27 gram atoms of Mn per mole of enzyme and had optical absorption shoulders around 470 nm and 600 nm. The Cd-, Cr- or Mn-substituted enzyme had no enzymatic activity. The reactivity of the four sulfhydryl groups and circular dichroism spectrum of the Cd-, Cr- or Mn-substituted enzyme were similar to the iron-reconstituted enzyme. These results indicate that these metals may bind to the same site as Fe in the enzyme.

Crystals of the superoxide dismutase were prepared by a method of vapor diffusion. The crystals grow as rectangular parallelipipeds. The crystals were space group $p2_1$ with unit cell dimensions a = 81.9 Å, b = 49.1 Å, c = 61.7 Å and $\beta = 106.8^o$.

INTRODUCTION

Superoxide dismutases (E.C. 1.15.1.1) have been found among oxygen-metabolizing organisms and some obligate anaerobes. Three distinct kinds of superoxide dismutase, i.e. Cu,Zn-, Mn- and Fe-containing superoxide dismutase, have been isolated.[1,2]

In our previous studies, an Fe-containing superoxide dismutase from *Pseudomonas ovalis* was purified and some of its properties were presented.[3] The enzyme had a molecular weight of 40,000 and was composed of two subunits. The iron content of the purified enzyme varied from 1.1 to 1.5 gram atoms per mole of enzyme. The crystalline Fe-superoxide dismutase was prepared and preliminary crystal data were reported.[4] The iron in the enzyme was removed reversibly by the alkaline

---

[*] Present address: Department of Chemistry, Wayne State University, Detroit, Michigan 48202, USA.

treatment method or the urea-treatment method. We have previously described some properties of the iron-free enzyme and the iron-reconstituted enzyme.[5,6] The question of whether the enzyme iron could be substituted by other metals arose. Some metal substitution studies on Fe- or Mn-superoxide dismutase have been reported. Ose and Fridovich[7,8] reported replacement of Mn in Mn-superoxide dismutase from E. coli with Co^{2+}, Zn^{2+}, Ni^{2+}, Fe^{2+} or Cu^{2+}. Brock and Harris[9] also reported on the metal substitution of Mn-superoxide dismutase from Bacillus stearothermophilus. They obtained Co^{2+}, Fe^{2+} or Cu^{2+} substituted superoxide dismutase. All of these metal substituted enzymes from Mn-superoxide dismutase had no enzymatic activity. In the case of Fe-superoxide dismutase, Puget et al.[10] reported replacmeent of Fe in the enzyme from Photobacterium leigonathi with Mn^{2+}, Zn^{2+}, Ni^{2+}, Co^{2+} or Cr^{2+} in a ratio of one gram atom of metal per mole of enzyme and Cu^{2+} as two gram atoms of metal per mole. These metal substituted enzymes had no enzymatic activity.[10]

Here, we wish to present both the preparation and some properties of Cd-, Cr- or Mn-substituted superoxide dismutase. The purpose of this substitution study is to know the specificity of metal incorporation into the enzyme, to see whether the metal substituted enzyme has catalytic activity, and to investigate the site of metal binding by using physico-chemical method applied to the metal substituted enzyme. We also wish to present recent development of x-ray crystallographic data on this iron-superoxide dismutase.

METHODS

Iron-superoxide dimsutase was isolated from Pseudomonas ovalis according to the procedure reported previously.[3] Protein concentration was determined by the microbiuret method of Itzhaki and Gill.[11] The concentrations of the native, apo and reconstituted enzyme could also be determined from the absorbance at 280 nm using extinction coefficients of 80,600 $M^{-1}cm^{-1}$, 77,880 $M^{-1}cm^{-1}$ and 80,400 $M^{-1}cm^{-1}$, respectively.[3,5] Cadmium, chromium, manganese, nickel and copper were measured by atomic absorption spectrometery using a Nippon Jarrell-Ash atomic absorption spectrometer with a frameless atomizer FLA-10. Iron was determined by the method of Massey[12] except that batho-phenanthroline was used as colorimetric reagent. Superoxide dismutase activity was measured by the method of McCord and Fridovich,[13] with a slight modification.[3] Reactivity of sulfhydryl groups of the enzyme was examined spectrophotometrically, as described by Ellman.[14]

Absorption spectra were obtained using a Hitachi 124 spectrophotometer or a Union Giken SM-401 spectrophotometer. CD spectra were measured with a JASCO J-20 automatic recording spectropolarimeter. CD measurements were carried out with a 0.5 cm light-path cell and a protein concentration of 30-40 μM from 250 nm to 340 nm and with a 0.063 cm light-path cell and a protein concentration of 10-12 μM from 200 to 250 nm.

Preparation of metal substituted superoxide dismutase. Metal substituted enzymes were prepared from apoenzyme by a method similar to the alkaline treatment method described previously.[5] The apoenzyme (10-20 mg) was incubated for 25 min at 30°C with 5 ml of 0.25 M sodium carbonate buffer (pH 11.0) containing 2 mM dithiothreitol and 10 mM $CdSO_4$, $CrCl_2$, $MnSO_4$, $(NH_4)Cr(SO_4)_2$, $NiSO_4$ or $CuSO_4$. The incubation was carried out under anaerobic conditions by evacuation and flushing with N_2 using a Thunberg tube. Then the solution was dialyzed overnight against 0.2 M sodium carbonate buffer (pH 9.2) containing 1 mM each of the metal salts and 0.5 mM dithiothrietol at 4°C under anaerobic condition. After further dialysis against 50 mM potassium phopshtae buffer (pH 7.8), containing 0.2 mM dithiothreitol, for 4 hr at 4°C, substituted enzyme was centrifuged at 15,000 x g for 20 min to eliminate the precipitate of the metal salt. Then the solution was dialyzed against 0.025 M potassium phosphate buffer (pH 7.8) for 4 hr at 4°C and recovered by DEAE-cellulose chromatography. Finally, the metal substituted enzyme was dialyzed against 50 mM potassium phosphate buffer (pH 7.8) containing 0.1 mM EDTA to eliminate non-specific binding metals.

Preparation of large single crystals. The enzyme was adjusted to 30 mg/ml protein concentration and dialyzed against distilled water for one day at 4°C before crystallization. Seed crystals were obtained by dialysis of the protein solution at 4°C against 50% saturated $(NH_4)_2SO_4$, pH 5.0 by adjustment with 0.05N H_2SO_4, followed by slow evaporation in a desiccator under reduced pressure for 2 days at 4°C. Large single crystals were prepared by the method of vapor diffusion. Three drops of the protein in 50% saturated $(NH_4)_2SO_4$, pH 5.0, were placed in a sealed chamber with a 2 ml reservoir of 63% saturated $(NH_4)_2SO_4$. Vapor phase equilibration at 4°C required about 10 days, during which time well-formed, large crystals grew.

X-ray crystallographic measurement. X-ray photographs were taken on a Buerger precession camera mounted on a conventional sealed-tube generator. Nickel filtered copper Kα X-rays were produced from a fine-focus tube run at 50kV and 25 mA. Photographs were taken at both 4°C and 20°C, for 12 hours, at a 6 cm crystal-to-film distance. The crystals were mounted in quartz capillary tubes and sealed in equilibrium with a small amount of mother liquor.

RESULTS

Substitution with various metal ions. An attempt to substitute the iron of the iron-superoxide dismutase with Cd^{2+}, Cr^{2+}, Mn^{2+}, Cr^{3+}, Ni^{2+} and Cu^{2+} was made. Cd^{2+}, Cr^{2+} and Mn^{2+} (as well as Fe^{2+}) were incorporated into the apoenzyme by using the method described here at a yield of 54%, 52%, and 67% respectively, while Cr^{3+}, Ni^{2+} and Cu^{2+} were not incorporated. Table 1 shows the metal content and specific activity of the Cd-, Cr- and Mn-substituted forms of the enzyme in comparison with that of the apoenzyme and Fe-reconstituted enzyme. The apoenzyme contained no significant amount of Cd, Cr, or Mn. Each metal substituted form of the enzyme contained about 1.2-1.3 gram atoms of metal per mole of enzyme but had no significant activity. From these results, it is concluded that Cd-, Cr or Mn-substituted enzyme has a tendency to contain only one metal per two subunits as in the case of the Fe-reconstituted enzyme and iron is needed specifically for the enzymatic activity.

TABLE 1

METAL CONTENT AND SPECIFIC ACTIVITY OF NATIVE, APO AND METAL SUBSTITUTED IRON-SUPEROXIDE DISMUTASE FROM PSEUDOMONAS OVALIS

Enzyme	Fe	Cr	Cd	Mn	Specific Activity (units/mg)[a]
	(gram atom per mole)[a]				
Native SOD	1.38 ± 0.03(3)				4.490 ± 480(4)
Apo-SOD	0.085 ± 0.02(3)	<0.025	<0.025	0	100 ± 23 (3)
Cd-SOD			1.16 ± 0.02(9)		15 ± 8 (5)
Cr-SOD		1.27 ± 0.13(6)			24 ± 10 (6)
Mn-SOD				1.27 ± 0.02(7)	22 ± 3 (7)

[a]Values are given as mean ± SD., with the number of determinations in parentheses.

Reactivity of sulfhydryl groups in the metal substituted enzyme. In previous studies, we found that the four sulfhydryl groups of the apoenzyme showed high reactivity with the Ellman reagent, and those of the native or Fe-reconstituted enzyme had little reactivity.[5] The reactivity of the four sulhydryl groups of the Cd-, Cr- or Mn-substituted enzyme with the reagent was shown in Figure 1. The sulfhydryl groups of the Cd-, Cr- or Mn-substituted enzyme had low reactivity like the groups of the Fe-reconstituted enzyme. This suggests that the sulfhydryl groups of the Cd-, Cr- or Mn-substituted enzyme are masked as in the native or Fe-reconstituted enzyme.

Absorption spectra. The Cd-substituted enzyme showed no absorption spectrum in the visible region. The visible absorption spectrum of the Cr-substituted enzyme is shown in Figure 2. The Cr-substituted enzyme had absorption peaks at 530 nm and 670 nm and a shoulder around 370 nm. The molar extinction coefficient was determined as 125 $M^{-1}cm^{-1}$ per chromium at 530 nm and 72 $M^{-1}cm^{-1}$ per chromium at 670 nm. The Mn-substituted enzyme had a broad visible absorption spectrum with a shoulder around 470-480 nm and 600 nm (Figure 3). The molar extinction coefficient was calculated at 56 $M^{-1}cm^{-1}$ per manganese at 480 nm. Interestingly this spectrum is very similar to that of other native Mn-superoxide dismutases,[15,16] especially Mycobacterium Mn-superoxide dismutases[15,16] or yeast mitochondrial Mn-superoxide dimutase.[17] The extinction coefficent of the Mn-substituted enzyme, however, shows a rather low value, being one third to one sixteenth of the value of native Mn-superoxide dismutases.

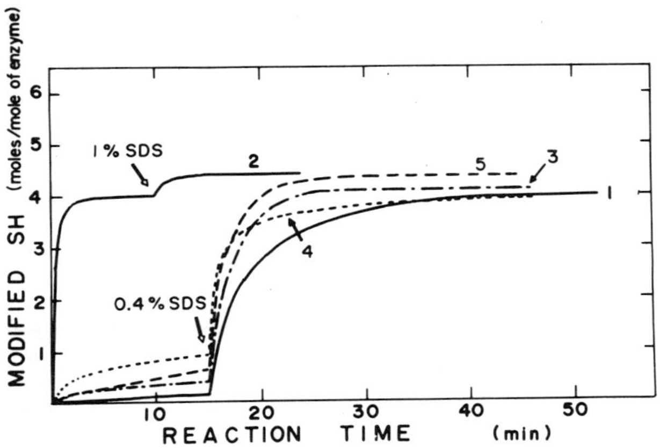

Fig. 1. Quantitative analysis of sulfhydryl groups of metal substituted enzyme by means of the Ellman reagent (Ref.14). Enzyme (0.25 - 0.3 mg) was allowed to react with the reagent in 1 ml of 100 mM sodium phosphate buffer (pH 8.0) containing 1 mM EDTA. The reaction was followed in terms of the absorption change at 412 nm. 1 (———), Fe-reconstituted enzyme; 2 (———), apoenzyme; 3 (—·—), Cd-substituted enzyme, 4 (-----), Cr-substituted enzyme; 5 (— — —), Mn-substituted enzyme. SDS, sodium dodecylsulfate.

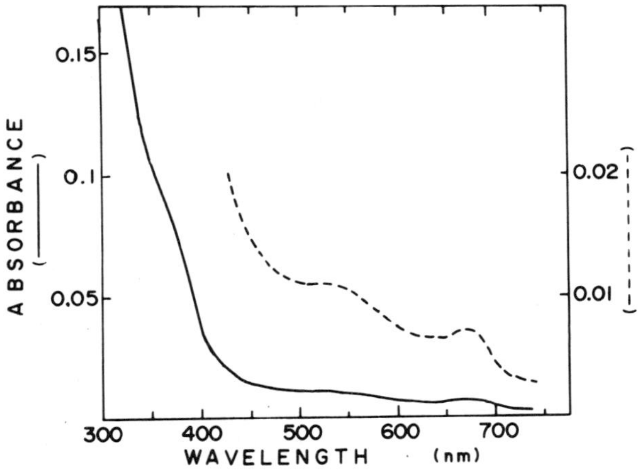

Fig. 2. Absorption spectrum of Cr-substituted enzyme. 6.9×10^{-5}M of the enzyme was used in 50 mM potassium phosphate buffer, pH 7.8.

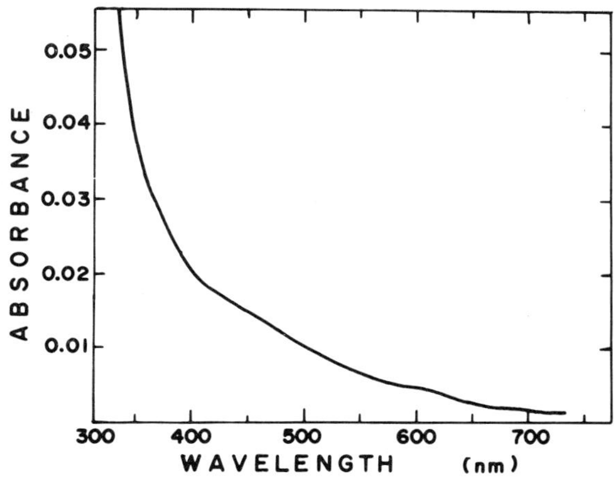

Fig. 3. Absorption spectrum of Mn-substituted enzyme. 1.87 x 10^{-4}M of the enzyme was used in 50 mM potassium phosphate buffer, pH 7.8.

The ultraviolet spectra of Cd-, Cr- or Mn-substituted forms of the enzyme below 300 nm were similar to that of the native enzyme.

CD spectra of the metal substituted enzyme. CD spectra were measured to compare the protein conformation of the Cd-, Cr- or Mn-substituted enzyme with that of the apoenzyme and the Fe-reconstituted enzyme. As shown in Figure 4, there is no significant difference between the Fe-reconstituted enzyme, Cd-, Cr- or Mn-substituted enzyme, and apoenzyme in the 200 to 250 nm region. Within experimental error, it is concluded that the gross conformation of the Cd-, Cr- or Mn-substituted enzyme seems to be nearly the same as that of the apoenzyme and the Fe-reconstituted enzyme. In the 250 to 340 nm region, only the apoenzyme showed a somewhat different spectrum, the other spectra being similar except around the 255 nm region. The Fe-reconstituted enzyme and the Cr-substituted enzyme had a small shoulder in this region, but apo-, Cr- and Mn-enzyme had no shoulder. It is not clear at present what causes this difference. From the CD results, it can be concluded that structure of aromatic amino acid residues of the Cd-, Cr- or Mn-substituted enzyme seems to be the same as in the Fe-reconstituted enzyme.

Crystallographic study. It is not difficult to grow large single crystals of the Fe-superoxide dismutase from Pseudomonas ovalis by vapor diffusion against 63% saturated $(NH_4)_2SO_4$. Large (1 mm x 1 mm x 0.4 mm) crystals, yellow-orange in color, form within a few days at 4°C. If no crystals are present after two weeks, seeding with a small single crystal grown by slow evaporation produces many large crystals. The crystals grow as rectangular parallelipipeds (Figure 5) with the real crystal axes along the points. X-ray precession photographs

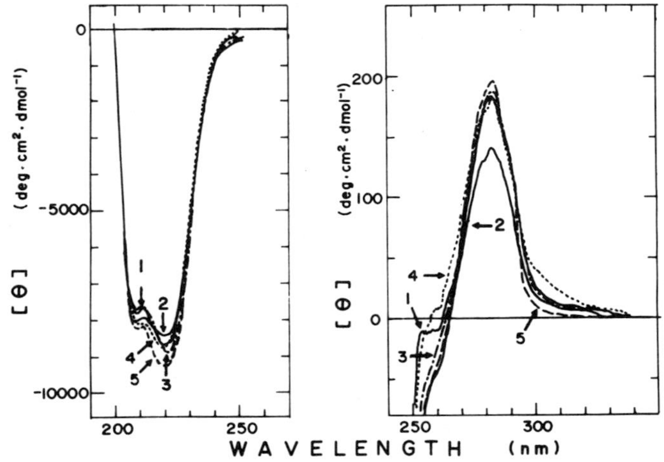

Fig. 4. CD spectra of metal substituted enzyme. All measurements were made in 50 mM potassium phosphate buffer, pH 7.8. The conditions are described in detail in the text. 1 (———), Fe-reconstituted enzyme; 2 (———), apoenzyme; 3 (—·—), Cd-substituted enzyme; 4 (------), Cr-substiututed enzyme; 5 (— — —), Mn-substituted enzyme.

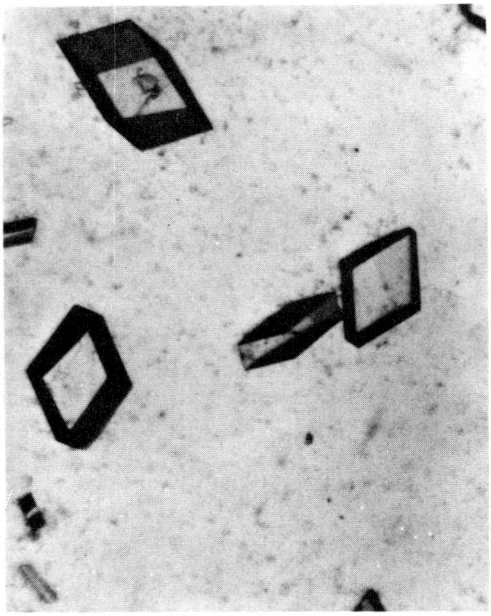

Fig. 5. Photomicrograph of crystals of the iron-containing superoxide dismutase from Pseudomonas ovalis. The longest dimension of the largest crystal in 1.4 mm.

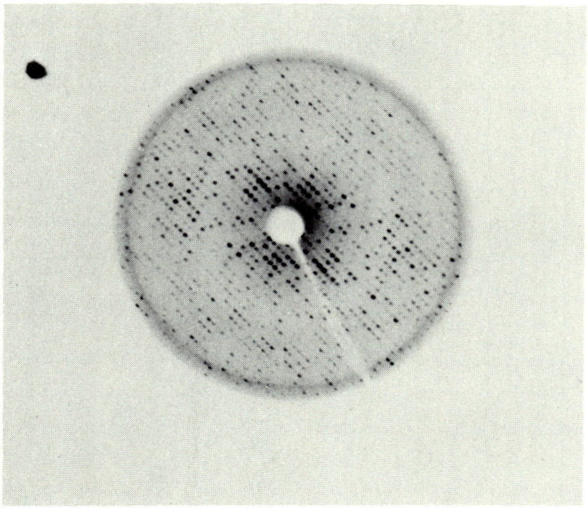

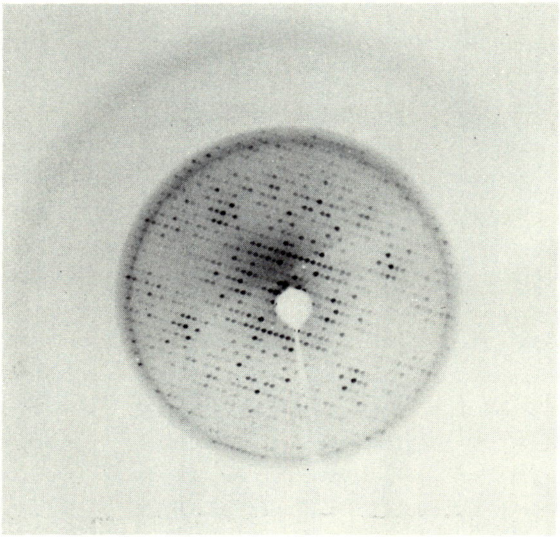

Fig. 6. Precession photographs of the (h0l) (top) and (hk0) (bottom) zones of the $p2_1$ space group Fe-superoxide crystals. The reflections furthest from the center of the photograph correspond to a resolution of 3 Å.

(Figure 6) of the two principal zones (h0l) and hk0) show no unusual symmetry and only the systematic absences k = 2n+1 where h,l = 0. The space groups is thus $p2_1$. The unit cell dimensions, measured both from photographs and from diffractometer data, are a = 81.90(5) Å, b = 49.10(5) Å, c = 61.70(5) Å with cell angles $\alpha = \gamma = 90^{o}$, $\beta = 106.80(6)^{o}$ (estimated standard deviations are given in parenthesis). The unit cell volume is 237,272 (300) $Å^3$, which, combined with an estimated molecular weight of 40,000 for the dimeric enzyme, gives a Vm (volume per unit mass) of 2.97 $Å^3$ per dalton assuming two dimers per unit cell. This value falls well within the range of commonly observed Vm values for proteins of this molecular weight,[18] which is not the case if one or four molecules per cell is assumed. The space-group symmetry rules out one molecule per cell in any case, since the two subunits of the dimer would then have to be related by the two-fold screw axis that relates that two asymmetric units in $p2_1$. A screw axis, being an infinite symmmetry element, cannot relate subunits in a finite molecule.

Assuming a partial specific volume of 0.74 cm^3g^{-1} protein, the solvent content of the crystals is calculated to be 59% by volume. Though not remarkable, this figure is relatively high for a protein crystal and may account for the relative ease with which the crystals may be shattered.

DISCUSSION

The data reported are the first detailed description of the properties of a metal substituted Fe-superoxide dismutase. The Cd-, Cr- or Mn-substituted enzyme contained only 1.2 - 1.3 gram atoms of metals per mole of enzyme (2 subunits) (Table 1). Since all metal substituted forms of the enzyme had no superoxide dismutase activity, iron is essential for the enzymatic activity at least among the examined metals. The Fe-reconstituted enzyme also contained about 1 gram atom of Fe per mole of enzyme.[5] From this evidence, it can be presumed that the apoenzyme has ability easily to combine only 1 metal atom per molecule with little tendency to bind an additional atom. Since Pseudomonas Fe-superoxide dismutase seems to be composed of identical subunit according to N-terminal sequence study[19] these results suggest the possibility that Fe-superoxide dismutase has negative cooperativity between the subunits for metal binding.

Although the coordination structure of the Cd-, Cr- or Mn-substiututed enzyme has not been finally determined, the following considerations and the binding stoichiometry of the metals support the conclusion that each metal occupies the same site as iron in the native enzyme. Thus, the sulfhydryl groups of the metal substituted enzyme have the same low reactivity as those of the native and the Fe-reconstiututed enzymes, in contrast to the high reactivity of the groups of the apoenzyme; and the spectra of the metal substituted enzyme show the same profile as the Fe-reconstituted enzyme in the 250-350 nm region which relfect the conformation of the aromatic amino acid residues.

The absorption spectrum of the Mn-substituted superoxide dismutase has bands around 470 nm and 600 nm (Fugure 2). This spectrum is similar to that of native Mn-superoxide dismutase.[15-17,20-25] Thus most Mn-superoxide dismutases have two absorption bands, one is around 450-480 nm with a molar extinction coefficient of 240 $M^{-1}cm^{-1}$ (Ref.14) to 900 $M^{-1}cm^{-1}$ (Refs.22,25) per Mn and the other is around 600 nm. Although molar extinction of the Mn-substituted enzyme in both the 470 nm and the 600 nm regions is less than one third of that of native Mn-superoxide dismutases, it can be presumed from the similarity of the absorption spectra that the Mn-substituted enzyme has similar ligand structure as native Mn-superoxide dismutases.

An earlier X-ray study of the same crystal form found the space group to be p2, with the same unit cell dimensions as reported here.[4] The assignment to this space group was based on the occurence of the (0,15,0) reflection, though all other (0,k,0) reflections of k = 2n+1 were absent. We have found that the Fe-superoxide dismutase crystals have a strong tendency to twin, usually in the $a\text{-}b^*$ plane, as though they were composed of many layers of thin plates. The twinning is often reflected in a splitting of the reflection profile on an ω-scan, but also produces extra reflections in severe cases. These extra reflections are generally of the form k = 2n+1, and we have grown cyrstals where the (0,9,0) and/or (0,15,0) reflections were present. We have also found specimens where reflections $(0,2\frac{1}{2},0)$ and $(0,12\frac{1}{2},0)$ were observable. The twinning is most severe for crystals grown by slow evaporation, as was the case in the original report,[14] and least severe for crystals grown by vapor diffusion. We can produce untwinned crystals with a success rate of about 50%; these show no k = odd reflections along the (0,k,0) direction. Proof that the correct space group is $p2_1$ has come from studies of the binding of heavy metal compounds to the crystalline enzyme to produce isomorphous derivatives for solution of the phase problem. The crystals were soaked in 85% $(NH_4)_2SO_4$, pH 5.0, plus 1 mM of various heavy atom-containing compounds. Precession photographs showed large, reproducible intensity changes for crystals soaked at 4°C in sodium platinum nitrite. An isomorphous difference Patterson map showed two strong peaks on the Harker section v = 0.5, where vectors are expected in space-group $p2_1$ (Figure 7). The Harker section at v = 0.0, corresponding to the symmetry of p2, was featureless. The positions of the cross-vectors also confirmed the true space group to be $p2_1$, as well as fixing the relative y-coordinates of the two sites.

One striking characteristic of the Fe-superoxide dimsutase crystals is their extreme sensitivity to both temperature and irradiation. In their normal mother liquor, 63% saturated $(NH_4)_2SO_4$, they cannot be warmed above 10°C without complete disordering. In 85% saturated $(NH_4)_2SO_4$, they are stable at room temperature if warmed slowly, but mechanical shock leads to complete loss of

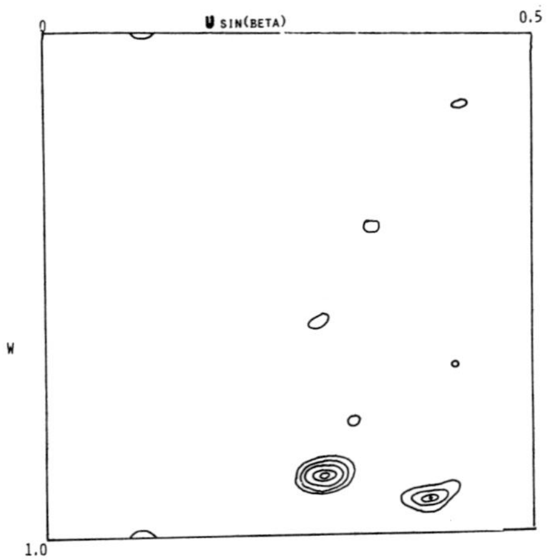

Fig. 7. Harker section (v = 0.5) of the isomorphous difference Patterson map of the platinum nitrite derivative at 4 Å resolution. Contours are at equal but arbitrary intervals.

the diffraction pattern. We have worked with them at 4°C or lower for this reason. At room temperature in 85% saturated $(NH_4)_2SO_4$, the crystals are extremely sensitive to X-rays, with 50% of the diffracted intensity being lost after 24 hours exposure. However, at -10°C, the radiation damage is much less severe, and the crystals withstand 180 hours of irradiation with only 10% loss in intensity. At this temperature a complete set of measurements to 2.9 Å resolution may be obtained from one crystal.

In summary, a method originally employed for Fe-reconstitution on the Fe-superoxide dismutase from *Pseudomonas ovalis*[5] is effective for metal substitution study of the enzyme. The resulting metal substituted forms of the enzyme are useful for characterization of the metal site. At low temperature (below 4°C), the monoclinic Fe-superoxide dismutase crystals diffract to at least 1.8 Å resolution. Though fragile, they are well-enough ordered for a high resolution structure determination by X-ray crystallography, Further characterization of the metal substituted enzyme and the elaboration of a 2.9 Å electron density map that should reveal the folding of the polypeptide chain and the environment of the iron atom are now in progress.

ACKOWLEDGEMENTS

We wish to thank Dr.J. Tobari for spectrophotometer facilities and Mr.E. Kobayashi for help in atomic absorption analyses. We are also grateful to Mr. David Roas, Ms. Anne Missavege and Dr. Phillip D. Martin for technical assistance. Part of this work was supported by a Scientific Research Grant from the Ministry of Education, Science and Culture of Japan (No. 278052). G.A. Petsko is an Alfred P. Sloan Foundation Fellow and holds a Research Career Development Award from the United States Public Health Service.

REFERENCES

1. Fridovich, I. (1974) Adv. Enzymol. 41, 35-97.
2. Fridovich, I. (1975) Annu. Rev. Biochem. 44, 147-159.
3. Yamakura, F. (1976) Biochim. Biophys. Acta, 426, 280-294.
4. Yamakura, F., Suzuki, K. and Mitsui, Y. (1976) j. Biol. Chem. 251, 4792-4793.
5. Yamakura, F. (1978) J. Biochem. Tokyo, 83, 849-857.
6. Yamakura, F. and Suzuki, K. (1976) Biochem. Biophys. Res. Commun. 72, 1108-1115.
7. Ose, D.E. and Fridovich, I. (1976) J. Biol. Chem. 251, 1217-1218.
8. Ose, D.E. and Fridovich, I. (1979) Arch. Biochem. Biophys. 194, 360-364.
9. Brock, C.J. and Harris, J.I. (1977) Biochem. Soc. Trans. 5, 1537-1539.
10. Puget, K., Lavelle, F., and Michelson, A.M. (1977) in Superoxide and Superoxide Dismutases, Michelson, A.M., McCord, J.M. and Fridovich, I., eds., Academic Press, New York, pp.139-150.
11. Itzhaki, R.F. and Gill, D.M. (1964) Anal. Biochem. 9, 401-410.
12. Massey, V. (1957) J. Biol. Chem. 299, 763-770.
13. McCord, J.M. and Fridovich, I. (1969) J. Biol. Chem. 244, 6049-6055.
14. Ellman, G.L. (1959) Arch. Biochem. Biophys. 82, 70-77.
15. Kusunose, M., Noda, Y., Ichihara, K. and Kusunose, E. (1976) Arch. Microbiol. 108, 65-73.
16. Ichihara, K., Kusunose, E., Kusunose, M. and Mori, T. (1977) J. Biochem. Tokyo, 81, 1427-1433.
17. Ravindranath, S.D. and Fridovich, I. (1975) J. Biol. Chem. 250, 6108-6112.
18. Matthews, B.W. (1968) J. Mol. Biol. 23, 491-497.
19. Yamakura, F. and Suzuki, K. (1978) Tanpakushitsu Kakusan Koso, 23, 221-231 (in Japanese).
20. Keele, B.B. Jr., McCord, J.M. and Fridovich, I. (1970) J. Biol. Chem. 245, 6176-6181.
21. Vance, P.G., Keele, B.B.,Jr. and Rajagopalan, .K.V. (1972) J. Biol. Chem. 247, 4782-4786.
22. Sato, S. and Harris, J.I. (1977) Eur. J. Biochem. 73, 373-381.
23. Weisiger, R.A. and Fridovich, I. (1973) J. Biol. Chem. 248, 3582-3592.
24. Misra, H.P. and Fridovich, I. (1977) J. Biol. Chem. 252, 6421-6423.
25. Sato, S. and Nakazawa, K. (1977) J. Biochem. Tokyo, 83, 1165-1171.

METAL BINDING PROPERTIES OF COPPER-ZINC SUPEROXIDE DISMUTASE

JOAN S. VALENTINE AND MICHAEL W. PANTOLIANO
Department of Chemistry, Busch Campus, Rutgers University, New Brunswick, New Jersey 08903, USA.

INTRODUCTION

The metalloprotein commonly named copper-zinc superoxide dismutase (Cu_2Zn_2SOD)* is particularly well-suited for studies of protein-metal ion-binding properties. We review here some previous work in this area and describe recent findings from our own laboratory concerning the pH-dependence of the metal-binding properties of this protein.

There is no general agreement about the exact function of Cu_2Zn_2SOD *in vivo*.[1-8] Between 1938, when it was first isolated,[9] and 1969, when its superoxide dismutase activity was discovered,[10] this protein was believed without direct evidence to be involved in copper metabolism. The presence of zinc in this protein was not reported until 1970.[11] Because this protein was known to contain copper and it was most frequently obtained from erythrocytes, it was commonly referred to as "erythrocuprein". In 1969, at the suggestion of McCord and Fridovich,[10] it was renamed "superoxide dismutase". However, the possibility still remains that this protein may be involved in zinc and/or copper metabolism or have some as yet unknown enzymatic function.

Whether or not metal ion binding is directly relevant to the major physiological role of Cu_2Zn_2SOD, this protein is ideally suited to studies of metal ion-binding properties. Copper-zinc SOD as isolated from bovine erythrocytes has a relatively low molecular weight (∿31,500) and is made up of two identical subunits[12,13] of known sequence.[14] Each subunit contains two metal ion binding sites[12,13] which, after isolation, contain one Cu^{2+} and one Zn^{2+}. The protein is readily isolated in large quantities and the holoprotein, partially metal depleted protein and apoprotein are unusually stable.[2] It has high solubility, a property which facilitates spectroscopic studies, and the native Cu and Zn ions are readily removed and replaced by a variety of metal ions (see below).

---

*Abbreviations: Cu_2Zn_2SOD, native bovine erythrocyte superoxide dismutase; X_2Y_2SOD, derivative of the native protein in which the metal ions X and Y have been substituted for Cu^{2+} and Zn^{2+} respectively; X_2E_2SOD, derivative in which the zinc site is vacant; E_2Y_2SOD, derivative in which the copper is vacant (E = empty). Metal ion oxidation states are 2+ unless denoted otherwise.

The X-ray crystal structure[12,13] shows that the metal binding sites on each subunit of Cu_2Zn_2SOD are very close to one another. In fact, the two metals bound to each subunit share a common bridging ligand, a deprotonated imidazole ring from the side chain of histidine 61.

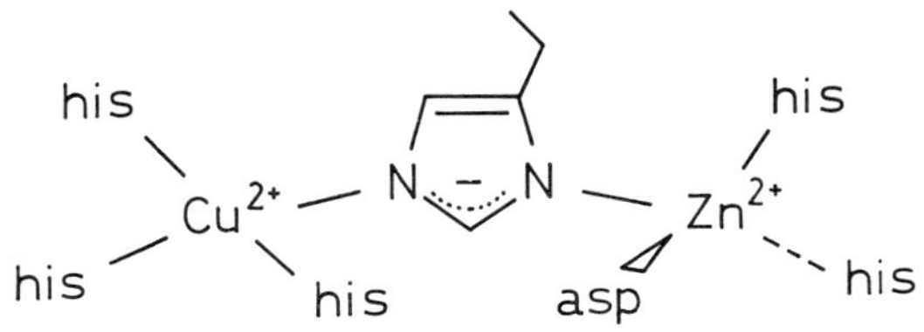

The remainder of the coordination sphere around Cu^{2+} consists of three additional histidyl imidazoles which complete a distorted square planar coordination geometry. In addition to histidine 61, the ligands around Zn^{2+} are two histidyl imidazoles and an aspartyl carboxylate forming an approximately tetrahedral coordination geometry. It can be confidently predicted from the crystal structure that occupancy of one of the two metal binding sites on a subunit will influence the metal binding properties of the remaining site. Since it is made up of two subunits, the apoprotein has four potential metal binding sites with several possible interactions between them. Thus occupancy of the copper site is predicted to affect binding of a metal ion to the zinc site, occupancy of the zinc site should affect binding to the copper site, and binding of metal ions to one subunit has been reported to affect binding of metal ions to the other subunit.[41,42] Recent work in our laboratory has been concerned with such properties of this protein. In the latter portion of this report, we describe the results of experiments which imply (1) that the Cu and Zn binding sites within each subunit are highly interactive, and (2) that the metal binding properties of this protein are markedly pH dependent.

Metal ion substitutions

(a) Methodology. The methodology to remove and replace metal ions in Cu_2Zn_2SOD was largely developed in the laboratories of Fridovich, Fee, Rotilio and Rajagopalan. Extensive characterization using spectroscopic and other methods established the nature of the substitutions in each case. The following scheme illustrates the various substitutions that have been carried out giving literature references which describe the procedures for substitution.

The most useful general method for metal ion substitution in this protein was discovered by Beem et al.[13,15] who found that the apoprotein could be successfully reconstituted with the native or other metal ions by first adding the metal ion intended for the copper site to the apoprotein at pH 3.8, followed by addition of the metal intended for the zinc site at pH 7.8. The rationale for changing pH was not known until more recently however.

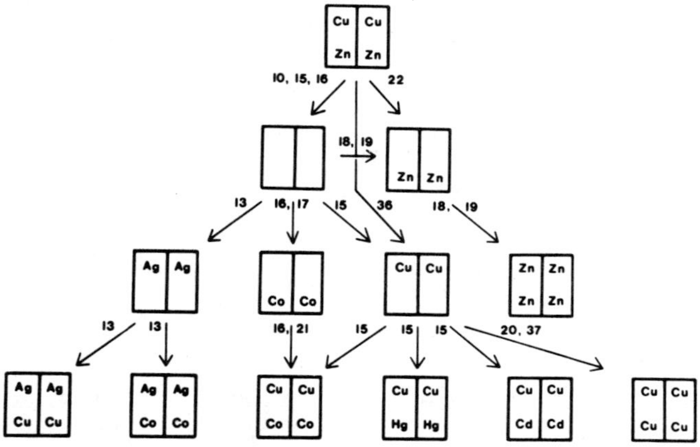

(b) Location of metals. From careful examination of ESR spectral data, Beem et al.[15] concluded that Cu^{2+} in Cu_2E_2SOD near neutral pH is bound to the protein in a manner similar to that of Cu^{2+} in Cu_2Zn_2SOD. They concluded furthermore that binding of Zn^{2+}, Hg^{2+}, or Cd^{2+} to Cu_2E_2SOD occurred at the native Zn site and that the resulting protein was very similar to native Cu_2Zn_2SOD as judged by the ESR signal of the protein bound Cu^{2+} ion.

In the case of E_2Zn_2SOD, Lippard et al.[18] concluded from examination of NMR data in the histidine N-H region and from protein chemical modification studies that addition of one equivalent of Zn^{2+} per subunit to the apoprotein at pH 6.0 caused Zn^{2+} to bind to the native Zn site and, moreover, that a conformational change occurred upon binding one Zn^{2+} per subunit which caused the entire metal binding region to become less accessible to the solvent. Lippard et al.[18] also observed NMR spectral changes upon addition of a second Zn^{2+} per subunit which they attributed to formation of Zn_2Zn_2SOD. Cass et al.[19] came to similar conclusions from observation of the histidine C-2 proton resonances upon addition of Zn^{2+} to apoprotein at pH 6.3.

Evidence that Co^{2+} is in the native Zn site in Cu_2Co_2SOD, E_2Co_2SOD and $Cu(I)_2Co_2SOD$ comes from visible and ESR spectral results of Fee[16] and Calabrese et al.[17,21] The visible spectrum that was obtained for Co^{2+} in these derivatives is clearly due to Co(II) in a non-centrosymmetric environment[23] since the extinction coefficient is quite large (∿600 $M^{-1}cm^{-1}$) (Refs.17,21).

Some particularly interesting metal substituted derivatives of Cu_2Zn_2SOD contain copper in the native zinc site. These are Cu_2Cu_2SOD, prepared by Fee and Briggs,[20] and $Ag(I)_2Cu_2SOD$, prepared by Beem et al.[13] In the case of

$Cu(II)_2Cu(II)_2SOD$, the magnetic interaction between the two Cu(II) ions results in a distinctive triplet ESR spectrum (see below). The ESR spectral properties of $Ag(I)_2CuSOD$ and $Cu(I)_2Cu(II)_2SOD$ are strikingly similar and indicate that Cu(II) in the native zinc site has considerable tetrahedral character.[13,20] This suggests the possibility that in the pH range near neutrality the zinc site is relatively inflexible and cannot readily adjust to the square planar geometry preferred by Cu(II) (Ref.24).

(c) Activities of metal-substituted derivatives. The superoxide dismutase activities measured for metal-substituted derivatives of Cu_2Zn_2SOD are summarized in Table 1. In general it appears that the only derivatives with substantial activity have Cu^{2+} bound to the native Cu site. The activity is close to that of Cu_2Zn_2SOD if the zinc site is occupied by a metal ion, but appears to be variable when the zinc site is unoccupied, i.e. in Cu_2E_2SOD (Table 1). Copper in the zinc site, in the derivatives $Ag(I)_2Cu_2SOD$ and Cu_2Cu_2SOD, apparently does not contribute to the activity. The variability of the activity of Cu_2E_2SOD is discussed below.

TABLE 1

RELATIVE ACTIVITY OF METAL-SUBSTITUTED DERIVATIVES OF COPPER-ZINC SUPEROXIDE DISMUTASE

Derivative	Activity (% of native)	Ref.
Cu_2Zn_2 (as isolated)	100	2, 10
Cu_2Zn_2 (reconstituted)	110	15
E_2E_2 (apo)	0, <0.1%	10, 15
E_2Co_2	0	16
E_2Zn_2	0	22
Cu_2E_2	20-80	15, 20
Cu_2Co_2	90	15
Cu_2Hg_2	90	15
Cu_2Cd_2	70	15
$Ag(I)_2Cu_2$	5	13
$Ag(I)_2Co_2$	trace	13
Cu_2Cu_2	100	20

(d) pH dependent properties of copper-zinc superoxide dismutase and derivatives. The superoxide dismutase activity of Cu_2Zn_2SOD is invariant over the pH range 5-9 (Refs.25,26) as are the visible,[27] NMR (ref.28) and ESR (Ref.27) spectral properties. Above pH 10, the ESR spectrum shows some differences from the spectrum below pH 10 and nitrogen superhyperfine splitting is observed. Also above pH 10, changes in proton nuclear relaxation rates have been observed[29,30] which have been attributed to binding of hydroxide to Cu^{2+} with concomittant cleavage of the bond between Cu^{2+} and the bridging histidine 61.[30] In contradiction to the above conclusion, electron spin echo experiments at pH 12.2 strongly suggest that the histidyl bridge between Cu^{2+} and Zn^{2+} remains intact at high pH.[31] The spectral changes that are observed at high pH have also been attributed to deprotonation of water complexed to Cu(II) (Ref.29).

Below pH 5, reversible changes in the ESR and visible spectra of Cu_2Zn_2SOD and Cu_2Co_2SOD have been attributed to cleavage of the bond between Zn^{2+} or Co^{2+} and histidine 61 combined with protonation of that residue.[32-35] Reversible changes in the visible and ESR spectra of E_2Co_2SOD (Ref.17) and in the NMR spectra of apo-SOD (Ref.33) have also been observed in the same pH range. In the latter case, it was concluded that the apoprotein has a compact conformation above pH∿5 but that a reversible transition occurs near pH 4 in which the protein adopts a random coil conformation. Sedimentation coefficients suggest furthermore that monomerization has occurred.[33]

Systematic studies of the effect of changing pH on the properties of other derivatives of copper-zinc SOD have not been previosuly reported. Such studies carried out in our laboratory are the subject of the remainder of this report.

Further studies on the nature of the low pH transition. The spectral changes that occur below pH 4.5 for Cu_2Zn_2SOD and Cu_2Co_2SOD have previously been ascribed to cleavage of the bond between histidine 61 and Zn^{2+} or Co^{2+}, but it has generally been assumed that the metal ion remains bound to the protein.[32-35]

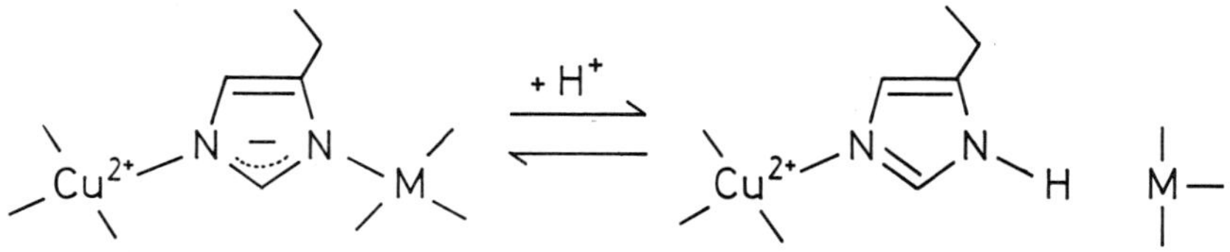

Our recent studies of some metal substituted derivatives of copper-zinc SOD have led us to a different conclusion.[36]

Figure 1 shows ESR spectra of Cu_2Zn_2SOD, Cu_2E_2SOD, and Cu_2Cu_2SOD in solutions of varying pH. We were struck first by the resemblance between the low pH ESR spectra of Cu_2Zn_2SOD and Cu_2E_2SOD (Figure 1, parts a and b). The visible spectra of these two derivatives are also identical at low pH.[36] This suggested the possibility that Zn^{2+} might be reversible released into solution at low pH. To test this possibility, we carried out the following experiment. A 0.5 mM solution of Cu_2Zn_2SOD was adjusted to pH 3.6 and then dialyzed against 0.1 M potassium phopshate at the same pH. Visible and ESR spectra were found to be only minimally changed by the dialysis procedure. Analysis of the metal ion

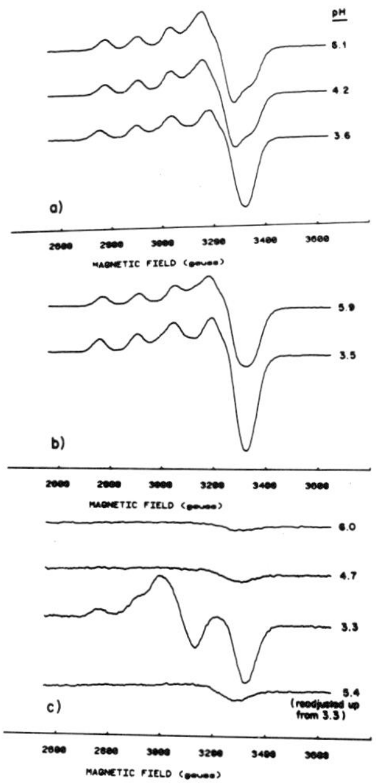

Fig. 1. The pH dependence of the ESR spectra of a) Cu_2Zn_2SOD, b) Cu_2E_2SOD and c) Cu_2Cu_2SOD at $30\pm2°C$. Ref. 36 for experimental details.

content by atomic absorption, however, indicated that 95% of the Zn and 5% of the Cu had been removed from the solution. Likewise, addition of one equivalent of Zn^{2+} per subunit to solutions of Cu_2E_2SOD at pH 3.6 caused only small changes in the spectra of the solutions, but raising the pH to 6.0 gave spectra identical to those of Cu_2Zn_2SOD rather than those of Cu_2E_2SOD at that pH. This experiment strongly suggests that Zn^{2+} is either very weakly bound or free in solution in solutions of Cu_2Zn_2SOD at low pH.

A similar reaction occurs for Cu_2Cu_2SOD. The changes in the ESR spectrum at 30°C (Figure 1, part c) illustrates this quite dramatically. We have found that Cu_2Cu_2SOD is ESR silent at this temperature.[37] This is undoubtedly due to proximity of the two Cu^{2+} ions bridged by imidazolate.

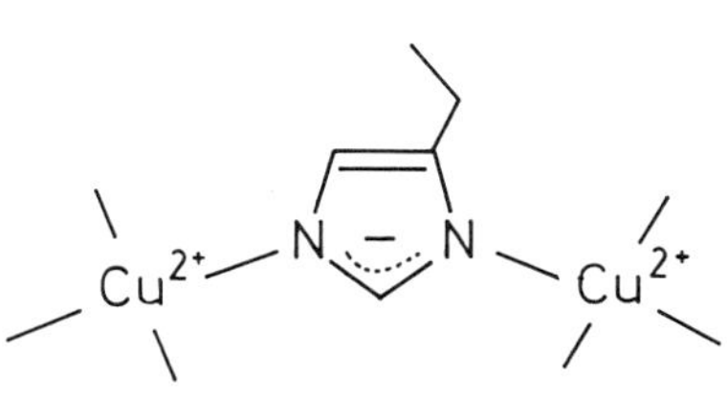

At low pH, the coppers apparently become uncoupled resulting in the spectrum shown in Figure 1 (part c), which is a superposition of the signal due to Cu_2E_2SOD and the isotropic signal due to free aqueous Cu^{2+}. Raising the pH causes the ESR spectrum to disappear again, establishing that the reaction is reversible.

We conclude from our results described above that the metal ion in the native Zn site of Cu_2Zn_2SOD and Cu_2Cu_2SOD is released in the range 4.5>pH>3.0 and is rebound when the pH is raised. It is interesting that the pH dependence of the properties of these derivatives is so similar and that Cu_2Co_2SOD (Ref.35), E_2Co_2SOD (Ref.17) and the apoprotein[33] also undergo reversible changes in the same pH region. It is possible that this behavior is due to similar conformational changes for all of these derivatives. The success of the reconstitution procedure of Beem et al.[13,15] where the metal ion intended for the native copper binding site is added at pH 3.8 presumably is related to the fact that only one strong metal binding site per subunit exists at that pH and competition with a second binding site in the subunit is thereby avoided.

pH dependent Cu^{2+} migration in Cu_2E_2SOD. Our investigation of the effects of varying the pH of solutions of Cu_2E_2SOD resulted from our studies of the variability of the superoxide dismutase activity of Cu_2E_2SOD (Table 1). The activity of preparations of Cu_2E_2SOD has variously been reported to be as low as 20% or as high as 80% that of Cu_2Zn_2SOD.[15,20] Our experiemnts have shown that much of this variability can be explained by the fact that the activity of Cu_2E_2SOD is pH dependent, unlike that of Cu_2Zn_2SOD.[25,26] Thus we find that the activity of Cu_2E_2SOD assayed at pH 8.7 is $\sim$35 ± 5% that of native protein while at pH 6.0 it is $\sim$80 ± 5% that of native protein.[37,38] In attempting to understand how the presence of Zn^{2+} in Cu_2Zn_2SOD causes the superoxide dismutase activity to be pH invariant over the range 5<pH<10, we have studied the ESR and visible spectral properties of Cu_2E_2SOD as a function of varying pH. Our results, summarized here, imply that a novel rearrangement has occurred in which Cu^{2+} has migrated from the Cu site of $\sim$40% of the subunits to form subunits containing Cu in both the native Cu and Zn sites.[37]

The changes in the 77°K ESR spectra of Cu_2E_2SOD as the pH is raised are shown in Figure 2. The spectrum at pH 9.3 is shown in Figure 2 (curve c) and, at higher gain, in Figure 3 (curve a). The resemblance of this spectrum to that of Cu_2Cu_2SOD, in which all metal ion sites contain Cu^{2+}, can be seen by comparing curves a) and b) of Figure 3. Particularly distinctive is the half-field transition at $\sim$1540 G which is characteristic of a magnetic triplet and is due to the magnetic interaction between the two Cu^{2+}(S = 1/2) ions. Very similar ESR spectra have been observed for low molecular weight binuclear imidazolate-bridged dicopper(II) complexes that have been structurally characterized by X-ray crystallography.[39] Visible spectra and ESR spectra at 30°C also show that

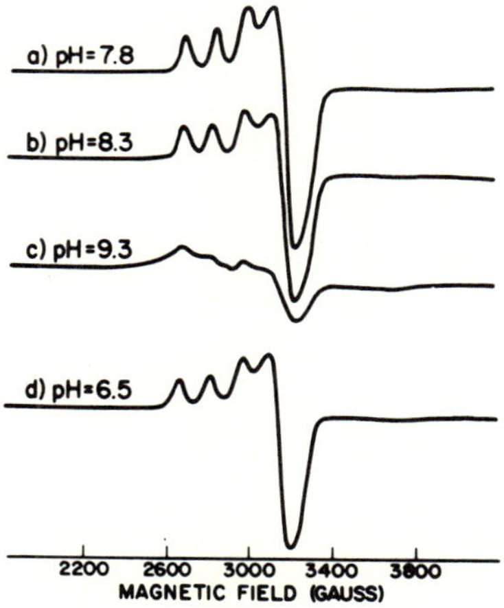

Fig. 2. The pH dependence of the ESR sepctra of zinc-free SOD at 77 K. See Ref.37 for experimental details.

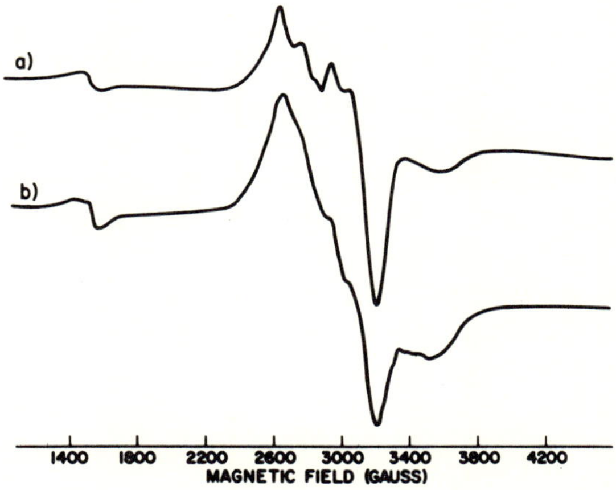

Fig. 3. a) ESR spectrum of zinc-free SOD at pH 9.3 and 77 K. This spectrum is the same as curve c) of Fig. 2 except that the gain was increased by a factor of 5. b) ESR spectrum of Cu_2Cu_2SOD at 77 K. See Ref. 37 for experimental details.

zinc-free SOD undergoes reversible changes as the pH is raised, consistent with the formation of subunits containing the imidazolate bridged dicopper(II) chromophore, such as is found in Cu_2Cu_2SOD. The midpoint of this transition was found to occur at pH 8.2.[37]

Our interpretation of this observation is that a reversible pH-dependent migration of copper(II) ion from one subunit to the zinc site of another subunit (not necessarily in the same protein dimer) occurs.

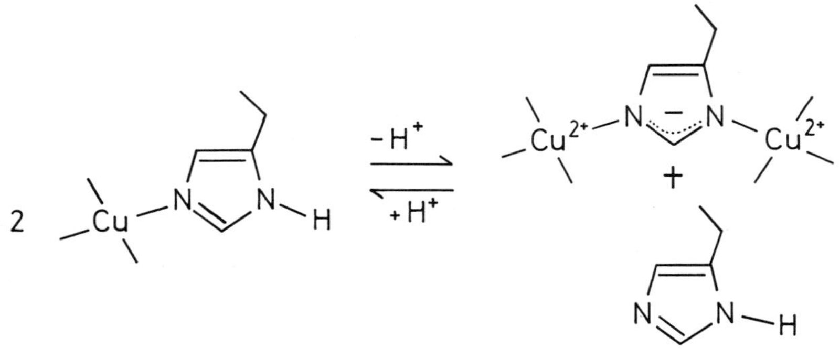

The migration of copper(II) from the copper site to the zinc site of some of the subunits explains why the superoxide dismutase activity is lower at high pH, since copper in the native site is known to be inactive (see above). It is also clear that some previous speculations concerning the role of zinc in determining the properties of Cu_2Zn_2SOD that have been based on studies of zinc-free SOD must be reexamined in light of these new findings. For example, the reported thermal instability of zinc-free SOD is based on experiments carried out at pH 7.8 (Ref.40). Our experiments indicate that 10% of the protein is present as the apoprotein at that pH and the apoprotein is well known to have less thermal stability than Cu_2Zn_2SOD (Ref.2).

Conclusions. The metal binding properties of copper-zinc superoxide dismutase are unique among well characterized metalloproteins. In fact, this protein is a particularly interesting system for the inorganic coordination chemist, apart from any consideration of its physiological function. It represents a unique type of binucleating ligand for which the metal ion affinities of the metal binidng sites can be regulated by varying the pH. Thus below pH 3, we find that the protein releases all metal ions. Between pH 3 and 4.5, it binds one metal ion either Cu^{2+} or Ag^{2+}, per subunit, probably but not necessarily at the native Cu site. Above pH 4.5, other metal ions will bind to the native zinc binding site, but if no other metal is added, above pH 8, a disproportionation occurs resulting in formation of equal amounts of binuclear subunits and apo subunits.

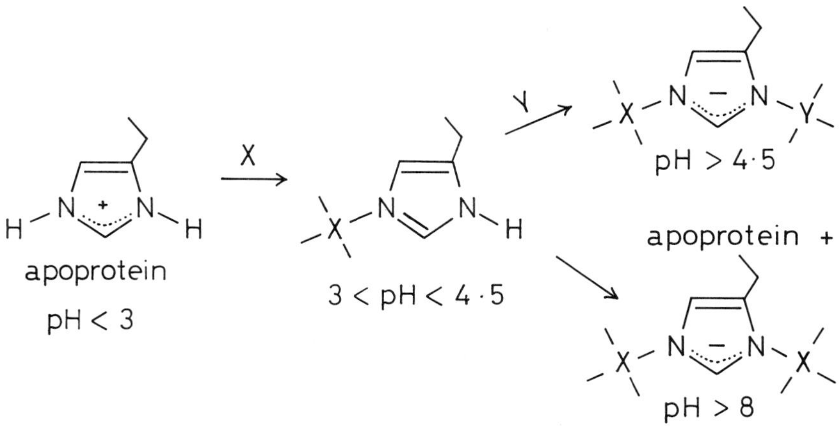

Similar low molecular weight synthetic ligands are not yet available for studies in inorganic coordination chemistry. Future work in our laboratory will be directed toward continued elucidation of the metal binding properties of this interesting biological ligand.

ACKNOWLEDGEMENTS

Financial support from the Charles and Johanna Busch Memorial Fund, U.S. Public Health Grant Service GM24759 and National Science Foundation Grant CHE78-08122 are gratefull acknowledged. M.W.P. is the recipient of a Johnson and Johnson Fellowship in the Biological Sciences (1978-79) and J.S.V. of a U.S.P.H.S. Research Career Development Award. J.S.V. also thanks the Rutgers Research Council for a grant for partial support of travel expenses.

REFERENCES

1. McCord, J.M. and Fridovich, I. (1977) in Superoxide and Superoxide Dismutases, Michelson, A.M., McCord, J.M. and Fridovich, I., eds., Academic Press, New York, pp.1-10.
2. Fridovich, I. (1974) Adv. Enzymol. 41, 35-97.
3. Fridovich, I. (1978) Science, 201, 875-880.
4. Fridovich, I. (1979) Adv. Inorg. Biochem. 1, 67-90.
5. Fee, J.A. and Valentine, J.S. (1977) in Superoxide and Superoxide Dismutases, Michelson, A.M., McCord, J.M., and Fridovich, I., eds., Academic Press, New York, pp. 19-60.
6. Valentine, J.S. (1979) in Oxygen: Clinical and Biochemical Aspects, Caughey, W.S., ed., Academic Press, New York, in press.
7. Fee, J.A. and McClune, G.K. (1978) in Mechanisms of Oxidizing Enzymes, Singer, T.P. and Ondarza, R.N., eds., Elsevier, New York, pp. 273-284.
8. Fee, J.A., Lees, A.C., Bloch, P.L. and Neidhardt, F.C. (1979) in Oxygen: Clinical and Biochemical Aspects, Caughey, W.S., ed., Academic Press, New York, in press.
9. Mann, T. and Keilin, D. (1938) Proc.Roy.Soc. London Ser.B. 126, 303-315.
10. McCord, J.M. and Fridovich, I. (1969) J. Biol. Chem. 244, 6049-6055.
11. Carrico, R.J. and Deutsch, H.F. (1970) J. Biol. Chem. 245, 723-727.

12. Richardson, J.S., Thomas, K.A., Rubin, B.H. and Richardson, D.C. (1975) Proc. Nat. Acad. Sci. USA, 72, 1349-1353.
13. Beem, K.M., Richardson, D.C. and Rajagopalan, K.V. (1977) Biochemistry, 16, 1930-1936.
14. Steinman, H.M., Naik, V.R., Abernethy, J.L. and Hill, R.L. (1974) J. Biol. Chem. 249, 7326-7338.
15. Beem, K.M., Rich, W.E. and Rajagopalan, K.V. (1974) J. Biol. Chem. 249, 7298-7305.
16. Fee, J.A. (1973) J. Biol. Chem. 248, 4229-4234.
17. Calabrese, L., Cocco, D., Morpugo, L., Mondovi, B. and Rotilio, G. (1976) Eur. J. Biochem. 64, 465-470.
18. Lippard, S.J., Burger, A.R., Ugurbil, K., Pantoliano, M.W. and Valentine, J.S. (1977) Biochemistry, 16, 1136-1141.
19. Cass, A.E.G., Hill, H.A.O., Bannister, J.V. and Bannister, W.H., (1979) Biochem. J. 177, 477-486.
20. Fee. J.A. and Briggs, R.G. (1975) Biochim. Biophys. Acta, 400, 439-450.
21. Calabrese, L., Rotilio, G., and Mondovi, B. (1972) Biochim. Biophys. Acta, 263, 827-829.
22. Rotilio, G., Calabrese, L., Bossa, F., Barra, D., Finazzi-Agro, A. and Mondovi, B. (1972) Biochemistry, 11, 2187-2187.
23. Cotton, F.A. and Wilkinson, G. (1972) Advanced Inorganic Chemistry, Interscience, New York, p. 881.
24. Cotton, F.A. and Wilkinson, G. (1972) Advanced Inorganic Chemistry, Interscience, New York, p. 912.
25. Rotilio, G., Bray, R.C. and Fielden, E.M. (1972) Biochim. Biophys. Acta, 268, 605-609.
26. Klug, D., Rabani, J. and Fridovich, I. (1972) J. Biol. Chem. 247, 4839-4842.
27. Rotilio, G., Finazzi-Agro, A., Calabrese, L., Bossa, F., Guerrieri, P. and Mondovi, B. (1971) Biochemistry, 10, 616-621.
28. Cass, A.E.G., Hilll, H.A.O., Smith, B.E., Bannister, J.V. and Bannister, W.H. (1977) Biochemistry, 16, 3061-3066.
29. Terenzi, M., Rigo, A., Franconi, C., Mondovi, B., Calabrese, L. and Rotilio, G. (1974) Biochim. Biophys. Acta, 351, 230-236.
30. Boden, N., Holmes, M.C. and KNowles, P.F. (1979) Biochem. J. 177, 303-309.
31. Fee, J.A. and Mims, W.B. (1979), personal communication.
32. Fee, J.A. (1977) in Superoxide and Superoxide Dismutases, Michelson, A.M., McCord, J.M. and Fridovich, I., eds., Academic Press, New York, pp. 173-192; and Rotilio, G. and Calabrese, L. (1977) ibid., pp. 193-198.
33. Fee, J.A. and Philips, W.D. (1975) Biochim. Biophys. Acta, 412, 26-38.
34. Rotilio, G., Morpurgo, L., Calabrese, L., Finazzi-Agro, A. and Mondovi, B. (1 (1977) in Metal-Ligand Interactions in Organic Chemistry and Biochemistry, Part 1, Pullman, B. and Goldblum, N. eds., D.Reidel Publ. Co., Dordrecht, Holland, pp. 243-253.
35. Calabrese, L., Cocco, D., Morpurgo, L., Mondovi, B. and Rotilio, G. (1975) FEBS Lett. 59, 29-31.
36. Pantoliano, M.W., McDonnell, P.J. and Valentine, J.S. (1979) J. Amer. Chem. Soc., in press.
37. Valentine, J.S., Pantoliano, M.W., McDonnell, P.J., Burger, A.R. and Lippard, S.J. (1979) Proc. Nat. Acad. Sci. USA, 76, 4245-4249.
38. Pantoliano, M.W. and Valentine, J.S. unpublished results.
39. O'Young, C.-L., Dewan, J.C., Lilienthal, H.R. and Lippard, S.J. (1978) J. Amer. Chem. Soc. 100, 7291-7300.
40. Forman, H.J. and Fridovich, I. (1973) J. Biol. Chem. 248, 2645-2649.
41. Rigo, A., Terenzi, M., Viglino, P., Calabrese, L., Cocco, D. and Rotilio, G. (1977) Biochem. J. 161, 31-35.
42. Rigo, A., Viglino, P., Calabrese, L., Cocco, D., and Rotilio, G. (1977) Biochem. J. 161, 27-30.

PREPARATION AND PRELIMINARY CHARACTERIZATION OF HYBRIDS OF BOVINE Cu,Zn SUPEROXIDE DISMUTASE WITH SELECTIVELY METAL-FREE OR METAL-SUBSTITUTED SUBUNITS

L. CALABRESE,[+] D. COCCO,[+] A. RIGO[++] AND G. ROTILIO[+]
[+]Institutes of Biological Chemistry and Applied Biochemistry, University of Rome, Rome, Italy; [++]Institute of Physical Chemistry, University of Venice, Venice, Italy

Scope of the experimental approach. A major problem in present research on Cu, Zn superoxide dismutases concerns the role of their dimeric structure in the catalytic mechanism. The isolation of active monomers, although already reported, relies on weak and largely circumstantial evidence.[1,2] Hybrids between active and inactive subunits have been shown to have half the specific activity of the native dimer[3] but the procedure for isolation of the hybrid and determination of protein concentration appears to be subject to a number of experimental errors. However the approach of using "hybrid" dimers is full of potentialities and deserves special attention in the case of Cu,Zn superoxide dismutases, where the presence of a double labelling on each subunits, that is the bimetal cluster typical of these protein, increases the probability for successful experimental record of cross-effects between metal centers and hence between subunits.

Cobalt substitution for the zinc in Cu,Zn superoxide dismutase has opened the area of "spectroscopic information transfer" between metal sites in these enzymes. In this case optical spectra of the cobalt have been shown to sense the valence changes of the adjacent copper on the same subunit and the nature of the spectral changes has demonstrated that the adjacent non-catalytic metal site is in the same environment in the copper-free protein and when the catalytically active copper becomes reduced in the catalytic cycle.[4] This result establishes that copper communicates to zinc its catalytic reduction via a protein change restricted to the region of the neighbor metal binding sites and this change has been identified as the coincident protonation of the copper-facing nitrogen of the imidazole between the two metals.

On the molecular scale, the question is whether information of this local change can be transferred to the other active site through the dimeric structure. One problem regards the magnitude of change, which is likely to be relatively small and restricted to the active site region; the second problem is that of the time scale, which must cope with the high catalytic constant of the enzyme. Two models are obviously available, that is, a statistical versus a cooperative situation. Evidence for cooperative interaction between the metal binding sites of each monomer has been inferred from steady state values of the redox state of the active copper ions during turnover,[5] from the equilibrium constants of binding of copper to the copper-free protein[6] and from the non-linear dependence of the catalytic constant on the extent of reconstitution with copper of the copper-free protein.[7] On the other hand the antithetical results mentioned above[1,3] require that additional evidence be provided by an entirely novel approach properly designed to answer the questions raised by previous experiments. In particular, preparation of hybrids between copper-containing and copper-free subunits, obtained without strong treatments such as concentrated urea or heating of monomers, may represent the ideal case of hybridization between active and inactive monomer. Furthermore inserting cobalt in the vacant region of copper could be useful, not only to "lock" the 1-Cu dimer against time-dependent changes such as metal migration, but to create novel "spectroscopic hybrids" as well, potentially able to record information transfer between the same sites on different monomers.

Preparation and properties of the "half-the-coppers" [(Cu,Zn)-(...,Zn)] dimer. Bovine Cu,Zn superoxide dismutase was made copper-free by the procedure described by Rigo et al.[6] To this protein 0.7 Cu(II) ions per dimer were added to maximize the yield of the "half-the-copper" species.[6] This species was actually separated from its mixture with copper-free and holo-proteins, as demonstrated by analytical polyacrylamide gel electrophoresis and by preparative column electrophoresis according to Whitehead et al.[8] using the conditions[7] of analytical gel-electrophoresis. In the latter case the elution was followed by recording the optical density at 260 and 280 nm, which takes advantage of different U.V. spectra of holo- and copper-free proteins.[9] As shown in Figure 1 the separation of the three species is obtained with fairly good resolution and allows the pooling of fractions relatively pure with respect to the 1-Cu dimer. They correspond to fractions having a 260 nm to 280 nm O.D. ratio of 1.4 and a copper content, as calculated by either double integration of the EPR signal or atomic absorption spectroscopy, of 1 copper atom per 32,000 daltons. This derivative is spectrally indistinguishable from the native 2-Cu dimer as far as line shape of optical and EPR spectra are concerned; also the extinction coefficient of the copper is the same, that is, $\varepsilon_{680\ nm} = 150\ M^{-1}\ cm^{-1}$. However prolonged standing at room temperature resulted in a slight but significant increase of the 2-Cu component in the electrophoretic pattern. It is not clear whether this is due to metal migration between subunits or to subunit exchange between 1-Cu dimers. The freshly isolated 1-Cu dimer quantitatively reconverts to the native molecule by addition of the full complement of Cu(II) ions, as judged by EPR and optical spectra and electrophoretic mobility.

Preparation and properties of the [(Cu,Zn)-(Co,Zn)] dimers. In a recent report from our laboratory it has been shown that, under certain

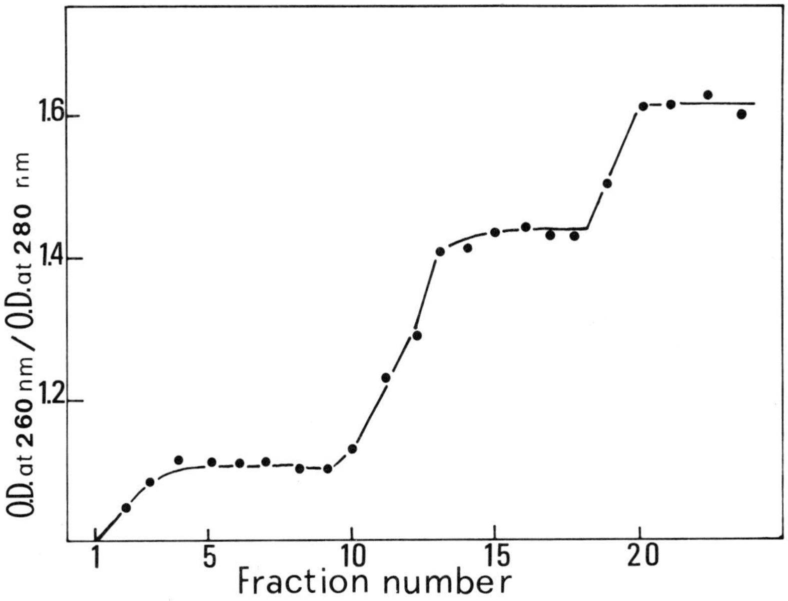

Fig. 1. Elution profile of the preparative column electrophoresis as followed by recording the 260 nm:280 nm O.D. ratio of the fractions. For details see text.

conditions of pH and buffering ions, cobalt binds in a slow process to superoxide dismutase which has zinc sites filled and copper sites vacant.[10] No evidence is against the assumption that this novel cobalt binding site is that occupied by copper in the native protein. According to the binding properties already described[10], slightly less than stoichiometric Co(II) ions (about 0.8 equiv.) were added to the 1-Cu protein in 0.1 M phosphate buffer, pH 7.4. The binding process was followed spectrophotometrically and when completion was reached the sample was extensively dialysed against the same buffer. After dialysis the metal content, as assayed by atomic absorption spectroscopy, was 1 copper and 0.7 cobalt per mole of protein. The electrophoretic mobility on polyacrylamide gel was identical to that of the 2-Cu dimers. The optical spectrum of this derivative is shown in Figure 2, where the spectrum of the (Cu,Co) species is also shown to point out the specific differences in band shape and absorption maxima, $\Delta\varepsilon_{606\ nm}\ M^{-1}\ cm^{-1} \simeq 330$, that is, the optical density referred to cobalt only is compatible with pseudotetrahedral geometry. The EPR spectra (Figure 3) differentiate the two types of cobalt derivatives even better: the [(Cu,Zn)-(Co,Zn)] spectrum (curve c) is characterized by a new type of high spin cobalt signal[10] that is very different as far as line

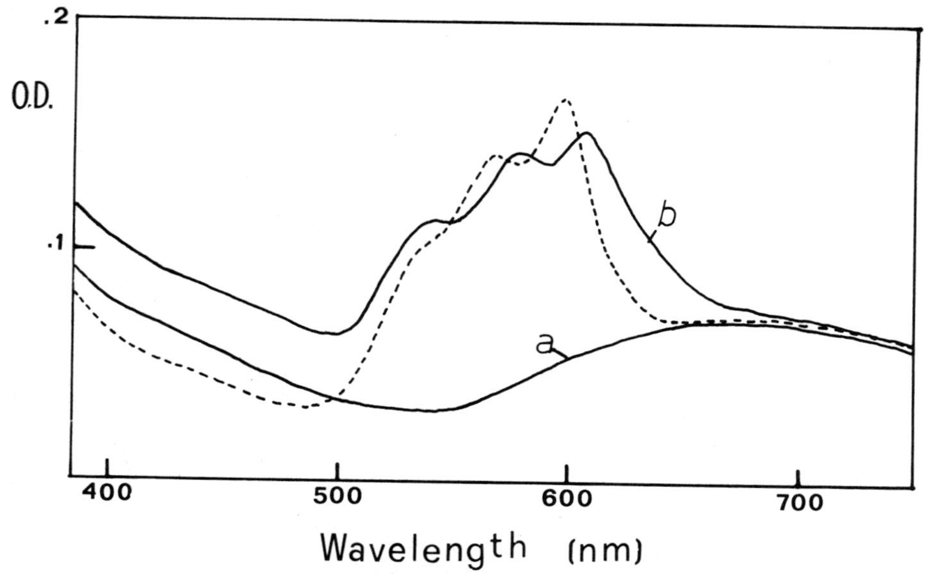

Fig. 2. Optical spectra of selectively metal-free and metal-substituted derivatives of bovine Cu,Zn superoxide dismutase. a) 0.45 mM 1-Cu dimer (0.46 mM copper); b) the same as a) after addition of 0.8 equiv. Co(II) and subsequent dialysis; dotted line: 0.25 mM (Cu,Co) derivative (0.45 mM copper, 0.28 mM cobalt). Buffer 0.1 M phosphate, pH 7.4.

shape and g values are concerned, from that of cobalt bound to the zinc site (curve a), although it is still consistent with pseudotetrahedral coordination. Moreover the absence of magnetic interaction between the two paramagnetic ions located in different subunits (curves b and c), while confirming that the novel cobalt binding site is likely to be the site occupied by copper in the native protein, makes both ions fully detectable by EPR and thus available for experiments of information transfer between subunits, which can take advantage of selective reactivity of the two ions. Preliminary unpublished experiments from this laboratory indicate that equilibrium reactions of the copper on one subunit do in fact modify the spectra of the cobalt on the other subunit.

Catalytic properties of the 1-Cu and the [(Cu,Zn)-Co,Zn)] dimers. Table 1 reports the catalytic constants of superoxide dismutase derivatives with different metal composition. The relevant data, which has been found to be reproducible in separate experiments, is that the specific activity of the 1-Cu dimer (k/protein in Table 1) is not much less than that of the controls reconstituted with the full complement

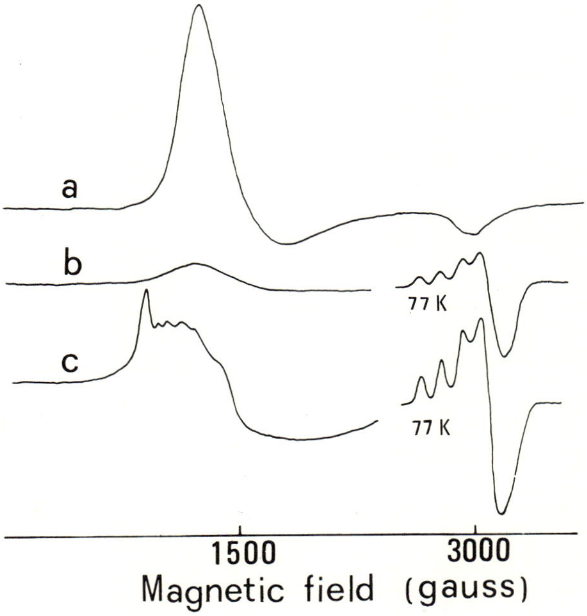

Fig. 3. EPR spectra of cobalt derivatives of bovine superoxide dismutase. a) 0.3 mM (...,Co) protein, 0.3 mM Co(II); b) 0.25 mM (Cu,Co) protein, 0.45 mM copper and 0.28 mM cobalt; c) 0.45 mM [(Cu,Zn)-(Co,Zn)] hybrid, 0.46 mM copper and 0.3 mM cobalt. The spectra were obtained with a Varian E-9 instrument equipped with an Air Products and Chemicals LT-3-110 liquid helium-transfer Cryo-Tip refrigerator with automatic temperature controller. The temperature was 10 K, except for the part of the spectrum marked 77 K, which was recorded at higher temperature to avoid saturation of the copper signal.

of copper. However, such figures are strongly influenced by the copper content of dimeric molecules, which is always less than 2 atoms even in the native protein,[11] and could affect any calculation[3] based only on protein determination. Therefore it appears to be more reliable to refer specific activities to the content of native copper, as evaluated by analysis of line shapes and double-integrated areas of EPR copper signals (k/copper in Table 1). By this parameter, the activity of the 1-Cu dimer is significantly higher than that of doubly occupied dimers. However, it levels down to the activity of controls after prolonged standing of samples at room temperature. These results confirm the observations made by Rigo et al.[7] on mixtures with different content of 1-Cu and 2-Cu dimers and are consistent with the assumption that copper on doubly metal-saturated dimers is less effective in the catalytic dismutation of superoxide, irrespective of the presence of copper or cobalt on the other subunit. On the other hand the decrease of enzymic activity with time in the 1-Cu samples can be reasonably related to the time-dependent changes of the electrophoretic pattern mentioned before, which leads to a relative increase of the supposedly less active 2-Cu species with time. Also intramolecular metal migration between subunits might occur, although it should be recalled that intermolecular metal exchange between copper-free and holo-proteins was not observed.[6]

TABLE 1

CATALYTIC CONSTANTS OF SUPEROXIDE DISMUTASE SAMPLES WITH DIFFERENT METAL COMPOSITION[a]

SOD species	Cu/protein[b]	k/protein[c]	k/Cu[c]
1-Cu dimer	1.1	1.4 (1.2)	1.3 (1.1)
1-Cu dimer + 0.6 equiv. Cu(II)	1.7	1.7 (1.7)	1.0 (1.0)
Copper-free dimer + 1.6 equiv. Cu(II)	1.6	1.6 (1.6)	1.0 (1.0)
1-Cu dimer + Co(II) (Co(II)/protein = 0.6)	1.1	1.1 (1.1)	1.0 (1.0)

[a]Catalytic constants (k) evaluated by polarographic method of Rigo et al. (Ref. 12) and expressed as M^{-1} sec^{-1} x 10^9.
[b]Determined by double integration of EPR signals against a Cu-EDTA standard.
[c]The values in parentheses were measured after 24 hr standing of the samples at 20°C

In conclusion these results leave two problems open. The first one is quantification of the activity of the 1-Cu dimer, which is made difficult by the not entirely complete separation of the supposedly more and less active species by electrophoresis. The second problem is that this effect can not be related to negative cooperativity during catalysis, but seems to be rather associated to conformational differences of the active site region in the singly- and doubly-occupied dimers. These differences are, however, not detected by EPR. Negative cooperativity seems not to exist in hybrids containing active and inactive monomers. In fact the results obtained with the (Cu-Co) hybrid confirm the result by Malinowski and Fridovich,[3] as it has been shown that binding of cobalt in the copper site produces inactive molecules.[10]

ACKNOWLEDGMENT

This work has been supported by C.N.R. grant No. 78.02261.04.

REFERENCES

1. Bannister, J.V., Anastasi, A. and Bannister, W.H. (1978) Biochem. Biophys. Res. Commun. 81, 469-472.
2. Bossa, F., Barra, D., Marmocchi, F., Martini, F., Rotilio, G., Bannister, J.V. and Bannister, W.H., this volume.
3. Malinowski, D.P. and Fridovich, I. (1979) Biochemistry, 18, 237-244.
4. McAdam, M.E., Fielden, E.M., Lavelle, F., Calabrese, L., Cocco, D. and Rotilio, G. (1977) Biochem. J. 167, 271-274.
5. Fielden, E.M., Roberts, P.B., Bray, R.C., Lowe, D.J., Mautner, G.N., Rotilio, G. and Calabrese, L. (1974) Biochem. J. 139, 49-60.
6. Rigo, A., Viglino, P., Calabrese, L., Cocco, D. and Rotilio, G. (1977) Biochem. J. 161, 27-30.
7. Rigo, A., Terenzi, M., Viglino, P., Calabrese, L., Cocco, D. and Rotilio, G. (1977) Biochem. J. 161, 31-35.
8. Whitehead, J.S., Kay, E., Lew, J.Y. and Shannon, L.M. (1971) Anal. Biochem. 40, 287-291.
9. Rotilio, G., Calabrese, L., Bossa, F., Barra, D., Finazzi Agro, A. and Mondovi, B. (1972) Biochemistry, 11, 2182-2187.
10. Calabrese, L., Cocco, D. and Desideri, A. (1979) FEBS Lett. 106, 142-144.
11. Fee, J.A. (1973) Biochim. Biophys. Acta, 295, 82-91.
12. Rigo, A. Viglino, P. and Rotilio, G. (1975) Anal. Biochem. 68, 1-8.

COMPARISON OF METHODS FOR DETERMINING THE PROSTHETIC METAL ION OF BACTERIAL SUPEROXIDE DISMUTASES IN CRUDE EXTRACTS

A.R. McEUEN[+], H.A.O. HILL[+], G.J. DRING[++] AND G.S. INGRAM[+++]
[+]Inorganic Chemistry Laboratory, University of Oxford, Oxford, OX1 3QR, England; [++]Unilever Research Laboratory, Colworth House, Sharnebrook, Bedfordshire, England; [+++]Unilever Research Laboratory, Port Sunlight, Merseyside, England.

INTRODUCTION

Three different types of superoxide dismutase (SOD) have been characterized on the basis of the prosthetic metal ion.[1,2] The copper- and zinc-containing enzyme, with two exceptions (Ref. 3, and P.M. Vignais et al., this volume), has been found exclusively in eukaryotes, the manganese-containing enzyme has been found in both prokaryotes and eukaryotes, and the iron-containing enzyme has been found predominantly in prokaryotes, though it has recently been reported in eukaryotic algae[4] (and higher plants; see M.L. Salin and S.M. Bridges, this volume). The manganese and iron superoxide dismutases are closely related on the basis of amino acid sequence homology,[5] while the cuprozinc enzymes form a phylogenetic class of their own.[5-7]

It is often very useful to be able to identify the prosthetic metal ion of a given SOD in a crude extract without having to purify the protein. Since cyanide inhibits CuZn-SOD but does not affect either Fe-SOD or Mn-SOD,[8] it has been employed in studying the SOD isozymes in eukaryotes[9] and has proved useful in determining the intracellular localization of the different isozymes.[10-18] For prokaryotes, three methods have been proposed for distinguishing between the SODs of this class of organisms: 1) growth of cells on media radiolabelled with ^{55}Fe (Ref. 19); 2) inhibition by hydrogen peroxide;[20] and 3) inhibition by azide.[21] During our studies of bacterial SODs, we have used the two inhibition methods on crude extracts of *Pseudomonas aeruginosa* and on the purified Mn-SOD of *Bacillus stearothermophilus*, and have used all three methods on crude extracts of *Actinomyces viscosus*. We present here an experimental comparison of these techniques and discuss the merits and limitations of each.

RESULTS

The superoxide dismutases of Actinomyces viscosus and Pseudomonas aeruginosa. When cell-free extracts of *A. viscosus* and *Ps. aeruginosa* were electrophoresed on 7.5% polyacrylamide gels at pH 8.3 (Ref. 22), staining with nitro blue tetrazolium (NBT) (Ref. 23) revealed two isozymes migrating towards the anode and

and none moving towards the cathode in each species (Figure 1). For convenience, the more slowly migrating isozyme from A. viscosus is labelled A_1 and the faster migrating one from the same species A_2. Similarly, for Ps. aeruginosa, the slower isozyme is P_1 and the faster one P_2. Densitometric scans showed that in most extracts the ratio of A_1:A_2 was approximately 80:20 and the ratio of P_1:P_2 was approximately 25:75. None of these four SODs were inhibited by 5 mM KCN.

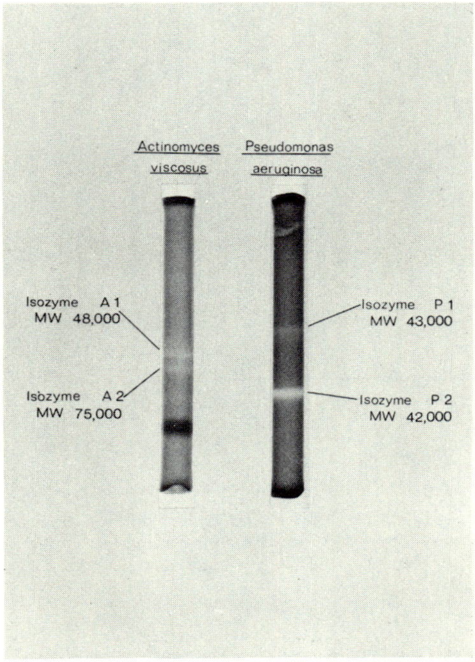

Fig. 1. Disc-electrophoretograms of unfractionated cell-free extracts of A. viscosus and Ps. aeruginosa. Electrophoresis was performed on 7.5% polyacrylamide gels (Ref. 22) which were stained for superoxide dismutase activity with nitro blue tetrazolium (Ref. 23). Molecular weights were determined by the method of Hedrick and Smith (Ref. 24).

Determination of molecular weights by the method of Hedrick and Smith[24] (using 5-10% polyacrylamide gels, NBT to localize the enzyme,[23] and comparison with several known standards) gave values of 48,000 for A_1, 75,000 for A_2, 43,000 for P_1 and 42,000 for P_2. By this technique the purified Mn-SOD from B. stearothermophilus had a molecular weight of 45,000, which is higher than the value of 40,000 determined by ultracentrifugation[25] but is very close to the value derived from the sequence data (see C.J. Brock and J.E. Walker, this volume). Thus, A_1, P_1 and P_2 are probably dimeric and A_2 is probably tetrameric. Since the larger isozyme, A_2, migrates more quickly than the smaller one, A_1, at gel concentrations less than 9% (Figure 2), it can not be an aggregated form of A_1.[22,24] At 9% polyacrylamide, A_1 and A_2 are coincident and at higher gel concentrations A_2 migrates more slowly than A_1.

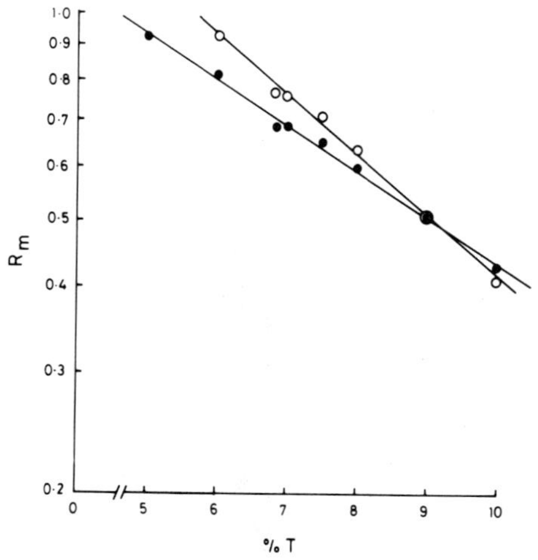

Fig. 2. Molecular weight determination by method of Hedrick and Smith (Ref. 24). Plot of the relative mobility (R_m) of the SOD isozymes of A. viscosus versus the gel concentration (%T). (●) A_1, (O) A_2. Each data point is the average of 2-4 trials.

Determination of the Prosthetic Metal Ions

A. Use of Radioisotopes. We adopted the procedure of Pile and Dougherty[19] with modifications. Cells of A. viscosus were grown on Brain Heart Infusion agar labelled with either ^{54}Mn or ^{55}Fe. Both unfractionated cell-free extracts and 60-80% saturated ammonium sulphate fractions were subjected to electrophoresis on polyacrylamide gels.[22] The gels were stained for enzyme activity with NBT[23] and then cut into 2mm thick slices and the radioactivity determined in each gel slice. In order to distinguish between a radiolabelled SOD and the coincidence of an unlabelled SOD with a different radioactive protein, the 60-80% saturated ammonium sulphate fraction was electrophoresed on gels of five different polyacrylamide concentrations. We did not examine Ps. aeruginosa by this technique, but Pile and Dougherty[19] did. They found a single band of SOD activity which corresponded in mobility to isozyme P_2 and concluded that it contained iron.

When extracts of ^{54}Mn-grown cells of A. viscosus were electrophoresed on 6.8% polyacrylamide gels, a peak of radioactivity was coincident with the gel slice

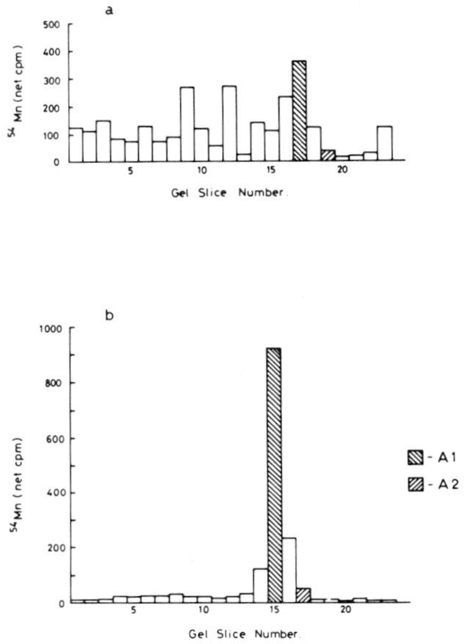

Fig. 3. Effect of partial purification of SOD on ^{54}Mn-labelled proteins from A. viscosus. Cells were grown on BHI agar containing ^{54}Mn. Samples were electrophoresed on 6.8% polyacrylamide gels, the gels were stained with nitro blue tetrazolium, and 2mm thick slices were cut, digested with H_2O_2 and assayed for radioactivity. Gel slices which contained SOD activity are marked by cross hatching. Numbering of slices is in the direction of migration; i.e. towards the anode. a) Crude extract, specific activity of SOD = 4 Units/mg, 4 Units SOD applied; b) 60-90% saturated ammonium sulphate fraction, specific activity of SOD = 37 Units/mg, 12 Units SOD applied.

containing isozyme A_1 in both the crude extract and the 60-80% saturated ammonium sulphate fraction (Figure 3). Even though most of the other Mn proteins were eliminated by this purification step, in three different preparations the ratio of ^{54}Mn in this particular gel slice to total enzyme activity remained nearly constant between the crude extract and the ammonium sulphate fraction. In addition to co-purifying with isozyme A_1, the ^{54}Mn-labelled protein co-electrophoresed with it at four other gel concentrations (Figure 4), indicating it had the same molecular weight and same charge at pH 8.3 as A_1. Since no similar behaviour was observed with extracts from cells grown on ^{55}Fe (Figures 5 and 6), A_1 and the observed ^{54}Mn-containing protein are almost certain to be the same.

In all radiolabelled cultures of A. viscosus, A_2 was very much the lesser of the two isozymes. As a result, the radioisotope experiments gave no clear indication of its constituent metal. For example, in Figure 4 (parts b and d) it is

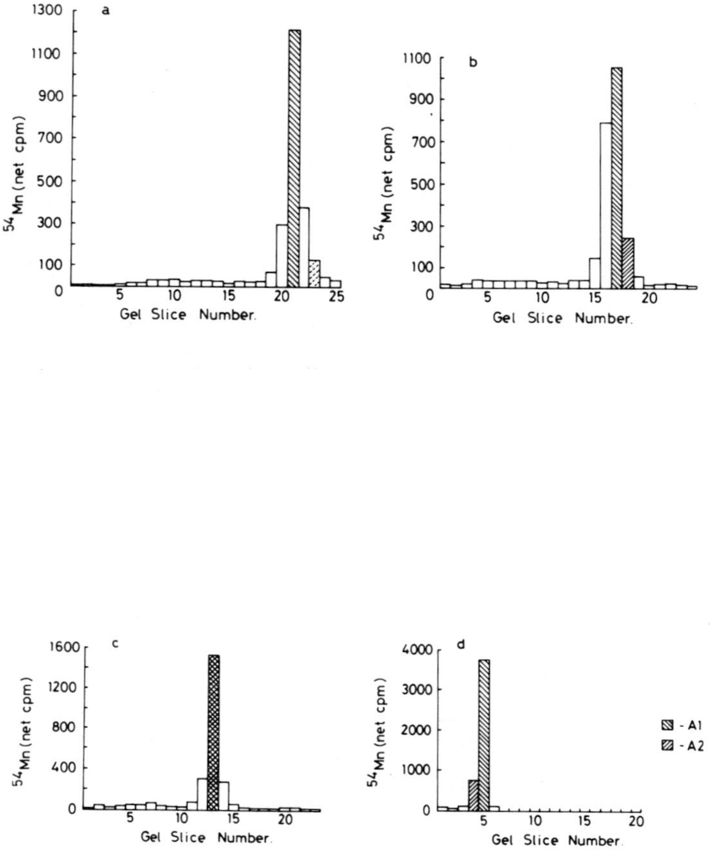

Fig. 4. Coincidence of ^{54}Mn-labelled protein from A. viscosus with isozyme A_1 at four different gel concentrations: a) 5.0%, b) 7.5%, c) 9.0%, d) 15.0%. The 60–80% saturated ammonium sulphate fraction was electrophoresed in each case. Gel slice 23 in a) probably contained A_2, but it was obscured by the presence of bromophenol blue. For further details, see legend to Fig. 3.

uncertain as to whether the radioactivity in the gel slice containing A_2 is due to the minor isozyme or whether it is merely diffusion of A_1 into the adjacent gel slices. The better resolution achieved in Figure 3 (part b) suggests that the latter possibility is the case. Also, ammonium sulphate fractionation more

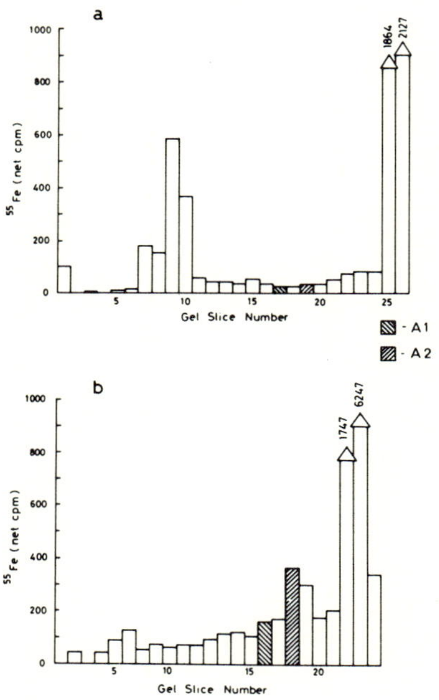

Fig. 5. Effect of partial purification of SOD on ^{55}Fe-labelled proteins of A. viscosus. Cells were grown on BHI agar labelled with ^{55}Fe and samples were electrophoresed on 6.8% gels. a) Crude extract, specific activity of SOD = 6 Units/mg, 4 Units SOD applied; b) 60-80% saturated ammonium sulphate fraction, specific activity of SOD = 43 Units/mg, 16 Units SOD applied. For further details, see legend to Fig. 3.

than halved the ratio of ^{54}Mn in the A_2-containing gel slice to the total SOD activity. The possibility still remains, however, that A_2 does contain Mn, but that it accounted for less than 10% of the total SOD and that in the crude extract it was coincident with another Mn protein. Electrophoresis of the ^{55}Fe-labelled extracts did not provide much information either. An ^{55}Fe-containing protein coincident with A_2 co-purified with it during ammonium sulphate fractionation (Figure 5), but electrophoresis on gels of different polyacrylamide concentration showed that A_2 and this Fe protein were not the same (Figure 6). There is still the possibility that A_2 is an Fe protein, but is so minor compared to other Fe proteins that it was lost in the background.

B. Azide. Misra and Fridovich[21] have recently reported differences between different SODs in their sensitivity to a reversible inhibition by azide. In their

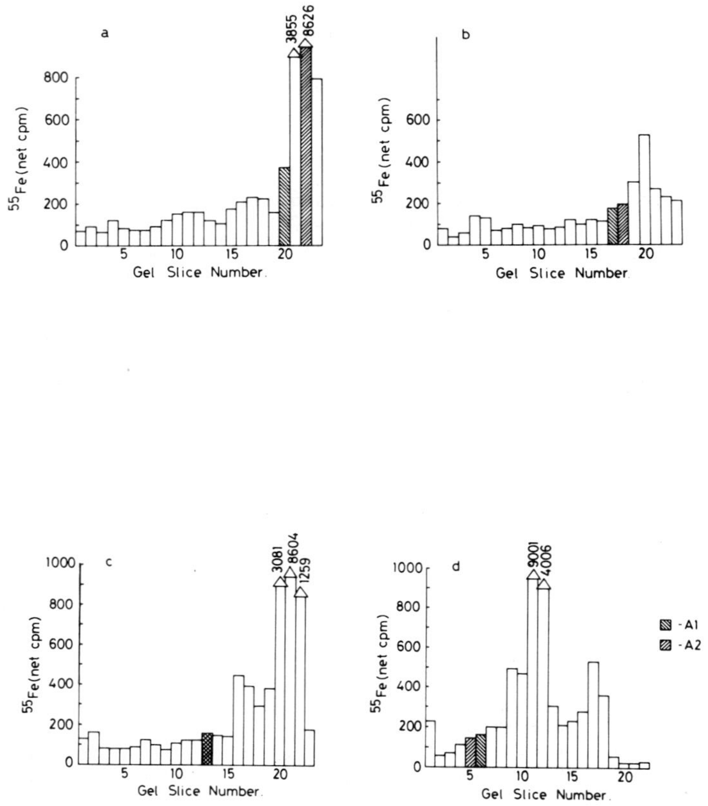

Fig. 6. Electrophoresis of the 60-80% saturated ammonium sulphate fraction containing ^{55}Fe-labelled proteins from *A. viscosus* on gels of different acrylamide concentrations: a) 5.0%, b) 7.5%, c(9.0%, d) 15.0%. For further details, see legend to Fig. 3.

examination of eight purified and well-characterized SODs from different sources they found that regardless of the source of the enzyme the response of a given SOD to azide was uniquely determined by the prosthetic metal ion. At pH 7.8, Fe-SODs were the most sensitive (50% inhibition by 4 mM NaN_3) followed by Mn-SODs

(50% inhibition by 20mM NaN_3) with CuZn-SODs the most resistant of all (32mM NaN_3 required for 50% inhibition). Crude extracts from cells with well-characterized SODs showed the same sensitivity to azide as the purified enzymes, so these authors proposed that azide could be used with crude extracts to determine the prosthetic metal ion of any SOD present.

TABLE 1

INHIBITION OF SUPEROXIDE DISMUTASE ACTIVITY BY AZIDE[a]

Sample	Percent Inhibition		
	4 mM Azide	10 mM Azide	20 mM Azide
B. stearothermophilus Mn SOD	-	30.2	49.0
Ps. aeruginosa (P_1:P_2 = 20:80)	54	-	-
Ps. aeruginosa (P_1:P_2 = 30:70)	47	-	-
A. viscosus (A_1:A_2 = 40:60)	-	50.6	-
A. viscosus (A_1:A_2 = 80:20)	-	19.0	34.6
Fe-SOD[a]	50	73	80
Mn-SOD[a]	18	30	50
CuZn-SOD[a]	-	10	26

[a]Method and data of Misra and Fridovich (Ref. 21)

Table 1 records the effects of azide on the purified Mn-SOD of B. stearothermophilus, on two different extracts of Ps. aeruginosa with different P_1:P_2 ratios, and on two different extracts of A. viscosus with different A_1:A_2 ratios. The purified enzyme from B. stearothermophilus behaved exactly as the Mn-SODs from Eschericia coli, Porphyridium cruentum and human liver.[21] The greater sensitivity of both Ps. aeruginosa extracts to azide suggests that the major isozyme (P_2) contains iron. The fact that the sample enriched in the minor isozyme was slightly more resistant to azide than the other sample suggests that P_1 is a Mn-SOD. For A. viscosus, A_2 is more sensitive to azide than either A_1 or the Mn-SOD from B. stearothermophilus, and so behaves as an Fe-SOD. A_1 is decidedly more resistant to azide than any purified Mn-SOD tested to date,[4,21] and may be as resistant as CuZn-SODs. The presence of A_2 makes it difficult to determine the azide sensitivity of A_1 alone since the sensitivity of a mixture of SODs is not a linear function of the composition of the mixture.[21]

C. Hydrogen Peroxide. Asada and co-workers[26] isolated both a Mn-SOD and an Fe-SOD from the blue-green alga Plectonema boryanum and found that although the

Fe-SOD was rapidly inactivated by H_2O_2, the Mn-SOD remained fully active even after 24 hr in 5mM H_2O_2. Britton and co-workers[20] have proposed a method to distinguish between these two prosthetic metals based on these findings. The method involves soaking electrophoretograms in a solution of 5mM H_2O_2, 0.1mM EDTA and 1mM KCN prior to staining with NBT. Cyanide is included to inhibit catalase. Comparison with control gels shows which SODs are sensitive to this treatment.

The purified Mn-SOD from B. stearothermophilus behaved as expected and was not inactivated by H_2O_2. The results with Ps. aeruginosa were identical to those of Britton and co-workers[20] in that P_1 was resistant to treatment while P_2 was obliterated by it. With A. viscosus, A_1 was inhibited and A_2 was not. Thus P_1 and A_2 behaved as Mn-SODs and P_2 and A_1 behaved as Fe-SODs.

DISCUSSION

The different methods for distinguishing between Fe-SODs and Mn-SODs present a consistent picture for the Mn-SOD from B. stearothermophilus and the two isozymes from Ps. aeruginosa, but present a contradictory one for the two isozymes from A. viscosus. The purified B. stearothermophilus enzyme responded to azide and H_2O_2 in a manner identical to other Mn-SODs.[20,21] Isozyme P_1 displayed the behaviour expected of a Mn-SOD in both these tests while the behaviour of P_2 was consistent with it being an Fe-SOD, a result in agreement with the radioisotope work of Pile and Dougherty.[19] However, A_1 behaved as a Mn-SOD with respect to the radioisotope studies, as a CuZn-SOD with azide, and as an Fe-SOD with H_2O_2. Isozyme A_2 responded to azide like an Fe-SOD and to H_2O_2 as a Mn-SOD, while the work with radioisotopes was inconclusive.

Disagreement between these different techniques is not unique to A. viscosus. The single isozyme from Bacillus megaterium was resistant to H_2O_2 (i.e., behaved as a Mn-SOD) (Ref. 20), but was inhibited 51% by 5.5mM NaN_3 (i.e., behaved as an Fe SOD) (Ref. 21). This enzyme has been isolated and shown to contain iron.[27] The single isozyme from Bacillus subtilis was coincident with a major peak of ^{55}Fe under all conditions of electrophoresis tested,[19] but it responded to H_2O_2 (Ref. 20) and azide[21] like a Mn-SOD. With such discrepancies apparent, it becomes necessary to examine the reliability of each technique.

The principal problem with the radioisotope incorporation method arises in distinguishing between a radiolabelled SOD and other radiolabelled proteins. This is relatively easy if the specific activity of the enzyme is high and the number or amounts of other Fe or Mn proteins is low, as was the case with A_1. However, if the specific activity of a particular isozyme is low, as with A_2, or there is a large number of other radioactive proteins, the task becomes difficult given the limited resolution of gel slicing. Perhaps the best example of interference from other radioactive proteins is the work on Corynebacterium pseudodiphtheriae.[19] When the extracts from this organism was electrophoresed, there was always a sizeable amount of ^{55}Fe (600-800 cpm) in each SOD-containing gel slice, an amount comparable to Fe-SODs in other organisms. However, the SOD was always part of a large band spanning twenty or more gel slices where the radioactivity in each gel slice varied from 500 to over 1200 cpm, thus masking any radioactivity which may have been due to the SOD itself. Complementary studies with ^{54}Mn might reveal that the principal isozyme contains Mn, but as the results with A_2 demonstrate, if the SOD fails to coincide consistently with a significant peak of ^{54}Mn, it does not necessarily follow that it possesses iron. So although the radioisotope method may be the most inherently reliable, the problem with interference from other proteins limits its usefulness. Also, the empirical reliability of this technique is unknown, for none of the SODs examined in this way have been isolated to purity to verify these predictions.

The empirical reliability of the azide inhibition method as presented by Misra and Fridovich[21] is very good. With eight purified SODs (CuZn-SOD from bovine, human, and chicken sources; Mn-SOD from human liver, E. coli and Porphyridium cruentum; and Fe-SOD from E. coli and Plectonema boryanum) and two crude extracts containing well-characterized enzymes (Mn-SOD from Streptococcus faecalis and Fe-SOD from B. megaterium) they found a completely consistent correlation of susceptibility to azide with prosthetic metal ion using two different assays for SOD. Kanematsu and Asada[4] have also applied this technique to eight purified SODs. The Fe-SODs from Chlorobium thiosulfatophilum, Chromatium vinosum, Azotobacter vinelandii, Rhizobium sp. and Plectonema boryanum were all inhibited 67-73% and the Mn-SOD from P. boryanum was inhibited 28% by 10mM NaN_3 at pH 7.8, in excellent agreement with Misra and Fridovich.[21] The Fe-SOD from Pseudomonas ovalis also fitted into this pattern.[28] However, the two Fe-SODs from Euglena gracilis were unusually resistant to azide (49-50% inhibition at 10mM NaN_3). So although this method has now proved consistent with eight Fe-SODs and six Mn-SODs, there are two, possibly three, known exceptions: the two Fe-SODs from Euglena gracilis and isozyme A_1 from A. viscosus, which in all probability is a Mn-SOD on the basis of the radioisotope incorporation work presented here. The exact agreement of our results with the Mn-SOD of B. stearothermophilus with the results of Misra and Fridovich[21] argues against any error in our technique being responsible for the anomalous results with A_1.

The empirical reliability of using H_2O_2 to determine the prosthetic metal ion is comparable to that of using azide. The Fe-SODs from P. boryanum,[26] Rhodopseudomonas sphaeroides,[29] Photobacterium leiognathi,[30] Chomatium vinosum,[4,31] Chlorobium thiosulfatophilum,[4] Brassica campestris (see M.L. Salin and S.M. Bridges, this volume), both Fe-SODs from E. coli,[20,32] and both Fe-SODs from Euglena gracilis[4] were all inactivated by H_2O_2, while the Mn-SODs from P. boryanum,[26] Spirulina platensis,[29] Pleurotus olearius,[33] B. stearothermophilus (Ref. 34 and this work), and Streptococcus faecalis[20] were all resistant. The exceptions to this pattern are the Fe-SOD from *B. megaterium*, which was not inhibited by H_2O_2 (Refs.20,27), and possibly A_1 from *A. viscosus*, which is probably a Mn-SOD but was inhibited. Also, there is controversy over whether the Mn-SOD from *E. coli* is inhibited by H_2O_2 (Refs. 20, 35).

That neither the azide nor the H_2O_2 inhibition techniques are fully reliable reflects the latitude of structure variation permissible within each class of SOD. In order for a metal ion to function solely as a highly efficient catalyst for the disproportionation of O_2^-, certain constraints must be put on its environment to produce the desired chemistry. In SOD, the protein portion of the molecule meets these requirements by not only providing and positioning the appropriate ligands around the metal ion, but also by controlling the access of solvent and solutes to it. A number of protein structures are compatible with these requirements. Azide inhibits by acting as a ligand to the prosthetic metal ion,[21,36] and H_2O_2 presumably inhibits by reacting with the reduced form of the enzyme to produce a reactive intermediate which then destroys essential amino acid residues.[30,37-39] The evidence presented and reviewed here demonstrates that the protein can vary sufficiently in structure to change the affinity of the prosthetic metal ion for azide or the reactivity towards H_2O_2 without significantly affecting the SOD activity.

After consideration of each of the three techniques used in this work, it should be possible to draw some conclusions about the four unknown SODs examined. Since both the azide and H_2O_2 methods predict that P_1 contains Mn and P_2 contains Fe, and since both methods have proved reliable with over 80% of the well-characterized SODs tested to date, it is safe to conclude that these predictions are correct. Studies with ^{55}Fe (Ref. 19), increase the certainty that P_2 contains Fe. A_1 is almost certainly a Mn-SOD because radioisotope incorporation is theoretically more reliable than either inhibition technique and interference from other ^{54}Mn-labelled proteins was minimal (Figure 4). For A_2, the choice is more difficult since both the azide and H_2O_2 methods have similar records of reliability. In the very similar situation with the single SOD of *B. megaterium*,[20,21] characterization of the purified enzyme[27] decided in favour of the prediction made by the azide technique. Thus, precedent favours A_2 being an Fe-SOD, but there is no *a priori* reason for favouring one inhibition technique over the other.

In view of the limitations which have been encountered with each of the three techniques, none of them can be relied upon to give the correct answer every time. If one seeks to distinguish between a Mn-SOD and an Fe-SOD in a cell-free extract, all three methods should be used, of which radiolabelling, when applicable, is inherently the most reliable. Indeed, complementary labelling with ^{54}Mn and ^{55}Fe promises to be a very powerful first technique. In the end, however, only isolation of the purified enzyme and chemical analysis can be absolutely reliable.

REFERENCES

1. Fridovich, I. (1975) Ann. Rev. Biochem. 44, 147-159.
2. Hill, H.A.O. (1978) in New Trends in Bioinorganic Chemistry, Frausto da Silva, J.R.R. and Williams, R.J.P., eds., Academic Press, London, pp. 173-208.
3. Puget, K. and Michelson, A.M. (1974) Biochem. Biophys. Res. Commun. 58, 830-838.
4. Kanematsu, S. and Asada, K. (1979) Arch. Biochem. Biophys. 195, 535-545.
5. Harris, J.I. and Steinman, H.M. (1977) in Superoxide and Superoxide Dismutases, Michelson, A.M., McCord, J.M. and Fridovich, I., eds., Academic Press, New York, pp. 225-230.
6. Steinman, H.M. (1978) J. Biol. Chem. 253, 8708-8720.
7. Barra, D., Martinie, F., Bossa, F., Rotilio, G., Bannister, J.V. and Bannister, W.H. (1978) Biochem. Biophys. Res. Commun. 81, 1195-1200.
8. McCord, J.M. (1976) Adv. Exp. Med. Biol. 74, 540-550.
9. Asada, K., Kanematsu, S. and Uchida, K. (1977) Arch. Biochem. Biophys. 179, 243-256.
10. Weisiger, R.A. and Fridovich, I. (1973) J. Biol. Chem. 248, 3582-3592.
11. Weisiger, R.A. and Fridovich, I. (1973) J. Biol. Chem. 248, 4793-4796.
12. Lumsden, J. and Hall, D.O. (1974) Biochem. Biophys. Res. Commun. 58, 35-41.
13. Tyler, D.D. (1975) Biochem. J. 147, 493-504.
14. Panchenko, L.F., Brusov, O.S., Gerasimov, A.M. and Loktaeva, J.D. (1975) FEBS Lett. 55, 84-87.
15. Arron, G.P., Henry, L., Palmer, J.M. and Hall, D.O. (1976) Biochem. Soc. Trans. 4, 618-620.
16. McCord, J.M., Boyle, J.A., Day, E.D., Jr., Rizzolo, L.J. and Salin, M.L. (1977) in Superoxide and Superoxide Dismutases, Michelson, A.M., McCord, J.M. and Fridovich, I., eds., Academic Press, New York, pp. 129-138.
17. DeRosa, G., Duncan, D.S., Keen, C.L. and Hurley, L.S. (1979) Biochim. Biophys. Acta, 566, 32-39.
18. Shatzman, A.R. and Kosman, D.J. (1979) J. Bacteriol. 137, 313-320.
19. Pile, J. and Dougherty, H.W. (1977) J. Gen. Appl. Microbiol. 23, 303-310.
20. Britton, L, Malinowski, D.P. and Fridovich, I. (1978) J. Bacteriol. 134, 229-236.
21. Misra, H.P. and Fridovich, I. (1978) Arch. Biochem. Biophys. 189, 317-322.
22. Maurer, H.R. (1971) Disc Electrophoresis and Related Techniques of Polyacrylamide Gel Electrophoresis, Walter de Gruyter, Berlin.
23. Beauchamp, C. and Fridovich, I. (1971) Anal. Biochem. 44, 276-287.
24. Hedrick, J.L. and Smith, A.J. (1968) Arch. Biochem. Biophys. 126, 155-164.
25. Bridgen, J., Harris, J.I. and Kolb, E. (1976) J. Mol. Biol. 105, 333-335.
26. Asada, K., Yoshikawa, K., Takahashi, H., Maeda, Y. and Enmanji, K. (1975) J. Biol. Chem. 250, 2801-2807.
27. Anastasi, A., Bannister, J.V. and Bannister, W.H. (1976) Int. J. Biochem. 7, 541-546.
28. Yamakura, F. (1976) Biochim. Biophys. Acta, 422, 280-294.
29. Lumsden, J., Cammack, R. and Hall, D.O. (1976) Biochim. Biophys. Acta, 438, 380-392.
30. Lavelle, F., McAdam, M.E., Fielden, E.M., Roberts, P.B., Puget, K. and Michelson, A.M. (1977) Biochem. J. 161, 3-11.
31. Kanematsu, S. and Asada, K. (1978) Arch. Biochem. Biophys. 185, 473-482.
32. Dougherty, H.W., Sadowski, S.J. and Baker, E.E. (1978) J. Biol. Chem. 253, 5220-5223.
33. Lavelle, F. and Michelson, A.M. (1975) Biochimie, 57, 375-381.
34. McAdam, M.E., Lavelle, F., Fox, R.A. and Fieldon, M. (1977) Biochem. J. 165, 81-87.
35. Keele, B.B., Jr., Giovagnoli, C. and Rotilio, G. (1975) Physiol. Chem. Phys. 7, 1-6.
36. Slykhouse, T.O. and Fee, J.A. (1976) J. Biol. Chem. 251, 5472-5477.
37. Bray, R.C., Cockle, S.A., Fielden, E.M., Roberts, P.B., Rotilio, G. and Calabrese, L. (1974) Biochem. J. 139, 43-48.
38. Hodgson, E.K. and Fridovich, I. (1975) Biochemistry, 14, 5294-5299.
39. Hodgson, E.K. and Fridovich, I. (1975) Biochemistry, 14, 5299-5303.

A COMPARISON OF THE ACTIVE SITE STRUCTURES OF BOVINE, HUMAN, SWORDFISH AND YEAST COPPER-ZINC SUPEROXIDE DISMUTASES

J.V. BANNISTER,[+] W.H. BANNISTER,[++] A.E.G. CASS,[+] H.A.O. HILL[+] AND J.T. JOHANSEN[+++]
+ Inorganic Chemistry Laboratory, University of Oxford, Oxford OX1 3QR, England; ++ Nuffield Department of Clinical Biochemistry, Radcliffe Infirmary, Oxford OX2 6HE, England; +++ Department of Chemistry, Carlsberg Laboratory, DK-2500 Copenhagen, Valby, Denmark

INTRODUCTION

The copper-zinc containing superoxide dismutases are very widely distributed enzymes that are found to be present in all eukaryotic cells. Bovine erythrocyte superoxide dismutase has been the most intensively studied enzyme of this class and both the amino acid sequence and 3Å crystal structure are known. Only the yeast, horse and human enzymes have also been sequenced (see reports, this volume) and no other crystal structures are known. Although the enzymes from the three mentioned species show considerable sequence homology it is not known how far the similarities in primary structure reflect the relationships between the three-dimensional structures of the active sites. We have been using high resolution ^{1}H nuclear magnetic resonance spectroscopy to study copper-zinc superoxide dismutases in solution[1-3] and because this method is sensitive to the relative orientations and distances of groups in the protein it can be used to probe the three-dimensional structures in solution.

An interesting feature of superoxide dismutase is that although its rate constant for substrate turnover is at the diffusion control limit, other copper complexes including the aquo copper(II) ion show comparable catalytic activity. This observation has been discussed and a role for the protein moiety described in some detail based on anion binding experiments (A.E.G. Cass and H.A.O. Hill, this volume). An additional reason for studying the structures of the enzymes from many different sources is to try and discern common features that have been retained during evolution.

In previous papers we have shown that ^{1}H nmr chemical shifts of the histidine C-2 protons are very sensitive to the environment of the active site in superoxide dismutases,[1,2] and to changes in that environment. We have also developed several criteria to distinguish resonances from non-ligand and ligand histidine residues; and furthermore to be able to assign resonances to groups in either the copper site or the zinc site. This paper presents a synthesis of our previous work with more recent studies to show how far the aim of using nmr to investigate structural homology may be realised.

RESULTS AND DISCUSSION

The most suitable form of the enzyme to compare is the copper(I)-zinc(II) derivative. This is because in the apo protein the histidine residues in the active site are very pH sensitive and show complex histidine-histidine interactions. Also the apo protein does not show superoxide dismutase activity. Copper(II)-zinc(II) superoxide dismutase is unsuitable because the unpaired electron on the copper ion is an efficient relaxation source and all of the protons within ∿13 Å are broadened beyond detection.[1]

A number of criteria have been used to assign the different types of C-2 protons in copper-zinc superoxide dismutases. Non-ligand histidine gives rise to sharp, singlet resonances in the chemical shift regions 7.5 - 9.0 ppm (C2(H)) and 6.5 - 7.5 ppm (C4(H)). These chemical shifts are pH dependent and the resonances shift upfield with increasing pH. In contrast to this behaviour, C-2 proton resonances from ligand residues have pH independent chemical shifts. Copper-ligand and zinc-ligand resonances may be distinguished by the observation that the former show a downfield shift on changing the medium from one containing perchlorate to one containing chloride ions. An alternative method for distinguishing between the different groups of histidine residues is based on the relative rates of deuterium exchange at the C-2 position. All of these criteria are self consistent.[1,2,4] Finally, the histidine residue at position 41 (a non-ligand) has an usually rapid rate of C-2 proton exchange.[5] This high rate is due to the local protein environment around this residue.

A complete summary of the C-2 proton assignments made in human, bovine, swordfish and yeast superoxide dismutases is shown in Table 1. The convolution difference spectra of the corresponding proteins are shown in Figure 1. Details of the assignments procedure for human bovine and yeast enzymes have previously been described.[1,3,4] In the case of the swordfish enzyme the methods used were the same.

Non Active Site Residues. Histidine 19 is only present in the bovine enzyme whilst the human enzyme contains a freely titrating surface histidine residue at position 108. The yeast enzyme has neither of these residues as determined from its nmr spectrum amd confirmed by sequence. Sequencing of the swordfish enzyme is being undertaken. The absence of a freely titrating group suggests that in this enzyme, histidine 19 has also been replaced. A histidine residue is also present at position 41 in the bovine and human enzymes and both proteins show a relatively rapidly exchanging proton in ca. 8.6 ppm. In contrast the yeast enzyme does not contain this residue and the corresponding resonance is absent from its nmr spectrum. Our results are consistent with the swordfish enzyme also having a histidine residue at position 41; furthermore the relatively rapid rate of deuteration of this residue implies that the region around this group is also conserved.

TABLE 1

COMPARISON OF HISTIDINE C-2 PROTON ASSIGNMENTS IN COPPER-ZINC SUPEROXIDE DISMUTASES[a]

Resonance	Species			
	Bovine	Human	Swordfish	Yeast
His 41	8.62 (1) 8.64	8.63 (1) N.D.	8.61 (1) 8.59	Not present
His 19	Titrates (5)	Not present His 108 (=6)	Not present	Not present
Cu ligand	8.58 (2) 8.30	8.53 (2) N.D.	8.53 (2) 8.29	8.54 (2) 8.20
Cu ligand	8.33 (4) 8.02	8.35 (4) N.D.	8.27 (4) 7.99	8.30(3) 7.88
Zn ligand	8.47 (3) 8.48	8.48 (3) N.D.	8.50 (3) 8.50	8.59 (1) 8.57
Zn ligand	7.72 (6) 7.73	7.76 (7) N.D.	7.73 (5) 7.74	7.76 (4) 7.75
Zn ligand	6.48 (10) 6.49	6.48 (8) N.D.	6.41 (6) 6.42	6.50 (5) 6.42

[a]Chemical shifts (ppm) in 1 M NaCl (upper value) and 1 M $NaClO_4$ (lower value), pH* 6.9, are given for each resonance together with the number shown in Fig.1 (in parentheses); N.D. = not determined.

Active Site Residues. The six active site histidines are all conserved in the primary structure of the bovine, human and yeast enzymes. Of these six three are ligands to the copper and three are ligands to the zinc, although one of the zinc ligands (histidine 61) is also coordinated to the copper. In our previous nmr studies we have also been able to identify three zinc- and two copper-ligand resonances in the spectrum of the reduced protein.

As can be seen in Table 1, the resonances of the zinc-ligand C2 protons occur at very similar chemical shifts except for one in the yeast enzyme which is significantly downfield shifted (8.59 ppm compared to 8.47 - 8.50 ppm) with respect to the enzyme from the other three sources. One zinc-ligand resonance in the yeast enzyme (6.50 ppm) shows a slight anion dependent downfield shift. These data suggest that the zinc site has retained a strongly conserved structure in the vertebrates; and has undergone some change between this group and the yeasts.

The ^{1}H nmr parameters of the copper-ligand resonances do not provide such a good guide to the structural conservation of the enzyme as they appear to be very responsive to the nature of the surrounding medium. This effect has been

(a)

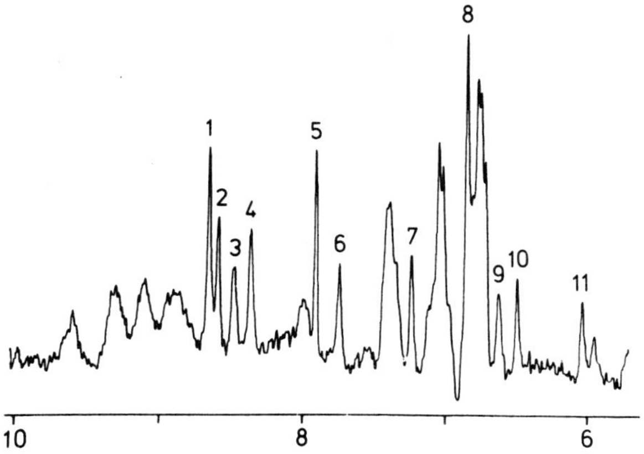
1
2
3
4
5
6
7
8
9
10
11
10
8
6

(b)

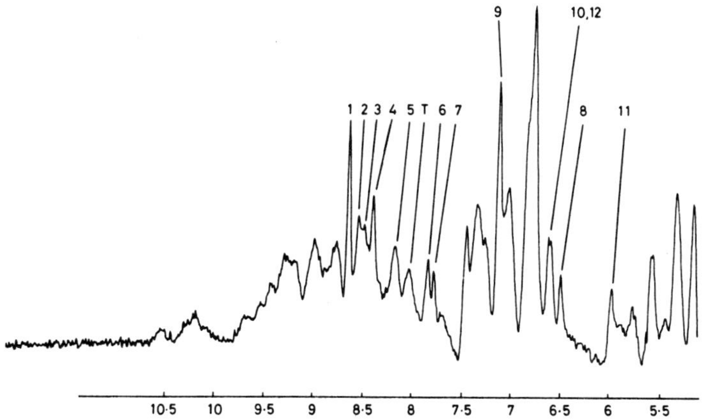
1 2 3 4 5 T 6 7
9
10,12
8
11
10·5
10
9·5
9
8·5
8
7·5
7
6·5
6
5·5

(c)

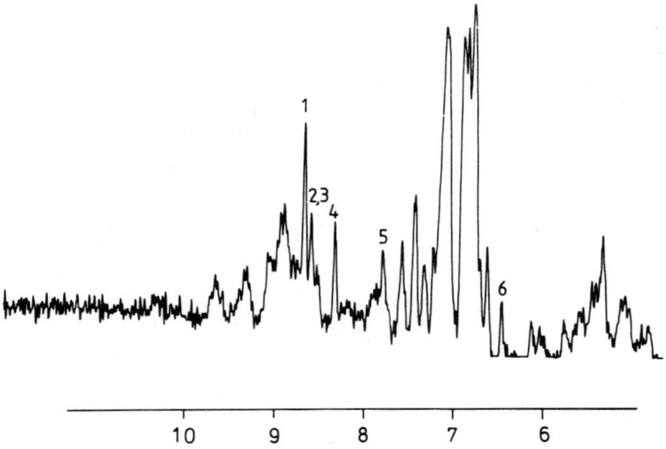

(d)

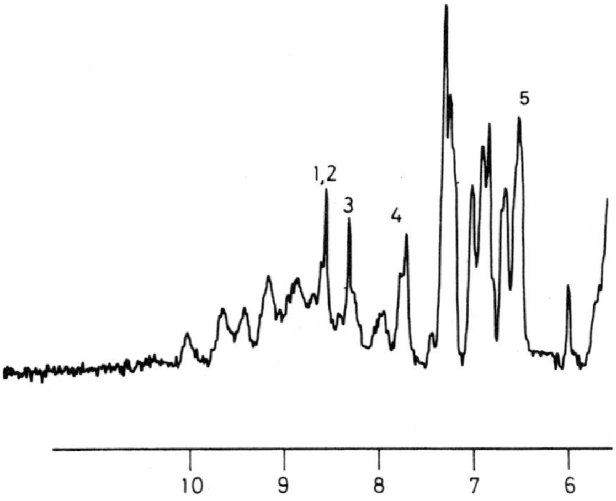

Fig.1. ^{1}H nmr convolution difference spectra of Cu(I)-Zn(II) superoxide dismutases. (a) Bovine, (b) Human, (c) Swordfish, (d) Yeast. Conditions: 3 mM protein in 1 M NaCl and 20 mM phosphate buffer at pH* 6.9 and 50°C. For numbering of resonances see Refs. 1-3 and Table 1. Abscissae: chemical shift (ppm).

studied in some detail for the bovine enzyme and it was concluded that the chemical shifts depended not only upon the coordination of groups to the copper ion but also upon the structure of the water in the active site (A.E.G. Cass and H.A.O. Hill, this volume).

A comparison of the chemical shifts of the copper ligand C2 proton resonances of the enzyme in solutions of 1 M chloride show some variation; 0.04 ppm for the low field copper ligand resonance and 0.08 ppm for the high field resonance. However, in the presence of 1 M perchlorate solution the shifts are much greater: 0.1 ppm and 0.14 ppm, respectively. Neither of these sets of data show a trend based on phylogeny. However, as the higher field resonance is always more affected on changing the medium a common mechanism may be involved.

These preliminary studies demonstrate that ^{1}H nmr spectroscopy is of use as a rapid and sensitive probe of protein structure, and we intend to expand this work to include enzymes from a variety of other species.

ACKNOWLEDGEMENTS

We thank the U.K. Science Research Council for a studentship to A.E.G.C., the Wellcome Trust and the Nuffield Foundation (J.V.B. and W.H.B.), and the Carlsberg Laboratory (J.T.J.) for support. This is a contribution from the Oxford Enzyme Group, of which H.A.O.H. is a member.

REFERENCES

1. Cass, A.E.G., Hill, H.A.O., Smith, B.E., Bannister, J.V. and Bannister, W.H. (1977) Biochemistry, 16, 3061-3066.
2. Cass, A.E.G., Hill, H.A.O., Hasemann, V. and Johansen, J.T. (1978) Carlsberg Res. Commun. 43, 439-450.
3. Hill, H.A.O., Lee, W.-K., Bannister, J.V. and Bannister, W.H. (1980) Biochem. J. 185, 245-252.
4. Cass, A.E.G., Hill, H.A.O., Bannister, J.V., Bannister, W.H., Hasemann, V. and Johansen, J.T. (1979) Biochem. J. 183, 127-132.
5. Cass, A.E.G., Hill, H.A.O., Smith, B.E., Bannister, J.V. and Bannister, W.H. (1977) Biochem. J. 165, 587-589.

ANION BINDING TO COPPER(I) SUPEROXIDE DISMUTASE: A HIGH RESOLUTION ^{1}H NUCLEAR MAGNETIC RESONANCE SPECTROSCOPIC STUDY

ANTHONY E.G. CASS AND H. ALLEN O. HILL
Inorganic Chemistry Laboratory, University of Oxford, Oxford, OX1 3QR
England

Nuclear magnetic resonance (nmr) spectroscopy is arguably the most powerful tool at the disposal of biochemists for studying proteins in solution.[1] The high resolution nmr spectrum of an enzyme provides information about both the structures and dynamics of the molecule and is often very responsive to small changes in either of these characteristics. In addition nmr spectroscopy reveals details about the protein at an atomic resolution although the sheer complexity of the spectrum does not readily allow full use to be made of the information available. A major problem in interpreting the ^{1}H nmr spectrum of a protein is that of assignment.[1] It is probably true to say that the step of assigning a particular resonance to a specific proton in the molecule is still the main barrier in the full exploitation of ^{1}H nmr spectroscopy in the study of enzymes. Despite this limitation much useful information can be gleaned at a qualitative level without the need for a complete assignment of the resonances.

A variety of different criteria have enabled us to assign several resonances in the ^{1}H nmr spectrum of bovine erythrocyte superoxide dismutase.[2,3] Especially useful have been the assignments which allow the non-ligand histidine residues, 19 and 41, to be distinguished from those involved in metal coordination and further to identify zinc-ligand and copper-ligand resonances. Figure 1 shows the aromatic region of the 270 MHz ^{1}H nmr spectrum of copper(I)-zinc(II) superoxide dismutase, the resonances have been artificially sharpened by the convolution difference method.[4] The assignments are summarized in Table 1.

An additional advantage of ^{1}H nmr spectroscopy in studying metallo proteins is that it is also responsive to changes occurring outside of the immediate coordination sphere of the metal ion. This property is especially useful when studying copper(I) centres in proteins as copper in this form is spectroscopically "silent" towards many of the commonly used methods. These two points, taken in conjunction with the assignment of resonances to residues in both the active site and elsewhere in the molecule, prompted us to study the binding of monovalent anions to copper(I) superoxide dismutase.

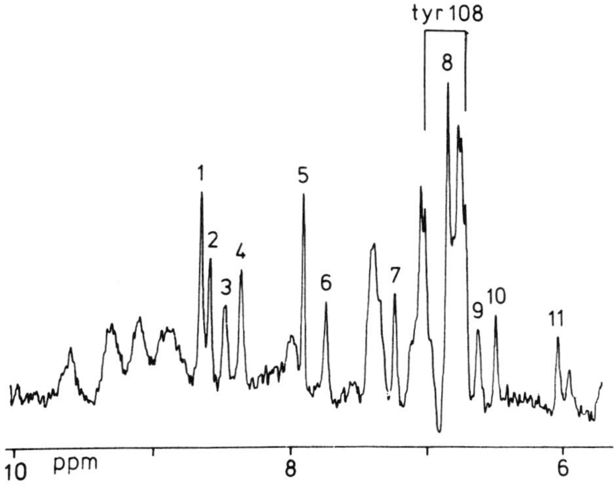

Fig. 1. Aromatic region of the 270 MHz ^{1}H nmr spectrum of copper(I)-zinc(II) bovine superoxide dismutase. Protein concentration is 3 mM in 1 M sodium chloride and 20 mM phosphate buffer at pH* 6.9 and temperature of 50°C. Assignments of the numbered peaks are presented in Table 1.

TABLE 1

ASSIGNMENT OF RESONANCES IN THE AROMATIC REGION OF THE NMR SPECTRUM OF COPPER(I)-ZINC(II) SUPEROXIDE DISMUTASE FROM BOVINE ERYTHROCYTES

Resonance	Assignment
1	Histidine 41 C2(H)
2	Copper Ligand C2(H)
3	Zinc Ligand C2(H)
4	Copper Ligand C2(H)
5	Histidine 19 C2(H)
6	Zinc Ligand C2(H)
7	Histidine 41 C4(H)
8	Histidine 19 C2(H)
9	Metal Ligand C4(H)
10	Zinc Ligand C2(H)
11	Metal Ligand C4(H)

The substrate of superoxide dismutase is a singly charged anion and the halides and pseudohalides have been studied as substrate analogues.[5] Cyanide has been the most intensively investigated[5,6] of the anions, partly because it binds tightly and partly because it causes large changes in the optical and electron paramagnetic resonance spectra of the copper(II) centre. A similar behaviour is also observed for azide binding. For both anions measurement of the water proton relaxation rates suggests that they directly coordinate to the copper(II) ion and displace an axially bound water molecule.[6,7] In our high resolution nmr studies cyanide and azide proved to be inconvenient for two reasons. One is that the rate of reoxidation by air in the presence of these ions is much more rapid than for the enzyme alone. A second more serious problem is that these ions are in slow exchange on the ^{1}H nmr time scale. Therefore, changes in intensity rather than chemical shift are observed. Both of these considerations led us to study halide ion binding although this required a much higher concentration of the relevant anions.

Nuclear Magnetic Resonance Results. ^{1}H nmr spectra of copper(I) bovine erythrocyte superoxide dismutase in the presence of 1 M chloride and 1 M perchlorate solutions are shown in Figures 2a and 2b, respectively. Comparison of these spectra reveals that two of the C2 proton resonances have been shifted downfield by the addition of chloride ions (resonances 2 and 4). It can be easily shown that under fast exchange conditions the shift (Δ) is related to the chloride concentration by:

$$\frac{1}{(Cl^-)} = \frac{\Delta_o}{K_d}\cdot\frac{1}{\Delta} - \frac{1}{K_d}$$

where Δ_o is the maximum shift when all of the enzyme is present as the chloride complex and K_d is the dissociation constant of that complex. A double reciprocal plot of the data from a chloride titration is shown in Figure 3, and from this curve and the corresponding one for the shift of resonance 4 dissociation constants, K_d values of 2.3 ± 0.2 M and 1.1 ± 0.2 M, and values of 0.9 ± 0.1 ppm and 0.6 ± 0.1 ppm, respectively, can be determined.

A spectrum was also measured of the copper(I) enzyme in the presence of fluoride, bromide and iodide (all at a concentration of 1M) and also at low ionic strength (20 mM phosphate). The measured shifts of C2 proton resonances under these conditions are shown in Table 2 and are also illustrated schematically in Figure 4 for a copper-ligand, a zinc-ligand and a non-ligand histidine residue.

Interpretation. Assignment of resonances 2 and 4 to C2 protons for two of the histidine residues acting as copper-ligands was made previously on the basis of differential rates of deuterium exchange.[3] The same procedure was used to assign resonances 3, 6 and 10 to the C2 protons of zinc-ligands. The above results on the effects of halide ions on these resonances are completely consistent with these assignments and with previous work which showed that anions bind at or close to

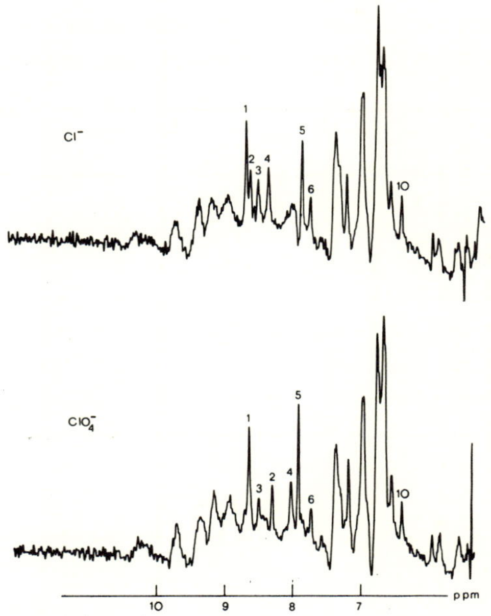

Fig. 2. Comparison of the spectra of bovine superoxide dismutase in the presence of chloride and perchlorate ions. The shifts of the two resonances from the copper ligands (2 and 4) are clearly seen.

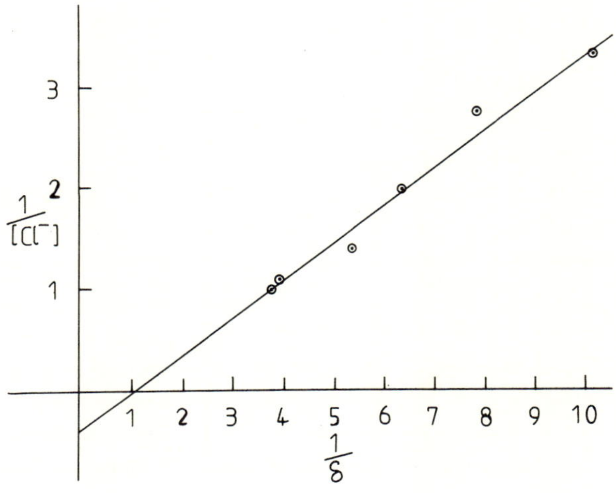

Fig. 3. Double reciprocal plot for chloride binding to bovine superoxide dismutase.

TABLE 2

CHEMICAL SHIFTS OF C2 PROTON RESONANCES IN BOVINE SUPEROXIDE DISMUTASE IN THE PRESENCE OF PERCHLORATE, HALIDES AND AT LOW IONIC STRENGTH

Anion	Resonance						
	1	2	3	4	5	6	10
ClO_4^-	8.64	8.30	8.48	8.02	-	7.73	6.48
F^-	8.62	8.37	8.49	8.10	-	7.73	6.49
Cl^-	8.64	8.58	8.47	8.33	-	7.72	6.49
Br^-	8.64	8.36	8.47	8.08	-	7.69	6.47
I^-	8.63	8.54	8.47	8.22	-	7.71	6.49
Low Ionic Strength	8.62	8.40	8.47	8.13	-	7.72	6.51

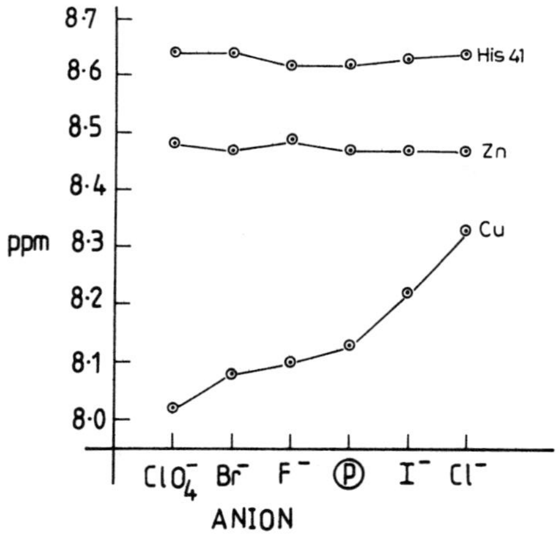

Fig. 4. Schematic representation of the effects of different anions on the chemical shifts of a non-ligand, zinc-ligand and copper-ligand C2(H) resonance. All the anions were at a concentration of 1 M except for the low ionic strength measurement which was made in 20 mM phosphate buffer.

the copper centre.[6,9] As none of the zinc-ligand resonances are perturbed by the addition of halide ions this site is well insulated from the copper site in the reduced enzyme. In particular the histidine residue interposed between the metal ions (histidine 61) must be coordinated only to the zinc ion in the reduced protein. Both of these results are in contrast to those obtained with the oxidised enzyme where spectroscopic evidence implies an interaction between the copper and zinc centres, possibly mediated by a bridging imidazolate anion from histidine 61 (Refs. 8, 10). The effect of halide ions on the superoxide dismutase from yeast is slightly different as in this case the resonance corresponding to resonance 10 shows a small but distinct downfield shift.[11] This result has been used to assign this resonance to histidine 61. Indeed the sensitivity of the residues in the copper site to external environment suggest that the copper ion may be very responsive to events occurring outside of its immediate coordination sphere.

As can be seen from Figure 4 and Table 2, iodide has a substantial effect on the resonance of the residues acting as copper ligands. This result is in contrast to the work of Rigo et al.[6] where it is claimed that the iodide ion is too large to penetrate the active site of the enzyme. In addition the effect of the remaining ions does not follow the order of binding constants determined by these authors. The perturbation caused by the anions seems to reflect not only coordination to the copper ion but also to depend on their ability to disrupt water structure, the so-called chaotropic effect.

A final feature to note in Figure 4 and Table 2 is that the effect of ionic strength is by no means simple. McAdam[10] has pointed out the dangers of trying to derive mechanistic information from the dependence of kinetic parameters on ionic strength. A similar note of caution should be sounded here, especially considering the high (1 M) ionic strength involved. However, even with this caveat we can make two observations at the qualitative level. The first is that the change in ionic strength is manifested only at the copper centre; those resonances assigned to the zinc-ligands, histidines 19 and 41, tyrosine 108 and the N-acetyl group are unaffected. This means that at least as judged by these groups the protein does not undergo any gross change in structure as a function of ionic strength. The second observation reinforces the point made above that the copper site is responsive even to influences that do not directly involve coordination.

This effect of ionic strength prompted us to consider the role of the protein bound water in the enzyme reaction, especially in relation to the source of protons in the reduction of the superoxide radical.

Enzyme Mechanism. Superoxide dismutase is a very effective catalyst for the disproportionation of O_2^- having a rate constant for substrate turnover of 3×10^9 $M^{-1}s^{-1}$ (Ref. 10). Other copper complexes, including the aquated Cu(II) ion are effective catalysts. This has led to a questioning of the role of the protein: Is its biological function to catalyze the disproportionation of O_2^-? Irrespective of the ultimate answer to this question, to base it on the premise that all enzymes must be inherently better catalysts than simple complexes and that the only biological property of enzymes that is important is their ability to act as catalysts reveals a fundamental misunderstanding of the properties of metalloproteins.

Though most enzymes are more effective catalysts than simple complexes this may only be a reflection of the lack of endeavour and/or originality on the part of those concerned with the synthesis of "models". In addition the substrate of most enzymes are complex molecules which require binding, often stereospecifically, to the protein. This requirement may be less important for "small" substrates. Therefore one might expect simple complexes to be effective catalysts for reactions involving, for example, O_2, O_2^- and H_2O_2. As far as metalloenzymes are concerned the protein must bind the transition metal tightly with a low rate constant for the "off" reaction and yet, in addition, allow the substrate free and easy access to the active site. Either of these requirements is easy to achieve with simple ligands, but not both in that they are usually mutually exclusive. Small complexes will often be toxic to the cell because they can release the metal ion which can then interfere with other biological processes. Further requirements including compartmentation and membrane binding can only be achieved through the protein part of the metalloenzyme. Therefore there is more required of an enzyme than just catalysis.

A number of properties of superoxide dismutase suggest that the role of the protein may be crucial in the apparent "simple" redox reactions involved in superoxide dismutation. The rate of water exchange at the copper centre is low, some 10-100 times than is observed in simple copper complexes, and less than the turnover rate of the enzyme.[7] An arginine residue at position 141, that is also present in the human and yeast enzymes, is essential for activity (D. Malinowski and I. Fridovich, this volume). In addition the results described above on the effects of anions suggest that the structure of the water in the active site may be important. In the mechanism described below these features are drawn together to try and provide a framework for the role of the protein in superoxide dismutase.

The slow rate of exchange of the copper-bound water in the enzyme suggests that the electron transfer reactions occur via an outer sphere process without the need for displacement of the water from the copper. Electron paramagnetic resonance spectroscopic studies of the superoxide ion[12] suggest that it will be hydrated, i.e. $O_2^- (H_2O_2)_n$. The water molecule coordinated to copper would also be expected to hydrogen bond to the superoxide ion. In the first step of the enzyme turnover the reaction is a simple outer sphere electron transfer from the superoxide ion to the copper(II) centre.

During the second step we suggest that the superoxide ion is bound between the copper centre and arginine 141, as suggested by Malinowksi and Fridovich (this volume) although in our mechanism there is a water molecule between the copper and the superoxide ions. The role of the arginine is to partially neutralise the charge on the superoxide ion and thus to aid electron transfer. Concomitantly with electron transfer the copper-bound water molecule and the guanidinium group of the arginine residue each donate a proton to the developing peroxide dianion. The copper(II)-bound hydroxide ion is then reprotonated by one of the water molecules in the active site to yield a free hydroxide ion. The guanadinium group is similarly reprotonated. At this point in the reaction scheme the hydroxide ion does not diffuse away but rather the ordered water structure in the cleft leading from the surface of the enzyme to the active site undergoes a reorganization of its hydrogen bond network to eject a hydroxide ion into solution. This process is essentially the same as that governing proton and hydroxide ion transport in ice, where these ions have a greater mobility than in water at 20°C. This mechanism is shown diagrammatically in Figure 5.

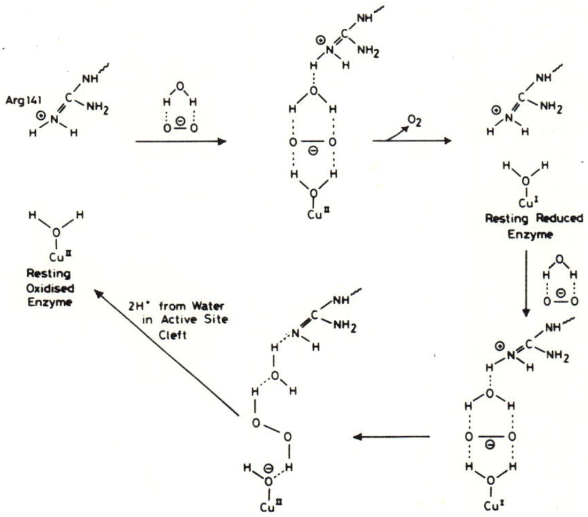

Fig. 5. Reaction scheme for copper-zinc superoxide dismutase.

The above discussion shows that the limitation on the reactivity of the copper ions in superoxide dismutase that are imposed by the protein in order to survive in a biological milieu are overcome by the structure of the protein active site.

ACKNOWLEDGEMENTS

A.E.G.C. wishes to thank the U.K. Science Research Council for a studentship, and the Wellcome Trust for a travel grant to present this work. Bovine erythrocyte superoxide dismutase was a gift from Dr. J.V. Bannister. This is a contribution from the Oxford Enzyme Group, of which H.A.O.H. is a member.

REFERENCES

1. Levine, B.A., Moore, G.R., Ratcliffe, R.G. and Williams, R.J.P. (1979) Int. Rev. Biochem. 24, 77-141.
2. Cass, A.E.G., Hill, H.A.O., Smith, B.E., Bannister, J.V. and Bannister, W.H. (1977) Biochemistry, 16, 3061-3066.
3. Cass, A.E.G., Hill, H.A.O., Hasemann, V., Johansen, J.T., Bannister, J.V. and Bannister, W.H. (1979) Biochem. J. 183, 127-132.
4. Campbell, I.D., Dobson, C.M., Williams, R.J.P. and Xavier, A.V. (1973) J. Magnet. Res. 11, 172-181.
5. Fielden, E.M., Roberts, P.B., Bray, R.C., Lowe, D.J., Mautner, G.N., Rotilio, G. and Calabrese, L. (1974) Biochem. J.139, 49-60.
6. Rigo, A., Stevanato, R., Viglino, P. and Rotilio, G. (1977) Biochem. Biophys. Res. Commun. 79, 776-783.
7. Boden, N., Holmes, M.C. and Knowles, P.F. (1979) Biochem. J. 177, 303-309.
8. Fee, J.A. and Gaber, B.P. (1972) J. Biol. Chem. 247, 60-65.
9. Fee, J.A. and Ward, R.L. (1976) Biochem. Biophys. Res. Commun. 71, 427-437.
10. McAdam, M.E. (1977) Biochem. J. 161, 697-699.
11. Cass, A.E.G., Hill, H.A.O., Hasemann, V. and Johansen, J.T. (1978) Carlsberg Res. Commun. 43, 439-449.
12. Green, M.R., Hill, H.A.O. and Turner, D.R. (1979) FEBS Lett. 103, 176-180.

AN ESSENTIAL ARGININE RESIDUE AT THE ACTIVE SITE OF COPPER-ZINC SUPEROXIDE DISMUTASES

DOUGLAS P. MALINOWSKI AND IRWIN FRIDOVICH
Department of Biochemistry, Duke University Medical Center, Durham, North Carolina 27710, USA.

ABSTRACT

Arginine-specific reagents cause an 80 → 90% inactivation of the bovine erythrocyte superoxide dismutase and a concomitant modification of one arginine residue per subunit. Reagents, whose attack on arginine is reversible, allowed the demonstration that reappearance of activity, during analysis, was paralleled by unmasking of the modified arginine. Strutcural analysis, utilizing tryptic fragments of cyanogen bromide peptides showed that the essential arginine was residue 141, which was known from X-ray crystallography to lie within 6Å of the copper at the active site. Copper and zinc containing superoxide dismutases, from wheat germ, chicken liver and several mammalian species, all exhibited an essential arginine residue, whereas the manganese-containing and the iron-containing enzymes from <u>Escherichia coli</u> did not. Two possible functions for the essential arginine residue in the catalytic mechanism are proposed.

INTRODUCTION

Bovine erthyrocyte superoxide dismutase (BESOD) catalyzes the reaction, $O_2^- + O_2^- + 2H^+ \rightarrow H_2O_2 + O_2$, and does so at a rate close to the diffusion limit, over the entire range of pH from 5 → 10.[1,2] The active site contains copper and zinc in close proximity, as determined by X-ray diffraction analysis[3,4] and the copper functions in catalysis by alternate reduction and reoxidation, during the successive encounters with O_2^- (Refs. 5,6). Several other copper proteins have been found devoid of this activity, as is the EDTA-copper complex.[7,8]

Hydrated Cu(II) is an effective catalyst of O_2^- dismutation, but it loses its activity as the pH is raised to the neutral and weakly alkaline range, due to conversion to $Cu(OH)_2$ (Refs. 9,10). If the formation of the hydroxide could somehow be prevented, then free Cu(II) might exhibit an activity comparable to that of the Cu(II) at the active site of the enzyme, over a broad range of pH. Even then, one would be hard put to explain the catalytic efficiency of the enzyme; since it has 130 times more surface area than does aquated Cu(II) and random collisions with the protein surface would render most collisions between the enzyme and O_2^- fruitless.[11] Furthermore, the isoelectric point of BESOD is 5.0 (Ref. 12) and the protein would acquire a progressively greater net negative charge, as the pH was increased from 5 → 10 and this would raise an electrostatic barrier to collision with O_2^-.

X-ray diffraction analysis has provided a possible explanation for the ability of BESOD to exhibit great catalytic efficiency over a broad range of pH. Thus, there is an arginine residue, number 141 in the linear sequence, which is situated within 6Å of the active site copper.[3,4] The positive charge on this arginine might serve to attract the anionic O_2^- into the active site. There is precedent for this concept, since arginine has been seen to play a role in binding anionic substrates in many enzymes. Alternatively one might propose that this arginine residue might facilitate proton conduction during the very rapid catalytic cycle of this enzyme. αβ-Diketones have frequently been used in group-specific attack upon arginine residues and we have employed several of these reagents in probing the essentiality of arginine 141 for the catalytic activity of BESOD.

MATERIALS AND METHODS

Copper-zinc superoxide dismutase (CuZn SOD) was prepared from bovine erythrocytes,[8] bovine liver,[13] chicken liver[14] and wheat germ,[15] while manganese-containing enzyme (MnSOD) was isolated from E.coli[16] and Streptococcus faecalis[17] and iron-containing enzyme (FeSOD) from E.coli.[18] Glyoxal, phenylglyoxal, 2,3-butanedione and 1,2-cyclohexanedione were from the Aldrich Chemical Co. TPCK-trypsin was from the Worthington Biochemical Corp.

SOD was assayed by the xanthine oxidase-cytochrome c method.[8] Small amounts of diketones, carried over from incubation mixtures, did not interfere with this assay; but borate, which is known to inhibit xanthine oxidase,[19] was a problem. Borate was prevented from inhibiting xanthine oxidase by incorporating 0.1 M D-mannitol into the assay mixtures, to scavenge the borate.

Addition of α,β-diketones to borate buffers caused some acidification which necessitated readjustment of pH with NaOH. Samples taken at intervals, for measurement of residual SOD activity, were diluted directly into the xanthine oxidase-cytochrome c assay mixture. Samples intended for amino acid analysis were delivered into 30% acetic acid to stabilize modified arginine. Protein content of diketone-modified SOD was determined by the Lowry method.[20]

Cleavage with CNBr was performed on samples which had been reduced and carboxymethylated[21] and then lyophilized, after desalting by gel exclusion on Biogel P2 in 20% acetic acid. The lyophilized samples were treated with 0.1 M CNBr in 70% formic acid for 24 hrs at 23°C. CNBr peptides were separated by gel exclusion on a 0.9 x 118 cm column of Sephadex G-75 in 20% acetic acid at 4 ml/hr. CNBr peptide (Refs. 22,23) was digested with 47 μg/ml TPCK-trypsin for 4 hrs at 37°C in a borate buffer at pH 8.0. Tryptic peptides were separated on thin layer cellulose plates.[22] Peptides were located with a ninhydrin spray while arginine or histidine-bearing peptides were located by spraying with phenanthraquinone[24] or with the Pauly reagent.[25]

Amino acid analyses were performed after 24 hr hydrolysis of 0.1 - 0.2 mg samples in 6.0 M HCl, 0.1% phenol, at 110°C. Hydrolysates were analyzed with the Beckman 120C and 119 automatic instruments by published procedures.[26,27] The cyclohexane-modified enzyme was hydrolyzed in the presence of 1% mercapto-acetic acid.[28]

RESULTS

Glyoxal. We began our studies with this reagent since it had already been seen, by McCord (personal communication), to inactivate BESOD. Incubation of BESOD with 40 mM glyoxal caused a progressive inactivation to a limit of 15-20% residual activity. Figure 1 illustrates the time course of this inactivation.

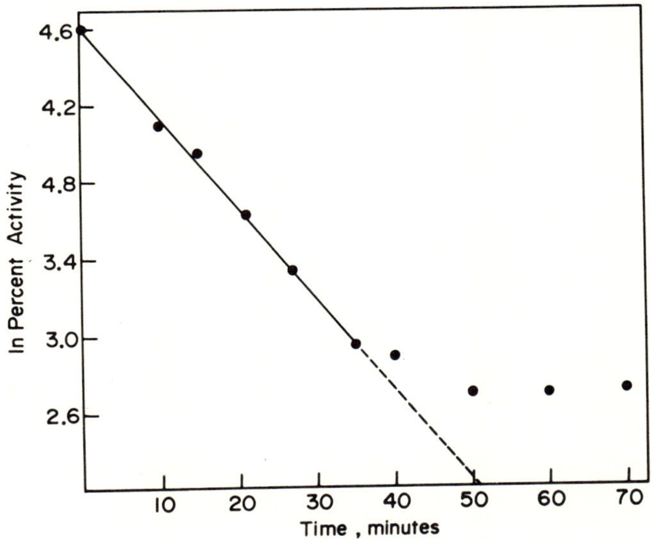

Fig. 1. Inactivation of bovine erythrocyte superoxide dismutase by glyoxal. Native BESOD (2 µL of 26 mg/ml) was added to 1.0 ml of 40 mM glyoxal, 0.125 M sodium bicarbonate, pH 8.0 (final concentrations, 1.7×10^{-6}M BESOD, 39.9 mM glyoxal). The samples was incubated at 25°C and aliquots were removed at the indicated times for determination of residual enzymatic activity. The natural logarithm of residual activity is here presented as a function of the time of incubation of the enzyme with the glyoxal.

The rate constant for this inactivation at pH 8.0 and at 25°C was 1.5 M^{-1} min^{-1} and the limit of 80-85% of inactivation was seen at all concentrations of glyoxal tested. Amino acid analysis of BESOD, inactivated with 50 mM glyoxal to the limit of 20% residual activity, demonstrated a loss of close to one arginine and one lysine per subunit. This is shown in Table 1. Because glyoxal modified lysine as well as arginine, its use was discontinued and other diketones were tested.

TABLE 1

AMINO ACID COMPOSITION OF BOVINE ERYTHROCYTE SUPEROXIDE DISMUTASE MODIFIED WITH DICARBONYL REAGENTS

	Native BESOD	Reagent	
		Glyoxal[a]	Phenylglyoxal[b]
Specific Activity (U/mg)	4000	800	386
Amino Acid		Residues per Subunit[c]	
Lysine	9.86	9.05	10.24
Histidine	7.47	7.21	7.52
Arginine	4.03	3.14	3.13
Aspartic acid	17.13	16.97	17.11
Threonine	11.69	11.47	11.43
Serine	7.55	7.80	7.66
Glutamic acid	11.04	10.87	10.93
Proline	6.38	6.45	6.46
Glycine	24.84	25.09	25.15
Alanine	8.96	9.09	8.95
Half-cystine	2.89	1.75	2.89
Valine	11.65	10.38	10.45
Methinone	0.82	1.55	0.54
Isoleucine	6.05	5.89	5.87
Leucine	8.12	7.99	7.79
Tyrosine	0.93	0.97	1.00
Phenylalanine	3.97	3.93	3.98

[a]Native BESOD incubated with 50 mM glyoxal, 0.125 M sodium bicarbonate, pH 8.0 for 60 min at 25°C.

[b]Native BESOD incubated with 19.7 mM phenylglyoxal, 0.125 M sodium bicarbonate, pH 8.0 for 60 min at 25°C.

[c]Composition normalized to (Asp + Glu + Gly + Ala) = 6.20 per subunit.

Phenylglyoxal. This reagent also inactivated BESOD to a limiting residual activity, in this case approximately 10% of the initial activity. The time course of this inactivation, at three different concentrations of phenylglyoxal, at pH 8.0, is shown in Figure 2. The rate of inactivation was 2.5 M^{-1} min^{-1} at pH 8.0 and was 4.0 M^{-1} min^{-1} at pH 9.0. The phenylglyoxal derivative of arginine has been reported to be stable in acidic, but not in neutral or alkaline media[29] and both 1:1 (Ref. 30) and 2:1 (Ref. 29) complexes of the reagent with this amino acid have been proposed. We noted that the arginine-phenylglyoxal product could be diluted into 30% acetic acid and then incubated for 72 hrs at 4°C in this medium without regenerating detectable free arginine. This provided confidence that enzyme, modified with phenylglyoxal, could safely be dialyzed against 30% acetic acid prior to amino acid analysis.

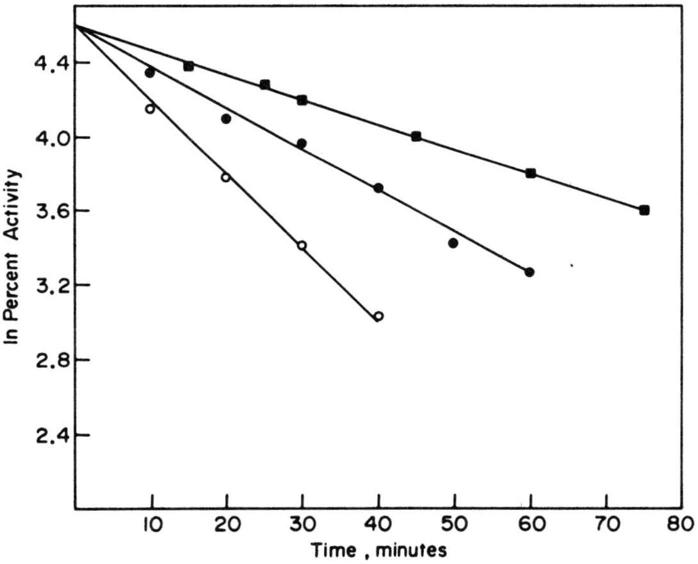

Fig. 2. Inactivation of bovine erythrocyte superoxide dismutase by phenylglyoxal. Native BESOD (1.57 μM, final concentration) was incubated in 1.0 ml of 0.125 M sodium bicarbonate, pH 8.0, 25°C with the following concentrations of phenylglyoxal: 5.1 mM(■); 10.2 mM (●) and 15.4 mM (0). The natural logarithm of residual activity is here presented as a function of time of incubation of the enzyme with phenylglyoxal.

BESOD was incubated with 19.7 mM phenylglyoxal for 1 hr at pH 8.0 and at 25°C. This enzyme, which exhibited 10% residual activity, was dialyzed against 30% acetic acid prior to lyophilization and amino acid analysis. The results of this analysis, shown in Table 1, revealed the loss of approximately one arginine residue per subunit and no significant changes in the recovery of any other amino acids. Inactivation by phenylglyoxal thus appears to be due to modification of a single arginine residue per subunit. This conclusion was strengthened by inactivating BESOD to different degrees and then analyzing for the extent of arginine modification. It is clear from the results in Figure 3 that loss of activity was a linear function of arginine modification with a limit at 1.0 arginine modified per subunit.

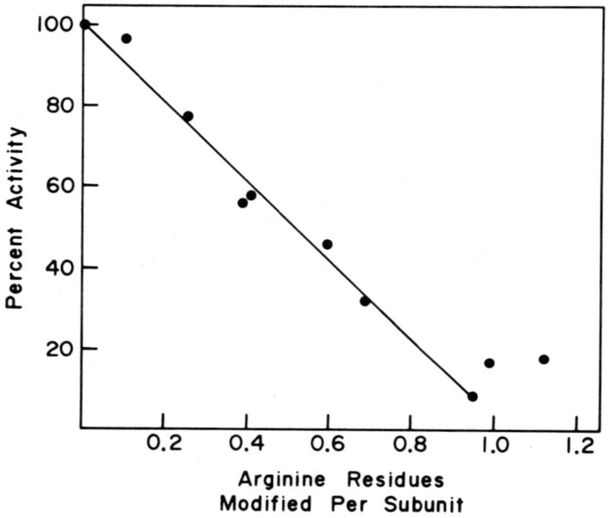

Fig. 3. Correlation of inactivation of BESOD with arginine modification by phenylglyoxal. The data represent the results of three separate experiments. In the first experiment, native BESOD (1.99 mg/ml) was incubated in 19.5 mM phenylglyoxal, 0.12 M sodium bicarbonate, pH 8.0, 25°C. At 10 min intervals, 5 μL were removed, diluted into 0.2 ml of 50 mM potassium phosphate, pH 7.8 and assayed for residual activity. In addition, 100 μL were removed, diluted with an equal volume of 60% acetic acid and then dialyzed against 30% acetic acid for amino acid analysis. In the second experiment, native BESOD (2 mg/ml) was incubated in 9.5 mM phenylglyoxal, 0.125 M sodium bicarbonate, pH 8.0, 25°C. At 13 min and 26 min, 10 μL were diluted into 0.4 ml of 50 mM potassium phosphate, pH 7.8 and assayed for resdiual activity. An additonal aliquot (50 μL) was removed, diluted into 1 ml of 30% acetic acid and dialyzed against the same solution. In the third experiment, native BESOD (1 mg/ml) was incubated in 10.2 mM phenylglyoxal, 0.125 M sodium bicarbonate, pH 8.0, 25°C. At 6.5 and 35 min, 10 μL and 15 μL, respectively, were diluted into 0.2 ml of 50 mM potassium phosphate, pH 7.8 and assayed for residual activity. An additional aliquot (100 μL) was removed, diluted into 1.0 ml of 30% acetic acid and dialyzed against the same solution at 4°C. Arginine modification was quantitated on the basis of 4.0 arginine residues per subunit in native BESOD.

Butanedione and cyclohexanedione plus borate. These reagents offer the great advantage of reversibility. Butanedione did inactivate BESOD when incubated with it at pH 9.0 in 50 mM borate and once again there was a limiting residual activity of approximately 10%. The rate of inactivation by this reagent at 25°C was 10.9 $M^{-1}min^{-1}$. Borate, which was itself without effect on BESOD, was needed to support the inactivation by butanedione and no inactivation was seen when the reagent was applied in a bicarbonate buffer.

Cyclohexanedione, whose behaviour is similar to that of butanedione,[28,31,32] also inactivated BESOD, when applied in 50 mM borate at pH 9.0 and the rate was 6.6 $M^{-1}min^{-1}$ at 25°C. The time course of inactivation at three concentrations of cyclohexanedione is shown in Figure 4. In this case the residual limiting activity was 18%. It has been shown,[28] that the arginine derivative, generated with cyclohexanedione plus borate, is stable in 30% acetic acid and is destroyed,

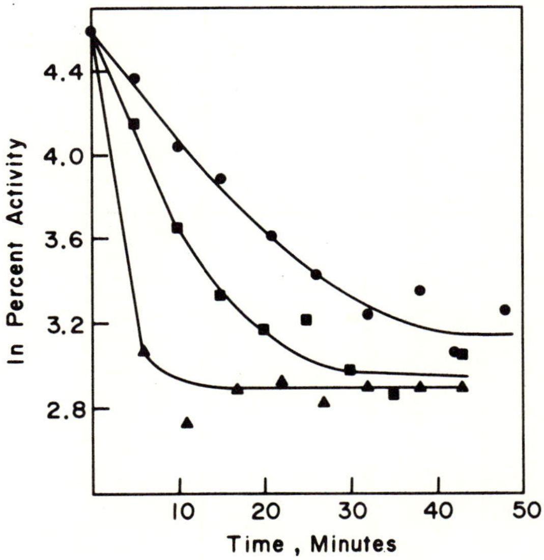

Fig. 4. Inactivation of native bovine erythrocyte superoxide dismutase by cyclohexandione. Native BESOD (1.57 μM, final concentration) was incubated in 1.0 ml of 50 mM sodium borate, pH 9.0, 25°C with the following concentrations of cyclohexanedione: 10.2 mM, (●); 20.5 mM, (■) and 60.5 mM, (▲). At the indicated times, aliquots were removed and assayed for residual activity.

without generating arginine, upon hydrolysis in 6N HCl in the presence of mercaptoacetic acid. Riordan,[31] similarly noted that the comparable butanedione adduct was stable at pH 2.2 and was destroyed in 1.0 N HCl. We found the butanedione-borate adduct of arginine stable in 30% acetic acid at 4°C for 72 hrs and noted no regeneration of arginine during subsequent acid hydrolysis. Amino acid analysis of BESOD, modified with these reagents, could therefore be undertaken without concern for the stability of the arginine adducts.

BESOD was not inactivated by 9.5 mM butanedione, when incubated for 45 min, in 50 mM bicarbonate at 25°C and pH 9.0, but it did suffer modification of 0.4 arginines per subunit. In contrast, when borate buffer was used, there was 77% inactivation and loss of 1.3 arginines per subunit. Essentially similar results were obtained with cyclohexanedione and these data are presented in Table 2. The half cystine data shown in this Table are not reliable since the protein was neither reduced and carboxymethylated nor treated with performic acid, prior to hydrolysis.[27] The non-integral values for arginine, seen with

TABLE 2

AMINO ACID COMPOSITION OF BOVINE ERYTHROCYTE SUPEROXIDE DISMUTASE INCATIVATED BY BUTANEDIONE AND CYCLOHEXANEDIONE[a]

	Sample				
	Native	BDO/Borate[b]	BDO/$NaHCO_3$[c]	CHD/Borate[d]	CHD/$NaHCO_3$[e]
Specific Activity (U/mg)	3500	801	3500	890	3070
Amino Acid	Residues per subunit				
Lysine	9.86	9.98	9.97	9.93	9.95
Histidine	7.47	7.41	7.54	7.34	7.48
Arginine	4.03	2.75	3.63	3.04	3.65
Aspartic acid	17.13	17.21	16.83	17.06	17.27
Threonine	11.69	11.33	11.90	11.40	11.39
Serine	7.55	7.59	7.61	7.61	7.63
Glutamic acid	11.04	11.08	11.22	11.13	11.12
Proline	6.38	5.38	5.81	6.30	5.85
Glycine	24.84	24.56	24.51	24.52	24.47
Alanine	8.96	8.98	9.08	9.03	8.95
Half-cystine	2.89	0.80	2.45	2.77	1.80
Valine	11.65	11.41	11.95	11.20	11.28
Methionine	0.82	0.76	0.56	0.75	0.60
Isoleucine	6.05	6.29	6.47	6.02	6.08
Leucine	8.12	8.06	8.15	7.94	7.95
Tyrosine	0.93	0.95	1.04	0.91	0.92
Phenylalanine	3.97	3.98	3.94	4.01	3.87

[a] All samples were hydrolyzed in 6 N HCl, 0.1% phenol for 24 hrs, in vacuo, at 110°C. Composition normalized to (Asp + Glu + Gly + Ala) = 62.0 per subunit. No significant difference was observed between hydrolysis of cyclohexanedione-modified BESOD in 6N HCl, 0.1% phenol and 6N HCl, 1% mercaptoacetic acid.

[b] Native BESOD (6.3 μM) was incubated in 9.5 mM butanedione (BDO), 50 mM sodium borate, pH 9.0, for 45 min at 25°C prior to activity measurements and termination of the reaction.

[c] Reaction conditions identical to those of (b) except that the buffer was 50 mM sodium bicarbonate, pH 9.0.

[d]Native BESOD (6.3 μM) was incubated in 20 mM cyclohexane dione (CHD), 50 mM sodium borate, pH 9.0 for 45 min at 25°C prior to activity measurements and termination of the reaction.

[e]Reaction conditions identical to those of (d) except that the buffer was 50 mM sodium bicarbonate, pH 9.0.

the modified proteins, were not a reflection of the use of 30% acetic acid to terminate the reaction, since similar results were obtained when 1.7 N HCl was used in place of the acetic aicd. Oligomerization of the diketones[33] was also not an apparent problem, since the use of freshly distilled butanedione did not change the results.

Reversibility. The correlation between loss of activity and modification of arginine would gain strength, if it could be followed in the reverse. BESOD, which had lost 80% of its activity and 1.3 arginines per subunit, as a consequence of exposure to butanedione plus borate, was restored to 93% of its original activity, when dialyzed against 50 mM potassium phosphate ± 0.1 mM EDTA at pH 7.8 and at 4°C. This dialysis also caused the re-appearance of one arginine per subunit. Similar reversibility of the effects of cyclohexanedione plus borate was also achieved and these results are represented in Table 3. The

TABLE 3

REVERSIBILITY OF BOVINE ERYTHROCYTE SUPEROXIDE DISMUTASE INACTIVATION AND ARGININE MODIFICATION BY BUTANEDIONE AND CYCLOHEXANEDIONE

	BDO-Non-dialyzed[a]	BDO-Dialyzed, KP_i[a]	BDO-Dialyzed, KP_i,EDTA[a]	CHD-Non-Dialyzed	CHD-Dialyzed, KP_i[b]	CHD-Dialyzed, KP_i,EDTA[b]
Specific Activity (U/mg)	607	3300	3475	1197	3565	3238
Amino Acid	Residues per subunit[c]					
Lysine	10.00	10.00	10.00	10.00	10.00	10.00
Histidine	7.40±0.02	7.61±0.02	7.56±0.00	7.52±0.06	7.56±0.01	7.56±0.03
Arginine	2.69±0.09	3.65±0.04	3.66±0.02	3.16±0.03	4.00±0.03	4.01±0.06

[a]Native BESOD (2.3×10^{-5}M) was incubated in 9.9 mM butanedione, 50 mM sodium borate, pH 9.0 for 45 min at 25°C. An aliquot (20 μL) was diluted into 0.2 ml of 50 mM sodium borate, pH 9.0 and assayed for resdiual activity. A second aliquot (0.25 ml) was diluted into 0.75 ml of 40% acetic acid and dialyzed against 30% acetic acid at 4°C prior to amino acid analysis. A third and fourth aliquot (1 ml each) were dialyzed against 50 mM potassium phosphate, pH 7.8 and 50 mM potassium phosphate, 10^{-4}M EDTA, pH 7.8 at 4°C for 24 hrs. The samples were assayed for activity and dialyzed against 30% acetic acid for amino acid dialysis.

[b]Reaction conditions identical to those of (a) except that 9.8 mM cyclohexandione was employed.

[c]Compositions normalized to Lys = 10.0 per subunit. Values represent the average of duplicate analyses ± 1 standard deviation.

correlation between activity and one arginine per subunit thus holds, both when the arginine is blocked by group-specific attack and when it is deblocked by dialysis. The fact that 0.1 mM EDTA, present during deblocking, did not prevent restoration of catalytic activity established that the effects of butanedione and cyclohexanedione did not depend upon labilization of copper. Borate, which is essential for the arginine-specific attack of butanedione and cyclohexanedione[31,34] could be replaced by phenylborate and by m-aminophenylborate.

Identification of the essential arginine residue. BESOD, treated with phenylglyoxal, showed modification of 1.2 arginines per subunit. Reduction and carboxymethylation reduced this to 0.8 per subunit. Partial reversal of phenylglyoxal modification of arginine during this treatment has also been noted by Weng et al.[35] Modification of arginine with butanedione plus borate proved more resistant to the conditions of reduction and carboxymethylation. Thus, BESOD treated with butanedione plus borate, to the limit of 13% residual activity, showed 1.6 arginines modified per subunit prior to reduction and carboxymethylation and 1.4 afterwards. Borate was maintained at 50 mM throughout, to gain maximal stability. A sample similarly inactivated with butanedione plus borate, was subsequently restored to 90% of its initial activity by dialysis against 50 mM potassium phosphate for 48 hrs at pH 7.8 and at 4°C and it then showed only 0.7 modified arginines per subunit. Both the modified and the modified-

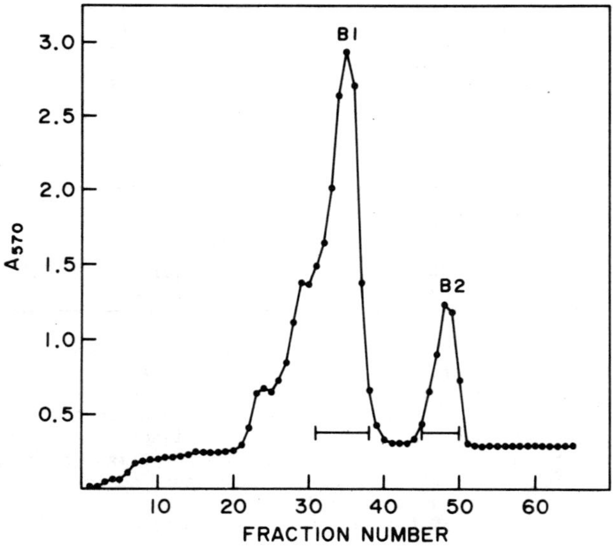

Fig. 5. Separation of cyanogen bromide peptides from reduced and carboxymethylated butanedione-modified BESOD. The sample was treated with cyanogen bromide and the resulting peptides were fractionated on a column of Sephadex G-75 equilibrated in 20% acetic acid as described in MATERIALS AND METHODS. Fractions 31-38 were pooled and identified as cyanogen bromide peptide B1 based upon amino analysis (Evans et al. (Ref.22)). Fractions 45-50 were pooled and identified as cyanogen bromide peptide B2 based upon amino acid analysis (Evans et al. (Ref.22)).

regernated enzymes were treated with cyanogen bromide after reductive carboxymethylation. The two CNBr peptides, B1 and B2, were isolated by chromatography, as shown in Figure 5. Identical elution profiles were obtained with CNBr peptides derived from native BESOD, modified BESOD and modified-regenerated BESOD.

Peptide B1 represents amino acid residues 1-114 and it contains two arginines, at positions 77 and 113. Peptide B2 represents residues 116-151, and also contains arginine, these at positions 126 and 141 (Refs. 22, 23). Amino acid analysis of peptides B1 from native, modified and modified-regenrated samples of BESOD showed that all contained two arginines. This is shown in Table 4.

TABLE 4

AMINO ACID COMPOSITION OF CYANOGEN BROMIDE PEPTIDE B1 FROM NATIVE AND BUTANEDIONE MODIFIED BOVINE ERYTHROCYTE SUPEROXIDE DISMUTASE

Amino Acid	RCM Native BESOD	RCM BDO-BESOD	RCM Regenerated BDO-BESOD
Lysine	7.00[a]	7.00	7.00
Histidine	6.58	6.88	6.59
Arginine	1.95	1.99	1.89
SCM cysteine	1.80[b]	2.03	2.06
Aspartic acid	13.60	13.35	13.47
Threonine	9.89	9.54	9.66
Serine	5.72	5.47	5.55
Glutamic acid	8.24	8.12	8.17
Proline	4.42	4.68	4.80
Glycine	18.36	18.17	18.51
Alanine	6.11	6.37	6.17
Valine	10.45	10.00	10.42
Isoleucine	5.07	4.87	5.13
Leucine	5.89	5.89	5.90
Tyrosine	0.82	0.88	0.79
Phenylalanine	4.17	3.90	3.96

[a]All basic compositions normalized to Lys = 7.0 as determined from the sequence (Evans et al. (Ref.22); Steinman et al. (Ref.23)).

[b]All compositions normalized to (Asp + Glu + Gly + Ala) = 46.0 as determined from the sequence (Evans et al. (Ref.22); Steinman et al. (Ref.23)).

This eliminates the possibility that the essential dione-modifiable arginine was either argine 77 or arginine 113. In contrast, peptide B2 from modified BESOD had only 0.5 arginines and the peptide from the modified-regenerated enzyme had 1.4 arginines. This is shown in Table 5 and establishes that the essential arginine was either arginine 126 or arginine 141.

TABLE 5

AMINO ACID COMPOSITION OF CYANOGEN BROMIDE PEPTIDE B2 FROM NATIVE AND BUTANEDIONE MODIFIED BOVINE ERYTHROCYTE SUPEROXIDE DISMUTASE

Amino Acid	RCM Native BESOD	RCM BDO-BESOD	RCM Regenerated BDO-BESOD
Lysine	3.00[a]	3.00	3.00
Histidine	0.98	0.98	0.96
Arginine	1.99	0.52	1.39
SCM Cysteine	0.95[b]	0.93	0.88
Aspartic acid	3.86	3.86	3.88
Threonine	1.89	1.86	1.92
Serine	1.81	1.77	1.81
Glutamic acid	2.88	2.86	2.89
Proline	0.96	1.00	0.90
Glycine	6.68	6.65	6.72
Alanine	2.88	2.92	2.84
Valine	1.59	1.68	1.57
Isoleucine	1.47	1.47	1.44
Leucine	1.93	1.94	1.93
Tyrosine	0	0	0
Phenylalanine	0	0	0

[a]All basic compositions normalized to Lys = 3.0 as determined from the sequence (Evans et al. (Ref.22); Steinman et al. (Ref. 23)).

[b]All compositions normalized to (Asp + Glu + Gly + Ala) = 16 as determined from the sequence (Evans et al. (Ref. 22);Steinman et al. (Ref. 23)).

The sequences of the CNBr peptide B2 of carboxymethlated BESOD and of its tryptic fragments are given in Figure 6. Peptides B2 from the native, modified and modified-regenerated enzymes were digested with trypsin and the peptide fragments, so produced, were analyzed by peptide mapping. Peptide spots were located with ninhydrin, arginine-containing spots with phenanthraquinone and histidine-containing spots with diazotized sulfanic acid. The results are summarized in Figures 7A, 7B and 7C. As shown in Figure 6, four tryptic peptides are expected. Of these, B2-A is the only one containing both histidine and arginine and it must therefore correspond to spot 2 in Figure 7A. B2-B contains neither arginine nor histidine and at pH 6.5 it would have a negative charge due to excess of glutamate over lysine. Spot 5 therefore corresponds with B2-B. Fragment B2-C contains arginine 141 and would be cationic. Spot 4 in Figure 7A therefore corresponds with B2-C. Fragment B2-D is the most hydrophobic of the tryptic fragments and would be neutral, because the positive charge on lysine 151 would be offset by the negative charge on the carboxymethyl cysteine 144. Spot 1 in Figure 7A therefore corresponds to B2-D. Spot 3 cannot be unequivocally identified, but was probably due to an undigested fragment containing residues 127-151, since it contained arginine, but not histidine.

116 126
Val-Val-His-Glu-Lys-Pro-Asp-Asp-Leu-Gly-Arg-Gly-Gly-Asn-

130 141
Glu-Glu-Ser-Thr-Lys-Thr-Gly-Asn-Ala-Gly-Ser-Arg-Leu-Ala-

144 151
CM Cys-Gly-Val-Ile-Gly-Ile-Ala-Lys-COOH

Trypsin

116 126
Val-Val-His-Glu-Lys-Pro-Asp-Asp-Leu-Gly-Arg B2-A

127 134
Gly-Gly-Asn-Glu-Glu-Ser-Thr-Lys B2-B

135 141
Thr-Gly-Asn-Ala-Gly-Ser-Arg B2-C

142 151
Leu-Ala-CM Cys-Gly-Val-Ile-Gly-Ile-Ala-Lys B2-D

Fig. 6. The amino acid sequence of the cyanogen bromide peptide B2 of bovine erythrocyte superoxide dismutase. Also indicated are the expected tryptic peptides of B2. CM Cys refers to S-carboxymethylcysteine. (Taken from Steinman et al. (Ref.23)).

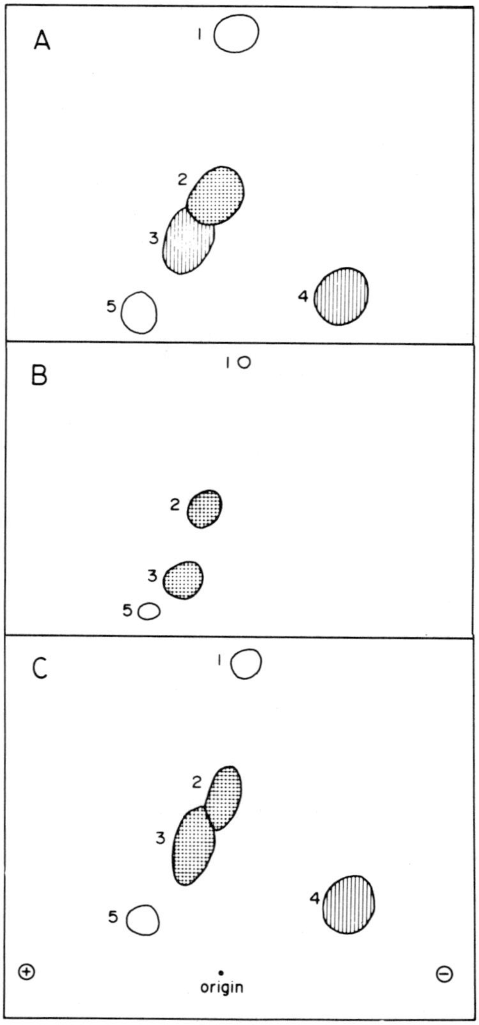

Fig. 7. Tryptic peptide maps of the cyanogen bromide peptide B2 from native and butanedione modified bovine erythrocyte superoxide dismutase. Digestion of the peptides by trypsin and subsequent mapping were performed as described in MATERIALS AND METHODS. Ninhydrin-positive spots are enclosed by solid lines, Phenanthraquinone-positive spots are cross-hatched vertically, and diazo-sulfanic acid-positive spots are cross-hatched horizontally. Tryptic digests of CNBr peptides B2 from reduced and carboxymethylated (A) native BESOD, (B) butanedione modified BESOD, (C) butanedione modified and regenerated BESOD.

The peptide map obtained from CNBr peptide B2 from modified enzyme, Figure 7B, showed smaller spots 1, 2 and 5 indicating that tryptic cleavage at arginine 126 was restricted and it lacked spot 4 entirely showing that tryptic cleavage at arginine 141 had been inhibited. This, in turn, means that arginine 141 had been the predominant site of modification, when the enzyme was treated with butanedione plus borate. Spot 3 in Figure 7B probably represents undigested B2 or a composite of B2 A-B plus B2 C-D. In the case of modified-regenerated enzyme, Figure 7C, spot 4 was again evident, with increased production of B2 D demonstrating deblocking of arginine 141, concomitant with the restoration of activity, during dialysis of the enzyme against phosphate buffer. Spot 3 in Figure 7C again probably represents undigested B2, while the other spots have the identities already discussed for the case of the native enzyme in Figure 7A.

It seems clear, from these results, that the essential arginine residue, whose chemical modification leads to loss of catalytic activity, is none other than arginine 141, already known, from X-ray crystallography, to lie in the active site region.

Survey of superoxide dismutase. Structural information, from X-ray crystallography, is thus far available only for one representative of the copper and zinc-containing superoxide dismutases and that is for BESOD (Refs.3,4). If the arginine, in the active site region, does indeed serve an important function in the catalytic process, one might expect this structural feature to be conserved during the evolution of CuZnSOD. In effect, one would expect CuZnSOD from a variety of sources to show comparable sensitivity to arginine-specific reagents. MnSOD and FeSOD from a variety of sources show considerable sequence homology, but appear unrelated to the CuZnSOD (Refs.26,37). Furthermore, MnSOD and FeSOD, unlike CuZnSOD, show diminished catalytic activity at elevated pH.[38] One might therefore venture to guess that MnSOD and FeSOD would probably not have an arginine residue in their active sites.

This reasoning prompted a survey of the effects of phenylglyoxal and of butanedione upon a wide range of superoxide dismutases. The results are shown in Table 6. It is clear that CuZnSOD from bovine liver, porcine liver, chicken liver and wheat germ (isoenzyme II) were all susceptible to inhibtion by phenylglyoxal and by butanedione plus borate. In each case the inhibition by butanedione plus borate was reversible and approximately one arginine per subunit was modified. In the case of the apo BESOD, essentially all four of the arginines per subunit were modified by phenylglyoxal, indicating that the native structure somehow masked all of the arginines, save the one in the active site region.

The possibility that the diketones might have exerted their effect by modifying the single thiol residue of BESOD can be eliminated since the crystal structure shows this thiol to be buried within the β-barrel.[3,4] Furthermore, the activity of BESOD is unaffected by p-chloromercuriphenyl sulfonate, much as we previously

TABLE 6

COMPARATIVE ARGININE MODIFICATION OF SUPEROXIDE DISMUTASE

Sample	Percent Activity			K_2 ($M^{-1}min^{-1}$)		Arginine Residues Modified per Subunit	
	PGO[a]	BDO[b]	BDO + Dialysis[c]	PGO	BDO	PGO	BDO
CuZnSOD							
Bovine Erythrocyte	10.2	12.0	95	2.4	10.8	1.0	1.6
Apo-Bovine Erythrocyte	-	-	-	-	-	3.6	-
Bovine Liver	12.9	15.0	89.9	2.0	9.2	1.0	-
Chicken Liver	16.0	31.5	86.1	5.7	5.0	1.1	1.5
Porcine Liver	16.3	14.1	100	1.9	9.2	1.3	-
Wheat Germ	6.2	4.6	94.3	-	11.5	1.1	-
MnSOD							
Escherichia coli	64	65.6	80.1	0.4	0.7	2.3	-
Streptococcus faecalis	63	100	92.4	0.3	-	1.7	-
FeSOD							
Escherichia coli	100	93	100	-	-	0.4	0.8

[a]Samples of superoxide dismutase were incubated in 15 mM phenylglyoxal (PGO) 0.125 M sodium bicarbonate, pH 8.0 at 25°C for 60 min. Aliquots were removed at 5 min intervals and assayed for residual activity. Second order rate constants were calculated without correction for residual activity of the samples, except in the case of the chicken liver enzyme, where the second order rate constant was corrected for the 16% residual activity of the sample. After 60 min incubation, acetic acid was added to 30% (v/v final concentration) and the samples were dialyzed against 30% acetic acid at 4°C prior to amino acid analysis.

[b]Samples of superoxide dismutase were incubated in 10 mM butanedione, 50 mM sodium borate, pH 9.0 at 25°C for 60 min, Aliquots were removed at 5 min intervals and assayed for residual activity. Second order rate constants were determined after correcting for the residual activity of each sample. When indicated, acetic acid was added to the modified samples after 60 min incubation to 30% final concentration and the samples were dialyzed against 30% acetic acid at 4°C prior to amino acid analysis.

[c]Samples of superoxide dismutase were incubated as described in (b) and then dialyzed against 50 mM potassium phosphate, pH 7.8 at 4°C for 24 hrs.

noted it to be resistant to iodoacetate.[39] In the case of the chicken liver CuZnSOD, which has a reactive thiol, modification of this thiol with the mercurial had no effect on enzymatic activity.[14] We have also tested p-chloromercuriphenyl sulfonate on the porcine and wheat germ enzymes and found it to have no effect. In the cases of MnSOD or FeSOD, there was some arginine modification, but little or no loss activity, suggesting that these enzymes, unlike the CuZnSOD, lack an essential arginine.

DISCUSSION

In toto, the evidence presented above strongly suggests that arginine 141 of BESOD, which lies within the solvent access channel to the active site and whose amidininium group is positioned within 6Å of the copper, is important for the catalytic activity of this enzyme. Thus, several αβ-diketones, known to exert group-specific attack upon arginine, caused loss of activity. None of the CuZnSOD enzymes studied were affected by thiol-specific reagents, thus eliminating the possibility that the diketones exerted their effects by modifying thiols. In the case of glyoxal, 1 arginine and 1 lysine were modified per subunit, but with phenylglyoxal, butanedione and cyclohexanedione, only arginine was modified. With phenylglyoxal the degree of inactivation was proportional to the extent of arginine modification, to a limit of one arginine per subunit at maximal inactivation. Both butanedione and cyclohexanedione give unstable diol adducts with the amidine portion of arginine, which can be stabilized with borate, which acts by adding onto adjacent hydroxyl groups. With these reagents, borate was essential for inactivation to occur and reactivation could be achieved by dialysis against phosphate buffer. Inactivation by butanedione or cyclohexanedione plus borate, correlated with modification of arginine and reactivation, by dialysis, correlated with deblocking one arginine per subunit.

There were several indications that the arginine, whose chemical modification caused inactivation, must lie close to the active site copper. The first of these was, of course, the inactivation itself. The second was the effect of arginine modification upon the visible absorption of the enzyme, which is entirely due to the Cu(II). This spectral modification was also seen with butanedione plus borate, but was most profound with phenylglyoxal. The third indication was a modest labilization of copper, seen only with the phenylglyoxal. Modification of the essential arginine with this reagent apparently perturbed the active site region more than did corresponding modification with glyoxal, butanedione or cyclohexanedione.

The most powerful evidence that the essential arginine was 141, was obtained by peptide analysis of native, modified and modified-regenerated BESOD. Cyanogen bromide B-2 alone bore the modified arginine and this immediately halved the range of possibilities from arginines 77, 113, 126 and 141 to only arginine 126 or arginine 141. Then, tryptic digestion of B-3 showed that the site of greatest modification was arginine 141. There was evidence of some modification also at arginine 126, and this was to be expected since butanedione plus borate modified somewhat more than one arginine per subunit. It appears that arginine 126 could to some extent be modified by butanedione, in the absence of borate, and that this could not be reversed by dialysis and was not associated with loss of catalytic activity.

BESOD retained some activity even after modification with arginine-specific reagents. The extent of this residual activity depended upon the specific reagent used, but never exceeded 20% of the native activity. It is possible, in the cases of butanedione and cyclohexandione, whose attack on arginine was supported by borate and was reversed by removal of borate, that some of this residual activity was actually a reflection of reversal, which occurred upon dilution into the assay mixture. This is not, however, a possibility when glyoxal or phenylglyoxal were used; since their effects on enzymatic activity were not reversible. It may be that the active site arginine is important in the electrostatic guidance of the incoming O_2^- to the catalytic center, but that such electrostatic guidance is not entirely essential for activity. If we may judge from the effect of arginine-modifying reagents, we would estimate that this arginine increased catalytic efficiency by approximately one order of magnitude. It is apparently a sufficiently important factor in this catalysis to have been retained during long periods of evolutionary change, as evidenced by the similar behavior of CuZnSOD from chicken, cow, pig and wheat.

Another possible role for the active site arginine involves proton conduction, which could be a limiting factor in catalysis, were a special facilitating mechanism not provided. We envision O_2^- binding, in a bridging position, between Cu(I) and the amidinium group of arginine 141, during the second half reaction of the catalytic cycle. Transfer of an electron from the Cu(I) and of a proton from the arginine would yield the leaving group, i.e. HO_2^-, which would protonate to H_2O_2 in free solution. The amidine group of arginine 141 would then recover a proton from the solvent and be ready for another cycle.

Superoxide dismutases based upon manganese or iron appeared not to possess an arginine uniquely important for activity. In these enzymes, the metal cycles between the trivalent and divalent states during catalysis and the charge on the metal itself may provide sufficient electrostatic effect to attract O_2^-. In contrast, the copper of CuZnSOD cycles between the divalent and monovalent states and the presence of an imidazole ligand bridging the copper to the zinc and an aspartate ligand to the zinc would practically eliminate residual positive charge, when the copper was in the univalent half of its catalytic cycle. It is tempting to speculate that the ligands to the manganese and to the iron in MnSOD and FeSOD, when identified, will be found not to diminish the charge on these metals. We might also anticipate that a lysine residue lies within the active site region of these enzymes and functions in catalysis, as does the arginine in CuZnSOD.

ACKNOWLEDGEMENTS

This work was supported by research grants from the National Institutes of Health Bethesda, Maryland (GM-10287); the United States Army Research Office, Research

Triangle Park, N.C. (DRXRO-PR-P-15319-L) and from the Merck, Sharp and Dohme Research Laboratories, Rahway, New Jersey.

REFERENCES

1. Klug, D., Rabani, J. and Fridovich, I. (1972) J. Biol. Chem. 247, 4839-4842.
2. Rotilio, G., Bray, R.C. and Fielden, E.M. (1972) Biochim. Biophys. Acta, 268, 605-609.
3. Richardson, J.S., Thomas, K.A. and Richardson, D.C. (1975) Biochem. Biophys. Res. Commun. 63, 986-992.
4. Richardson, J.S., Thomas, K.A., Rubin, B.H. and Richardson, D.C. (1975) Proc. Nat. Acad. Sci. USA, 72, 1349-1353.
5. Klug, D., Fridovich, I. and Rabani, J. (1973) J. Amer. Chem. Soc. 95, 2786-2790.
6. Fielden, E.M., Roberts, P.B., Bray, R.C. and Rotilio, G. (1973) Biochem. Soc. Trans. 1, 52-53.
7. Asada, K., Kanematsu, S., Takahashi, M. and Kona, Y. (1976) Adv. Exp. Med. Biol. 74, 551-564.
8. McCord, J.M. and Fridovich, I. (1969) J. Biol. Chem. 244, 6049-6055.
9. Rabani, J., Klug-Roth, D. and Lilie, J. (1973) J. Phys. Chem. 77, 1169-1175.
10. Koppenol, W.H., Van Buren, K.J.H., Butler, J. and Braams, R. (1976) Biochim. Biophys. Acta, 449, 157-168.
11. Koppenol, W.H. (1979) in Third International Symposium on Oxidases and Related Redox Systems, King, T.E., Mason, H. and Morrison, M., eds., Albany, New York, in press.
12. Bannister, J., Bannsiter, W. and Wood, E. (1971) Eur. J. Biochem. 18, 178-186.
13. Abernethy, J.L., Steinman, H.M. and Hill, R.L. (1974) J. Biol. Chem. 249, 7339-7347.
14. Weisiger, R.A. and Fridovich, I. (1973) J. Biol. Chem. 248, 3582-3592.
15. Beauchamp, C. and Fridovich, I. (1973) Biochim. Biophys. Acta, 317, 50-64.
16. Keele, B.B., Jr., McCord, J.M. and Fridovich, I. (1970) J. Biol. Chem. 245, 6176-6181.
17. Britton, L., Malinowski, D.P. and Fridovich, I. (1978) J. Bacteriol. 134, 229-236.
18. Yost, F.J. and Fridovich, I. (1973) J. Biol. Chem. 248, 4905-4908.
19. Rousch, A. and Norris, E.R. (1950) Arch. Biochem. Biophys. 29, 344-347.
20. Lowry, O.H., Rosebrough, M.J., Farr, A.L. and Randall, R.J. (1951) J. Biol. Chem. 193, 265-275.
21. Crestfield, A.M., Moore, S. and Stein, W.H. (1963) J. Biol. Chem. 238, 622-627.
22. Evans, H.J., Steinman, H.M. and Hill, R.L. (1974) J. Biol. Chem. 249, 7315-7325.
23. Steinman, H.M., Naik, V.R., Abernethy, J.L. and Hill, R.L. (1974) J. Biol. Chem. 249, 7326-7338.
24. Easley, C.W. (1965) Biochim. Biophys. Acta, 107, 386-388.
25. Easley, C.W., Zegers, B.J.M. and DeVijlder, M. (1969) Biochim. Biophys. Acta, 175, 211-213.
26. Moore, S., Spackman, D.H. and Stein, W.H. (1958) Anal. Chem. 30, 1185-1206.
27. Moore, S. and Stein, W.H. (1963) Methods Enzymol. 6, 819-831.
28. Patthy, L. and Smith, E.L. (1975) J. Biol. Chem. 250, 557-564.
29. Takahashi, K. (1968) J. Biol. Chem. 243, 6171-6179.
30. Borders, C.L. and Riordan, J.F. (1975) Biochemistry, 14, 4699-4704.
31. Riordan, J.F. (1973) Biochemistry, 12, 3915-3923.
32. Patthy, L. and Smith, E.L. (1975) J. Biol. Chem. 250, 565-569.
33. Yankeelov, J.A. (1970) Biochemistry, 9, 2433-2439.
34. Daemen, F.J.M. and Riordan, J.F. (1974) Biochemistry, 13, 2865-2871.
35. Weng, L., Heinrikson, R.L. and Westley, J. (1978) J. Biol. Chem. 253, 8109-8119.
36. Steinman, H.M. and Hill, R.L. (1973) Proc. Nat. Acad. Sci. USA, 70, 3725-3729.
37. Bridgen, J., Harris, J.I. and Northrop, F. (1975) FEBS Lett. 49, 392-395.
38. Forman, H.J. and Fridovich, I. (1973) Arch. Biochem. Biophys. 159, 396-400.
39. Malinowski, D.P. and Fridovich, I. (1979) Biochemistry, submitted for publication.

POLYCYCLIC PHENOLS IN ALKALINE SOLUTION: STABLE SUBSTRATES FOR SUPEROXIDE DISMUTASE?

CHRISTIAN AUCLAIR
Laboratoire d'enzymologie des cellules sanguines, INSERM U160, CNRS ERA 573, Hôpital Beaujon, 92118 Clichy, France

INTRODUCTION

Superoxide dismutases (SODs) (E.C. 1.15.1.1.) are enzymes which catalyse the dismutation of superoxide anion (O_2^-) according to the following reaction:

$$O_2^- + O_2^- + 2\,H^+ \longrightarrow H_2O_2 + O_2 \qquad (1)$$

This reaction also proceeds spontaneously in aqueous solution and SODs are unique among enzymes in that their substrate is an unstable free radical.[1] This unusual feature complicates the measurement of the catalytic activities of SODs. Assays for measuring the activity of these enzymes usually consist in observing the inhibition of a reaction in which O_2^- is a transient intermediate. The extent of inhibition is taken as a measure of the catalytic activity of the enzyme (for general review, see Ref.2). Such methods are indirect and do not permit to determine a unit of enzyme as usually defined. Procedures for direct determination of the catalytic activity of SODs require the generation of high concentration of O_2^- which is achieved either by pulse radiolysis[3] or by the decomposition of KO_2 in alkaline aqueous solution.[4] However, such procedures involve the generation of high concentration of H_2O_2 which strongly interacts with the enzyme.[5] Moreover, in all proposed assays, saturation of the enzyme with the substrate is never approached and consequently, the affinity of the enzyme for its substrate cannot be determined. In this situation, suitable conditions for the study of SOD activity according to classical methods used in enzymology, require the discovery of a stable substrate for the enzyme. Among the possible candidates for such function are oxygenated complex with oxygen in the superoxide state according to the general structure: $X \ldots O_2^-$.

Such a structure may be achieved through a charge transfer process between an organic molecule with an electron releasing group and molecular oxygen. Since the phenolate form of various phenols is known to react with electron acceptors according to a charge transfer process, we have studied the ability of some polycyclic phenols to react with molecular oxygen according to the following reaction:

$$R - O^- + O_2 \longrightarrow [R - O^{\bullet} \ldots O_2^-]^- \qquad (2)$$

Such an oxygenated complex, as pointed out above, may possibly react with SOD to break down according to a dismutation process as follows:

$$2\ [R\text{-}O^{\bullet}\ \ldots\ O_2^{\bullet}]^- + 2\ H^+ \xrightarrow{SOD} R = O + R\text{-}O^- + H_2O_2 + O_2 \qquad (3)$$

The present report describes the results of investigations performed with phenolic compounds derivated from hydroxyl indole, namely: 5-hydroxyindole (5-OH-I) (A in Figure 1), 6-hydroxy 1-4, dimethyl carbazole (6-OH-C; B in Figure 1), and 9-hydroxy 5-11, dimethyl, 6-H-pyrido (4-3b) carbazole* (9-OH-PC; C in Figure 1). Under suitable experimental conditions, these compounds generate a stable oxygenated complex which acts as a stable substrate for SOD, suitable for a direct spectrophotometric assay.

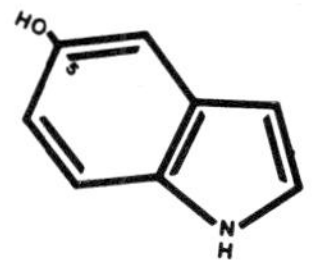

A

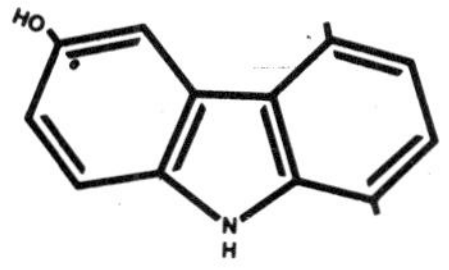

B

Fig. 1. Formula of hydroxy-indole derivatives.

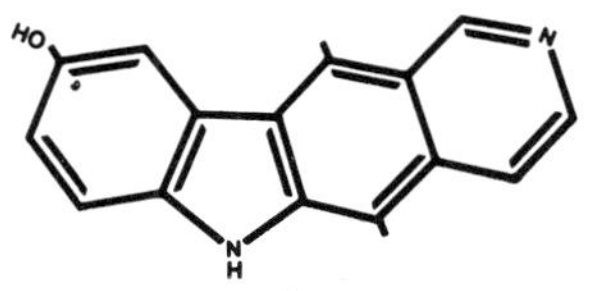

C

*Pyridocarbazole derivatives are molecules endowed with antitumor activity and presently under clinical trials in the treatment of human cancers under the name of Ellipticines.

MATERIALS AND METHODS

6-Hydroxy, 1-4, dimethyl carbazole, and 9-hydroxy, 5-11, dimethyl, 6-H pyrido (4-3b) carbazole were synthesized according to Dalton et al.[6] and were a gift of Dr. Dat Xuong (ICNS, Gif sur Yvette, France). 5-Hydroxyindole was obtained from Ega-Chemie (West Germany). Salicylate-copper complex (bis-(salicyclato)-copper(II) was prepared according to de Alvare et al.[7]

Catalase was obtained from Boehringer (West Germany) and Cu-Zn SOD was purified from human erythrocytes according to McCord et Fridovich except that CM-52 was used instead of CM-32 in the last step of purification.

Ultraviolet and visible absorption spectra were performed on a Aminco DW-2 spectrophotometer and e.p.r. spectra on a Bruker B-ER 420 apparatus. Oxygen consumption was measured with an oxygen electrode from Yellow Springs Instruments connected to a Gilson oxygraph. H_2O_2 concentration was measured by the amount of O_2 released in the presence of catalase.

RESULTS AND DISCUSSION

Interaction between molecular oxygen and polycyclic phenols derived from indole in alkaline medium. A number of phenols are known to undergo a spontaneous oxidation at alkaline pH (Ref.8). This property has been extensively studied with diphenols[9] and it has been established that the initial step of the oxidation process involves the one electron transfer from the phenolate form of an OH group on molecular oxygen yielding the phenoxy radical and O_2^- as follows:

$$R\text{-}\bar{O}\!: + O_2 \rightleftharpoons R\,O^\bullet + O_2^- \qquad (4)$$

It has, moreover, been pointed out[10] that in such reaction, O_2^- is reversibly generated with a pre-existing charge transfer complex between molecular oxygen and the electron donor according to the following sequence:

$$R\text{-}\bar{O}\!: + O_2 \rightleftharpoons [R - O: \ldots O_2] \rightleftharpoons R\text{-}O^\bullet + O_2^- \qquad (5)$$

The univalent electron transfer from the organic molecule to molecular oxygen should generate the biradical form of the oxygenated intermediate as indicated in reaction:

$$[R - O: \ldots O_2]^- \longleftrightarrow [R - O^\bullet \ldots O_2^-]^- \qquad (6)$$

The biradical complex with O_2 under superoxide state may react with SOD according to reaction (3). If O_2^- does not act as chain propagating species in the reaction, SOD is thus expected to accelerate the autoxidation of the electron donor. Such a reaction may provide a basis for a direct assay for the measurement of SOD activity. Conditions required for a convenient assay are stability of the oxygenated intermediate in the absence of any catalyst and the easy detection of the two-electron oxidized product of the oxidizable phenolic compound. According to this view, polycyclic phenols having an indole moiety in their structure may be oxidized into quinone-imine derivatives (Figure 2) by abstraction of two electrons and two protons.

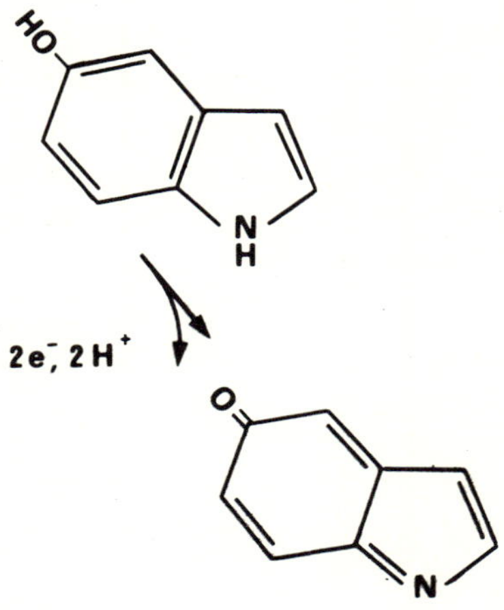

Fig. 2. Two-electron oxidation of the hydroxyindole ring.

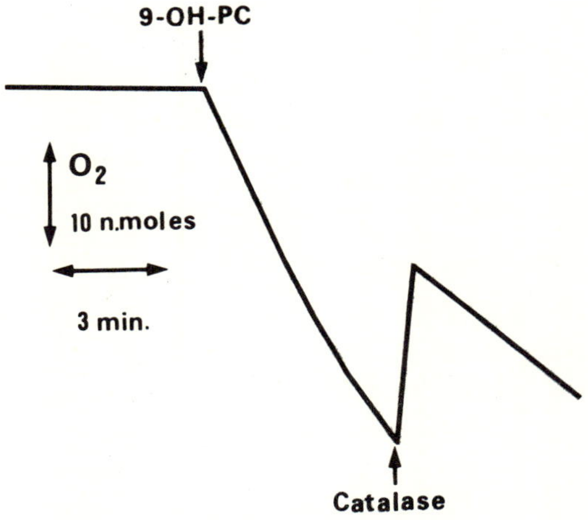

Fig. 3. Oxygen consumption associated with the autoxidation of 9-OH-PC. Oxygen consumption was measured with a Clark oxygen electrode at 37°C in 1.0 ml 0.05 M carbonate (pH 11.2) containing 10^{-4}M EDTA. Where indicated by arrows, 0.1 mM 9-OH-PC and 10 μg catalase were added.

The quinone-imine derivatives so formed are easily detectable spectrophotometrically in the 500 nm range. Typical data are indicated in Figure 3.

We performed a series of experiments in alkaline DMSO (pH 14). Under these conditions the phenolate anion of the polycyclic phenols is formed. Addition of indole derivatives such as 5-OH-I, 6-OH-C and 9-OH-PC to air saturated DMSO containing 50 μM NaOH results in the oxygen consumption illustrated in Figure 4. Furthermore, addition of salicylate copper complex, an hydrophobic copper chelate endowed with SOD activity[11] results in a release of oxygen consistent with the proposed mechanism of the reaction. The generation of free radicals as products of the reaction between the phenolic compounds and O_2 may be investigated by e.p.r. spectroscopy. Experiments performed at room temperatures indicate

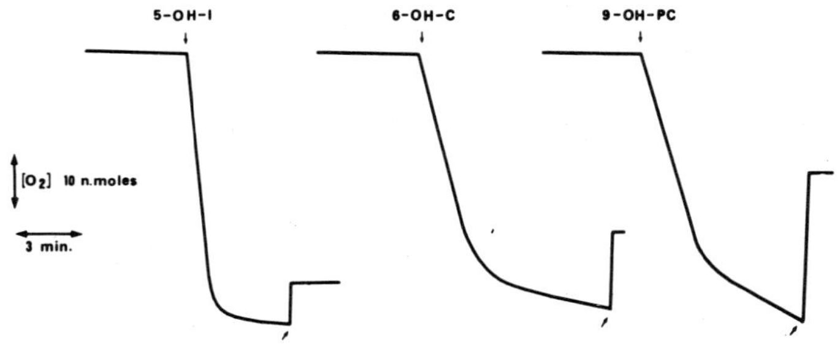

Fig. 4. Oxygen consumption associated with the generation of oxygenated complex of hydroxyindole derivatives. Oxygen consumption was measured as described in Figure 3 in DMSO containing 50 μM NaOH (pH 14) and 0.55 M H_2O. Phenolic compounds were added at a concentration of 0.1 mM. Arrows at the lower part of the traces indicate the addition of 10 mM Cu(II)-Sal_2 complex dissolved in DMSO.

that, in alkaline DMSO, the three phenols tested spontaneously generate radical species in the presence of oxygen (Figure 5). The e.p.r. spectra observed with 5-OH-I indicate, that in alkaline DMSO, this compound readily generates a stable free radical related to a phenoxy radical (Figure 5, upper traces). In contrast, 6-OH-C generates under similar conditions an unstable radical which exhibits an unresolved spectrum (Figure 5, middle trace, left hand side). This spectrum may be that of the biradical derived from 6-OH-C, as indicated in reaction (6). This complex is unstable and the e.p.r. spectrum recorded again after 30 min (Figure 5, middle trace, right-hand side) indicates that the signal of the oxygenated complex has been replaced by a signal exhibiting hyperfine structure related to the phenoxy free radical of 6-OH-C. 9-OH-PC gives rise to a stable oxygenated complex (Figure 5, lower trace). It is of interest to consider that the increase in the number of conjugated double bounds of the organic molecules is correlated with the increase of the stability of the oxygenated complex.

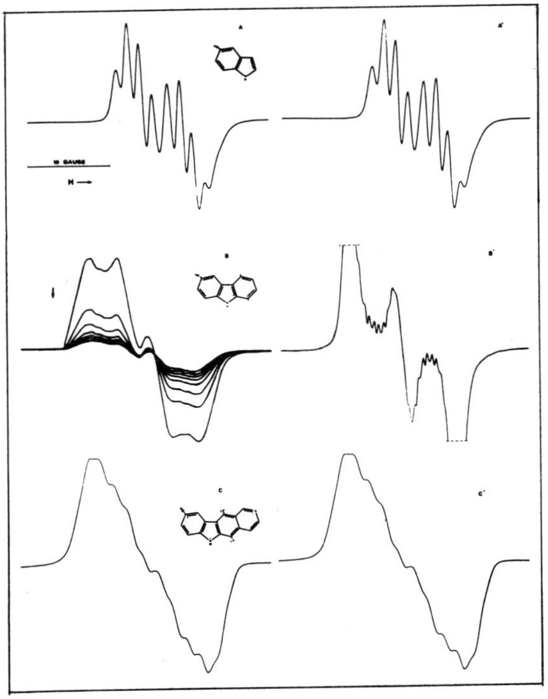

Fig. 5. E.p.r. spectra of hydroxyindole derivatives in air saturated alkaline DMSO. E.p.r. spectra were measured at room temperature with a microwave frequency of 9.753 GHz, 1 Gauss modulation and 250 mW microwave power. Assay mixtures were composed of DMSO containing 50 μM NaOH (pH 14) and 0.55 M H_2O. Phenolic compounds were added at a concentration of 1 mM. Spectra in A, B and C were recorded just after the addition of the phenols in alkaline DMSO whereas spectra in A' B' and C' were recorded at 30 min. afterwards. In B (left-hand side) the spectrum of 6-OH-C was recorded at 2 min. intervals. g values were estimated to be 2.0046 ± 0.0002 for the oxygenated complex and 2.0050 ± 0.0002 for the phenoxy free radicals.

It may be concluded from these experiments that the phenolate anion of hydroxyindole derivatives can transfer one electron to molecular oxygen. In aprotic solvants the reactions might be summarized as follows:

$$R = 5\text{-OH-I} \quad RO^{\bar{\cdot}} + O_2 \longrightarrow [RO^{\cdot} \ldots O_2^{\cdot}]^{-} \xrightarrow{\text{fast}} RO^{\cdot} + O_2^{\bar{\cdot}} \quad (7)$$

$$R = 6\text{-OH-C} \quad RO^{\bar{\cdot}} + O_2 \longrightarrow [RO^{\cdot} \ldots O_2^{\cdot}]^{-} \xrightarrow{\text{slow}} RO^{\cdot} + O_2^{\bar{\cdot}} \quad (8)$$

$$R = 9\text{-OH-PC} \quad RO^{\bar{\cdot}} + O_2 \longrightarrow [RO^{\cdot} \ldots O_2^{\cdot}] \not\longrightarrow RO^{\cdot} + O_2^{\bar{\cdot}} \quad (9)$$

In alkaline aqueous solution, an adequte supply of available protons permits the disproportionation of the transient intermediates. Accordingly, addition of 6-OH-C or 9-OH-PC to medium containing carbonate (pH 11.2) results in a spontaneous oxidation of these compounds into red colored products identified as quinoneimine derivatives. However, 5-OH-I was not oxidized. Experiments performed with 9-OH-PC indicate that during the autoxidation process, moelcular oxygen is effectively reduced to H_2O_2 (Figure 3). Moreover, the autoxidation rate depends on both the oxygen concentration and the pH of the solution (Figure 6). Lowering the pH from 11.2 to 10.2 results in a decrease of the oxidation rate suggesting that the concentration of phenolate anion is rate limiting. It is of interest to investigate the occurrence of the oxygenated

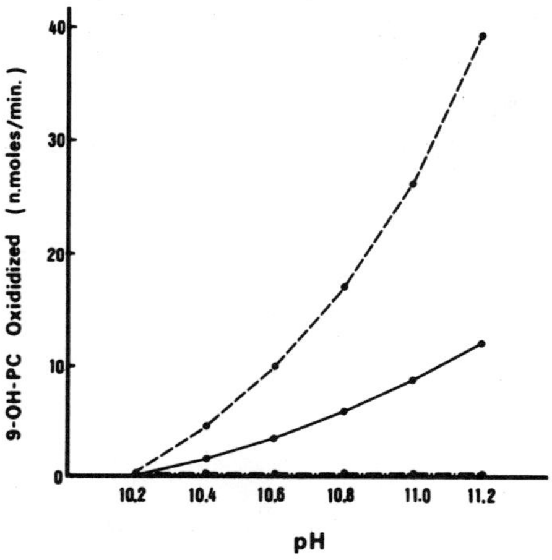

Fig. 6. Autoxidation of 9-OH-PC; effects of oxygen concentration and pH. 9-OH-PC oxidation rate was taken as the initial velocity of increase in absorbance at 515 nm using a molar extinction of 4,400 $M^{-1}cm^{-1}$. Assay media were composed of 0.05 M Na_2CO_3, 10^{-4}M EDTA and 10^{-4}M 9-OH-PC. The pH were adjusted by addition of 5N HCl. (●——●) indicates air saturated, (●·—·—●·) nitrogen saturated, and (●---●) oxygen saturated medium.

intermediates during the autoxidation reaction in aqueous solution, Unfortunately the poor solubility of the phenols in carbonate medium does not permit concentrations sufficient for e.p.r. experiments to be obtained. However, the interaction between the phenolate of 9-OH-PC and oxygen is shown by the appearance of a difference UV spectrum between 9-OH-PC under aerobic and anserobic conditions (Figure 7). Comparison of the spectra obtained in carbonate solution (pH 10.2) with those obtained in alklaine DMSO (pH 14) indicates that similar species were generated in each mixture. Such data were not obtained with 6-OH-C. It is reasonable to assume that the autoxidation of 9-OH-PC in carbonate medium occurs through the generation of the biradical. At pH 10.2 a slight amount of the oxygenated complex is generated without further oxidation of the phenolic compound.

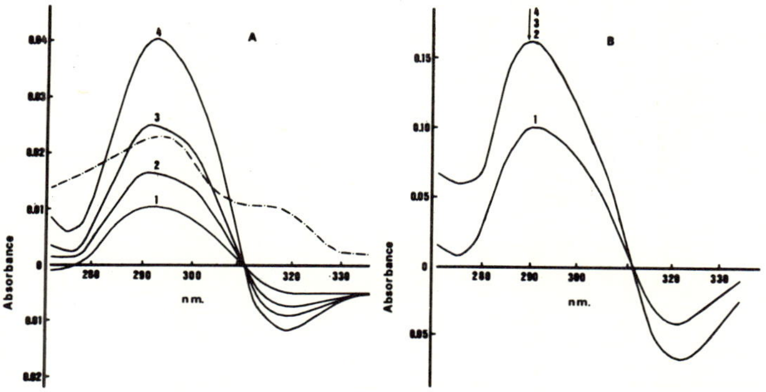

Fig. 7. Difference spectra of 9-OH-PC in oxygenated and deoxygenated mixtures A, carbonate medium (pH 10.2); B, alkaline DMSO (pH 14). In both experiments the sample cuvette contained 0.050 mM 9-OH-PC in nitrogen saturated medium. The reference cuvette contained 0.050 mM 9-OH-PC in air saturated (curve 1) and oxygen saturated medium (curves 2, 3, 4). Curves 2, 3 and 4 were recorded just after the introduction of O_2, 5 min. and 15 min. afterwards respectively. The dotted line in A indicates the difference spectra obtained after the addition of 1 µg SOD in both deoxygenated and oxygenated mixtures.

Effect of superoxide disumyase on the autoxidation of 9-OH-PC in carbonate medium. The above investigations indicate that among the phenolic compounds tested, only 9-OH-PC generates a stable oxygenated complex both in alkaline DMSO and in carbonate medium (pH 10.2). In agreement with this hypothesis (see reactions (3) and (6)), addition of SOD in carbonate medium results in a strong increase in the autoxidation rate of 9-OH-PC as measured by oxygen consumption and generation of the quinone-imine (Figure 8). At pH 10.2, where 9-OH-PC is not spontaneously oxidized, SOD promotes the aerobic oxidation of this compound. Under these conditions, the oxygenated complex of 9-OH-PC may be considered as a substrate for SOD. SOD inhibits the autoxidation of 6-OH-C indicating that O_2^- is generated free in solution and acts as chain propagating species as observed duirng the autoxidation of diphenols. Stability of the oxygenating complex, as in alkaline DMSO, is thus required to provide a convenient substrate for SOD.

Under suitable experimental conditions, the SOD catalyzed 9-OH-PC oxidation, is linear with respect of incubation time (Figure 9,part A) and the oxidation

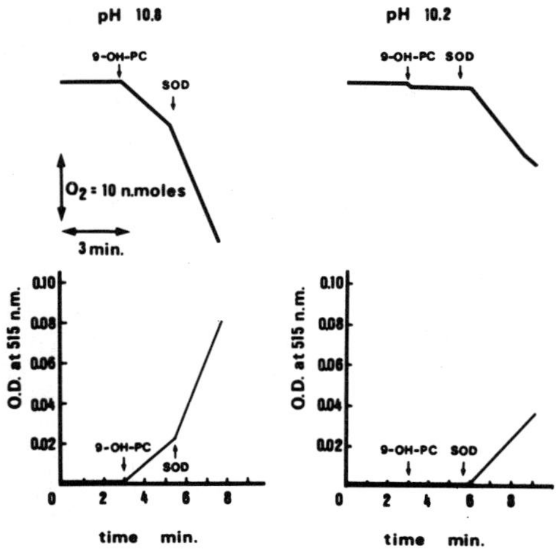

Fig. 8. The effect of SOD on 9-OH-PC oxidation and associated oxygen consumption. All experiments were carried out as described in Figure 2. Where indicated, SOD was added at a final concentration of 1 µg/ml.

rate is linear with respect of SOD concentration (Figure 9, part B). SOD (0.1 µg, about 0.03 nmoles) is easily detected. For higher concentrations of SOD, oxygen tension is the limiting factor. In the specified conditions the assay is thus suitable for samples containing between 0.05 and 0.5 µg Cu-Zn SOD. As expected, cyanide strongly inhibits the catalytic activity of the Cu-Zn SOD used in the experiments. From these experiments it may be concluded that the aerobic oxidation of the polycyclic conditions for the determination of the enzymatic activity of SOD as usually defined for measurement of enzymatic activity.

ACKNOWLEDGEMENTS

We are grateful to Dr. Gianotti (ICNS, Gif sur Yvette, France) for his kind guidance on e.p.r. studies.

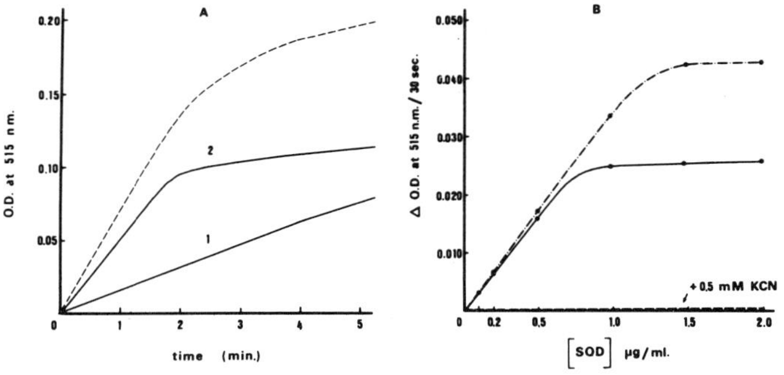

Fig. 9. Effects of incubation time and SOD concentration on the SOD-catalysed 9-OH-PC oxidation. Reactions were carried out in 0.05 M Na_2CO_3 containing 10^{-4}M EDTA and 0.050 mM 9-OH-PC. A: solid lines indicate the time-course of 9-OH-PC oxidation in the presence of 0.2 μg/ml SOD (curve 1) and 0.5 μg/ml SOD (curve 2); dashed line indicates the time course of 9-OH-PC oxidation in oxygen saturated medium and in the presence of 0.5 μg/ml SOD. B: solid line indicates air saturated and (—·—·—·—·—) line, oxygen saturated medium.

REFERENCES

1. McCord, J.M. and Fridovich, I. (1969) J. Biol. Chem. 244, 6049-6055.
2. Fridovich, I. (1975) Annu. Rev. Biochem. 44, 147-159.
3. Rotilio, G., Bray, R.C. and Fielden, E.M. (1972) Biochim. Biophys. Acta, 268, 605-609.
4. Marklund, S. (1976) J. Biol. Chem. 251, 7504-7507.
5. Hodgson, E.K. and Fridovich, I. (1975) Biochemistry, 14, 5294-5303.
6. Dalton, L.K., Demerac, S., Elmes, B.C., Loder, J.W., Swan, J.M. and Teitei, T. (1967) Aust. J. Chem. 20, 2715-2727.
7. Alvare De, L.R., Goda, K. and Kimura, T. (1976) Biochem. Biophys. Res. Commun. 69, 687-694.
8. Hawley, M.D., Tatawawadi, S.V., Piekarski, S. and Adams, R.N. (1967) J. Amer. Chem. Soc. 89, 447.
9. Cohen, G. and Heikkila, R.E. (1974) J. Biol. Chem. 249, 2447-2452.
10. Da Silva Araujo, J., Bechara, E.J.H., Zinner, K. and Cilento, G. (1970) Rec. Chem. Progr. 31, 189.
11. Sorenson, J.R.J. (1978) Progr. Med. Chem. 15, 211-260.

TRANSIENT REACTION OF 2Cu,2Zn-SUPEROXIDE DISMUTASE WITH A SUPEROXIDE GENERATING SUBSTRATE

LUTZ M. SCHUBOTZ, MAGED YOUNES AND ULRICH WESER
Anorganische Biohcemie, Physiologisch-chemisches Institut der Universität Tübingen, D-7400 Tübingen, W.Germany

Many transition metals are known to react with excited oxygen species.[1] Among these metals copper deserves special attention as it is widely distributed in biochemical systems. Copper is involved in reversible oxygenation reaction, the catalysis of superoxide dismutation, peroxide metabolism and the production of water.[2] Out of these reactions the catalysis of superoxide dismutation has gained considerable interest.[3] The 2Cu2Zn-superoxide dismutase accelerates the spontaneous superoxide dismutation to a rate of 2.6 x $10^9 M^{-1} s^{-1}$ (Ref.3). This functional phenomenon is also exerted by many low-molecular-weight copper chelates in which the metal is bound to tyrosine, lysine,[4,5] salicylates,[6,7] penicillamine[8] or indomethacine.[9] The latter low-molecular-weight copper complexes are progressively used to study the involvement of superoxide ions in those metabolic pathways where the native enzyme is unable to reach the compartmented sections of oxygen activation.[10-13]

It has to be emphasized that there is no difference in the reaction of the protein-bound copper and the low-molecular-weight copper chelates when superoxide is available in the free state or in the protonated form. However, no experimental details are known about the transient state during the generation of superoxide radicals.[14] For example, this could mean that superoxide is still bound to a biochemical substrate and is not yet liberated into solution. A convenient system to examine this task was to study autoxidizing adrenaline in more detail. Autoxidizing adrenaline can be specifically inhibited by superoxide dismutase.[15] In fact, this assay is routinely used for the detection of superoxide in biological material.[16] Earlier observations from our laboratory showed that low-molecular-weight copper chelates react differently in this assay compared to native 2Cu2Zn-superoxide dismutase.[17] For an explanation of this phenomenon we were led by the hypothesis that the specific inhibition by the native enzyme may be attributed to the formation of a ternary complex between the activated substrate and the enzyme.[18] This reaction was reinvestigated employing biochemical methods and circular dichroism.

Autoxidizing Adrenaline. The autoxidation of adrenaline at pH 10 is suggested to proceed in the generally accepted reaction scheme (Figure 1). Four proposed intermediary radical components including several semiquinone radicals appear to be generated in the course of the reaction. The total of 4 electrons is transferred until adrenochrome is formed. Apart from the semiquinone radicals, superoxide and direct electron transport are probably involved.[23] As a consequence added 2Cu2Zn-superoxide dismutase will be initially reduced. This reaction could take place by either reaction intermediate. Decision of these possibilities is a major open question. The autoxidation of adrenaline is illustrated in

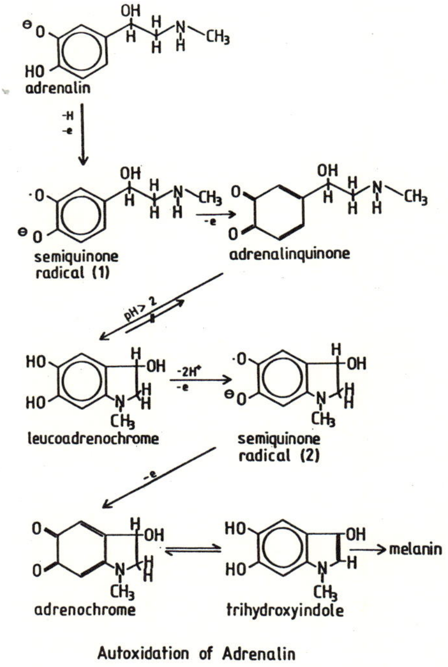

Fig. 1. Autoxidation reactions of adrenaline.

Figure 2. Autoxidizing adrenaline monitored at 480 nm passes a maximum after 5 minutes. This inidicates that adrenochrome will be converted into the end product melanin after some time. In the presence of 2Cu2Zn-superoxide dismutase the formation of adrenochrome is significantly diminished. Ultraviolet irradiation or heating of the native enzyme both produce control values. The same is seen after the addition of the apo-protein. Surprisingly the superoxide dismutative active copper tyrosine complex accelerates the formation of adrenochrome. For the enzyme mediated specific inhibition the need of both protein and coordinated copper can definitely be concluded. This can additionally be shown by inhibiting the native enzyme with cyanide. The cyanide-exposed 2Cu2Zn-superoxide dismutase dramatically accelerates the autoxidation of adrenaline. Cyanide alone did not affect the autoxidation of adrenaline. Chelation of the vicinal hydroxyl groups of the aromatic ring by borate stopped the autoxidation reaction completely. Additions of copper either as copper complexes or 2Cu2Zn-superoxide dismutase had no effect.

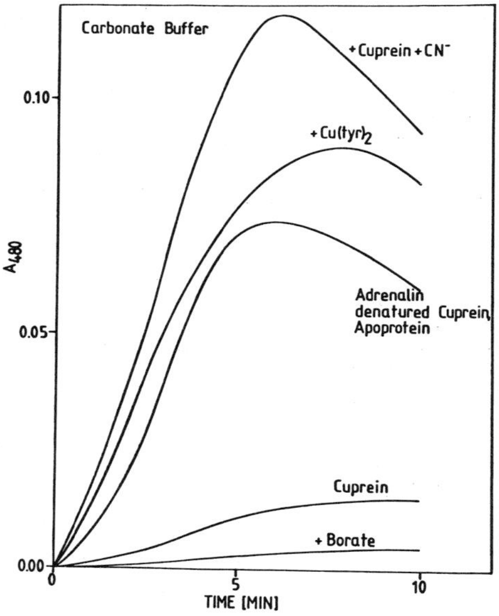

Fig. 2. Autoxidation of adrenaline in carbonate buffer. The reaction was followed by monitoring adrenochrome formation at 480 nm and 25°C in a Pye Unicam SP 1800 spectrophotometer. 1 cm light path cuvettes contained a total of 1 ml composed of 850 µl 0.1 M carbonate buffer pH 10, 100 µl water or the indicated addition, and 50 µl 0.01 M adrenalin pH 2. Additions were made to give the final concentrations: adrenaline 0.5 mM, 2Cu2Zn-superoxide dismutase, denatured 2Cu2Zn-superoxide dismutase, apo-2Cu2Zn-superoxide dismutase and 2Cu2Zn-superoxide dismutase 1 µM, copper-tyrosine 0.1 mM, 2Cu2Zn-superoxide dismutase and cyanide 50 µM and 5 mM, 2Cu2Zn-superoxide dismutase and borate 50 µM and 5 mM. Cuprein in this and subsequent Figs. denotes 2Cu2Zn-superoxide dismutase.

Chiroptical properties. The chiroptical properties of adrenaline in the presence and absence of micromolar concentrations of 2Cu2Zn-superoxide dismutase were markedly different. The adrenaline autoxidation products displayed Cotton extrema at 290 nm (+), 330 nm (-) and 370 nm (+). The magnitude of the ellipticity values averaged around 300 deg. $cm^2.dmol^{-1}$. In the presence of 2Cu2Zn-superoxide dismutase, new Cotton bands appearing at 400 nm (-) and a detectable blue-shift of the Cotton band at 330 nm to 317 nm as well as a shift of the 370 nm band was seen. The magnitude of the Cotton bands increased by 30% (Figure 3).

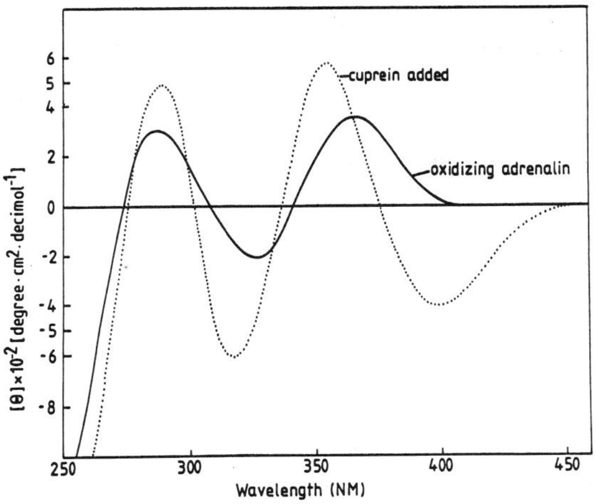

Fig. 3. Circular dichroism of autoxidizing adrenaline. Circular dichroism was measured within 20 minutes of the start of the reaction on a Jasco J-20 A spectropolarimeter employing 1 cm light path quartz cells. Autoxidizing adrenaline 0.5 mM in 0.1 M carbonate pH 10 (———) 2Cu2Zn-superoxide dismutase added to give 1 μM (·········). The incubation conditions were those described in the legend to Fig.2.

It is needless to say that these changes observed in the dichroic properties indicate that a reaction between 2Cu2Zn-superoxide dismutase and oxidizing adrenaline should have occurred. This observation is supported by further circular dichroism measurements at higher 2Cu2Zn-superoxide dismutase concentrations. 50 μM concentrations of 2Cu2Zn-superoxide dismutase in the presence of a tenfold excess of adrenaline were used to study possible conformational changes of the protein moiety. Within ten minutes oxidizing adrenaline affects the Cotton bands of native 2Cu2Zn-superoxide dismutase significantly (Figure 4). A positive Cotton extremum at 280 nm with a shoulder at 295 nm and at 365 nm appears. Two negative Cotton bands are detectable at 325 nm and at 385 nm. Neither of these new Cotton bands can be assigned to autoxidizing adrenaline only. The contribution of the substrate is negligible. It may be of interest to realize that in the region between 200 nm and 250 nm no detectable changes of the protein conformation were seen. This might raise the hypothesis that the conformational change is limited to the active site. It is further worthwhile to mention that only 50% of the protein-bound copper was reduced during the first ten minutes of the reaction. Whether or not this partial reduction can be assigned to the possible cooperativity between the subunits of the protein as earlier concluded awaits further investigation.[24]

Although these measurements only tentatively give an indication that a (ternary) complex is formed, it was of interest to examine the role of the active site metals. For this purpose the circular dichroic changes of partially

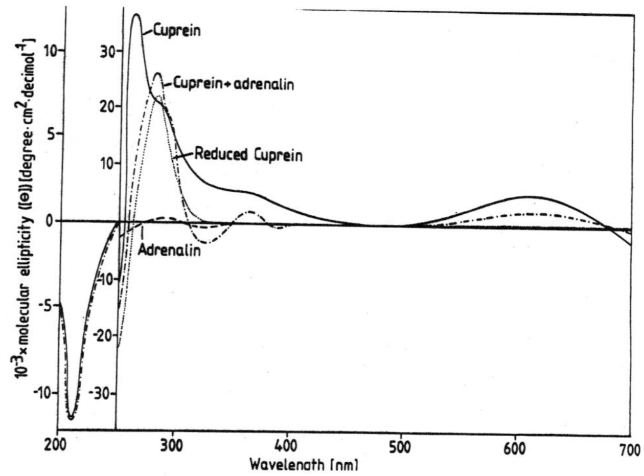

Fig. 4. Circular dichroism of 2Cu2Zn-superoxide dismutase. 50 µM 2Cu2Zn-superoxide dismutase in 0.1 M carbonate buffer pH 10 (———), in the presence of 0.5 mM adrenaline (—·—·—), of 0.5 mM adrenaline alone (-----), and of 50 µM 2Cu2Zn-superoxide dismutase reduced with excess dithionite. Spectra were taken within 10 minutes on a Jasco J-20 A spectropolarimeter employing 1 cm light path quartz cells above and 0.01 cm cells below 250 nm. Incubation conditions were the same as in the legend to Fig.2.

reconstituted apo-2Cu2Zn-superoxide dismutases in the presence and absence of adrenaline were examined. The differently reconstituted apo-2Cu2Zn-superoxide dismutases were prepared according to procedures already published.[25,26] Compared to the native protein, in the presence of oxidizing adrenaline the 2Zn-apoprotein showed only a positive Cotton band at the position of the shoulder at 290 nm (Figure 5). The 2Cu-apoprotein showed a marked positive Cotton band at 285 nm without the shoulder at 295 nm given by the native protein. Another major change is a large negative Cotton extremum at 325 nm in place of the weak band given by a more pronounced negative band near 390 nm (Figure 5).

The enzyme denatured by heat as well as by destruction of the histidines with ultraviolet light did not appear to react with oxidizing adrenaline. The observed dichroic properties represent the superposition of the spectrum of adrenaline oxidation products over the spectrum of denatured protein (Figure 6).

Significance of the catecholamine side chain. The addition of 2Cu2Zn-superoxide dismutase to borate chelated adrenaline did not change the circular dichroic properties of the protein moiety. It should be recalled that borate chelates the 3,4-benzene hydroxyl groups yielding tetrahedrally coordinated boron-adrenaline complexes of 1:1 and 1:2 stoichiometry with the observed consequence of a totally inhibited autoxidation.[27] The identical circular dichroic properties supported the proposal of a substantial contribution of the catecholamine side chain to any ternary complex formation. This proposal was pursued by measuring

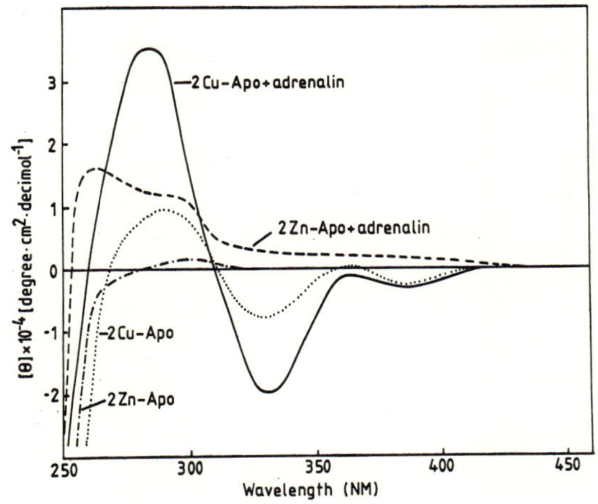

Fig. 5. Circular dichroism of partially reconstituted apo-superoxide dismutase. 50 μM 2Zn-aposuperoxide dismutase (–·–·–·), 50 μM 2Cu-aposuperoxide dismutase (·······) in 0.1 M carbonate buffer pH 10. 50 μM 2Cu-apoprotein and 0.5 mM adrenaline (———) within 10 minutes of the start of the reaction. Incubation conditions were identical to those given in the legend to Fig.2.

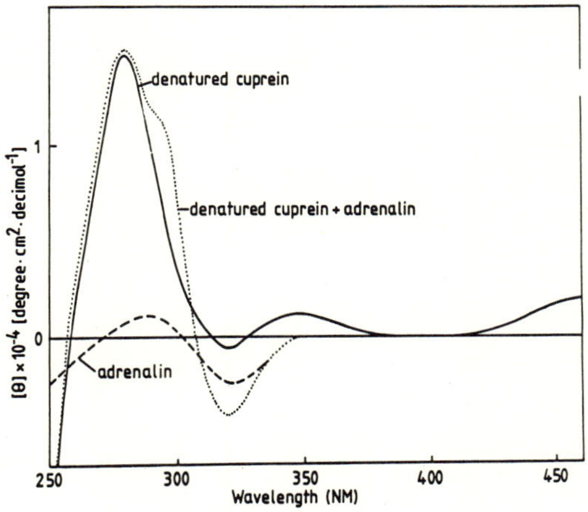

Fig. 6. Circular dichroism of superoxide dismutase devoid of activity. 50 μM denatured superoxide dismutase (———) and in the presence of 0.5 mM adrenaline (·······), and of 0.5 mM adrenaline alone (----). Spectra were recorded after 10 minutes of the autoxidation reaction in 0.1 M carbonate buffer pH 10. The incubation conditions were identical to those given in the legend to Fig.2. 1 h ultraviolet irradiation (35 W, Sterisol type 5012 Hanau, UV-lamp) was sufficient to destroy the active sites of superoxide dismutase.

the oxidation of 1,2-dihydroxybenzene, i.e. a catechol with no side chain. The autoxidation of 1,2-dihydroxybenzene could not be inhibited by 2Cu2Zn-superoxide dismutase. In fact, the same reactivity was observed regardless whatever copper-complex was used, i.e. low-molecular-weight copper chelates or the native copper enzyme. These experiments support the suggestion that the side chain is involved in the binding of 2Cu2Zn-superoxide dismutase to adrenaline. This consideration was further supported by X-ray photoelectron spectrometric measurements. The binding energy value of the adrenaline nitrogen 1s electron is unusually low (400.3 eV). In contrast, dihydroxyphenylalanine and noradrenaline have values corresponding to higher binding energies (401.0 eV and 400.9 eV). It should be pointed out that only adrenaline is specifically inhibited by 2Cu2Zn-superoxide dismutase while the other two catecholamines are not. It may be attractive to conclude that a specific interaction of the side chain nitorgen atom with 2Cu2Zn-superoxide dismutase can be deduced. This would imply, that the binding is strongest in the case of adrenaline which would increase the chances of a stable complex formation with 2Cu2Zn-superoxide dismutase (Table 1).

TABLE 1

AUTOXIDIZING CATECHOLAMINES IN THE PRESENCE OF COPPER-ZINC SUPEROXIDE DISMUTASE[a]

Catecholamine 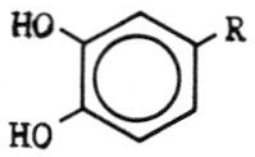	R	Rate of autoxidation of 0.5 mM solution at pH 10 (ΔA/min)	Concentration of superoxide dismutase at which inhibition is detectable. (nM)	Inhibition of the autoxidation by 100nM superoxide dismutase (%)
Dihydroxyphenyl-alanine	H COOH -C-C-NH_2 H H	0.045 (455 nm)	10	22
Noradrenaline	OH H -C-C-NH_2 H H	0.033 (455 nm)	50	19
Adrenaline	OH H -C-C-N-CH_3 H H H	0.028 (400 nm)	0.5	95

[a]Autoxidations were monitored at the wavelength inidcated in 1 cm light path cuvettes at 25°C using a Pye Unicam SP 1800 spectrophotometer. A total volume of 1 ml contained 850 µl 0.1 M carbonate buffer pH 10, and 100 µl water or additions to yield the desired final concentrations. The reactions were started by the addition of 50 µl 10 mM catecholamine stabilized at pH 2.

ACKNOWLEDGEMENT

Parts of this study were generously aided by DFG grant (We 401/14) donated to U.W.

REFERENCES

1. Hamilton, G.A. (1974) in Molecular Mechanisms of Oxygen Activation, Hayaishi, O., ed., Academic Press, New York, pp. 405-451.
2. Weser, U., Schubotz, L.M. and Younes, M. (1979) in Copper in the Environment, Nriagu, J.O., ed., Wiley Interscience, New York, in press.
3. Michelson, A.M., McCord, J.M. and Fridovich, I. (1977) Superoxide and Superoxide Dismutases, Academic Press, New York.
4. Brigelius, R., Spöttl, R., Bors, W., Lengfelder, E., Saran, M. and Weser, U. (1974) FEBS Lett. 47, 72-75.
5. Brigelius, R., Hartmann, H,-J., Bors, W., Lengfelder, E., Saran, M. and Weserm U. (1975) Z. Physiol. Chem. 356, 739-745.
6. Richter, C., Azzi, A., Younes, M. and Weser, U. (1978) Bioinorg. Chem. 8, 201-213.
7. Younes, M., Lengfelder, E., Zienau, S. and Weser, U. (1978) Biochem. Biophys. Res. Commun. 81, 576-580.
8. Lengfelder, E., Fuchs, C., Younes, M. and Weser, U. (1979) Biochim. Biophys. Acta, 567, 492-502.
9. Sellinger, K.-H. and Weser, U. (1979) submitted for publication.
10. Richter, C., Azzi, A., Weser, U. and Wendel, A. (1977) J. Biol. Chem. 252, 644-647.
11. Younes, M. and Weser, U. (1979) Biochim. Biophys. Acta, 526, 644-647
12. Myllylä, R., Schubotz, L.M., Weser, U. and Kivirikko, K.I. (1979) Biochem. Biophys. Res. Commun. in press.
13. Richter, C., Azzi, A. and Wendel, A. (1976) FEBS Lett. 64, 332-337.
14. Paschen, W. and Weser, U. (1975) Z. Physiol. Chem. 356, 727-737.
15. Misra, H.P. and Fridovich, I. (1972) J. Biol. Chem. 247, 3170-3175.
16. Weser, U. and Schubotz, L.M. (1978) Bioinorg. Chem. 9, 505-519
17. Younes, M. and Weser, U. (1976) FEBS Lett. 71, 87-90.
18. Weser, U. (1973) Structure and Bonding, 17, 1-65.
19. Bors, W., Michel, C., Saran. M. and Lengfelder, E. (1978) Biochim. Biophys. Acta, 540, 162-172.
20. Heacock, R.A. (1959) Chem. Rev. 59, 181-237.
21. Harrison, W.H., Whistler, W.W. and Hill, B.J. (1968) Biochemistry, 7, 3089-3094.
22. Graham, D.G. (1978) Molec. Pharmacol. 14, 633-643.
23. Schubotz, L.M. and Weser, U. (1979) submitted for publication.
24. Fielden, M.E., Roberts, P.B., Bray, R.C., Lowe, D.J., Mautner, G.N., Rotilio, G., and Calabrese, L. (1974) Biochem. J. 139, 49-60.
25. Weser, U., Barth, G., Djerassi, C., Hartmann, H.-J., Krauss, P., Voelcker, G., Voelter, W., and Voetshc, W. (1972) Biochim. Biophys. Acta, 278, 28-44.
26. Rotilio, G., Calabrese, L., Bossa, F., Finazzi-Agro, A. and Mondovi, B. (1972) Biochemistry, 11, 2182-2192.
27. Weser, U. (1968) Z. Physiol. Chem. 349, 982-988.

CATALYSIS OF SUPEROXIDE DISMUTATION BY ANTI-INFLAMMATORY ACTIVE COPPER SALICYLATES

MAGED YOUNES,[+] EDMUND LENGFELDER,[++] CHRISTOPH RICHTER,[+++] LUTZ M. SCHUBOTZ[++++] AND ULRICH WESER[++++]
+ Institut für Toxikologie, Medizinische Hochschule Lübeck, D-200 Lübeck, W. Germany; ++ Strahlenbiologisches Institut der Universität München, D-8000 München, W. Germany; +++ Laboratorium für Biochemie, ETH Zürich, CH-8006 Zürich, Switzerland; ++++ Physiologisch-chemisches Institut der Universität Tübingen, D-700 Tübingen, W.Germany.

INTRODUCTION

Activated oxygen species, such as O_2^- and OH-radicals, are thought to play a role in inflammation. Several authors reported that leukocytes and macrophages were able to produce and release superoxide anion radicals.[1-5] The superoxide production in macrophages could be stimulated in the presence of Pt(IV) ions,[16] phosphate, pyrophosphate or adenine nucleotides,[7] and is suppressed when several anti-inflammatory agents are present.[8]

McCord suggested that $^{\cdot}OH$ radicals, produced via the reaction of O_2^- with H_2O_2, were responsible for inflammatory arthritis and the associated degradation of synovial fluid.[9] His suggestion was supported by Oyanagui's finding that 2Cu,2Zn-superoxide dismutase administrated intravenously inhibited the prostaglandin phase swelling of carrageenan foot-edema.[10] Further support resulted from the discovery of an anti-inflammatory active protein, orgotein, which turned out to be cuprein (2Cu,2Zn-superoxide dismutase).[11-13]

Sorenson showed that the anti-inflammatory action of the Cu complexes of several antiphlogistic drugs, including different salicylates, was much higher than that of their parent compounds.[14,15] The complexes were less toxic than the ligands and possessed, in addition, an antiulcer activity. Bearing in mind that the serum copper level is increased during arthritic diseases,[16] he developed the hypothesis that anti-inflammatory agents exerted their therapeutic action via their copper complexes formed _in vivo_.

Work done in our laboratory revealed that low molecular weight copper chelates showed nearly the same activity towards pulse radiolytically produced superoxide radicals as native cuprein.[17,18] They were also able to inhibit the reduction of both cytochrome _c_[19] and nitroblue tetrazolium.[20] $Cu(tyr)_2$ could also suppress microsomal dealkylation reactions[21,22] as well as the diamine oxidase reaction[23] and exert its superoxide dismutative action, thus, in more complex systems.

It was of interest to study the ability of the copper complexes of four anti-inflammatory active salicylates to react with superoxide radicals from several sources and to try, thus, to formulate the mechanism of the therapeutic action of these chelates.

MATERIALS AND METHODS

Chemicals and enzymes. Catalase, xanthine oxidase, xanthine and beef heart cytochrome *c* were obtained from Boehringer, Mannheim; potassium superoxide from K & K, Hollywood; nitroblue tetrazolium chloride and sodium p-aminosalicylate from Serva, Heidelberg; diisopropylsalicylic acid from EGA-Chemie, Steinheim; ethyliodide, umbelliferone, glutathione and adrenaline from Fluka, Buchs; [$^{3}H_3$]-benzo-[α]-pyrene from Amersham-Buchler, Braunschweig; and [^{14}C]-arachidonic acid from Technosa AG, Pully, Lausanne. All other chemicals were purchased from Merck, Darmstadt.

Xanthine oxidase was further purified by gel chromatography (Sephadex G-75). Cu,Zn-superoxide dismutase was isolated as given in Ref. 24. The Cu(II) chelates of acetylsalicylate and diisopropylsalicylate were synthetized according to the procedure given in Ref. 14. Cu(II)-salicylate was prepared as described in Ref. 25 and Cu(II)-p-aminosalicylate was prepared in a similar manner. 7-ethoxycoumarin was prepared according to Ullrich and Weber.[26]

Structural studies. Elemental analyses (C,H,N) were carried out using a Perkin-Elmer 240 unit. Copper was estimated by means of a Perkin-Elmer 400 S atomic absorption spectrometer equipped with a HGA 76B calculator. X-ray photoelectron spectra were run on a Varian V-IEE 15 high resolution electron spectrometer equipped with a 620 L on-line computer (8K) as described earlier.[27-29]

Assays of superoxide dismutative action. Indirect assays were carried out using a Unicam SP 1800 spectrophotometer. The inhibition of cytochrome *c* reduction was assayed as given in Ref. 30. The nitroblue tetrazolium reduction assays were carried out using both the xanthine-xanthine oxidase and the potassium superoxide systems.[20] In pulse radiolysis experiments a Febetron pulse radiolysis unit was used as described in previous work.[17,18,31,32]

Cytochrome P-450-dependent reactions. Rat liver microsomes of methylcholanthrene or phenobarbital induced animals were isolated according to the procedure of Remmer et al.[33] Benzo-[α]-pyrene hydroxylase and 17-ethoxycoumarin deethylase activities were measured as described in Refs. 21, 22, 34 and 35.

Prostaglandin sythetase. Prostaglandin synthetase activity was assayed according to White and Glassman[36] employing microsomes from bull seminal vesicles[37] diluted to give a final concentration of 0.3 mg protein per 100 µl assay volume. The incubation time was 30 min, the temperature was maintained at 37°C.

RESULTS

Structural studies. Elemental analyses were carried out for all four chelates and were in accordance with $C_{14}H_{18}O_{10}Cu$ for the salicylate, $C_{26}H_{34}Cu$ for the diisopropylsalicylate, $C_{36}H_{28}O_{16}Cu_2$ for the acetylsalicylate and $C_{14}H_{20}N_2O_{10}Cu$ for the p-amino-salicylate complex. Thus, in all cases, a 1:2 (Cu:ligand) chelate was detected.

It was of interest to characterize the nature of the chelated copper with regard to its oxidation state as well as its coordination prior to the functional studies. X-ray photoelectron spectra of the four chelates under investigation were run. Apart from the main signal of the Cu $2p^{3/2}$ core electrons around 934 eV, well resolved satellites were seen near 943 eV (Figure 1). Such satellites are specific for the presence of Cu(II) (Refs. 27, 28, 38, 39). In these cases the satellites were split indicating that double-bonded oxygen is involved in the coordination of Cu(II) (Ref. 28). The fact that in the case of Cu-p-aminosalicylate no satellite splitting is seen could be interpreted either in terms of double-bonded oxygen not coordinating to copper or the splitting not being resolved. Thus, the X-ray photoelectron spectrometric data revealed that the chelates employed are well defined Cu(II) complexes.

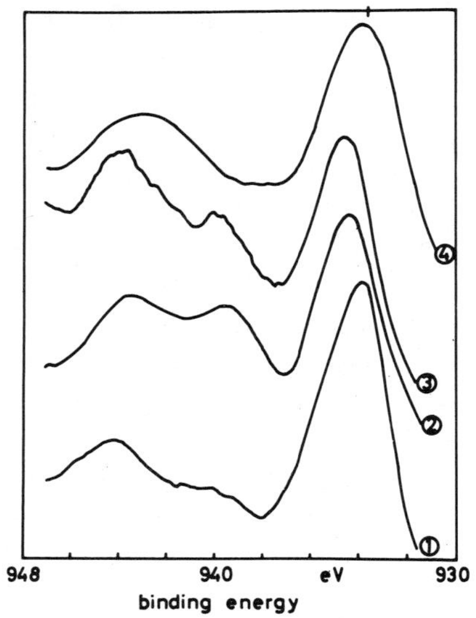

Fig. 1. X-ray photoelectron spectra of the Cu $2p^{3/2}$ level of (1) Cu(II)-salicylate, (2) Cu(II)-acetylsalicylate, (3) Cu(II)-diisopropylsalicylate and (4) Cu(II)-p-amino-salicylate. The recording conditions were: work function 6.0 eV; analyzer energy 100 eV; sweep width 20 eV; sweep time 20 sec; number of channels 200; pressure range 1-2 μ Torr; number of scans 10.

Indirect assays of superoxide dismutase. Using xanthine and xanthine oxidase as source of reducing radicals the reduction of cytochrome *c* was followed spectrophotometrically at 550 nm in the presence of different amounts of Cu(II) chelates as well as in their absence. The increase in absorbance per minute (ΔA_{550}/min) was determined in each case, and the rate of increase in the absence of all chelates was taken as the value for 100% activity. The reduction rates of cytochrome *c* in the presence of different amounts of Cu(II) were calculated in percent of this value and plotted against the logarithm of the copper concentration. The plots obtained were linear allowing a simple determination of the amount of chelated copper required to yield a 50% inhibition of the reaction (Table 1).

TABLE 1

INHIBITION OF CYTOCHROME c REDUCTION BY Cu(II)-SALICYLATES[a]

Cu(II)-chelate	Concentration of chelated Cu(II) yielding 50% inhibition (μM)	Cuprein-Cu: chelate-Cu
2Cu,2Zn-superoxide dismutase (cuprein)	0.002	-
Cu(diisopropylsalicylate)$_2$	0.860	1 : 430
Cu(acetylsalicylate)$_2$	6.400	1 : 3200
Cu(p-aminosalicylate)$_2$	8.600	1 : 4300
Cu(salicylate)$_2$	10.500	1 : 5250

[a]The assay was performed in a total volume of 0.76 ml containing: HEPES buffer, 50 mM, pH 7.8; xanthine, 0.33 mM; cytochrome *c* ox., 27 μM; catalase, 320 U; xanthine oxidase, 0.02 U. The time course of absorption at 550 nm was recorded and the rate of cytochrome *c* reduction per minute determined.

In the nitroblue tetrazolium reduction assay by the xanthine-xanthine oxidase system, the formazan formation was recorded at 540 nm in the absence and in the presence of different amounts of the copper chelates. The increase in absorbance per minute (ΔA_{540}/min) was determined in each case. Taking ΔA_{540}/min in the absence of Cu(II) chelates as the value for 100% activity, the concentration of the respective chelated copper yielding 50% inhibition was determined by plotting the percentage of activity against the logarithm of the concentration as described in Table 2.

To avoid uncontrolled side reactions of the components required in the xanthine-xanthine oxidase system, a saturated potassium superoxide solution in dimethylsulphoxide was used as the reducing agent of nitroblue tetrazolium. Again, the values for 50% inhibition were determined graphically (Table 3).

TABLE 2

THE INHIBITORY ACTION OF COPPER SALICYLATES ON THE XANTHINE-XANTHINE OXIDASE MEDIATED REDUCTION OF NITROBLUE TETRAZOLIUM[a]

Cu(II)-chelate	Concentration of chelated Cu(II) yielding 50% inhibition (μM)	Cuprein-Cu: chelate-Cu
2Cu,2Zn-superoxide dismutase (cuprein)	0.04	-
Cu(salicylate)$_2$	16.00	1 : 400
Cu(acetylsalicylate)$_2$	23.00	1 : 575
Cu(p-aminosalicylate)$_2$	28.00	1 : 700
Cu(diisopropylsalicylate)$_2$	73.00[b]	1 : 1825

[a]The assay mixture (100 mM phosphate buffer, pH 7.8; 0.14 mg/ml gelatine; 0.342 mg/ml nitroblue tetrazolium chloride; 0.14 mM xanthine; 0.2 U/ml catalase; 16.8 U/ml xanthine oxidase) had a total volume of 3 ml. The rate of formazan formation per minute was determined spectrophotometrically at 540 nm.
[b]Due to the low solubility, the value yielding 50% inhibition had to be extrapolated.

TABLE 3

THE INHIBITORY EFFECT OF COPPER SALICYLATES ON THE REDUCTION OF NITROBLUE TETRAZOLIUM BY POTASSIUM SUPEROXIDE[a]

Cu(II)-chelate	Concentration of chelated Cu(II) yielding 50% inhibition (μM)	Cuprein-Cu: chelate-Cu
2Cu,2Zn-superoxide dismutase (cuprein)	0.06	-
Cu(salicylate)$_2$	1.30	1 : 22
Cu(acetylsalicylate)$_2$	2.15	1 : 36
Cu(diisopropylsalicylate)$_2$	2.85	1 : 48
Cu(p-aminosalicylate)$_2$	3.00	1 : 50

[a]In a total volume of 3 ml the assay mixture contained: 100 mM phosphate buffer, pH 7.8; 0.16 mg/ml nitroblue tetrazolium chloride; 0.24 U/ml catalase. The reaction was initiated by the addition of 100 μl of saturated potassium superoxide solution in dimethylsulphoxide.

The fact that the complexes do not show the same degree of inhibition in the different assay systems is not surprising, as these reactions are all only in part due to the action of superoxide. Thus, for example, the discrepancy between the inhibitory action of Cu(II)-salicylate on cytochrome c-reduction on the one hand and its effect on formazan formation on the other, is only an apparent one. Only in the case of the KO_2-mediated reduction of nitroblue tetrazolium was a strictly inorganic system employed which ensured that the reducing agents can only be O_2^- and/or its direct interconversion products. Here the highest inhibitory activities of the chelates were observed.

The ligands alone did not inhibit any of the reactions investigated. And, although hydrated Cu(II) ions are good superoxide dismutases,[17,18] the fact that the chelates are very stable (the stability constant of Cu(II)-salicylate, for example, being 18.5 M^{-1} (Ref. 40)) and that free Cu(II) ions would be directly coordinated to any of the species present in the complex systems studied, rule out the possibility that the observed superoxide dismutative actions are due to Cu^{2+} dissociated from the chelates. Furthermore, the complexes did not influence the reaction of xanthine oxidase itself as could be seen when the enzymic oxidation of xanthine to urate was followed spectrophotometrically at 293 nm.

Pulse radiolysis. The ability of the Cu(II) chelates of salicylate, acetylsalicylate, p-aminosalicylate and diisopropylsalicylate to catalyze the spontaneous dismutation of superoxide anion radicals was assayed directly at pH 7.5 using pulse radiolytically produced O_2^- in the presence of formate to convert $^{\bullet}OH$-radicals to superoxide.[41,42]

The catalyzed dismutation reactions showed pseudo first order characteristics. The pseudo first order rate constants of the various reactions were determined by plotting lnA (the natural logarithm of the absorption of O_2^- at 250 mm) against time, as is demonstrated for the dismutation of superoxide in the presence of 5 μM Cu-salicylate (Figure 2). The second order rate constants were calculated by dividing the pseudo first order rate constants by the respective concentration of chelated copper. Values ranging from 0.79 to 2.40 x 10^9 $M^{-1}s^{-1}$ were found for the four Cu(II)-chelates (Table 4). This shows that these complexes react with O_2^- as rapidly and effectively as native superoxide dismutase and other low molecular weight copper complexes.[17,18,43,44]

The spontaneous dismutation of superoxide followed second order kinetics. The second order rate constant was 1.06 x 10^7 $M^{-1}s^{-1}$ at pH 7.15. To ensure that the acceleration of the disproportionation of O_2^- in the presence of the copper chelates was due to a real catalysis of the spontaneous dismutation and not to a direct reaction of superoxide with chelated Cu(II), the amounts of oxygen and H_2O_2 produced after the addition of the same amount of a saturated KO_2 solution in dimethylsulphoxide to a buffer solution in the absence and in the presence of either copper chelate as well as in the presence of native 2Cu,2Zn-superoxide dismutase were determined. In all cases the same amount of O_2 and H_2O_2 was detected, indicating that a real dismutation reaction had taken place.

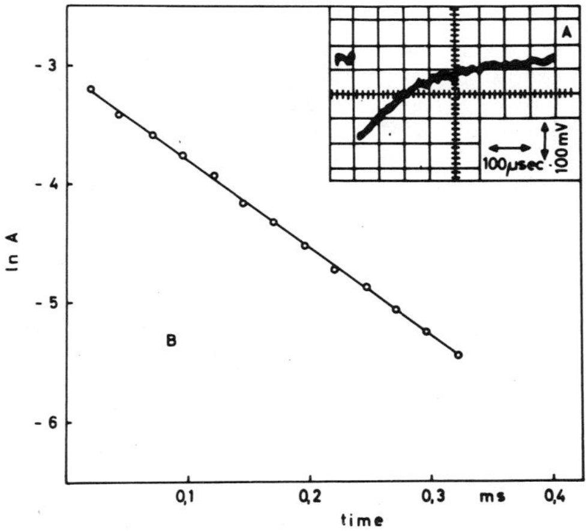

Fig. 2. The decay of pulse radiolytically produced superoxide in the presence of 5.0 μM $Cu(salicylate)_2$ (insertion A) and the pseudo first order plot of this reaction (B). The decay was observed at 250 nm employing a Tectronix 7704 oscilloscope equipped with a Polaroid camera. Experimental conditions were as described in Refs. 17, 18 and 31.

TABLE 4

SECOND ORDER RATE CONSTANTS DETERMINED FOR THE REACTIONS OF SUPEROXIDE WITH ANTI-INFLAMMATORY ACTIVE COPPER CHELATES[a]

Copper chelate	Concentration range (μM)	$K_{250} \times 10^{-9}$ ($M^{-1}s^{-1}$)
2Cu,2Zn-superoxide dismutase	1.0 - 5.0	1.30 ± 0.10 (per subunit)
$Cu(diisopropylsalicylate)_2$	2.2 - 5.4	2.40 ± 0.12
$Cu(salicylate)_2$	1.0 - 5.0	1.64 ± 0.15
$Cu(acetylsalicylate)_2$	1.1 - 8.4	0.96 ± 0.20
$Cu(p\text{-}aminosalicylate)_2$	2.0 - 10.0	0.79 ± 0.16

[a]Mean values of the reactions at different O_2^- (4-36 μM) and Cu concentrations are given.

Inhibition of the benzo-[α]-pyrene hydroxylase and the 7-ethoxycoumarin deethylase reactions. The hydroxylation of [^{3}H]-benzo-[α]-pyrene by methyl-cholanthrene-induced rat liver microsomes could be suppressed by the presence of the Cu(II) complexes of the four salicylates investigated (Table 5). Cu(II)-acetylsalicylate was the most active agent, while the other chelates had smaller inhibiting effect.

TABLE 5

INHIBITION OF METHYLCHOLANTHRENE-INDUCED RAT MICROSOMAL HYDROXYLATION OF [^{3}H]-BENZO-[α]-PYRENE BY DIFFERENT COPPER SALICYLATES[a]

Cu(II)-chelate	nmoles of benzo-[α]-pyrene metabolized/min/mg of microsomal protein	% inhibition
none	0.500	-
Cu(acetylsalicylate)$_2$	0.085	83
Cu(diisopropylsalicylate)$_2$	0.330	34
Cu(salicylate)$_2$	0.370	26
Cu(p-aminosalicylate)$_2$	0.375	25

[a]Experimental conditions are given in Ref. 31. The amount of hydroxylated benzo-[α]-pyrene in the total absence of the Cu(II) complexes was 0.5 nmoles/min/mg protein and was considered the value for 100% activity. The concentration of the respective Cu-chelates was 30 μM of copper in all cases.

Using phenobarbital-induced rat liver microsomes the 7-ethoxycoumarin deethylation was followed fluorimetrically in a way similar to that described by Ullrich and Weber.[26] The acetylsalicylate-copper complex was again the most effective inhibitor. The inhibition depended upon the concentration of the chelate in the assay mixture (Figure 3). At a copper concentration of 80 μM, the reaction was inhibited by 70%. At this same Cu concentration the salicylate complex exhibited 30% inhibition, while both the p-amino-salicylate and the diisopropylsalicylate chelates suppressed the deethylation reaction by 25%.

Inhibitory action on prostaglandin biosynthesis. As significant evidence has accumulated suggesting that prostaglandins play a key role in the development of inflammatory response,[45,46] and as it could be demonstrated that acetylsalicylate, indomethacin and other anti-inflammatory agents were able to inhibit the prostaglandin synthetase reaction,[49-52] it was of interest to study the effect of our four copper salicylates on prostaglandin biosynthesis.

Using a simple assay allowing the determination of the total prostaglandin biosynthesis reaction following the addition of [^{14}C]-arachidonic acid, we found slight, though significant, inhibitory effects of Cu-salicylate, Cu-acetysalicylate and Cu-p-aminosalicylate. The ligands alone exhibited a much lower inhibition (Table 6). In the case of the diisopropylsalicylate chelate the copper complex showed the same degree of inhibition as the ligand alone.

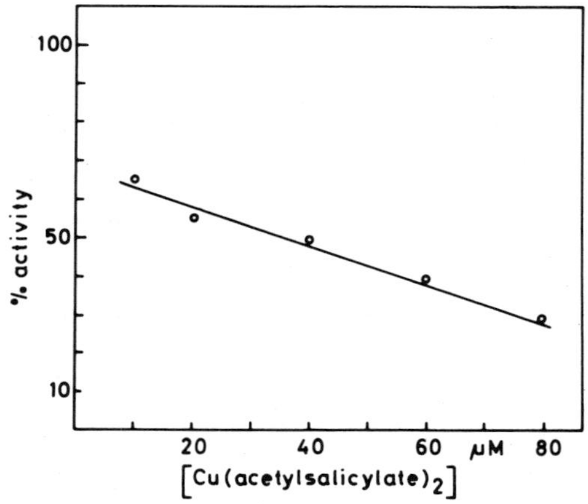

Fig. 3. Inhibition of 7-ethoxycoumarin dealkylation by Cu(acetylsalicylate)$_2$. The reaction was carried out as given in Ref. 31. The value for 100% activity (in the total absence of Cu-acetylsalicylate) was 1.1 nmoles of dealkylated ethoxycoumarin/min/mg protein.

TABLE 6

ACTION OF ANTI-INFLAMMATORY AND SUPEROXIDE DISMUTASE ACTIVE COPPER SALICYLATES AND THEIR PARENT COMPOUNDS ON PROSTAGLANDIN BIOSYNTHESIS IN VITRO[a]

Cu-chelate	% inhibition	Ligand	% inhibition
Cu(Salicylate)$_2$	24	Na-salicylate	6
Cu(acetylsalicylate)$_2$	17	acetylsalicylic acid	10
Cu(p-aminosalicylate)$_2$	32	Na-p-aminosalicylate	17
Cu(diisopropyl-salicylate)$_2$	10	Na-diisopropylsalicylate	10

[a]In a total assay volume of 100 μl bull seminal microsomes were used yielding a final concentration of 3 mg protein/ml. The inhibitor concentrations were 100 μM throughout. After the addition of [^{14}C]-arachidonic acid (9 μM, 1.1 x 10^6 cpm), the mixture was incubated at 37°C for 30 min. Following the incubation, 5 μl of unlabelled prostaglandin solution (0.2 mg/ml in ethanol) was added as carrier, then each assay mixture was transferred onto a Bio-Sil column and eluted as described in Ref. 36. Product formation was estimated using a Packard liquid scintillation counter. The 100% value corresponded to 24,000 cpm.

DISCUSSION

The fact that the copper chelates of the salicylates investigated, which all exhibited a higher anti-inflammatory activity than the ligands alone,[14,15] possess a well-defined superoxide dismutase activity is of great interest with regard to their mechanism of action. All four chelates were able to react with superoxide in an enzymic and in a strictly inorganic system as well as with pulse radiolytically produced O_2^-. They were also able to react with this oxygen radical in complex systems, as could be shown by their ability to inhibit the activation of oxygen by cytochrome P-450 during hepatic microsomal hydroxylation and dealkylation reactions. These findings confirm that superoxide and/or its possible interconversion products, such as $^{\cdot}OH$ and $^1\Delta_g\ O_2$, might play a role in inflammatory response and that the copper chelates exert their anti-inflammatory action by scavenging these radicals.

It has been suggested that during prostaglandin biosynthesis O_2^- is produced.[10] The Cu-salicylates inhibited prostaglandin biosynthesis to various extents. As prostaglandins are known to be mediators of inflammatory reactions, part of the therapeutic action of the Cu(II) complexes might be due to an inhibition of the prostaglandin sythetase system, possibly by scavenging oxygen radicals needed for the biosynthesis. These findings are in agreement with those of Lee and Lands[50] and Maddox,[51] who found that Cu^{2+} ions inhibited the biosynthesis of the pro-inflammatory prostaglandin PGE_2 while increasing the synthesis of the anti-inflammatory prostaglandin $PGF_{2\alpha}$. It should be kept in mind, however, that acetylsalicylic acid, for example, inhibits the protaglandin synthetase complex by direct acetylation.[52]

Although we are far from understanding the nature of inflammatory response, the role of activated oxygen species as direct as well as indirect mediators should be taken into consideration. In this context, the hypothesis that anti-inflammatory agents exert their therapeutic action through their copper complexes formed *in vivo*, which scavenge superoxide, appears to be a plausible one.

REFERENCES

1. Allen, R.C., Yevich, S.J., Orth, R.W. and Steele, R.H. (1974) Biochem. Biophys. Res. Commun. 64, 760-767.
2. Babior, B.M., Kipnes, R.S. and Curnutte, J.T. (1973) J. Clin. Invest. 52, 741-744.
3. Curnutte, J.T. and Babior, B.M. (1974) J. Clin Invest. 53, 1662-1672.
4. Salin, M.L. and McCord, J.M. (1977) in Superoxide and Superoxide Dismutases, Michelson, A.M., McCord, J.M. and Fridovich, I., eds., Academic Press, New York, pp. 257-270.
5. Ohmori, H., Komoritya, K., Azuma, A., Kurozumi, S. and Hashimoto, Y. (1979) Biochem. Pharmacol. 28, 333-334.
6. Oyanagui, Y. (1977) Biochem. Pharmacol. 26, 473-476.
7. Oyanagui, Y. (1977) Agent. Action. 7, 125-132.
8. Oyanagui, Y. (1976) Biochem. Pharmacol. 25, 1473-1480.
9. McCord, J.J. (1974) Science, 185, 529-531.
10. Oyanagui, Y. (1976) Biochem. Pharmacol. 25, 1465-1472.
11. Carson, S., Vogin, E.E., Huber, W. and Schulte, T.L. (1973) Toxicol. Appl. Pharmacol. 26, 184-202.
12. Huber, W. and Saifer, M.G.P. (1977) in Superoxide and Superoxide Dismutases, Michelson, A.M., McCord, J.M. and Fridovich, I., eds., Academic Press, New York, pp. 517-536.

13. Menander-Huber, K.B. and Huber, W. (1977) in Superoxide and Superoxide Dismutases, Michelson, A.M., McCord, J.M. and Fridovich, I., eds., Academic Press, New York, pp. 537-549.
14. Sorenson, J.R.J. (1976) J. Med. Chem. 19, 135-148.
15. Sorenson, J.R.J. (1976) Inflammation, 1, 317-333.
16. Underwood, E.J. (1971) Trace Elements in Human and Animal Nutrition, 3rd ed., Academic Press, New York, p. 51.
17. Brigelius, R., Spöttl, R., Bors, W., Lengfelder, E., Saran, M. and Weser, U. (1974) FEBS Lett. 47, 72-75.
18. Brigelius, R., Hartmann, H.-J., Bors, W., Lengfelder, E., Saran, M. and Weser, U. (1975) Z. Physiol. Chem. 356, 739-745.
19. Joester, K.-E., Jung, G., Weber, U. and Weser, U. (1972) FEBS Lett. 25, 25-28.
20. Younes, M. and Weser, U. (1976) FEBS Lett. 61, 209-212.
21. Richter, C., Azzi, A. and Wendel, A. (1976) FEBS Lett. 64, 332-337.
22. Richter, C., Azzi, A., Weser, U. and Wendel, A. (1977) J. Biol. Chem. 252, 5061-5066.
23. Younes, M. and Weser, U. (1978) Biochim. Biophys. Acta, 526, 644-647.
24. Weser, U., Bunnenberg, E., Cammack, R., Djerassi, C., Flohé, L., Thomas, G. and Voelter, W. (1971) Biochim. Biophys. Acta, 203-213.
25. De Alvare, L.R., Goda, K. and Kimura, T. (1976) Biochem. Biophys. Res. Commun. 69, 687-694.
26. Ullrich, V. and Weber, P. (1972) Z. Physiol. Chem. 353, 203-213.
27. Rupp, H. and Weser, U. (1976) Bioinorg. Chem. 6, 45-59.
28. Rupp, H. and Weser, U. (1976) Biochim. Biophys. Acta, 446, 151-165.
29. Younes, M. and Weser, U. (1977) Biochem. Biophys. Res. Commun. 78, 1247-1253.
30. Weser, U. and Voelcker, G. (1972) FEBS Lett. 22, 15-22.
31. Younes, M., Lengfelder, E., Zienau, S. and Weser, U. (1978) Biochem. Biophys. Res. Commun. 81, 576-580.
32. Lengfelder, E., Fuchs, C., Younes, M. and Weser, U. (1979) Biochim. Biophys. Acta, 567, 492-502.
33. Remmer, H., Greim, H., Schenkmann, J.B. and Estabrook, R.W. (1979) Methods Enzymol. 10, 703-708.
34. De Pierre, J.W., Moron, M.S., Johannesen, K.A.M. and Ernster, L. (1975) Anal. Biochem. 63, 470-484.
35. Weser, U., Richter, C., Wendel, A. and Younes, M. (1978) Bioinorg. Chem. 8, 201-213.
36. White, H.L. and Glassman, A.T. (1974) Prostaglandins, 7, 123-129.
37. Auggard, E., Machinsky, F.M. and Samuelson, B. (1969) Science, 163, 479-480.
38. Frost, D.C., Ishitani, A. and McDowell, C.A. (1972) Mol. Phys. 24, 861-877.
39. Frost, D.C., McDowell, C.A. and Tapping, R.L. (1975) J. Electron Spectrosc. 6, 347-355.
40. Weitzel, G. and Schaeg, W. (1964) in Biochemisches Taschenbuch, Vol. 11, Rauen, H.M. ed., Springer Verlag, Berlin, pp. 150-159.
41. Behar, D., Czapski, G., Rabani, J., Dorfman, L.M. and Schwarz, H.A. (1970) J. Phys. Chem. 74, 3209-3213.
42. Draganic, I.G., Nenadovic, M.T. and Draganic, Z.D. (1969) J. Phys. Chem. 73, 2564-2571.
43. Klug, D. Rabani, J. and Fridovich, I. (1972) J. Biol. Chem. 247, 4839-4842.
44. Fielden, E.M., Roberts, P.B., Bray, R.C., Lowe, D.J., Mautner, G.N., Rotilio, G. and Calabrese, L. (1974) Biochem. J. 139, 49-60.
45. Crunkhorn, P. and Willis, A.L. (1971) Brit. J. Pharmacol. 41, 507-512.
46. Crunkhorn, P. and Willis, A.L. (1971) Brit. J. Pharmacol. 41, 49-56.
47. Vane, J.R. (1971) Nature, 231, 232-235.
48. Flower, R.J. (1975) Nature, 253, 88-90.
49. Pong, S.-S. and Levine, L. (1976) J. Pharmacol Exp. Ther. 196, 226-230.
50. Lee, R.E. and Lands, W.E.M. (1972) Biochim. Biophys. Acta, 260, 203-211.
51. Maddox, I.S. (1973) Biochim. Biophys. Acta, 306, 74-81.
52. Roth, G.J., Stanford, N. and Majerus, P.W. (1975) Proc. Nat. Acad. Sci. USA, 72, 3073-3076.

SUPEROXIDE DISMUTATION BY Cu-PENICILLAMINE

E. LENGFELDER[+], E.F. ELSTNER[++], M. YOUNES[+++] AND U. WESER[+++]
[+]Strahlenbiologisches Institut, Universität München, D-8000 München 2, W. Germany; [++]Institut für Botanik and Mikrobiologie, Technische Universität München, D-8000 München 2, W. Germany; [+++]Anorganische Biochemie, Physiologisch-Chemisches Institut der Universität Tübingen, D-7400 Tübingen, W. Germany

Superoxide radicals are known to be produced by a number of biological reactions, probably in all oxygen metabolozing cells.[1,2] There is very much evidence that oxygen radicals, such as superoxide and ˙OH radical, play a role in inflammatory processes as demonstrated by the results of many experiments *in vitro*[5-8] and *in vivo*.[9-12]

The functional phenomenon of superoxide dismutation is not restricted to the protein-bound copper in haemocuprein,[3] i.e., superoxide dismutase.[4] The aquo-complex of Cu(II) and the copper chelates of some amino acids[13-15] and salicylates[16,17] exert the same catalytic acitivity. Moreover, $Cu(Tyr)_2$ (Ref. 18) and the Cu-salicylates[16] have been successfully employed to examine the generation of superoxide ions in those metabolic sites to which the large superoxide dismutase molecules has no access. It was deduced that all chelated copper is capable of reacting with bound or free superoxide in essentially the same way. One important advantage of the active copper in the enzyme superoxide dismutase is its higher resistance to EDTA treatment.

Recently, Sorenson showed that copper complexes of a great number of antiarthritic drugs are more active as anti-inflammatory agents than their parent compounds.[19] As the serum level of copper is increased during arthritic diseases,[20] Sorenson proposed that anti-inflammatory drugs coordinate copper ions *in vivo*, thus forming the active forms of such therapeutic agents. In the case of D-penicillamine, however, Sorenson observed that the red-violet copper complex showed an anti-inflammatory activity while the parent compound was inactive at the initial screening doses used in the three test models of inflammation he investigated.[19] Due to its anti-inflammatory activity, D-pencilillamine is of value in the treatment of rheumatoid arthritis.[21] Because of the high stability constant of the copper pencillamine complex[22] D-penicillamine easily coordinates to copper ions even *in vivo* as can be judged by its high cupriuric activity.[23]

To improve possible applications of low molecular weight copper chelates provided with superoxide dismutase activity for analytical and/or therapeutic purposes, we searched for a copper complex with high resistance to strong chelators. We here review the properties of the copper penicillamine complex as such a compound with high superoxide dismutative activity. In spite of about fifteen systems[24] which have been described, assaying superoxide dismutase activity is less than straight forward. Evidence is presented that the way copper penicillamine acts in catalyzing superoxide dismutation is different from that of other copper chelates or of the native enzymes. Moreover, the advantage of low molecular weight compared to that of the enzyme in studying superoxide dependent reactions connected with cell particles is demonstrated.

Structure and properties of the Cu-penicillamine complex. At physiological pH, Cu(II) and D-penicillamine react to form a deep purple complex, which is stable for a long time in solution. The complex is a mixed valence anionic Cu(I)Cu(II) cluster. Its structure has been studied intensively[25,27] and was finally determined by Birker and Freeman[28] by X-ray diffraction to be $(Cu^{II}{}_6Cu^{I}{}_8(\text{D-penicillamine})_{12}Cl)^{5-}$. Depending on the preparation procedure the proportions of Na^+ and Cu^{2+} as counterions varies. The molecular weight has been determined to be between 2200 (Refs. 25, 29) and 2600 (Refs. 26-28), depending on the type of the counterions and the state of hydration. X-ray photoelectron spectra recorded by Rupp and Weser show the presence of a large proportion of Cu(II) in the complex.[30] The stability of the complex was demonstrated by the same authors. Chelex-100 did not remove copper from the complex as shown by X-ray photoelectron spectroscopy of samples which have been treated with Chelex-100, desalted on Sephadex G-10, again treated with Chelex-100 and rechromatographed on Sephadex G-10.

Determination of the superoxide dismutative activity of Cu-penicillamine by indirect assay methods. In the xanthine/xanthine oxidase-cytochrome *c* assay system,[4,31,39] Cu-penicillamine exhibits high SOD activity, as shown in Figure 1. 8.6 nmol Cu-penicillamine itself does not show any inhibitory effects in the same test system even at concentrations two orders of magnitude higher. It has also been demonstrated[32] that Cu-penicillamine does not change the rate of oxidation of xanthine to urate by xanthine oxidase indicating that the enzymic activity of the xanthine oxidase is not influenced.

Cu-penicillamine shows SOD activity using the nitroblue tetrazolium test with both xanthine/xanthine oxidase and KO_2 as superoxide generating systems. With KO_2 the SOD activity of Cu-penicillamine proves to be constant in the presence of a 20-fold excess of cyanide.[33]

Cu-penicillamine is SOD-active in hydroxylamine assay system of Elstner and Heupel.[34] Addition of $CuSO_4$ to this test system stimulates nitrite (i.e. reaction product) formation. The Cu(II)-dependent stimulation of nitrite formation, however, is clearly diminished in the presence of EDTA, while the SOD activity of the complex is not affected.[35] In this test 21 nmol Cu-penicillamine exhibit the activity of 1 SOD unit.

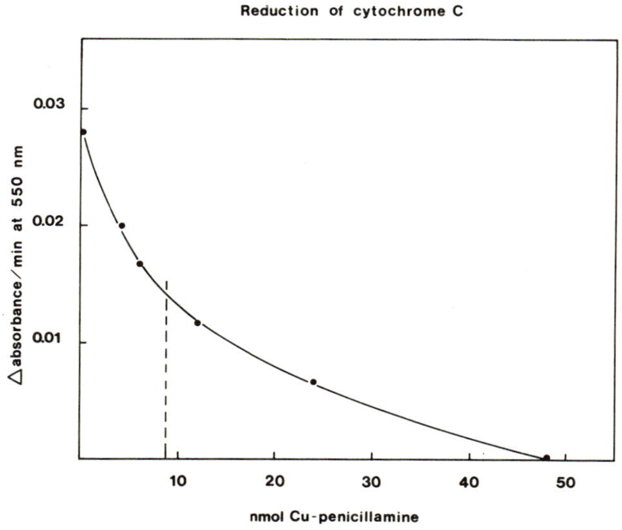

Fig. 1. Determination of the superoxide dismutative activity of Cu-penicillamine by the cytochrome c reduction assay.

Crocin, the digentiobiose ester of the carotenoid acid crocetin, has been used as a model compound for studies of carotene bleaching by superoxide radicals from the xanthine/xanthine oxidase system. Crocin destruction can be monitored by the decrease in absorption at 440 nm. A comparison of the inhibitory effects of superoxide dismutases and Cu-penicillamine in this system is presented in Figure 2. The superoxide dismutases tested were isolated from dry green peas (Cu-Zn) or *Euglena gracilis* (Fe) (Ref. 35). Cu-penicillamine shows high SOD activity, 11 nmol of the complex has the equivalent activity of 1 unit Cu-Zn SOD and 8.9 nmol has the equivalent activity of 1 unit Fe SOD. When 10^{-4} M KCN is added to this test system, the inhibitory effect of Cu-Zn SOD is considerably reduced, as is to be expected. The activities of both Fe SOD and Cu-penicillamine remain nearly unchanged. From the cyanide-insensitivity of Cu-penicillamine it follows that its superoxide dismutative activity cannot be explained satisfactory by a change of valence of copper alone, as is assumed for the action of Cu-Zn SOD[36] and for copper complexes of peptides with dismutative activity, which are know to be Cu(II) chelates[14,15,33] and to be inhibited by KCN.

Pulse radiolysis. In the direct and, therefore, less objectionable[37,38] assay

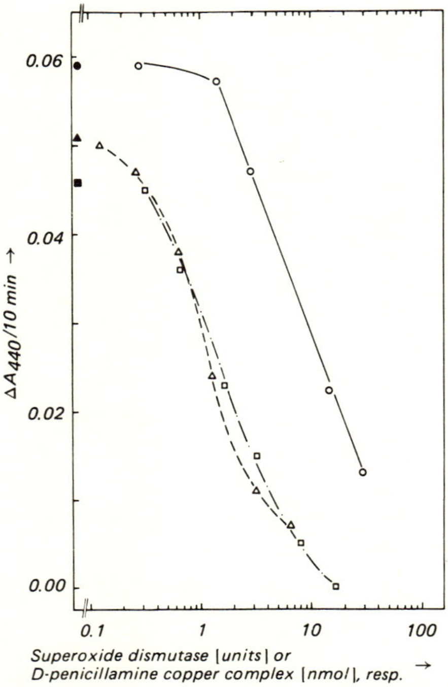

Fig. 2. Inhibitory effects of Cu-penicillamine ○——○ , Cu-Zn superoxide dismutase △---△ and Fe superoxide dismutase □-·-·-□ on crocin destruction by xanthine/xanthine oxidase. Taken from Ref. 35.

Fig. 3. The second order rate constant (k_2) of the reaction between Cu-penicillamine and O_2^- at different concentrations of the copper chelate, measured by pulse radiolysis. Oxygen saturated solutions in triply distilled and pyrolized water containing 10^{-2} M formate were employed. The pH was adjusted to 7.0. Taken from Ref. 29.

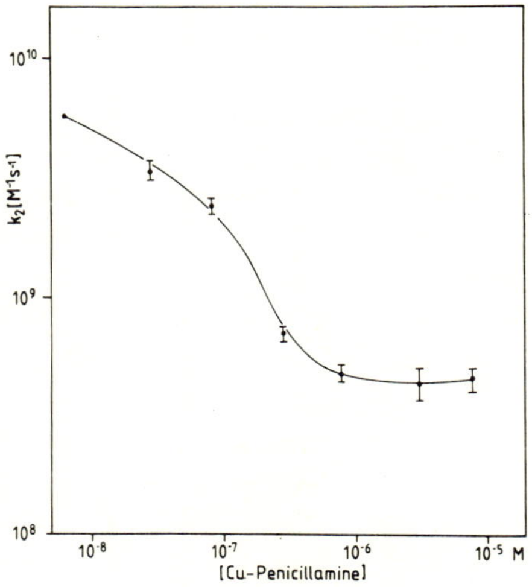

method of pulse radiolysis Cu-penicillamine exhibits marked SOD activity.[29] In a pure aqueous system the rate constants of the reactions with superoxide determined at various concentrations of Cu-penicillamine are between $6 \times 10^9\ M^{-1}sec^{-1}$ (10^{-8} M Cu-penicillamine) and $4.5 \pm 0.5 \times 10^8\ M^{-1}sec^{-1}$ (10^{-5} M Cu-penicillamine), as shown in Figure 3. It should be pointed out that Cu-penicillamine catalyzed superoxide dismutation proceeds with essentially the same velocity as observed with many known superoxide dismutases ($k_2 = 2.4 \times 10^9$ to $5.5 \times 10^8\ M^{-1}sec^{-1}$) (Refs. 40-45). The significant rise of k_2 at 10^{-7} - 10^{-8} M concentrations of Cu-penicillamine can be attributed to the dissociation of extraneous Cu(II) from the outer sphere of the cluster. This disturbance in reactivity can be countered by the addition of EDTA. This strong chelator is usually added to the different native superoxide dismutases to avoid uncontrolled side reactions of unspecifically bound transition metals. EDTA alone effects the O_2^- decay. From 10^{-7} M on to higher concentrations EDTA suppresses the spontaneous superoxide dismutation. However, in the presence of 10^{-8} - 10^{-6} M Cu-penicillamine the catalysis of superoxide dismutation remains constant at about $2.0 \times 10^8\ M^{-1}sec^{-1}$ (Ref. 29).

Unlike the strong pH-dependent rate of the spontaneous dismutation reaction, variation from pH 3 to pH 9 of the Cu-penicillamine solutions results in the rate constants in the range of the usual enzymic catalyzed dismutation. However, some acceleration can be observed at the pK of the HO_2/O_2^- system. Addition of EDTA has no effect on the rate constants of the Cu-penicillamine catalyzed dismutation reaction at different pH values (Figure 4).

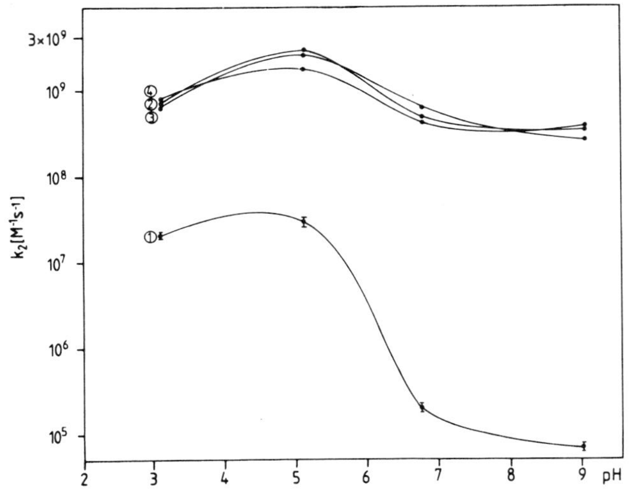

Fig. 4. pH-dependence of the rate constant of the reaction between pulse-radiolytically generated superoxide and Cu-penicillamine in the presence of 10^{-6} M EDTA. Cu-penicillamine was present in the concentrations: (1) 0; (2) 4.8×10^{-7} M; (3) 9.1×10^{-7} M; (4) 4.8×10^{-6} M. Taken from Ref. 29.

Cyanide, when it is present in equimolar concentrations, does not inhibit the dismutative action of Cu-penicillamine on pulse-radiolytically generated radicals (Figure 5). Higher concentrations result in slightly reduced reaction rates. This is attributed to cyanide-superoxide interaction, because free cyanide reduces the rate of spontaneous dismutation considerably. The lack of cyanide inhibition in the dismutative action of Cu-penicillamine, which is contrasted by the strong inhibition of the enzymic activity of native Cu-Zn superoxide dismutase, suggests that a different mode of action has to be considered.

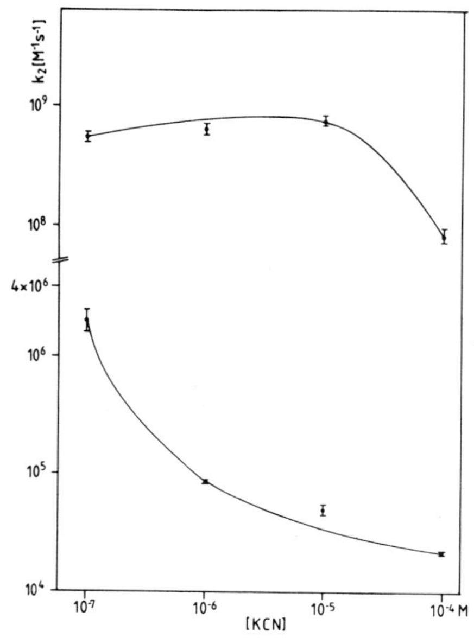

Fig. 5. Effect of added cyanide on the rate constant of the reaction of pulse-radiolytically generated superoxide with Cu-penicillamine. Lower curve: spontaneous decay of superoxide; upper curve: decay of O_2^- in the presence of 9.1 x 10^{-6} M Cu-penicillamine. Taken from Ref. 291.

Originally, the $(Cu_{14}(D\text{-penicillamine})_{12}Cl)^{5-}$ complex employed was assigned a Cu(I)/Cu(II) mixed valence as deduced from X-ray diffraction studies.[27,28] Six out of the fourteen Cu ions are present in the square planar environment commonly found for Cu(II). The high absorption in the blue region allowed the proposal of some similarity with the type 1 copper chromophores. The phenomenon that S → Cu charge transfer transitions do occur in both the blue copper proteins and the Cu-penicillamine complex implicates that the actual electronic state of the copper should be considerably affected. As also shown in earlier work[25,30,32] X-ray photoelectron spectrometric measurements of the Cu $2p_{3/2}$-level gives a binding energy value of 933.4 eV and no satellite is detectable. Thus it can be concluded that regardless of the molecular geometry essentially all copper in the Cu-penicillamine complex is present in the $3d^{10}$ electronic state, having the actual electronic state of +1 as defined by Jørgensen.[46] Further details of the redox state of copper in the Cu-penicillamine complex have been discussed in Ref. 29. From X-ray photoelectron spectrometric studies it is suggested that throughout the reaction of Cu-penicillamine with O_2^- the copper remains in the $3d^{10}$(Cu(I)) state. Thus the actual electron transfer ought to take place according to:

$$Cu(II)^{-}SR \rightleftharpoons Cu(I)^{\cdot}SR$$

$$Cu(I)^{\cdot}SR + O_2^- \longrightarrow Cu(I)^{-}SR + O_2$$

$$Cu(I)^{-}SR + O_2^- \longrightarrow Cu(II)^{\cdot}SR + O_2^{2-}$$

Provided this reaction mode is favoured, added cyanide should not influence the catalysis of superoxide dismutation. Cyanide would only bind to the copper, if at all, leaving the redox active sulphur radical unaffected.

Advantage of Cu-penicillamine in studying superoxide dependent reactions. The involvement of superoxide radicals in biochemical reactions is most often demonstrated by the inhibitory effect of the added enzyme superoxide dismutase. As long as the site of generation or action of superoxide dismutase turns out to be an efficient inhibitor. If the site of interest is surrounded by a barrier like a membrane, as is the case with cells or cell particles, the access of the enzyme is commonly limited, e.g. by molecular size. A small molecular weight compound with high solubility might diffuse much easier.

Chloroplasts produce ethylene from methionine in the presence of pyridoxal phosphate and ferredoxin.[47] In this reaction both O_2^- and H_2O_2 are formed.[48-50] The production of ethylene is strongly increased by the addition of superoxide dismutase as well as by Cu-penicillamine and the concomitant increase of ethylene formation. In this system 1.8 x 10^{-10} mole Cu-penicillamine has the same stimulating effect on ethylene formation as 1 unit Fe SOD. The amounts of Cu-penicillamine, which show equivalent effects as 1 unit of the natural enzyme in the indirect assay systems described above, are in the range of 8.6 nmol (cytochrome c reduction) to 21 nmol (hydroxylamine oxidation). Thus less than a fourtieth part of Cu-penicillamine proves to be as active as 1 unit Fe SOD in the production of ethylene by chloroplasts.

It has to be assumed that an enzyme like superoxide dismutase is unable to come close to the active site of photosystem I due to molecular size. Therefore it will act from the "outside" upon those superoxide radicals, which are generated

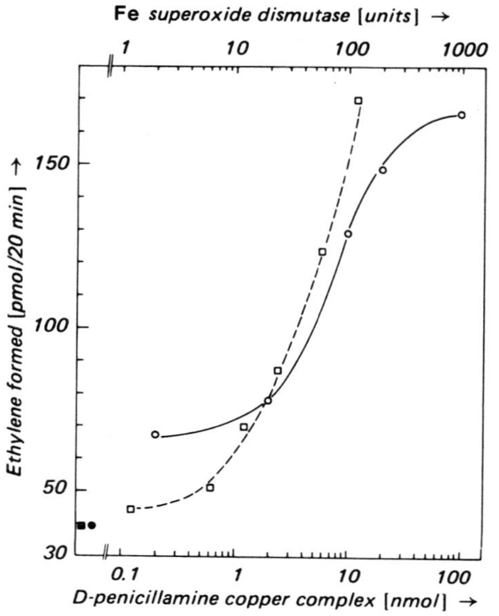

Fig. 6. Rates of ethylene production by isolated chlorplasts as stimulated by either Cu-penicillamine O———O or Fe superoxide dismutase □---□ . Taken from Ref. 35.

by the primary acceptor of photosystem I. The fact that Cu-penicallamine stimulates ethylene production from methionine to a higher extent as compared to Cu-Zn SOD (Ref. 51) suggests a better accessibility for Cu-penicillamine to the invagination of the thylakoids at the original ferredoxin site.

To conclude, the stability of the complex in the presence of strong chelators and KCN, its applicability in a wide pH-range and its molecular size demonstrate the advantage of Cu-penicillamine in the catalysis of superoxide dismutation, where other compounds, even the natural enzyme, may fail.

REFERENCES

1. Fridovich, I. (1972) Account. Chem. Res. 5, 321-326.
2. Fridovich, I. (1975) Annu. Rev. Biochem. 44, 147-149.
3. Mann, T. and Keilin, D. (1938) Proc. Roy. Soc. Ser. B, 126, 303-315.
4. McCord, J.M. and Fridovich, I. (1969) J. Biol. Chem. 244, 6049-6055.
5. Salin, M.L. and McCord, J.M. (1975) J. Clin. Invest. 56, 1319-1323.
6. Johnston, R.B. and Lehmeyer, J.E. (1977) in Superoxide and Superoxide Dismutases, Michelson, A.M., McCord, J.M. and Fridovich, I., eds., Academic Press, New York, pp. 291-305.
7. Salin, M.L. and McCord, J.M. (1977) in Superoxide and Superoxide Dismutases, Michelson, A.M., McCord, J.M. and Fridovich, I., eds., Academic Press, New York, pp. 257-270.

8. McCord, J.M. (1974) Science, 185, 529-531.
9. Oyanagui, Y. (1976) Biochem. Pharmacol. 25, 1465-1472.
10. Lund-Olesen, K. and Menander, K. (1974) Curr. Ther. Res. 16, 706-717.
11. Marberger, H., Huber, W., Bartsch, G., Schulte, T. and Swoboda, P. (1974) Int. Urol. Nephrol. 6, 61-74.
12. Menander-Huber, K.B. and Huber, W. (1973) in Superoxide and Superoxide Dismutases, Michelson, A.M., McCord, J.M. and Fridovich, I., eds., Academic Press, New York, pp. 537-549.
13. Joester, K.-E., Jung, G. and Weber, U. (1972) FEBS Lett. 25, 25-28.
14. Brigelius, R., Spöttl, R., Bors, W., Lengfelder, E., Saran, M. and Weser, U. (1974) FEBS Lett. 47, 72-75.
15. Brigelius, R., Hartmann, J.-J., Bors, W., Saran, M., Lengfelder, E. and Weser, U. (1975) Z. Physiol. Chem. 356, 739-745.
16. Weser, U., Richter, C., Wendel, A. and Younes, M. (1978) Bioinorg. Chem. 8, 201-213.
17. Younes, M., Lengfelder, E., Zienau, S. and Weser, U. (1978) Biochem. Biophys. Res. Commun. 81, 576-580.
18. Richter, C., Azzi, A., Weser, U. and Wendel, A. (1977) J. Biol. Chem. 252, 5061-5066.
19. Sorenson, J.R.J. (1976) J. Med. Chem. 19, 135-148.
20. Underwood, E.J. (1971) Trace Elements in Human and Animal Nutrition, 3rd ed., Academic Press, New York, p. 57.
21. Kreysel, H.-W. (1977) in D-Penicillamin, Schattauer Verlag, Stuttgart.
22. Henschler, D. (1975) in Pharmakologie und Toxikologie, Forth, W., Henschler, D. and Rummel, W., eds., Bibliographisches Institut, Mannheim, pp. 542-546.
23. Walshe, J.M. (1966) in The Biochemistry of Copper, Peisach, J., Aisen, P. and Blumberg, W.E., eds., Academic Press, New York, pp. 475-498.
24. Bors, W., Saran, M., Lengfelder, E., Michel, C., Fuchs, C. and Frenzel, C. (1978) Photchem. Photbiol. 28, 629-638.
25. Rupp, H. and Weser, U. (1976) Biochim. Biophys. Acta, 446, 151-165.
26. Wright, J.R. and Frieden, E. (1975) Bioinorg. Chem. 4, 163-175.
27. Birker, P.J. and Freeman, H.C. (1976) J. Chem. Soc. D, 164, 312-313.
28. Birker, P.J. and Freeman, H.C. (1977) J. Amer. Chem. Soc. 99, 6890-6899.
29. Lengfelder, E., Fuchs, C., Younes, M. and Weser, U. (1979) Biochim. Biophys. Acta, 567, 492-502.
30. Rupp, H. and Weser, U. (1976) Biochem. Biophys. Res. Commun. 72, 223-229.
31. Weser, U. and Voelcker, G. (1972) FEBS Lett. 22, 15-22.
32. Younes, M. and Weser, U. (1977) Biochem. Biophys. Res. Commun. 78, 1247-1253.
33. Younes, M. and Weser, U. (1976) FEBS Lett. 61, 209-212.
34. Elstner, E.F. and Heupel, A. (1976) Anal. Biochem. 70, 616-620.
35. Lengfelder, E. and Elstner, E.F. (1978) Hoppe-Seylers Z. Physiol. Chem. 359, 751.
36. Fielden, E.M., Roberts, P.B., Bray, R.C., Lowe, D.J., Mautner, G.N., Rotilio, G. and Calabrese, L. (1974), Biochem. J. 139, 49-60.
37. McCord, J.M., Crapo, J.D. and Fridovich, I. (1977) in Superoxide and Superoxide Dismutases, Michelson, A.M., McCord, J.M. and Fridovich, I. es., Academic Press, New York, pp. 11-17.
38. Lengfelder, E. and Elstner, E.F. (1979) Z. Naturforsch. 34, 374-380.
39. Azzi, A., Montecocco, C. and Richter, C. (1975) Biochem. Biophys. Res. Commun. 65, 597-603.
40. Rotilio, G., Bray, R.C. and Fielden, E.M. (1972) Biochim. Biophys. Acta, 268, 605-609.
41. Klug, D., Rabani, J. and Fridovich, I. (1973) J. Biol. Chem. 247, 4839-4842.
42. Klug-Roth, D., Fridovich, I. and Rabani, J. (1972) J. Amer. Chem. Soc. 95, 2786-2790.
43. Bannister, J.V., Bannister, W.H., Bray, R.C., Fielden, E.M., Roberts, P.B. and Rotilio, G. (1973) FEBS Lett. 32, 303-306.
44. Pick, M., Rabani, J., Yost, F. and Fridovich, I. (1974) J. Amer. Chem. Soc. 96, 7329-7333.
45. Lavalle, F., McAdam, M.E., Fielden, E.M., Roberts, P.B., Puget, K. and Michelson, A.M. (1977) Biochem. J. 161, 3-11.

46. Jørgensen, C.K. (1974) in Advance in Quantum Chemistry Vol. 8, Academic Press, New York, pp. 137-182.
47. Konze, J.R. and Elstner, E.F. (1976) FEBS Lett. 66, 8-11.
48. Allen, J.F. and Hall, D.O. (1973) Biochem. Biophys. Res. Commun. 52, 856-862.
49. Elstner, E.F. and Kramer, R. (1973) Biochim. Biophys. Acta, 314, 340-353.
50. Elstner, E.F. and Konze, J.R. (1974) FEBS Lett. 45, 18-21.
51. Elstner, E.F., Saran, M., Bors, W. and Lengfelder, E. (1978) Eur. J. Biochem. 89, 61-66.

PULSE RADIOLYSIS INVESTIGATION OF THE INTERACTION OF BOVINE SUPEROXIDE DISMUTASE WITH ORGANIC FREE RADICALS

P. O'NEILL AND E. MARTIN FIELDEN
Physics Division, Institute of Cancer Research, Sutton, Surrey, SM2 5PX, England

ABSTRACT

The reactions of superoxide dismutase (SOD) with $HO_2^{\cdot}/O_2^-$ and with the peroxyl radical derived from 2-methyl-2-propanol have been investigated using pulse radiolysis. The rate constant for reaction of $HO_2^{\cdot}/O_2^-$ under turn-over conditions was found to decrease on lowering the pH (<5.3). The treatment used to obtain the best fit to the data involves reaction of $HO_2^{\cdot}$ with SOD and an equilibrium between an active and inactive form of the enzyme. In the case of 2-methyl-2-hydroxypropyl-peroxyl radicals, no interaction with SOD was observed; the rate constant for this reaction was estimated to be $<10^8$ dm^3 mol^{-1} s^{-1}.

INTRODUCTION

Bovine superoxide dismutase (SOD) has been widely studied since it was first isolated (for reviews, see Fridovich[1,2]). The catalytic mechanism of copper and zinc containing superoxide dismutase (E.C. 1.15.1.1) has been shown[3,4] to involve alternate oxidation and reduction of the copper by O_2^- where $k_1 \approx k_2 \approx (2.0\text{-}3.7) \times 10^9$ dm^3 mol^{-1} s^{-1}.

$$Cu(II) + O_2^- \xrightarrow{k_1} O_2 + Cu(I) \quad (1)$$

$$Cu(I) + O_2^- \xrightarrow{k_2} O_2^{2-} + Cu(II) \quad (2)$$

Enzyme activity has been demonstrated[5,6] to be independent of pH in the range 5-9 whereas at pH<5 the activity is reduced. It was suggested[6] that the enzyme reacts with O_2^- ($pK_a \sim 4.7$, Ref. 7)) rather than with $HO_2^{\cdot}$. It should also be noted that transitions, which are reversible, involving Cu^{2+} of SOD occur[8] between pH 3 and 5.

The experiments reported below were performed to reinvestigate the possible reaction of $HO_2^{\cdot}$ with SOD and to extend the study to include possible reactions of the enzyme with organic peroxyl radicals derived from 2-methyl-2-propanol.

METHODS

The pulse radiolysis facilities used have previously been described.[9] All solutions were prepared from triply distilled water and, in the majority of cases, contained 3×10^{-2} mol dm^{-3} formate in order to convert $^{\bullet}OH$ radicals and H atoms, produced in the radiolysis of water, into $CO_2^{\bullet -}$. It should be noted that using 3×10^{-2} mol dm^{-3} formate the catalytic rate constant for dismutation should be independent of ionic strength.[10] Since solutions were saturated with O_2, all the radicals produced on irradiation of water were converted into O_2^-.

$$H_2O \rightsquigarrow H,\ OH,\ e_{aq}^- \quad (3)$$

$$e_{aq}^- + O_2 \longrightarrow O_2^- \quad (4)$$

$$OH(H) + HCO_2^- \longrightarrow CO_2^{\bullet -} + H_2O(H_2) \quad (5)$$

$$CO_2^{\bullet -} + O_2 \longrightarrow CO_2 + O_2^- \quad (6)$$

The enzyme was assayed as previously described[4] and the catalytic rate constant for this batch of enzyme was determined as 3.0×10^9 dm^3 mol^{-1} s^{-1}.

RESULTS

Effects of variation of pH. The dose per pulse used produced 1.6 or 3×10^{-6} mol dm^{-3} O_2^-. All optical measurements were performed at 250 nm where both $HO_2^{\bullet}$ and O_2^- are known to absorb. In Figure 1 the rate constants of removal of optical absorption at 250 nm in the presence of SOD at various pH values are presented. The rate of decay of O_2^- ($HO_2^{\bullet}$) was found to follow first order kinetics and to be proportional to the concentration of SOD. The rate constants in Figure 1 have been corrected for the "natural" rate of decay of O_2^- ($HO_2^{\bullet}$) by measuring the rate of change of optical absorption at 250 nm in the absence of SOD. The initial slopes of the first order plots were used to apply the correction. Similar changes of rate constant on pH have previously been reported.[6] On repetitive pulsing ($[O_2^-]$ per pulse $= 3 \times 10^{-6}$ mol dm^{-3}) the kinetics of decay did not change during at least 7 pulses. No appreciable change in the rate constants (<10%) at different pH values was observed when the experiments were repeated in the presence of 10^{-4} mol dm^{-3} EDTA.

In order to check that the activity of the enzyme is not permanently altered by acidic conditions (pH ≥3.8) the following controls were performed. The rate of dismutation at pH 3.8 was determined in an aqueous solution of SOD containing 3×10^{-2} mol dm^{-3} formate and saturated with O_2. The pH of the solution was then adjusted to 9.0 (left for 20 min on re-adjustment of pH) and the rate of dismutation re-determined. The rate measured at pH 9.0 is in agreement with that measured at this pH without initial acidification of the system. This observation is consistent with the reversibility of the transformation as previously demonstrated.[8] It was not possible to go to lower pH values since, at pH <3, further transitions occur[8] which result in inactivation of the enzyme.

The dependence of the rate constant for dismutation of O_2^- ($HO_2^{\bullet}$) in the presence of SOD on pH is presented in Figure 1. The dashed and broken curves in Figure 1

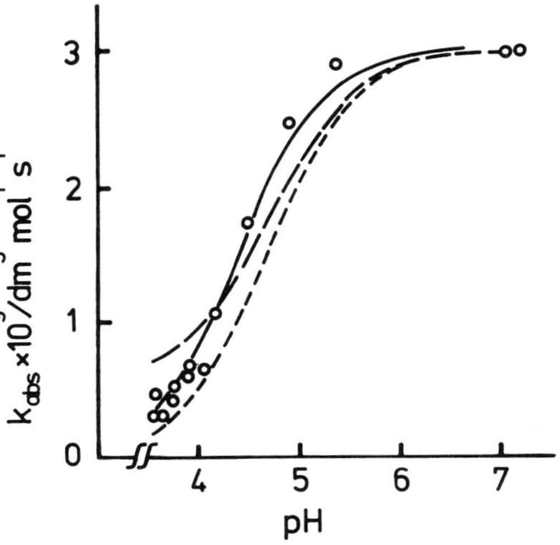

Fig. 1. Dependence of rate constants (k_{obs}) on pH determined at 250 nm on pulse irradiation of oxygenated, aqueous solutions containing 2-4 x 10^{-7} mol dm^{-3} SOD and 3 x 10^{-2} mol dm^{-3} formate/formic acid. Solid, dashed and broken lines determined using equations (12) and (13), respectively (see text).

are based on the following reaction scheme:

$$O_2^- + H^+ \underset{}{\overset{K}{\rightleftharpoons}} HO_2^{\cdot} \tag{7}$$

$$O_2^- + Cu(II) \xrightarrow{k_1} Cu(I) + O_2 \tag{8}$$

$$O_2^- + Cu(I) \xrightarrow{k_2} Cu(II) + O_2^{2-} \tag{9}$$

$$HO_2^{\cdot} + Cu(II) \xrightarrow{k_3} Cu(I) + O_2 + H^+ \tag{10}$$

$$HO_2^{\cdot} + Cu(I) \xrightarrow{k_4} Cu(II) + HO_2^- \tag{11}$$

The dashed line in Figure 1 has been fitted using equation (12) assuming K = 2 x 10^{-5} mol dm^{-3} (Ref. 3), $k_1 = k_2 = 3.0 \times 10^9$ dm^3 mol^{-1} s^{-1}, and $k_3 = k_4 = 0$, i.e., assuming that $HO_2^{\cdot}$ does not react with the enzyme.

$$k_{obs} = \frac{K\,k_1}{K+[H^+]} \tag{12}$$

If reaction (10) and (11) do occur and assuming $k_3 = k_4$ then, using steady state principles, equation (13) is derived.

$$k_{obs} = \frac{K\,k_1}{K+[H^+]} + \frac{k_3[H^+]}{K+[H^+]} \tag{13}$$

Using the known values for K and k_1, values for k_3 were calculated at different H^+ concentrations using equation (13). The mean value determined for $k_3(k_4)$ is $5.4 \times 10^8\ dm^3\ mol^{-1}\ s^{-1}$. Using this value and those for K and k_1, the broken line in Figure 1 was determined using equation (13) and the assumption stated above. It is apparent that using other values for k_3/k_4 will only improve the fit at one end of the curve at the expense of the fit at the other end.

An alternative reaction scheme based on reactions (7)-(11) and equilibrium (14), in which the active form of the enzyme (SOD_A) is converted into an inactive form (SOD_I), has been used to obtain the solid line in Figure 1.

$$SOD_A + H^+ \xrightarrow{K'} SOD_I \qquad (14)$$

This fit is based on an equation analogous to equation (12) and the assumptions that the rate constants for reaction of the oxidised and reduced forms of SOD_A with O_2^- or $HO_2^{\cdot}$ are the same, i.e., $k_1 = k_2 = k_3 = 3.0 \times 10^9\ dm^3\ mol^{-1}\ s^{-1}$, and $K' = 4.0 \times 10^{-5}\ mol\ dm^{-3}$ (estimated from the dependence of k_{obs} on pH).

The latter treatment affords a reasonable fit of the experimental data. In this treatment it is assumed that $HO_2^{\cdot}$ may react catalytically with the active form of SOD and that the variation of k_{obs} on pH is due to the existence of equilibrium (14). An equilibrium between an active and inactive form of the enzyme is consistent with the experimental finding that the activity of the enzyme is not permanently altered by acidic conditions. It is of interest that no marked pH dependence (pH$\sim$3-9) was observed[11] in the superoxide dismutative action of Cu-penicillamine. However, an alternative mechanism has been postulated[11] in the latter system whereby the redox-system involved in the reaction with superoxide was suggested to be $RS^{\cdot}/RS^-$. These redox reactions would not be applicable in the case of bovine SOD.

The reaction of SOD with the peroxyl radical from 2-methyl-2-propanol (reaction (15)) was studied to obtain information concerning possible reaction of SOD with organic peroxyl radicals. This system was chosen since

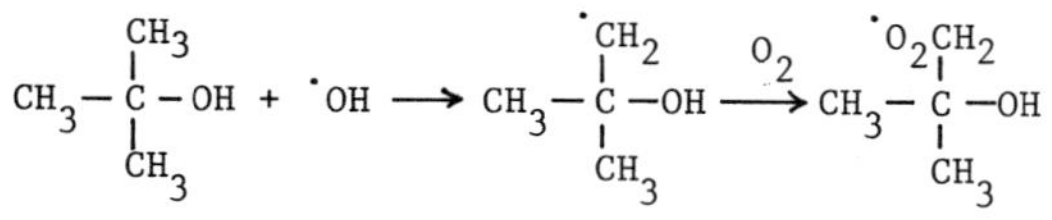
(15)

β-hydroxyperoxyl radicals, derived from 2-methyl-2-propanol, do not eliminate[12,13] O_2^- in an OH^- catalysed reaction whereas α-hydroxyperoxyl radicals eliminate[13] O_2^-. Furthermore, 2-methly-2-hydroxypropylperoxyl radicals have been shown[12] to undergo bimolecular reactions with production of O_2^- (The radiation chemical yield of O_2^- in units of molecules $(100\ eV)^{-1}$, $G(O_2^-)$ is $\approx$ 0.6-0.7 in oxygenated aqueous solution, i.e.$\sim$ 22% of peroxyl radicals from 2-methyl-2-propanol yield O_2^-).[12]

On pulse irradiation of a $4.5 \times 10^{-6}\ mol\ dm^{-3}$ SOD solution containing 0.5 mol dm^{-3} 2-methyl-2-propanol and saturated with O_2 at pH 9.0, the absorption at 680 nm decreases during the first 1 ms and there is no further change during the next 10 ms (percent loss of absorption is ~53%). At the end of this period O_2^- and the peroxyl radicals, formed in reaction (15), should have reacted since the lifetime of the peroxyl radicals cannot be[12] > 100 μs at the dose used of 3.6 krads. At this dose $[O_2^-] \geqslant 1.1 \times 10^{-5}$ mol dm^{-3} and therefore the enzyme reacts under turnover conditions since $[O_2^-] \gg [SOD]$. It is suggested that if SOD reacts catalytically with the 2-methyl-2-hydroxypropylperoxyl radicals also present, then the rates of reaction with the oxidised and reduced forms of SOD should be fast due to the "natural" lifetime[12] of the peroxyl radicals of < 100 μs under the conditions used. It has previously been shown[12] that bimolecular reactions of 2-methyl-2-hydroxypropylperoxyl radicals results in production of O_2^-,

$$2(Me)_2 - \overset{\overset{\bullet O_2CH_2}{|}}{C} - OH \longrightarrow O_2^- + \text{products} \qquad (16)$$

If the peroxyl radicals disappear only as shown in reaction (16) then the O_2^- produced will not cause further optical changes at 680 nm since the oxidised and the reduced forms of the enzyme are in equilibrium under turnover conditons.

In order to check the reaction of SOD (4.5×10^{-6} mol dm^{-3} Cu^{2+} centres) in the system containing 2-methyl-2-propanol (0.5 mol dm^{-3}) under non-turnover conditions (~ 170 rads), the optical absorption changes at 680 nm with time were monitored. The concentrations of 2-methyl-2-hydroxypropylperoxyl radicals and O_2^- initially produced are ~ 4.8×10^{-7} mol dm^{-3}. In Figure 2 a typical absorbance-time trace is shown for the optical changes at 680 nm. At the dose used, the total percent loss of absorption at 680 nm corresponds to ~ 15%. The initial reduction of absorption (~ 1 ms) is due to reaction with SOD of O_2^-, produced on scavenging $H^\bullet$ and e_{aq}^- by O_2. The subsequent loss of absorption over several msec is suggested to be due to reaction of SOD with O_2^- produced[12] in reaction (15). This is based on the following considerations.

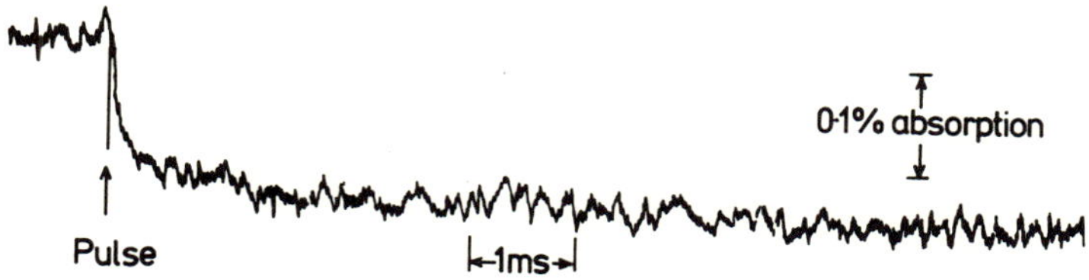

Fig. 2. Optical absorption changes with time monitored at 680 nm on pulse irradiation of an oxygenated, aqueous solution containing 4.5×10^{-6} mol dm^{-3} SOD and 0.5 mol dm^{-3} 2-methyl-2-propanol. Dose = 0.17 krad/pulse.

The yield of loss of absorption, G(-SOD) ~ 3.7, i.e. 6.3×10^{-7} mol dm^{-3} at the dose used, is in reasonable agreement with the total yield of O_2^- formed in reactions (4), (16) and (17) [$G(O_2^-) = G(e_{aq}^-) + G(H^{\cdot}) + G(O_2^-$; reaction (16)) ~ 3.8-3.9; at the dose used this is equivalent to a concentration of O_2^- of ~ 6.5×10^{-7} mol dm^{-3}]. If the peroxyl radicals had reacted with SOD at a rate similar to

$$H^{\cdot} + O_2 \longrightarrow HO_2^{\cdot} \rightleftharpoons H^+ + O_2^- \qquad (17)$$

that with O_2^- then the extent of removal of optical absorption at 680 nm would have been greater (i.e. G(-SOD) ~ 5-6; equivalent to $[O_2^-]$ of $8.5 - 10.2 \times 10^{-7}$ mol dm^{-3}). This would have been readily observed experimentally. Therefore, the rate constant for reaction of 2-methyl-2-hydroxypropyl-peroxyl radicals with SOD, if this reaction occurs, is estimated to be $< 10^8$ dm^3 mol^{-1} s^{-1} since the lifetime of the peroxyl radicals in the absence of SOD is ~ 3 msec (Ref. 12) at the dose used.

At present it cannot be ruled out that other organic peroxyl radicals undergo reaction with SOD; however, peroxyl radicals derived from 2-methyl-2-propanol react only slowly, if at all, with SOD. In order to study the reaction of SOD with other peroxyl radicals, e.g., from lipids, more information is required concerning the chemistry of the peroxyl radicals. In contrast, it is inferred that the reaction of $HO_2^{\cdot}$ with SOD may occur. It is of interest that SOD has been shown[14] to react with other types of organic radicals. The radical anions derived from nitroaromatic compounds, e.g. p-nitroacetophenone, are capable of reacting with bovine SOD but not in a catalytic manner. Whether this interaction involved reduction of the Cu centres is, as yet, not certain.

ACKNOWLEDGEMENTS

We thank Mr. J. Currant for technical assistance and Dr. M. McAdam and Mr. R. Bugden for the preparation of the enzyme. This work was funded by the MRC/CRC joint committee.

REFERENCES

1. Fridovich, I. (1974) Adv. Enzymol. 41, 35-97.
2. Fridovich, I. (1975) Annu. Rev. Biochem. 44, 147-159.
3. Klug-Roth, D., Fridovich, I. and Rabani, J. (1973) J. Amer. Chem. Soc. 95, 2786-2790.
4. Fielden, E.M., Roberts, P.B., Bray, R.C., Lowe, D.J., Mautner, G.N., Rotilio, G. and Calabrese, L. (1974) Biochem. J. 139, 49-60.
5. Rotilio, G., Bray, R.C. and Fielden, E.M. (1972) Biochim. Biophys. Acta, 268, 605-609.
6. Klug, D., Rabani, J. and Fridovich, I. (1972) J. Biol. Chem. 247, 4839-4842.
7. Behar,D., Czapski, G., Rabani, J., Dorfman, L.M. and Schwarz, H.A. (1970) J. Phys. Chem. 74, 3209-3213.
8. Fee, J.A. and Phillips, W.D. (1975) Biochim. Biophys. Acta, 412, 26-38.
Fee, J.A. (1977) in Superoxide and Superoxide Dismutases, Michelson, A.M., McCord, J.M. and Fridovich, I., eds., Academic Press, New York,pp. 173-192.

9. McAdam, M.E., Fox, R.A., Lavelle, F. and Fielden, E.M. (1977) Biochem. J. 165, 71-79.
10. McAdam, M.E. (1977) Biochem. J. 161, 697-699.
11. Lengfelder, E., Fuchs, C., Younes, M. and Weser, U. (1979) Biochim. Biophys. Acta, 567, 492-502.
12. Schuchmann, M.N. and von Sonntag, C. (1979) J. Phys. Chem. 83, 780-784.
13. Rabani, J., Klug-Roth, D. and Henglein, A. (1974) J. Phys. Chem. 78, 2089-2093; Bothe, E., Behrens, G. and Schulte-Frohlinde, D. (1977) Z. Naturforsch. 32b, 886-889.
14. Wardman, P., submitted to the meeting of the Association of Radiation Research, Jan. 1979.

POTENTIATING EFFECT OF CATALASE ON INHIBITION BY SUPEROXIDE DISMUTASE OF 6-HYDROXYDOPAMINE AUTOXIDATION

STEPHEN GENE SULLIVAN, LAURA SUDARSKY AND ARNOLD STERN
Department of Pharmacology, New York University School of Medicine,
New York 10016, USA

INTRODUCTION

6-Hydroxydopamine is a catecholamine analog which accumulates in catecholamine containing neurons and causes degeneration of nerve terminals when injected into animals.[1-3] 6-Hydroxydopamine autoxidizes at neutral pH forming a red colored compound thought to be the p-quinone of 6-hydroxydopamine.[4] Intermediates and products of autoxidation implicated in the cytotoxic actions of 6-hydroxydopamine include various quinones, indoles, and related compounds, hydrogen peroxide, superoxide, hydroxyl radical, and singlet oxygen.[4-14] Of these intermediates and products, superoxide has the unique role of catalyzing 6-hydroxydopamine autoxidation.[8] Hydrogen peroxide will oxidize 6-hydroxydopamine with formation of p-quinone but hydrogen peroxide is not catalytic in the autoxidation.[15]

Superoxide and hydrogen peroxide are both products of 6-hydroxydopamine autoxidation with potential cytotoxic actions but their role in the kinetics of autoxidation is not fully characterized. We have studied the effects of superoxide dismutase and catalase on p-quinone formation from 6-hydroxydopamine to determine the role of superoxide and hydrogen peroxide in 6-hydroxydopamine autoxidation and to determine the potential protective action of these enzymes against the cytotoxic effects of 6-hydroxydopamine.

RESULTS

Autoxidation of 6-hydroxydopamine at pH 7.4 and the effect of superoxide dismutase. Autoxidation of 6-hydroxydopamine hydrobromide (Aldrich Chemical Company, Milwaukee, Wisconsin) was studied by following the formation of the p-quinone derivative of 6-hydroxydopamine[4] at pH 7.4. The formation of the p-quinone was observed as an increase in absorbance at 490 nm using a Cary 14 spectrophotometer. A stock solution of 10^{-2} M 6-hydroxydopamine was prepared in 10^{-3} M KCl, pH 2.0. 6-hydroxydopamine is stable at pH 2.0 and autoxidation proceeds very slowly.[5] Autoxidation was initiated by injecting 40 µl of 10^{-2} M 6-hydroxydopamine stock solution into a sample cuvette containing 1.5 ml 10^{-4} M EDTA, 0.1 M NaH_2PO_4, pH 7.4. The solution in the sample cuvette was mixed continuously by a small magnetic stirrer. The absorbance at 490 nm was measured against a reference cuvette containing all additions except 6-hydroxydopamine. As can be seen in Figure 1, (curve a) formation of 6-hydroxydopamine p-quinone proceeds at a linear rate and then reaches a plateau. The addition of superoxide dismutase (Figure 1, curves b to d) gives rise to sigmoidal kinetics in the formation of the p-quinone. An initial lag period is followed by a linear phase of p-quinone formation and then by a plateau. The higher the concentration of superoxide dismutase the longer the lag phase, the lower the rate of p-quinone formation in the linear phase, and the longer the total time elapsed before reaching the plateau. Superoxide dismutase has no effect on the absolute amount of p-quinone formed. The final color of approximately 0.6 absorbance units was reached in about 15 minutes in the presence of 10 μg/ml superoxide dismutase (Figure 1, trace d). The time course of p-quinone formation in in the presence of superoxide dismutase mimics the kinetics of 6-hydroxydopamine autoxidation at low pH (Ref. 5). The results in Figure 1 are similar to those of Heikkila and Cohen[8] and confirms that superoxide is catalytic in the autoxidation of 6-hydroxydopamine. The relationship between increasing concentrations of superoxide dismutase and increasing length of the lag period indicates that superoxide is important in the events before color formation.

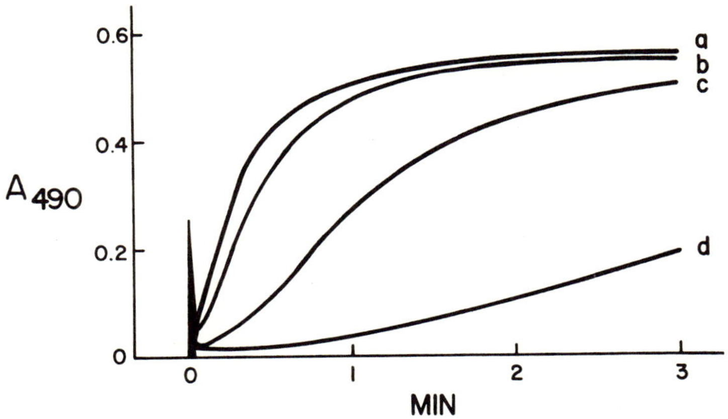

Fig. 1. The effect of superoxide dismutase on the autoxidation of 6-hydroxydopamine. Autoxidation of 2.8×10^{-4} M 6-hydroxydopamine in 10^{-4} M EDTA, 0.1 M NaH_2PO_4, pH 7.4 was followed by measuring changes in absorbance at 490 nm. 6-hydroxydopamine was added to the solution at zero time. Concentration of superoxide dismutase (3000 units/mg, Truett Laboratories, Dallas, Texas) in μg/ml: a, none; b, 0.67; c, 2.0; d, 10.0.

<u>The effects of catalase on 6-hydroxydopamine autoxidation</u>. The effects of increasing concentrations of catalase on 6-hydroxydopamine autoxidation are

shown in Figure 2. Catalase decreases the rate of the linear phase of p-quinone formation but this effect is saturable. There is no change in the absolute amount of p-quinone formed and no appearance of a lag period in p-quinone formation. These results indicate that hydrogen peroxide contributes to p-quinone formation but is not necessary for catalysis autoxidation. Since no lag period in p-quinone formation occurs in the presence of catalase, hydrogen peroxide is probably not involved in the events before p-quinone formation.

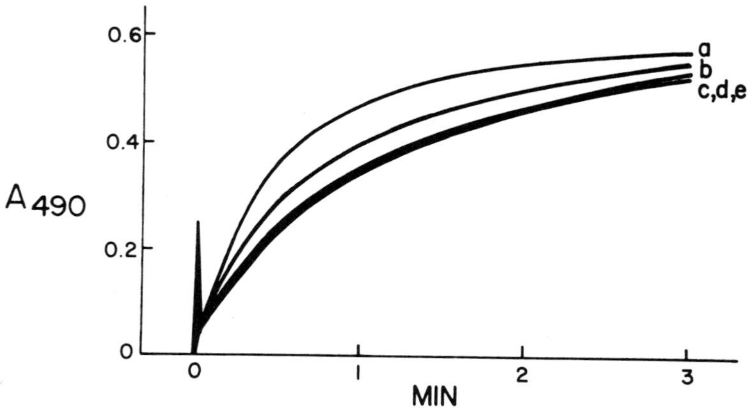

Fig. 2. The effect of catalase on the autoxidation of 6-hydroxydopamine. Concentration of catalase (63,863 units/mg, Calibiochem-Behring Corp., La Jolla, California) in μg/ml: a, none; b, 2.8; c, 7.0; d, 14.0; e, 28.0. Other conditions as in Fig. 1.

The influence of a saturating concentration of catalase on inhibition of 6-hydroxydopamine autoxidation by superoxide dismutase. The maximum inhibitory effect of catalase alone on p-quinone formation is shown in Figure 3, (curve b). Additions of varying amounts of superoxide dismutase (identical to those in Figure 1, curves b to d) shows that catalase greatly potentiates the inhibitory actions of superoxide dismutase on p-quinone formation (Figure 3, curves c to e). Doubling the concentration of catalase used in Figure 3 had no further effect on inhibition in the presence of superoxide dismutase. The potentiating effect of catalase on inhibition of p-quinone formation by superoxide dismutase appears as a further decrease in the rate of the linear phase. Superoxide dismutase still causes a lengthening of the lag phase with little influence of catalase. Combining superoxide dismutase and catalase had no effect on the end point of p-quinone formation. The final color of approximately 0.6 absorbance units was reached in about 30 minutes in the presence of 10.0 μg/ml superoxide dismutase

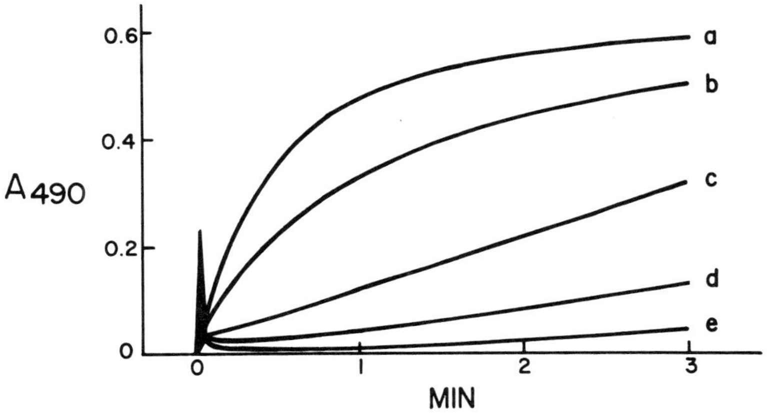

Fig. 3. The effect of superoxide dismutase on the autoxidation of 6-hydroxydopamine in the presence of a saturating amount of catalase. Concentration of superoxide dismutase in µg/ml: a and b, none; c, 0.67; d, 2.0; e, 10.0. Concentration of catalase in µg/ml: a, none; b, c, d and e, 14.0. Other conditions as in Fig. 1.

and 14.0 µg/ml catalase (Figure 3, curve e). These results show that the potentiating effect of catalase on superoxide dismutase inhibition of p-quinone formation is reflected in a decrease in the rate of the linear phase without changing the length of the lag period or the final color formed.

Nature of the colored product. Absorption spectra of the colored product formed by 6-hydroxydopamine autoxidation had maxima at 270 nm and 490 nm in agreement with Graham.[4] Absorption spectra were identical whether autoxidation was carried out with no additions, in the presence of either catalase or superoxide dismutase, or a mixture of catalase and superoxide dismutase. These spectra have been identified with the p-quinone of 6-hydroxydopamine.[4] Our results indicate that superoxide and hydrogen peroxide affect the kinetics but not the final product of 6-hydroxydopamine autoxidation.

The effect of mixtures of superoxide dismutase and catalase on the rate of the linear phase of p-quinone formation. Mixtures of catalase and superoxide dismutase were studied for their combined effect on the rate of the linear phase, to determine their potential protective effect against 6-hydroxydopamine cytotoxicity. Figure 4, part A shows the potentiating effect of a saturating amount of catalase (14 µg/ml) on the rate of p-quinone formation. Saturating catalase causes 47% inhibition of p-quinone formation in the absence of superoxide dismutase (Figure 4, part A, 0 µg/ml SOD). By comparison, p-quinone formation in a solution containing 0.5 µg/ml superoxide dismutase is 75% inhibited by the addition of saturating catalase (Figure 4, part A, 0.5 µg/ml SOD).

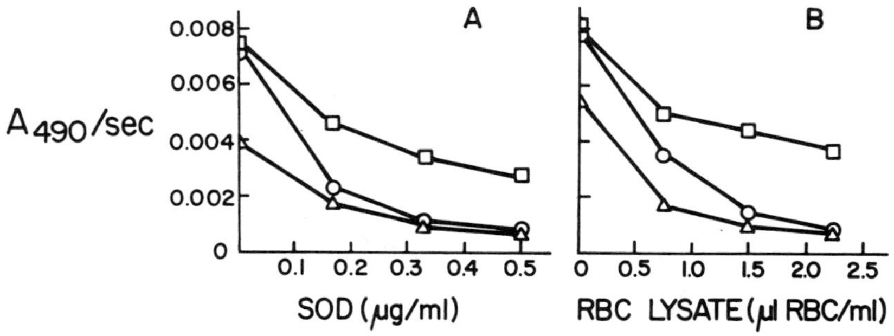

Fig. 4. The effects of mixtures of superoxide dismutase and catalase on the rate of autoxidation of 6-hydroxydopamine. Studies were carried out with purified enzymes (A) or with red cell lysates (B). The rate of autoxidation was measured as the change in absorbance at 490 nm per second during the linear phase of color formation. Autoxidation of 6-hydroxydopamine was carried out as described in Fig. 1. For the preparation of red cell lysates freshly drawn adult human blood was washed three times in phosphate-buffer saline (9 parts, 0.9% NaCl; 1 part, 0.1 M KH_2PO_4, pH 7.4). A 25% (v/v) suspension of red cells was prepared in phosphate-buffered saline. 0.5 ml of this suspension was lysed in 4.5 ml H_2O and added to 15 ml 10^{-4} M EDTA, 0.1 M NaH_2PO_4, pH 7.4 to give a red cell lysate stock of 6.25 μl RBC/ml. Additions in A: □, none; ○, 8.4 μg catalase per μg superoxide dismutase; Δ, 14.0 μg/ml catalase. Additions in B: □, 3.3 x 10^{-5} M azide; ○, none; Δ, 14.0 μg/ml catalase.

When the amount of added catalase is kept constant relative to superoxide dismutase (8.4 μg catalase per μg superoxide dismutase) the potentiation of inhibition by catalase (Figure 4, part A, circles) quickly approaches that seen with a saturating amount of catalase (Figure 4, part A, triangles). In the presence of 0.17 μg/ml superoxide dismutase, 1.4 μg/ml catalase inhibits p-quinone formation by 50%, whereas 14.0 μg/ml catalase (ten-fold) increases the inhibition to only 60%. Lysates of biological material would be expected to follow the pattern of adding a constant mixture of μg catalase per μg superoxide dismutase. This is shown with human red cell lysates in Figure 4, part B (circles). To prove that the shape of the lysate concentration curve is dependent only on endogenous superoxide dismutase and catalase, two types of experiments were carried out. In one experiment, a saturating amount of pure catalase was added to lysates, thereby eliminating the effects of the endogenous catalase. The resulting rate of autoxidation versus concentration curve (Figure 4, part B, triangles) mimics the curve obtained with pure superoxide dismutase plus saturating catalase (Figure 4, Part A, triangles). In another experiment, endogenous catalase activity was eliminated by specific inhibition with azide; 3.3 x 10^{-5} M sodium azide was used since this concentration was shown to completely inhibit catalase activity while having no effect on 6-hydroxydopamine autoxidation or on superoxide dismutase activity. The addition of azide to lysates

(Figure 4, part B, squares) gives a result which mimics the curve obtained with pure superoxide dismutase alone (Figure 4, part A, squares). Therefore, the effect of red cell lysate concentration on the rate of 6-hydroxydopamine autoxidation can be completely explained by the endogenous superoxide dismutase and catalase. The results in Figure 4 demonstrate that the presence of catalase in biological systems may greatly potentiate the action of superoxide dismutase in inhibiting 6-hydroxydopamine autoxidation.

DISCUSSION

Our data indicate that superoxide is important in initial events of 6-hydroxydopamine autoxidation leading to p-quinone formation. Once a steady state has been reached, observable as the linear phase of p-quinone formation, both superoxide and hydrogen peroxide participate in p-quinone formation. In the presence of superoxide dismutase, oxidation of 6-hydroxydopamine by hydrogen peroxide may become quantitatively more important in the face of decreased availability of superoxide. This would explain the potentiating effect of catalase on superoxide dismutase inhibition of p-quinone formation. Another possibility is that catalase protects superoxide dismutase from inactivation by hydrogen peroxide.[16,17]

Hydroxyl radical is a product of 6-hydroxydopamine autoxidation[10] and its formation is dependent on the presence of superoxide and hydrogen peroxide. Superoxide dismutase and catalase lower the concentration of superoxide and hydrogen peroxide and thereby secondarily block formation of hydroxyl radical. If hydroxyl radical were necessary for p-quinone formation, then the potentiating effect of catalase on superoxide dismutase inhibition of p-quinone formation might result from blocking the formation of hydroxyl radical. This is unlikely since scavengers of hydroxyl radicals, including ethanol, have no effect on autoxidation of 6-hydroxydopamine (R.E. Heikkila, personal communication). In our system, concentrations of ethanol up to 5×10^{-2} M had no effect on formation of p-quinone in the presence or absence of superoxide dismutase and catalase, confirming that hydroxyl radical is not involved in p-quinone formation.

Measurement of inhibition of autoxidation of 6-hydroxydopamine has been used as an assay for superoxide dismutase activity.[18] Our results show that this technique may underestimate superoxide dismutase activity in cellular systems in which catalase is partially or totally inhibited. To correct for this, measurements should be undertaken with either saturating amounts of catalase or an amount of azide that specifically inhibits catalase in the lysate. Polarographic[19] and chemical methods[20] for determining superoxide dismutase activity may be influenced by catalase when hydrogen peroxide is present. Chemical methods involving a reduction process such as cytochrome c reduction by superoxide may result in an overestimation of superoxide dismutase activity in cellular prepations that have partially or totally inhibited catalase. Cellular studies that have reported decreased superoxide dismutase activity, especially where catalase activity was or might be inhibited, should be reassessed in light of these results. Examples include reports of decreased superoxide dismutase activity in rabbit eyes after feeding 3-aminotriazole,[21] in red cells incubated with 1,4-naphthoquinone-2-sulfonic acid,[22] and the general finding of decreased superoxide dismutase in cancer cells.[23] Reports using indirect assays to determine relative superoxide dismutase content of various tissues or age groups should also be reassessed.

It has been suggested that since superoxide catalyzes 6-hydroxydopamine autoxidation, local concentrations of superoxide dismutase may be important in protecting cells exposed to 6-hydroxydopamine.[8] Our data show that the combined presence in tissues of superoxide dismutase and catalase or other peroxidatic activity may be more effective than superoxide dismutase alone in protection against 6-hydroxydopamine cytotoxicity.

ACKNOWLEDGMENTS

This work was supported by a grant-in-aid from the New York Heart Association and by Grant 19532 from the National Institutes of Health. We thanks Drs. F.S. Cabbat, G. Cohen and R.E. Heikkila for helpful discussions and advice during the preparation of this manuscript.

REFERENCES

1. Cohen, G., Heikkila, R.E., Allis, B., Cabbat, F., Dembiec, D., MacNamee, D., Mytilineou, C. and Winston, B. (1976) J.Pharmacol.Exp.Ther. 199, 336-352.
2. Kostrzewa, R.M. and Jacobowitz, D.M. (1974) Pharmacol.Rev. 26, 199-288.
3. Sachs, C. and Jonsson, G. (1975) Biochem.Pharmacol. 24, 1-8.
4. Graham, D.G. (1978) Mol.Pharmacol. 14, 633-643.
5. Saner, A. and Thoenen, H. (1971) Mol.Pharmacol. 7, 147-154.
6. Heikkila, R. and Cohen, G. (1971) Science, 172, 1257-1258.
7. Heikkila, R. and Cohen, G. (1972) Mol.Pharmacol. 8, 241-248.
8. Heikkila, R.E. and Cohen G. (1973) Science, 181, 456-457.
9. Blank, C.L., Kissinger, P.T. and Adams, R.N. (1972) Eur.J.Pharmacol. 19, 391-394.
10. Cohen, G. and Heikkila, R.E. (1974) J.Biol.Chem. 249, 2447-2452.
11. Rotman, A., Daly, J.W. and Creveling, C.R. (1976) Mol.Pharmacol. 12, 887-899.
12. Heikkila, R.E. and Cabbat, F.S. (1977) Res.Commun.Chem.Pathol.Pharm. 17, 649-662.
13. Borg, D.C., Schaich, K.M., Elmore, J.J. and Bell, J.A. (1978) Photochem. Photobiol. 28, 887-907.
14. Graham, D.G., Tiffany, S.M., Bell, W.R. and Gutknecht, W.F. (1978) Mol. Pharmacol. 14, 644-653.
15. Liang, Y.O., Wightman, R.M. and Adams, R.N. (1976) Eur.J.Pharmacol. 36, 455-458.
16. Bray, R.C., Cockle, S.A., Fielden, E.M., Roberts, P.B., Rotilio, G. and Calabrese, L. (1974) Biochem.J. 139, 43-48.
17. Hodgson, E.K. and Fridovich, I. (1975) Biochemistry, 14, 5294-5299.

18. Heikkila, R.E., Cabbat, F.S. and Cohen, G. (1976) J.Biol.Chem. 251, 2182-2185.
19. Rigo, A., Tomat, R. and Rotilio, G. (1974) Electroanal.Chem. Interfacial Electrochem. 57, 291-296.
20. Fridovich, I. (1975) Annu.Rev.Biochem. 44, 147-159.
21. Bhuyan, K.C. and Bhuyan, D.K. (1978) Biochim.Biophys. Acta, 542, 28-38.
22. McMahon, S. and Stern, A. (1979) Biochim.Biophys. Acta, 566, 253-258.
23. Oberley, L.W. and Buettner, G.R. (1979) Cancer Res. 39, 1141-1149.

SUPEROXIDE DISMUTASE: POTENTIAL FOR A WIDER ROLE AS AN INHIBITOR OF FREE RADICAL REACTIONS

CHRISTINE C. WINTERBOURN
Department of Clinical Biochemistry, Christchurch Hospital, Christchurch, New Zealand

ABSTRACT

A mechanism whereby superoxide dismutase (SOD) inhibits reactions of free radicals that can act as precursors of superoxide is described. The reaction studied was the reduction of methaemoglobin by a range of semiquinone radicals. In air as well as N_2, the semiquinones and not superoxide caused the reduction. However, in air, the reactions of all the semiquinones were inhibited by SOD. An explanation for these observations is that because the semiquinone radicals can react reversibly with O_2,

$$\text{Semiquinone} + O_2 \rightleftharpoons \text{quinone} + O_2^-$$

Superoxide dismutase, by catalysing the removal of superoxide, can displace this equilibrium to the right and thus inhibit competing reactions of the semiquinones. The reaction of methaemoglobin with semiquinones with a wide variety of redox potentials were inhibited. Conditions under which SOD should be capable of inhibiting other free radical reactions are outlined, and it is suggested that a number of reactions that SOD has been shown to influence could be due to radical precursors of superoxide.

INTRODUCTION

Current opinion generally supports the proposal that superoxide itself is not a particularly damaging species, and that the major protective role of superoxide dismutase (SOD) is to prevent superoxide going on to produce other forms of active oxygen, such as hydroxyl radicals, that can undergo far more deleterious reactions.[1-4] However, during an investigation of the oxidation of oxyhaemoglobin by menadione, we observed that SOD markedly accelerated the reaction, and this effect could not be explained in terms of SOD inhibiting a reaction of superoxide or a subsequent product.[5] Further studies showed that SOD can prevent reduction of methaemoglobin by the menadione semquinone, via an O_2-dependent mechanism.[6] These findings, as well as providing an explanation for the oxyhaemoglobin reaction, demonstrate how SOD, by reacting with superoxide, can inhibit a reaction of a free radical precursor of the superoxide.

The apparent requirement for this mechanism to operate is that the free radical ($R^{\cdot -}$) can react reversibly with O_2, reaction (1):

$$R^{\cdot -} + O_2 \rightleftharpoons R + O_2^{\cdot -} \quad (1)$$

Superoxide dismutase, by removing superoxide, displaces this equilibrium to the right, lowering the concentration of the free radical $R^{\cdot -}$, and therefore the extent to which it can undergo competing reactions. Removal of free radical precursors of superoxide by SOD could in theory be a general mechanism with important biological implications. In many reactions involving superoxide, the source of superoxide is the reaction of a free radical, frequently a semiquinone, with O_2 (e.g.,Refs.7,8). It may be that the effect of SOD in these situations is to prevent other reactions of these radicals.

METHODS

To investigate how a range of radical reactions could be affected by SOD, six quinones with quite different redox potentials have been investigated. Reduction of methaemoglobin by the semiquinones was found to be a very suitable reaction to study, because reduction by superoxide is slow, with known rate constant[9] and can readily be distinguished from the faster reactions of the semiquinones. Two experimental approaches have been used. In the first, semiquinones were produced by UV-irradiating isopropanol-containing solutions of the quinones in the presence of methaemoglobin. The rates of reduction of methaemoglobin in air and N_2, and in the presence and absence of SOD were compared. The other approach was to generate superoxide from xanthine and xanthine oxidase, and measure the rate of methaemoglobin reduction in the presence of each quinone. Catalase was present to eliminate effects due to H_2O_2, and the exposed (β-93) sulphydryl groups of the methaemoglobin were blocked with iodoacetamide to prevent their reacting with the quinones to produce semiquinones.

RESULTS

The quinones studied are listed in Table 1, along with their redox potentials for one electron reduction and equilibrium constants for reaction 1. All the quinones, on irradiating with 366 nm UV light in the presence of isopropanol, caused the reduction of methaemoglobin (Table 2). There was no reduction in the dark, or if methaemoglobin was irradiated in the absence of quinone. The rate varied considerably for the different quinones and was largely dependent on the amount of 366 nm light absorbed by each.

The rate of reduction of methaemoglobin by some of the quinones was little affected or unaffected by the presence of air, although with others it was decreased by up to 60% (Table 2). In air, but not N_2, all the reactions were inhibited by SOD (Table 3). Complete or almost complete inhibition was possible with the four quinones with highest redox potential, but with anthraquninone-

TABLE 1

LIST OF THE QUINONES STUDIED, THEIR REDOX POTENTIALS FOR ONE ELECTRON REDUCTION, AND EQUILIBRIUM CONSTANTS FOR REACTION (1)

Quinone	E^o	K	Ref
1,4-naphthoquinone-2-sulphonate (NQS)	-0.06	0.017	9
2,5-dimethylbenzoquinone (DMB)	-0.07	0.02	9
1,4-naphthoquinone (NQ)	-0.10	0.07	11[a]
menadione (MD)	-0.20	3.0	9
duroquinone (DQ)	-0.24	12	9
9,10-anthraquinone-2-sulphonate (AQS)	-0.39	5000	10[a]
O_2 (IM)	-0.17		9[b]

[a] Half of standard 2 electron reduction potential

[b] Normalized from 1 atmosphere to IM O_2

TABLE 2

REDUCTION OF METHAEMOGLOBIN BY UV-IRRADIATED QUINONES[a]

Quinone	Rate of methaemoglobin reduction (ΔμM/2m)	Rate in air / Rate in N_2 (%)[b]
1,4-naphthoquinone-2-sulphonate	21	95-100
2,5-diemthylbenzoquinone	1.3	80-100
naphthoquinone	23	80-100
menadione	11	45-70
duroquinone	0.7	40-50
9,10-anthraquinone-2-sulphonate	2.0	50-70

[a]The methaemoglobin had its β-93 sulphydryl groups blocked with iodoacetamide[5], and was prepared by adding ferricyanide to oxyhaemoglobin that had been purified by ion exchange chromatography.[5] Solutions contained 32 μM methaemoglobin and 50 μM quinone in 0.02 M sodium phosphate buffer containing catalase (20 μg/ml) and isopropanol (4 ml/100 ml). Irradiation was with 366 nm light. Concentration changes were calculated from ΔA_{577}, using 11.9 as the difference in mM extinction coefficients. Concentrations are expressed in terms of haem groups.

[b]Solutions were bubbled with N_2 and the tubes sealed. Air was reintroduced before recording A_{577}.

2-sulphonate only 45% inhibition could be obtained, and there was also a slight decrease with duroquinone. The amount of SOD needed for half-maximum inhibition also varied and was greatest for dimethylbenzoquinone and naphthoquinone-2-sulphonate.

Superoxide generated from xanthine by xanthine oxidase slowly reduced methaemoglobin, and the reaction is inhibited by SOD.[9] Addition of each of the six quinones to this system gave a marked increase in rate of reduction. With 50 μM quinone this varied from a 4-fold increase with anthraquinone-6-sulphonate to about 20-fold with menadione and naphthoquinone (Figure 1). With all the quinones the reaction could be fully inhibited by SOD. The quinones

TABLE 3

EFFECT OF SUPEROXIDE DISMUTASE ON THE REDUCTION OF METHAEMOGLOBIN BY UV-IRRADIATED QUINONES[a]

	Max. inhibition by SOD (%)	SOD concn. for half max. inhibition (μg/ml)
1,4-naphthoquinone-2-sulphonate	96	1.26
2,5-diemthylbenzoquinone	96	2.70
1,4-naphthoquinone	98	0.51
menadione	100	0.38
duroquinone	83	0.38
9,10-anthraquinone-2-sulphonate	45	<0.40

[a]Experiemtns were carried out under the conditions described in Table 2. Solutions contained 32 μM methaemoglobin, 50 μM quinone and variable amounts of SOD, and were irradiated for 2-15 min (depending on the reaction rate of each quinone). Values were determined from double reciprocal plots of per-cent inhibition *versus* SOD concentration, which were linear.

with highest redox potential required the most SOD for maximum inhibition, or, as shown in Figure 1, the amount of inhibition by a given concentration of SOD increased as the redox potential of the quinone decreased. This trend is the same as seen with the photochemically generated semiquinones.

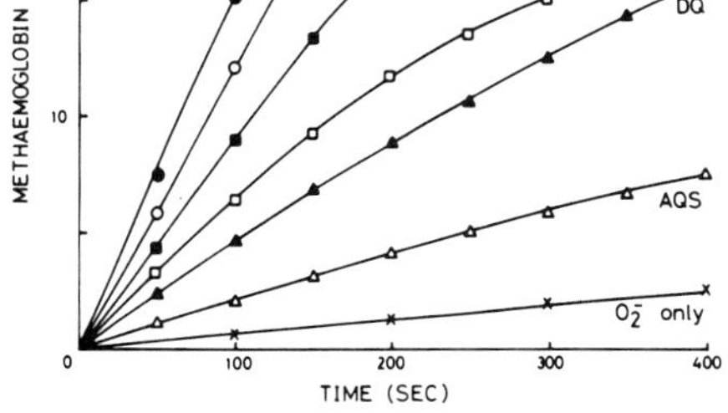

Fig. 1. Rate of reduction of methaemoglobin by superoxide generated from xanthine and xanthine oxidase in the presence of different quinones, and the effect of SOD. Reaction mixtures contained methaemoglobin (32 μM), quinone (50 μM), catalase (20 μM), xanthine (0.67 mM) and xanthine oxidase (0.008 U/ml) in pH 7.4 phosphate buffer. A_{577} was continuously recorded and from this the change in methaemoglobin concentration calculated. Abbreviations for the quinones are given in Table 1.

DISCUSSION

Superoxide dismutase inhibited the reaction of methaemoglobin with all the quinones investigated. The degree of inhibition was greatest with quinones such as menadione for which the equilibrium constant for reaction (1) is near unity, but this function of SOD applied for quinones with a wide range of redox potentials. In the presence of maximum SOD, anthraquinone-6-sulphonate was the only quinone that reacted to any extent with methaemoglobin, although for the least electro negative quinones, higher concentrations of SOD were required to prevent the reaction and this could limit its influence.

Methaemoglobin reduction by semiquinones is one example of a free radical reaction that can be inhibited by SOD. It would seem reasonable that other reactions of either semiquinone or other radicals, which may be more biologically significant, could be similarly affected. Analysis of the factors determining the reactivity of methaemoglobin with the different semiquinones gives some indication of the likelihood of SOD inhibiting other free radical reactions.

The main steps involved in the overall reaction between quinone, oxygen and methaemoglobin (shown in Figure 2) are likely to be the reactions of both the semiquinone and superoxide with methaemoglobin (reactions (2) and (3)), dismutation of both radicals (reactions (4) and (5)), and a reversible reaction between the semiquinone and O_2 (reaction (1) and (-1)). The first set of

Fig. 2. Main steps in the reaction between methaemoglobin, superoxide, and a quinone. Abbreviations: Q, quinone; $SQ^{\cdot -}$, semiquinone; Hb^{3+}, methaemoglobin; Hb^{2+}, reduced haemoglobin.

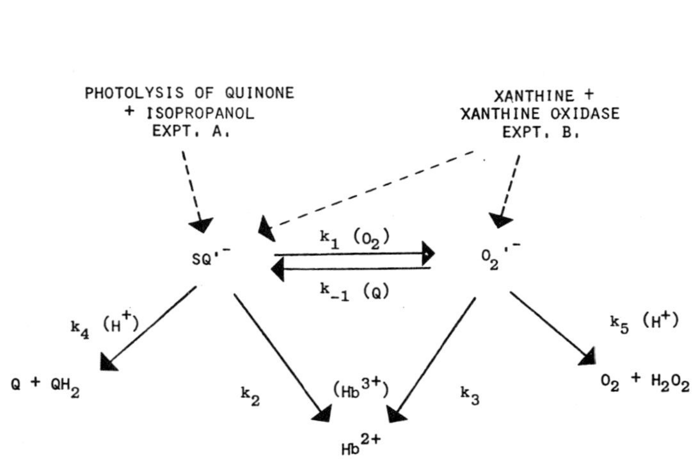

experiments described were initiated by producing the semiquinone. With xanthine and xanthine oxidase, it appears from studies with cytochrome c (C.C. Winterbourn, unpublished observations) that at least some of the quinones were acting as a substrate for the enzyme to produce the semiquinone directly. In the second set of experiments, therefore, production of semiquinone radicals as well as superoxide by the enzyme system was probably the initial event. Otherwise the reaction steps in the two experiments should be the same. Several points are apparent from the results (see Figure 2):

(a) Since SOD was effective only in the presence of O_2, there is no evidence that it reacted directly with the semiquinone radicals, but rather that it performed its usual function, to accelerate dismutation of superoxide (reaction (5)). P. Wardman (personal communication) also has been unable to show any catalysis of semiquinone radical dismutation by SOD.

(b) When superoxide is generated in the presence of methaemoglobin alone, only reactions (3) and (5) should be significant, and methaemoglobin reduction is slow because reaction (5) is favoured[9]. In the presence of quinone the other steps become involved. The increase in rate observed in each case implies that relative to spontaneous dismutation, reaction with methaemoglobin is much more favourable for the semiquinones than for superoxide.

(c) Reactions (2) and (4) should be the only significant reactions involving semiquinone amd methaemoglobin in the absence of O_2. In air, reaction (1) and hence reactions (3) and (5) could also occur. If this pathway were of major significance, and most of the semiquinone radicals were converted to superoxide, a much lower rate of methaemoglobin reduction in air would be expected. However, with three of the quinones the rate was scarely affected by air, and with the others it was decreased to no less than about a half. This suggests that in air as well as N_2 it was the semiquinone that reacted with methaemoglobin.

(d) In inhibiting methaemoglobin reduction by the photolysed quinones in air, SOD was not therefore inhibting a direct reaction of superoxide. The results can be explained, however, if the SOD, by increasing the rate of reaction (5), prevented the reverse of reaction (1), i.e. reaction (-1), if it normally predominates and allows only limited net production of superoxide. The semiquinone would then only be involved in reactions (1) and (2),and if reaction (1) is faster, methaemoglobin reduction should be inhibited.

The finding that SOD does inhibit methaemoglobin reduction in air is an indication that reaction (-1) must contribute to the overall reaction. Where complete or almost complete inhibition was attainable with maximum SOD, as observed with the top four quinones in Table 1, the contribution of this back reaction should be high. Lesser inhibition, as observed with anthraquinone-6-sulphonate, should indicate that this reaction is less significant. This type of sequence would be expected from the equilibrium constants listed in Table 1.

In addition, the rate of methaemoglobin reduction in air/SOD compared with that in N_2 should be a measure of the relative rates of reactions (1) and (2). As $k_1:k_2$ increases, so should the difference between the rates in N_2 and air/SOD.

The amount of SOD required for half maximum inhibition was also determined and should be an approximate measure of the relative rates of reaction (5) (SOD-catalysed) and reaction (-1), the faster the reaction of superoxide with the quinone, the more SOD being required to inhibit it. In both sets of experiments, the amount of SOD required was greatest for the quinones at the top of Table 3, and decreased down the list. The rate constants for the reactions of some of the quinones with superoxide have been measured[11,12] and do decrease in approximately this order.

Using this information, it can be postulated that the radical reactions that could be inhibited by SOD by the O_2-dependent mechanism described should meet the following criteria:

(a) For a reaction to occur it must be faster than the spontaneous dismutation of the radical. This condition is favoured by continuous low radical concentrations, which is likely to be the case in biological systems.

(b) The reaction must be much faster in comparison with radical dismutation than is the case for the corresponding reactions of superoxide.

(c) The reaction must be slower than the reaction of the radical with O_2.

(d) However, the reaction of the radical with O_2 must not be so fast compared with the reverse reaction that there is effectively no equilibrium. If this were the case, O_2 alone could inhibit other reactions of the radical and there should be no additional requirement for SOD. However, rather surprisingly, even with anthraquinone-2-sulphonate, which has an equilibrium constant of about 5000, O_2 was only partially effective at inhibiting methaemoglobin reduction.

(3) The rate constant for the reaction of superoxide with the parent compound of the radical must be considerably less than the rate constant for SOD-catalysed dismutation of superoxide. Otherwise excessively high SOD levels would be required for inhibition.

These conditions do not appear to be very restrictive, and an examination of relevant rate constants may well show that SOD could inhibit a wide range of reactions of precursor radicals of superoxide. It is noteworthy that for many reactions that SOD has been shown to affect, the findings are compatible with the reaction due to either superoxide or a precursor radical. For example, recent studies in this laboratory have indicated that in the presence of various quinones SOD-dependent cytochrome c reduction is due to the semiquinone radical. In many instances where superoxide production is monitored by cytochrome c reduction, a semiquinone or other free radical is the precursor of the superoxide, and in these systems it may also react directly with cytochrome c.

The apparent lack of reactions of superoxide that could cause significant biological damage provides something of a stumbling block in attempts to explain an essential role for SOD. This could be overcome if the range of reactions inhibited by SOD could be extended to include those of some organic free radicals. Many of these are far more reactive than superoxide, and can undergo biologically deleterious reactions.

ACKNOWLEDGEMENTS

This work was supported by a grant from the Medical Research Council of New Zealand. I would like to thank Dr. H.C. Sutton for helpful discussion.

REFERENCES

1. Fee, J.A. and Valentine, J.S. (1977) in Superoxide and Superoxide Dismutases, Michelson, A.M., McCord, J.M. and Fridovich, I., eds., Academic Press, New York, pp.19-60.
2. Halliwell, B. (1978) Cell Biol. Rep. 2, 113-128.
3. Babior, B.M. (1978) N. Engl. J. Med. 298, 657-666.
4. McCord, J.M. and Day, E.D. (1978) FEBS Lett. 86, 138-142.
5. Winterbourn, C.C., French, J.K. and Claridge, R.F.C. (1979) Biochem. J. 179, 665-673.
6. Winterbourn, C.C., French, J.K. and Claridge, R.F.C. (1978) FEBS Lett. 94, 269-272.
7. Goodman, J. and Hochstein, P. (1977) Biochem. Biophys. Res. Commun. 77, 797-803.
8. Bus, J.S., Aust, S.D. and Gibson, J.E. (1974) Biochem. Biophys. Res. Commun. 58, 749-755.
9. Sutton, H.C., Roberts, P.B. and Winterbourn, C.C. (1976) Biochem. J. 155, 503-510.
10. Ilan, Y.A., Czapski, G. and Meisel, D. (1976) Biochim. Biophys. Acta, 430, 209-224.
11. Rao, P.S. and Hayon, E. (1975) J. Phys. Chem. 79, 397-402.
12. Patel, K.B. and Willson, R.L. (1973) J. Chem. Soc., Faraday Trans. I, 69, 814-825.

SUPEROXIDE DISMUTASE ACTIVITY IN THE FUNCTIONING CHLOROPLAST

CHRISTINE H. FOYER AND DAVID HALL
Department of Plant Sciences, University of London King's College,
London SE24 9JF, England

INTRODUCTION

The utilisation of water as the primary donor for photosynethetic electron transport results in the formation of molecular oxygen within the chloroplast. As a consequence of this a high concentration of oxygen is present within the chloroplast in the light,[1] and this provides a potentially toxic environment which, in the absence of protective countermeasures, would prove most hazardous. The mechanisms of oxygen toxicity are many[2] ranging from the interaction of the triplet ground state with essential -SH groups of enzymes[3] to the production of free radicals such as the hydroxyl free radical $^{\bullet}OH$, lipid peroxides and excited state singlet oxygen ($^1\Delta_g$) which, becuase of their reactivity, can cause extensive damage to the functioning chloroplasts.[2,3] These highly active oxygen derivatives are produced by such processes as photodynamic action via excited chlorophyll, and by chain reactions involving less reactive oxygen derivatives such as the superoxide free radical anion, O_2^-, and hydrogen peroxide.

Mechanisms of superoxide production within the chloroplasts. The univalent reduction of oxygen to produce superoxide appears to be a common, if undesirable, process in chloroplast metabolism. Figure 1 summarises the mechanisms of O_2^- production in the photosynthesising chloroplast. Electron flow to oxygen via

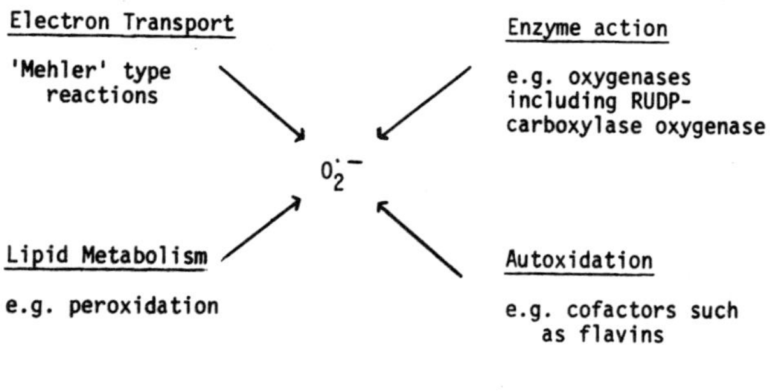

Fig. 1. Systems generating superoxide in the active chloroplast.

the autoxidation of certain of the redox carriers of the PSI electron transport chain occurs at a low but constant rate.[4,5] It has been suggested that this process serves to prevent over-reduction of the electron transport system by 'poising' the electron carriers of the cyclic pathway while still remaining a coupled system leading to ATP synthesis.[6,7] Illuminated chloroplasts donate electrons to oxygen in single electron steps to produce superoxide and this initiates a Mehler-type reaction.[8] The production of superoxide from photosystem I (PSI) has been measured in isolated thylakoids by various means such as EPR[9] and the oxidation of compounds such as sulphite[10] and adrenaline.[11] The autoxidisable electron carrier responsible for the electron donation to oxygen is suggested to be P-430 (Ref.12) which is analogous to the iron sulphur centres 'A' and 'B' (Ref.13).

It has been shown that several mono and dioxygenases produce superoxide as a product of enzyme activity. Direct evidence has now been presented for the production of superoxide during the oxygenase reaction of ribulose-1,5-bisphosphate carboxylase.[14] This reaction in conjunction with electron transport to oxygen and the oxidation of autoxidisable compounds such as flavins, results in the liberation of superoxide in the chloroplast stroma. The production of superoxide as an intermediate in the water-splitting reaction of photosystem II (PSII) has been suggested.[15] It is evident from kinetic experiments that the oxygen evolving system, Y, accumulates four oxidising equivalents, one with each turnover of the PSII reaction centre.[16,17] Four sequential single electron transfers are required to promote the liberation of oxygen and some changes in the redox state of bound manganese, during this charge accumulation, have been observed.[18,19] Thus it has been suggested that the water-splitting system, Y, is a manganoprotein to which the intermediates of photosynthetic water cleavage are bound.[15] The univalent four-step oxidation process would result in the production of intermediate oxidation states which are termed 'cryptosuperoxide', 'cryptohydroperoxide' and 'cryptohydroxyl' complexes. However, there is no evidence for the production of superoxide on the inner side of the thylakoid membrane and it is likely that if superoxide is produced it is stabilised as an inner sphere complex with the manganese and therefore would be tightly bound.

In order to prevent damage to the molecular structure of the chloroplast (e.g. lipid peroxidation mediated by singlet oxygen formed as a result of unchecked chain reactions involving superoxide) the superoxide produced must be removed immediately and this is efficiently accomplished by the activity of superoxide dismutase (SOD). We give here an experimental appraisal of SOD in the chloroplast.

RESULTS AND DISCUSSION

Localization of Cu-Zn superoxide dismutase in the chloroplast stroma. It has been known for some time that superoxide dismutase, which catalyses the rapid, safe dismutation of superoxide, is associated with chloroplast fractions.[20,21] In fact (type A) chloroplasts prepared from the leaves of higher plants are found to contain both a cyanide sensitive (copper-zinc) SOD and a cyanide insensitive (Mn) superoxide dismutase[22]. In the past this has led to conflicting reports concerning the compartmentalisation of these enzymes in the green leaf.[20,23,24] We therefore, carried out subcellular fractionation experiments[22] to determine the amount and nature of the SOD activity present in the chloroplasts of green leaves. Intact chloroplasts were prepared from spinach and lettuce by the method of Walker[25] and assayed for SOD activity as described by Halliwell[26]; chloroplasts were assayed both in the intact state and after bursting by osmotic shock. Washing these chloroplast fractions in ion-free sorbitol medium[27] removes contamination by broken chloroplasts and other organelles and increases the yield of intact chloroplasts. Figure 2 shows the result of a typical experiment when intact chloroplasts were prepared and washed and assayed for SOD activity after each wash. $NADP^+$ glyceraldehyde-3-phosphate dehydrogenase (G3PD) was used as a chloroplast stromal enzyme marker. From Figure 2 it can be seen that unless the chloroplasts are disrupted by osmotic

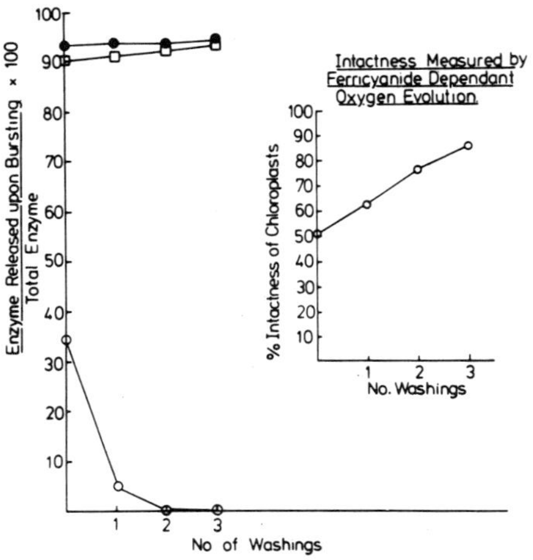

Fig. 2. The distribution of cyanide-senstive and cyanide-insensitive SOD activities in spinach chloroplasts. Type A chloroplasts were prepared by the method of Walker (Ref.25) and washed three times in cation-free medium (Ref.27). Chloroplasts were assayed after each wash for cyanide-senstive (●—●) and cyanide-insensitive SOD (○—○) by including 2.0 mM KCN, pH 8.0 in the reaction medium, and for NADP-glyceraldehyde-3-phosphate dehydrogenase (□—□). The chloroplasts were assayed in the intact state by including 0.33 M sucrose in the reaction medium, and after disruption by osmotic shock.

shock only low cyanide-sensitive SOD and G3PD activities can be measured. Over 90% of the total amount of these enzymes present is released from the chloroplast upon bursting. In the initial unwashed chloroplast fractions there is also some release of the cyanide-insensitive enzyme when the chloroplasts are subjected to osmotic shock. In the washed chloroplast fractions, however, no release of cyanide-insensitive SOD is observed upon bursting. Since the cyanide-sensitive SOD shows a similar latency pattern to the G3PD, which is known to be a chloroplast stromal enzyme,[28] these results show that cyanide-sensitive SOD is located within the chloroplast. In contrast, the pattern of latency of the cyanide-insensitive SOD is totally different to that of the G3PD and in washed chloroplast fractions there is no release of this enzyme when the chloroplasts are ruptured. The cyanide-insensitive SOD is therefore located outside the chloroplast. From these results it is not possible to determine whether the enzyme is non-specifically adsorbed onto the organelle during isolation or located on the chloroplast envelope. It is however possible to calculate from these experiments that the majority of the total leaf Cu-Zn SOD is located within the chloroplasts. The superoxide dismutase activity of the chloroplast fractions was also observed on polyacrylamide gels stained for SOD activity by the photochemical method.[29] Figure 3 shows that only one cyanide-sensitive SOD band is observed when disrupted, washed spinach chloroplast fractions are stained for activity after disc gel electrophoresis.

When intact chloroplast fractions were subjected to osmotic shock and the immediately returned to isotonic medium 25-35% of the original SOD activity remained with the thylakoid fraction. The possibility that some of the chloroplast SOD is membrane bound is now considered.

<u>SOD activity associated with PSI</u>. The finding that some of the Cu-Zn SOD remains associated with the thylakoid membrane has led to several proposals that there is a membrane-bound SOD (Refs.24,30). Values for the amount of SOD which remains in the thylakoid fraction vary from 30% (Ref.14) to 80% (Ref. 24) of the amount present in the original chloroplast fractions. The discrepancy in these values probably arises from the different methods of chloroplast preparation used in each case, and as there is no value given for the percent intactness of the original chloroplast fraction it is difficult to compare these results. We have found that some of the Cu-Zn SOD does remain with the thylakoids when they are pelleted after bursting the chloroplasts. The activity remaining is approximately 50 units/mg chlorophyll for spinach chloroplasts and is in agreement with the results of Asada et al.[20] We have, however, found that this SOD can be removed by washing the thylakoids two to three times with hypotonic medium (e.g. 20 mM potassium phosphate buffer, pH 7.8) but is only gradually removed by isotonic washing (e.g. 0.33 M sorbitol, 50 mM HEPES, pH 7.5). There is some evidence that the SOD associated with the intact thylakoids is concentrated

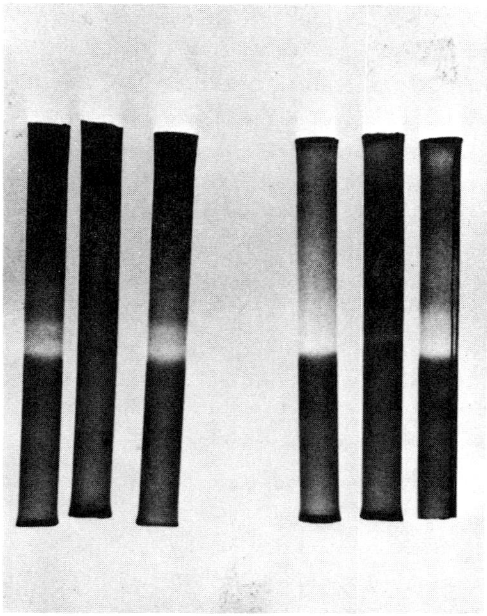

Fig. 3. Polyacrylamide gels (7.5%) showing stromal and PSI enriched fractions prepared from type A spinach chloroplasts and stained for SOD activity (Ref.29) in the presence or absence of 2 mM KCN, pH 8.0. The three gels on the right contain samples of chloroplast stroma, the three gels on the left contain PSI enriched samples. The middle gel in each case was treated with 2 mM KCN, pH 8.0 during the staining procedure. Approximately 20 units of SOD activity was added to each gel.

in the region of PSI (Ref.23) and that preparations of PSI enriched vesicles have a fourfold enrichment in SOD activity. This data is very intersting as a functional requirement for SOD activity at the level of PSI might well be expected. However, the French press/differential centrifugation method used to prepare these complexes produces fractions still heavily contaminated with other membrane polypeptides. We have been unable to repeat this experiment. The enzyme found with the thylakoid fraction is cyanide-sensitive and similar in properties to the stromal enzyme. Figure 3 shows the pattern of SOD activity produced when membrane fractions enriched in PSI are stained for SOD activity after disc gel electrophoresis. A previous report[24] has shown that SOD activity is present in PSI enriched fractions produced by digitonin treatment. Intact chloroplasts were prepared and washed once in ion-free medium.[20] The final chloroplast pellet was dispersed in water to burst the chloroplasts and release the stroma. The thylakoid membranes were then pelleted and incubated with 2 ml 10% (w/v) digitonin for 20 minutes. The mixture was then centrifuged at 12,000 g for 10 minutes to pellet the large grana stacks. The supernatant is enriched in stroma lamellae and PSI. Figure 3 shows the results of such an experiment.

In both the stroma and PSI enriched fractions there is a discrete band of cyanide-insensitive SOD activity. This is probably due to the presence of mitchondria which are a usual contaminant of intact chloroplast preparations. On both the stromal gels and the PSI gels there is only one band of cyanide-senstive SOD activity. The PSI enriched fraction show bands of chlorophyll-protein at the top of the gel, and most of the protein applied entered the gel. The band of cyanide-sensitive SOD is in an equivalent position to that on the stromal gel. There is no apparent SOD activity on the top of the gels with the chlorophyll-protein complexes and so it appears that there is no SOD activity bound to these complexes. Therefore the SOD activity is also removed from the membranes by mild detergent treatment. It is possible that there is some localization of Cu-Zn SOD at the level of PSI where it would be most useful. It may be that the enzyme is able to associate with the thylakoid in this region but it appears doubtful that it is truly membrane bound.

SOD activity associated with PSII. Manganese is an essential component for photosynthesis, the process of oxygen evolution being, apparently, totally dependent on the presence of bound manganese in the chloroplast thylakoids.[31] Estimations of the content of this metal in the thylakoid membranes of higher plants give a value of 1 atom of manganese per 50 to 100 chlorophyll molecules.[32,33] Since there are Mn-SODs known from several organisms[34,35] and Mn^{++} is well known to react with superoxide[36] the possibility that thylakoid manganese may play some role in protecting the molecular structure of the membrane from the deleterious effects of superoxide has been considered.[34,37] Recently two Mn-SODs have been found to be associated with high molecular weight protein complexes from two photosynthetic organisms, *Dunaliella tertiolecta* and *Phaeodactylum tricornutum* (Dr. K. Wegmann; Dr. E. Holdsworth - personal communications). These complexes have features which suggest that they originated from PSII.[38,39] We have, however, found no evidence for the presence of an enzymic Mn-SOD in the thylakoid membrane of higher plants.

Free Mn^{++} has been found to be an inhibitor of lipid peroxidation in membranes[4] and will inhibit superoxide-mediated reactions in a manner similar to SOD; therefore some of the thylakoid manganese may afford protection against superoxide. This possibility has been investigated in the past[37,41] and a procedure was reported for the preparation of sub-thylakoid particles which would inhibit the superoxide-mediated reduction of nitro blue tetrazolium. These particles displayed properties which suggested that they consisted of the light-harvesting chlorophyll a/b protein complex[42] plus some of the components of the PSII electron transport chain. The SOD activity of these particles was insensitive to cyanide and was considered to be a consequence of the presence of manganese. This work received some criticism[24] and no localisation of the activity was attempted. We therefore prepared the sub-thylakoid particles as described by by Lumsden and Hall[37] and found that the procedure gave a very low yield and a large number of polypeptide bands were visible when the gels were stained for protein after SDS gel electrophoresis. We modified the procedure to concentrate the active component and remove as much contamination as possible. The procedure we developed is shown in Figure 4 and is slightly modified from that which we originally reported.[43] The resultant preparation was found to possess a cyanide-insensitive SOD activity (Figure 5) which was totally inactivated by

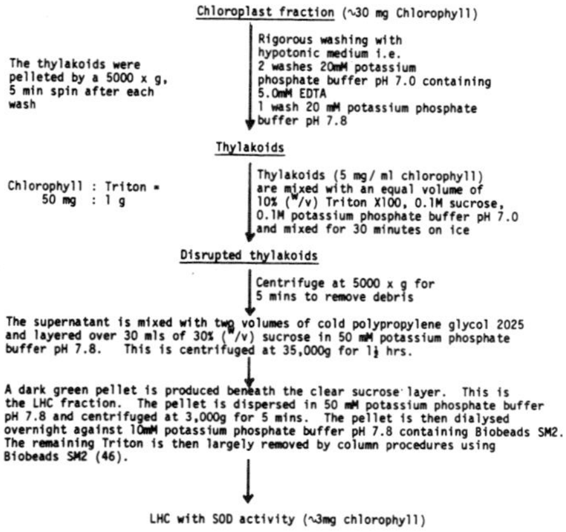

Fig. 4. The preparation procedure of sub-thylakoid particles which possess SOD activity

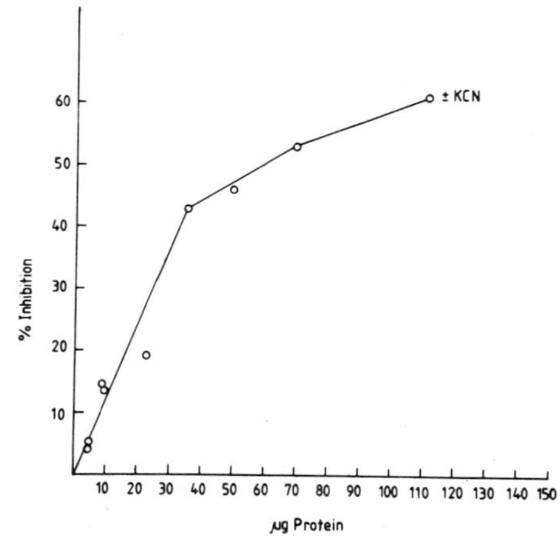

Fig. 5. The SOD activity of the LHC fraction prepared as described in Fig.4. SOD activity was assayed as in Ref.26 except that EDTA was omitted.

by washing with 0.8 M Tris-HCl, pH 8.0 or in the presence of 3.0 mM EDTA and partically inactivated by heating at 50°C for 5 minutes. The preparations were found to consistently contain bound manganese at concentrations of between 10-70 Mn/mg protein. The absorption spectrum of the final preparation is shown in Figure 6 and shows that there is a high concentration of chlorophyll b and the chlorophyll a/b ratio was approximately 1.2. The elucidation of the polypeptide

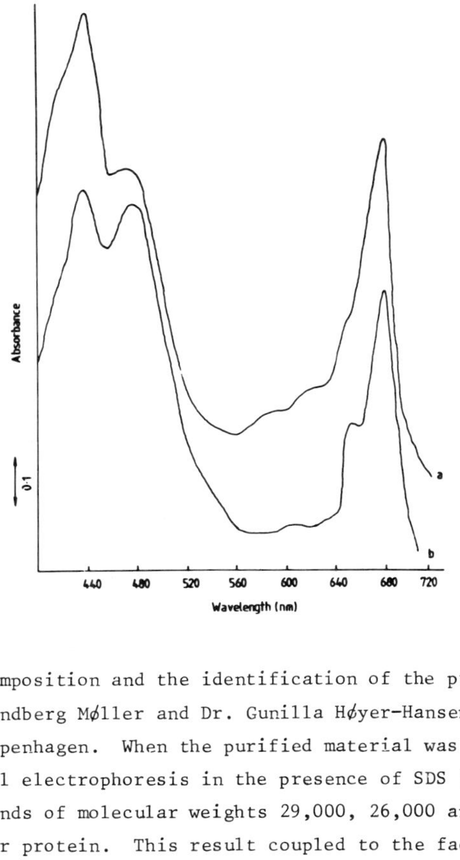

Fig. 6. The room temperature absorption spectrum of washed spinach thylakoids (a) and the freshly prepared LHC fraction (b). The samples were dispersed in 0.25 M sucrose, 50 mM potassium phosphate buffer, pH 7.5.

composition and the identification of the protein was carried out by Dr. Birger Lindberg Møller and Dr. Gunilla Høyer-Hansen at the Carlsberg laboratory, Copenhagen. When the purified material was subjected to polyacrylamide slab gel electrophoresis in the presence of SDS (Ref. 44) three main polypeptide bands of molecular weights 29,000, 26,000 and 23,000 were evident after staining for protein. This result coupled to the facts that the purified samples had a low chlorophyll a/b ratio and were shown both by us and by Dr.B. Lindberg Møller to be without photochemical activity and not to contain cytochromes, led us to conclude that the preparations consisted largely of the light-harvesting chlorophyll a/b protein complex (LHC). Dr. B. Lindberg Møller also prepared pure LHC from barley by a method[45] slightly modified from that previously described.[46] The purified LHC was dialysed overnight and largely depleted of Triton X-100 using Biobeads SM2 (Ref.47) to avoid interference by the detergent in the SOD assay (Refs.30,43). The LHC purified by this method was also found to contain manganese and to have superoxide dismutase activity similar to that present in the samples purified by the method described in Figure 4 (Ref.43). The activity is not due to free Mn^{++} as the characteristic Mn^{++} EPR spectrum is absent from the freshly prepared samples and appears only after incubation of the LHC with concentrated HCl. The LHC was always found to contain bound manganese even when prepared from thylakoids which had been subjected to rigorous EDTA washing before detergent treatment.[46] It appears that it is this manganese which is responsible for the SOD activity since treatments which remove the manganese, e.g. washing with 0.8 M Tris-HCl, pH 8.0, also destroy the SOD activity. Leaves grown under intermittent light (1 msec flash every 15 minutes) were found to produce no LHC in their thylakoids, as detectable by polyacrylamide slab gel electrophoresis. When chloroplasts prepared from these leaves are subjected to the LHC preparation procedure, only a small white pellet is produced which contains little protein ($<$ 1 mg from 50 mg thylakoid protein) and no manganese. This pellet has no SOD activity.

SUMMARY

It is apparent from the data presented here that the chloroplast is equipped with large quantities of SOD activity (approximately 200 units/mg chlorophyll). This, in itself, is an indication that the chloroplast is constantly faced with the problem of superoxide production. Superoxide dismutase and secondary protective agents such as ascorbic acid[48] and glutathione[3] provide an adequate defence against the build up of superoxide in the chloroplast under normal conditions. The SOD of the chloroplast stroma is of the Cu-Zn type and this can associate with the thylakoid membrane. The Mn-SOD is located outside the chloroplast. However, the internal structure of the thylakoid may be protected against superoxide by the presence of manganese bound to the light-harvesting chlorophyll a/b protein complex.

ACKNOWLEDGEMENT

This work is supported by the U.K. Science Research Council.

REFERENCES

1. Steiger, H.M. and Beck, E. (1977) Plant. Physiol. 60, 903-906.
2. Halliwell, B. (1978) Cell. Biol. Int. Rep. 2, 113-128.
3. Haugaard, N. (1968) Physiol. Rev. 48, 311-345.
4. Egneus, H., Heber, U., Mathiessen, U. and Kirk, M.R. (1975) Biochim. Biophys. Acta. 408, 252-268.
5. Forti, G. and Jagendorf, A.F. (1961) Biochim. Biophys. Acta, 54, 322.330
6. Heber, U., Egneus, H., Hanck, U., Jensen, M. and Koster, J.F. (1978) Planta, 143, 41-49.
7. Grant, B.R. and Whatley, F.R. (1964) Biochemistry of Chloroplasts, 2, 505-521.
8. Mehler, A.H. (1951) Arch. Biochem. Biophys. 33, 65-77.

9. Harbour, J.R. and Bolton, J.R. (1975) Biochem. Biophys. Res. Commun. 64, 803-807.
10. Asada, K. and Kiso, K. (1973) Eur. J. Biochem. 33, 253-257.
11. Asada, K. and Kiso, K. (1973) Agr. Biol. Chem. Tokyo, 37, 453-454.
12. Asada, K. and Nakano, Y. (1978) Photochem. Photobiol. 28, 917-920.
13. Ke, B. and Beinert, H. (1973) Biochim. Biophys. Acta, 305, 689-693.
14. Bhagwat, A.S. and Sane, P.V. (1978) Biochem. Biophys. Res. Commun. 84, 865-873.
15. Renger, G. (1978) in Photosynthetic Oxygen Evolution, Metzner, H., eds., Academic Press, New York, pp.229-248.
16. Kok, B., Forbush, B. and McGlobin, M. (1979) Photochem. Photobiol. 30, 309-329.
17. Slabas, A.R. and Evans, M.C.W. (1977) Nature, 270, 169-171.
18. Wydrzynski, T., Zumbulyadis, N., Schmidt, P.G., Glutowsky, H.S. and Govindjee. (1976) Proc. Nat. Acad. Sci. USA, 73, 1196-1198
19. Wydrzynski, T., Zumbulyadis, N., Schmidt, P.G. and Govindjee. (1975) Biochim. Biophys. Acta 408, 349-354.
20. Asada, K., Urano, M. and Takahashi, M. (1973) Eur. J. Biochem. 36, 257-266.
21. Allen, J.F. and Hall, D.O. (1973) Biochem. Biophys. Res. Commun. 52, 856-862.
22. Jackson, C., Dench, J., Moore, A.L., Halliwell, B., Foyer, C.H. and Hall, D.O. (1978) Eur. J. Biochem. 91, 339-344.
23. Halliwell, B. (1975) Eur. J. Biochem. 55, 355-360
24. Elstner, E.F. and Heupel, A. (1975) Planta, 123, 145-154.
25. Walker, D.A. (1971) Methods Enzymol. 23A, 211-220.
26. Halliwell, B. (1975) FEBS Lett. 56, 34-38.
27. Nakatani, H.-Y. and Barber, J. (1977) Biochim. Biophys. Acta, 461, 510-512.
28. Halliwell, P. (1978) Progr. Biophys. Mol. Biol. 33, 1-54.
29. Weisiger, R.A. and Fridovich, I. (1973) J. Biol. Chem. 248, 3582-3592.
30. Van Ginkel, G. and Brown, J.S. (1978) FEBS Lett. 94, 284-286.
31. Cheniae, G.M. (1970) Annu. Rev. Plant Physiol. 21, 467-498.
32. Anderson, J.M., Boardman, N.K. and David, D.J. (1964) Biochem. Biophys. Res. Commun. 17, 685-689.
33. Henriques, F. and Park, R.B. (1976) Biochim. Biophys. Acta, 430, 312-320.
34. Asada, K., Kanematsu, S., Takahashi, M. and Kono, Y. (1976) in Iron and Copper Proteins, Yasunobu, K.T., Mower, H.F. and Hayaishi, O., eds., Plenum Press, New York, pp.551-564.
35. Hall, D.O. (1977) Adv. Chem. Ser. 162, 227-250.
36. Bielski, B.H.J. and Chan, P.C. (1978) J. Amer. Chem. Soc. 100, 1920-1921.
37. Lumsden, J. and Hall, D.O. (1975) Biochim. Biophys. Res. Commun. 64, 595-602.
38. Von Kameke, E. and Wegmann, K. (1978) in Photosynthetic Oxygen Evolution, Metzner, H., ed., Academic Press, New York, pp.371-380.
39. Holdworth, E.S. and Arshad, J.H. (1977) Arch. Biochem. Biophys. 183, 361-373.
40. Panyim, S. and Chalkley, R. (1972) J. Biol. Chem. 246, 7557-7564.
41. Lumsden, J. and Hall, D.O. (1974) Biochem. Biophys. Res. Commun. 58, 35-41.
42. Thornber, J.P. (1975) Annu. Rev. Plant Physiol. 26, 127-158.
43. Foyer, C.H. and Hall, D.O. (1979) FEBS Lett. 101, 324-328.
44. Høyer-Hansen, G. and Simpson, D.J. (1977) Carlsberg Res. Commun. 42, 379-389.
45. Simpson, D.J., Møller, B.L. and Høyer-Hansen, G. (1978) in Chloroplast Development, Akoyūnoglou, G. et al., eds., Elsevier North Holland Biomedial Press, Amsterdam, pp.507-512.
46. Burke, J.J., Ditto, C.L. and Arntzen, C.J. (1978) Arch. Biochem. Biophys. 187, 252-263.
47. Holloway, P.W. (1973) Anal. Biochem. 53, 304-308.
48. Foyer, C.H. and Halliwell, B. (1976) Planta, 133, 21-25.

MECHANISMS AND PRODUCTS OF PHOTOSYNTHETIC OXYGEN ACTIVATION

ERICH F. ELSTNER
Institut für Botanik und Mikrobiologie, Technische Universität,
München, 8000 München 2, W.Germany.

Oxygen is not only a photosynthetic product in algae and higher plants, but also an electron acceptor of photosynthetic electron transport (PET) in vivo[76,79] and in isolated, intact chloroplasts.[22,61] As recently reviewed in detail,[3] isolated chloroplast lamellae can photoreduce oxygen monovalently yielding O_2^- which in turn undergoes secondary reactions as SOD catalyzed dismutation[4,16,27,58] or reaction with other compounds.[35] In the late sixties, shortly before McCord and Fridovich's first report on superoxide dismutase (SOD) appeared,[70] puzzling effects in isolated chloroplasts concerning the oxygen metabolism in the presence of ascorbate were explained by assuming O_2^- as a photosynthetic product, stimulating effects of cyanide were interpreted by the function of a cyanoperoxidase.[24,72] In the following years, with the aid of SOD, catalase, relatively unspecific non-proteinaceous inhibitors and several more or less sophisticated detector reactions, different mechanisms, products and physiological implications of photosynthetic oxygen reduction have been described and assessed (see Ref.41). We recently presented models for reactions observed in vivo, especially concerning herbicidal activities. Since the reviews by Allen,[3] and Halliwell,[51] no dramatic changes in the view of the physiology of photosynthetic oxygen reduction during steady state PET can be reported. This contribution therefore will mainly deal with model reactions of photosynthetic oxygen activation and oxygen toxicity emerging from altered oxygen metabolism under changed physiological conditions or under the influence of certain important herbicides.

(1) Photosynthetic electron transport (PET) and function of herbicides interfering with PET. Photosynthetic CO_2 fixation is a dark reaction, driven by ATP and NADPH as products of PET. Two photoacts drive PET, reducing NADP and synthesizing ATP via a pH gradient through the membraneous coupling factor.[87]

Four types of artificial effectors influence photosynthetic product (ATP, NADPH) formation namely, (a) inhibitors of PET, (b) acceptors of PET, (c) uncouplers of PET, and (d) inhibitors of photosynthetic phosphorylation. Types (a) and (b) are relevant for the present discussion. Type (a) blocks PET and thus energy dissipation in the light traps inducing photobleaching and unsaturated fatty acid oxidation and is represented by the well known herbicides 3-(3,4-dichlorophenyl)-1,1-dimethylurea (DCMU), atrazine, simazine and many others, inhibiting PET at the reducing side of photosystem II (PS II). Type (b), represented by several low potential dyes of the bipyridylium family (e.g. paraquat, PQ = methylviologen), substitute for ferredoxin and NADP at the reducing side of photosystem I (PS I) (see Scheme 1 below) These compounds dissipate the photosynthetic reducing power, producing O_2^- (see Ref.19).

Scheme 1: Photosynthetic electron transport

H_2O --> PS II --> PA II --> plastoquinone --> plastocyanin, cytochrome f --> PS I --> PA I --> ferredoxin --> NADP-ferredoxin-reductase --> NADP (PA II, PA I = acceptors of photosystems II and I)

(2) Photosynthetic oxygen reduction

(I) Physiological aspects. Photosynthetic oxygen reduction is a physiological reaction and has been demonstrated to occur in algae[50,76,79] as well as in intact chloroplasts of higher plants;[22,61] H_2O_2 has been determined as the product. Photosynthetic oxygen reduction may have a function in the energy balance of chloroplasts ("flexible coupling", Ref.57), where oxygen reduction is switched on as soon as the NADPH/NADP ratio reaches a value of about 0.8 (Refs.29,30). The products of the oxygen reduction (in isolated chloroplast lamellae, O_2^-, H_2O_2 and a Fenton-type oxidant can be determined[3,38] may also have physiological functions for certain biosyntheses in chloroplasts, for example, of hydroxylated aromates or unsaturated fatty acids (see Ref.41). Most of the produced O_2^- is certainly detoxified inside the chloroplasts by the thylakoid-bound SOD (Cu-Zn type[4,16,29,58]) by reduced ferredoxin,[2] ascorbate[1,24,25,44] and yet other compounds containing SH- or o-dihydroxy groups (see Refs. 3,51). Since the chloroplasts apparently lack catalase, excess H_2O_2 is deomposed by ascorbate peroxidases yielding dehydroascorbate, which in turn is reduced by NADPH via glutathione-glutathione reductase (see Ref.51).

Two mechanisms for photosynthetic oxygen reduction have been proposed:

(a) a two step mechanism, involving reduced ferredoxin, fd_{red}, (Ref.2):

$$fd_{red} + O_2 \longrightarrow fd_{ox} + O_2^-$$

$$fd_{red} + O_2^- + 2\,H^+ \longrightarrow H_2O_2 + fd_{ox}$$

(b) a three step mechanism, involving an oxygen reducing factor (ORF, Ref.26) which is released from chloroplast lamellae upon heat treatment:

$$ORF\,H_2 + O_2^- + H^+ \longrightarrow ORFH^{\cdot} + H_2O$$

$$ORFH^{\cdot} + O_2 \longrightarrow ORF + O_2^- + H^+$$

$$ORF + 2\,fd_{red} + 2\,H^+ \longrightarrow ORFH_2 + 2\,fd_{ox}$$

In analogy to reactions of certain o-diphenols (e.g. 3-hydroxytyramine[26,27,33]) ORF has been expected to be represented by (a) compound(s) structurally similar to caffeic acid. We recently identified an (the?) oxygen reducing compound from the supernatant of heat-treated chloroplast as a flavin derivative (riboflavin; Ref.45). Since isolated NADP-ferredoxin reductase(a flavin enzyme) has no ORF-activity but heat-treated NADP-ferredoxin reductase is active, our proposed[26] oxygen reducing factor (ORF), which already has mislead several groups ("primary acceptor" of PS I, see Refs.84,87), might well be an artifact. On the other hand, there might exist flavins bound to other components of PS I as reported for photosynthetic bacteria.[10] We have, however, no indication for the function of a flavin other than NADP-ferredoxin reductase in PET; thus Allen's scheme[3] at the moment is likely to represent the physiological reactions operating during photosynthetic oxygen reduction.

(ii) Artificial oxygen reducing sites in isolated chloroplast lamellae. Isolated chloroplast lamellae lack the enzymes of CO_2 fixation but have an intact PET coupled to phosphorylation.[87] As recently outline.[36,37,40] there is evidence for at least four different mechanisms of light-dependent oxygen activation in isolated chloroplast lamellae:

(A) Photosystem II reactions. In the presence of μM 2,5-dibromo-3-ethyl-6-isopropyl-p-benzoquinone (DBMIB) which inhibits PET between plastoquinone and plastocyanin (see Scheme 1), 2,3-dimethyl-5,6-methylenedioxy-p-benzoquinone (DIMEB) can accept electrons from plastoquinone ($E_o' = 0$ V) and transfer to oxygen, where H_2O_2 but not O_2^- is produced[36,37] as determined with NADH peroxidase or with the hydroxylamine method.[31] DCMU, which blocks PET between PA II and plastoquinone (see Scheme 1) inhibits this two-electron transfer; SOD has no influence on H_2O_2 formation. DBMIB itself in concentrations above 2×10^{-5}M can also act as two-electron donor for oxygen, yielding H_2O_2 and no O_2^-.

(B) Photosystem I reactions. The reducing side of PS I (PA in Scheme 1) provides a strong reductant with a redox potential of lower than -0.5 V. Therefore, in contrast to PS II with an electron acceptor with an E_o' of about -0.05 V at its reducing side (PA II in Scheme I), PS I can reduce oxygen monovalently (E_o' for O_2/O_2^- = -0.33 V; Ref.59). Low potential compounds like anthraquinone sulfonic acid (E_o' = -0.2 V), PQ (E_o' = -0.44 V), triquat (E_o' = -0.55 V) and many others produce O_2^- and thus oxidize hydroxylamine, yielding nitrite.[24,29,32]

Oxygen uptake by illuminated spinach chloroplasts is also stimulated by o-diphenols as 3-hydroxytyramine or caffeic acid. Depending on the concentrations, 3-hydroxytyramine catalyzes both O_2^- and H_2O_2 production, while caffeic acid (and other o-diphenols) only yield H_2O_2 at all tested concentrations. The mechanisms of these oxygen activating reactions have been written as follows:

(c) o-diphenol + O_2^- --> semiquinone + H_2O_2

semiquinone + O_2 --> o-quinone + O_2^-

o-quinone + 2 e + 2 H^+ --> o-diphenol

starter reaction: O_2 + e --> O_2^- (e = PA I_{red})

(d) in the case of 3-hydroxytyramine which in contrast to caffeic acid is a good substrate for chloroplast lamellar polyphenoloxidase, the product at low 3-hydroxytyramine concentrations is O_2^-, at high concentrations only H_2O_2:

$$\text{o-diphenol} + 1/2\ O_2 \xrightarrow{\text{phenoloxidase}} \text{o-quinone} + H_2O$$

$$\text{o-quinone} + e + H^+ \longrightarrow \text{semiquinone}$$

$$\text{semiquinone} + O_2 \longrightarrow \text{o-quinone} + O_2^-$$

At high 3-hydroxytyramine concentrations, not enzyme bound o-diphenol reacts with O_2^- producing H_2O_2. This reaction may play a role in wounded leaves of certain plants, for example, sugar beet.[35] Oxygen uptake and H_2O_2 production catalyzed by o-diphenols is inhibited by SOD and by DBMIB, characterizing this H_2O_2 production as PS I reaction transferring two times one electron to oxygen.[37,38]

Some years ago we though to have good reason to assume that the free OH radical (OH$^•$) is a good product of photosynthetic oxygen reduction since ethylene formation from methional was strongly inhibited by catalase and partially inhibited by SOD [28]– results which have been suggested to be indicative for the "Haber-Weiss" reaction.[14] Recent results[38] on ethylene formation from methionine in the presence of pyridoxalphosphate[64] and on photobleaching of p-nitrosodimethylaniline,[39] however, are in contrast to the above suggestion. It is more likely that a "Fenton"-type "reduced H_2O_2" is the oxidant for methionine (CH_3-S-R) yielding ethylene:

(e) 4 fd + 3 e ---> 4 fd_{red}

$$2\ fd_{red} + O_2 + 2\ H^+ \longrightarrow H_2O_2 + 2\ fd_{ox}$$

$$fd_{red} + H_2O_2 \longrightarrow (fd_{red}\text{-}H_2O_2)$$

$$fd_{red} - H_2O_2 \longrightarrow \begin{pmatrix} fd_{ox} = \begin{matrix} OH^\bullet \\ OH^- \end{matrix} \end{pmatrix}$$

$$\left(2\ fd_{ox} = \begin{matrix} OH^\bullet \\ OH^- \end{matrix}\right) + CH_3\text{-S-R} \rightarrow (fd_{ox} - 2\ OH^-) + CH_3\text{-}S^+\text{-R}$$

$$(fd_{ox} - 2\ OH^-) \longrightarrow fd_{ox} + 2\ OH^-$$

$$CH_3\text{-}S^+\text{-R} \longrightarrow C_2H_4 + \text{other products}$$

This reaction is, however, not identical with the physiological production of the plant hormone ethylene. The biosynthetic way of ethylene production seems to involve S-adenosyl methionine (SAM) and 1-amino cylcopropane-1-carbonic acid (enzymatically derived from SAM) as direct precursor of ethylene.[5]

(C) Involvement of pteridines in oxygen activation. Fuller et al.[48] and very recently Iwai et al.[60] reported on the involvement of pteridines in photosynthetic reactions. Other groups have also looked for unconjugated pteridines in higher plants, but without success (Ref.80, H. Ziegler, personal cummunication, Elstner, unpublished results). The search for unconjugated pteridines was successful, however, in the case of Euglena gracilis, where the pteridine-like compounds described[32] could be identified as a new type of N-2-substituted pteridines. These compounds (in agreement with reports by Fridovich,[47] and Nishikimi[73] produce O_2^- upon autoxidation after a ferredoxin-dependent reduction by PS I of isolated chloroplast lamellae.[35] Böhme et al.[11] studied these "Euglenapterines" which apparently are 2'3'-cyclic-and 3'-phosphate esters of 2-N-dimet-ylamino-4-hydroxy-6-trihydroxypropylpteridine. Meanwhile we have also good evidence that E. gracilis, in contrast to higher plants is able to directly hydroxylate phenylnlanine in a reaction which may be similar to the one in animals, requiring unconjugated pteridines.[63] The O_2^--producing sequence which involves Euglenapterin can be written:

(f) fd_{red} + Euglenapterin ---> $Euglenapterin_{red}$

$Euglenapterin_{red} + O_2$ ---> $Euglenapterin_{ox} + O_2^-$

As exceptional as these pteridines of E. gracilis is its SOD: it is an iron-containing enzyme.[68] Since we do not yet know the compartmentations of either SOD or Euglenapterin it is difficult to assess physiological functions (fatty acid desaturation[13,17] or phenylalanine hydroxylation[63]). Both unconjugated pteridin and Fe-SOD might indicate that E. gracilis chloroplasts originally were symbiotic cyanobacteria which "learned" to synthesize chlorophyll b.

(3) Photorespiration and desctructive photooxidations.

(I) Photorespiration. Dependent on environmental parameters, green plants (of the C_3-type) lose part of their fixed carbon during "photorespiration". Andres and Lorimer[6] recently asked the question: "Photorespiration - still unavoidable?" In the view of these authors photorespiration is the inevitable consequence of the active site chemistry of ribulose bisphosphate carboxylase - oxygenase. This bifunctional enzyme either carboxylates ribulose bisphosphate yielding two molecules of phosphoglycerate, or oxidizes its substrate to form phosphoglycerate and phosphoglycollate. Phosphoglycollate is further metabolised in the "photorespiratory cycle", which is located in different compartments of the cell, including peroxisomes, mitochondria and chloroplasts.[6]

In another view, photorespiratory CO_2-production serves as an "energy valve" protecting photosynthesizing plants[66,74] and bacteria from photodamage, since under high light, CO_2 concentration in the atmosphere is the over-all rate limiting factor in photosynthesis and thus dissipation of trapped light energy is a problem. There is circumstantial evidence that photorespiration cannot only be seen as a dissipative process since CO_2 losses inside the cell are at least partially corrected by protecting effects of CO_2 (see below).

(II). Destructive photooxidations. Bleaching of photosynthetic pigments connected with unsaturated fatty acid oxidation and degradation of membraneous cell compartments are characteristic symptons of oxidative plant damage induced by certain chemicals, physical "stress" conditions and infections. Among the chemicals, the activities of certain herbicides as DCMU or paraquat have been shown to be dependent on both light and oxygen: the mechanism of these destructive photooxidations have been studied by several workers (see Ref.19).

(A) Assays. Destructive photooxidations can be studied by (a) following morphological changes of subcellular structures[52], (b) measuring the decrease in photosynthetic pigments as carotenes or chlorophyll (bleaching effects[53,77]), (c) activities of the PET-system or enzyme connected with photosynthetic carbon metabolism,[8,12,77] or (d) determining the products of lipid oxidation.[19,55,56]

As shown earlier for CCl_4-treated animals and tissue homogenates[81] gas chromatographic ethane determination is a simple and non-destructive method for determining light dependent and herbicide-induced lipid peroxidation[40,42,43] in addition to the well-known thiobarbituric acid reaction.[75] Besides ethane, other hydrocarbon products as ethylene[34,65,78] and many others (see Ref.90) have been described. We recently showed that the diatom *Phaeodactylum tricornutum* produces hexanal at slightly acidic pH in a light dependent reaction, competing with ethane formation: ethane formation is strongly stimulated, hexanal formation is strongly inhibited after the addition of α-linolenic acid. Gas formation correlates with pigment bleaching (both chlorophyll and certain carotenoids are concerned) and is inhibited by the radical scavenger propylgallate. These biochemical changes are accompanied by characteristic changes of the cell wall and the outer membrane of the cell. Thus, ethane (and in the above case also hexanal) determination proves a valuable tool for following pathological, oxidative lipid metabolism.[83]

In vitro crocin bleaching[18,40,71,91] p-nitrosodimethylaniline bleaching, hydroxylamine oxidation[30], ethylene formation from methionine[38,64] and NADH-peroxidation have been used for studying photosynthetic oxygen reduction. In some cases, penicillamine copper as a low molecular weight "SOD" (Refs.68,69) has been used as an aid to systems where proteinaceous SOD is of less value due to its size.[40]

(B) Differentiation of mechanism of herbicidal activities. As already reported by Dodge and cowrokers,[19,52,53,77] the symptoms after herbicide treatments may be very similar for different herbicides although the interactions with cellular mechanisms are different: PQ as an autooxidizable electron acceptor induces bleaching effects and unsaturated fatty acid oxidation in a similar manner as the PET-blocker DCMU. Dodge[19] and Ridley[82] explain the DCMU effects by an oxygen activation as consequences of limited energy dissipation while the PQ effects are explained as due to dissipation of the reducing power (PQ instead of NADP) in addition to the production of deleterious O_2^- radicals.[19] Comparing DCMU and PQ by measuring chlorophyll bleaching and ethane formation in E. gracilis cells, a correlation between "malondialdehyde" formation, ethane production and chlorophyll bleaching with both herbicides has been observed.[43] The mechanisms studied in cell-free systems and with model reactions are in general agreement with the view of Dodge[19]: ethane formation and chlorophyll bleaching are enhanced by both PQ and DCMU. The DCMU effect is reduced in the presence of bicarbonate as regulator of electron flow at a site close to or identical with the DCMU binding site in PET (Ref.49); 10^{-5} M DBMIB and 10^{-5} M DCMU have similar effects[87]; 10^{-4} M DBMIB, in contrast to 10^{-4} M DCMU, strongly reduces chlorphyll bleaching and ethane formation due to divalent oxygen reduction driven only by PS II (Refs.36,37), thus relieving the energy transfer block set by lower DBMIB concentrations. This is a good indication that DCMU- and DBMIB-induced chlorophyll bleaching and ethane formation are PS II reactions. Using E. gracilis homogenates, the mechanisms of chlorophyll (chl) bleaching and ethane formation have been shown to include oxygen reduction yielding H_2O_2 as oneproduct, dependent on α-linolenic acid (α-lin) but independent of an intact PET. The carotenoid crocin inhibits both chlorophyll bleaching and ethane formation in E. gracilis homogenates. We may thus write:

(g)	chl	$\xrightarrow{h\nu}$	chl "	(chl " = excited chl)
	chl " + O_2	$\longrightarrow$	chl + $^1O_2$	
	chl " + α-lin	$\longrightarrow$	chl + α-lin$^{\cdot}$	
	α-lin + O_2	$\longrightarrow$	α-lin-OO$^{\cdot}$	
	α-lin-OO$^{\cdot}$ + chl	$\longrightarrow$	chl$^+$ + α-linOOH	
or				
	α-lin + $^1O_2$	$\longrightarrow$	α-lin-OOH	

---

chl$^+$	$\longrightarrow$	bleached products
α-lin-OOH	$\longrightarrow$	ethane and other products

This sequence is in agreement with reports by Dodge[19], Dumelin and Tappel,[21] Foote,[46] Cobern et al.[15] and Tera and Matsushita.[86]

The oxygen activating step in PQ-mediated photobleaching and ethane formation is diferent to the DCMU-induced reactions. As disucssed by several authors (see Refs.3,19), monovalent oxygen reduction by PQ is dependent on the function of both PS II and PS I, and inhibited by DCMU. As recently shown, intact flax leaves floating on PQ solution under low light intensities (3 Klux) evolve ethane and become bleached. Both ethane formation and chlorophyll are reduced in the presence of a penicillamine-copper complex[68] or by DCMU. (Ref.40). At high intensities (35 Klux), the activity of low potential compounds with redox potentials between -0.2 V and -0.55 V is merely dependent on the pigment system of PS I. This can be shown with digitonin-fragmented chloroplasts with no PS II-activity[23,24,89] but exhibiting PS I - activity at the expense of artifical electron donors as ascorbate. The autoxidation and O_2^- production of the reducing side of of PS I of these particles at the expense of enzymatically reduced plastocyanin of cytochrome c_{552} is enhanced by low potential compounds as anthraquinone-sulfonic acid (E_o' = -0.2 V) or triquat (E_o' = -0.55 V) (Ref.24). As shown in Table 1, ethane formation and chlorophyll bleaching in digitonin fragmented

TABLE 1

EFFECTS OF SEVERAL LOW POTENTIAL COMPOUNDS OF THE BIPYRIDYLIUM TYPE ON ETHANE FORMATION AND CHLOROPHYLL BLEACHING IN DIGITONIN FRAGMENTED CHLOROPLASTS[a]

Compound added	Redox potential (E_o', V)	Ethane formed (pmol/mg chl)	chlorophyll bleached (%)
Control	-	1.02	55
PQ	-0.44	3.8	90
CGA 44361	-0.45	2.5	100
CGA 82222	-0.51	2.1	75
CGA 82225	-0.75	0.8	70
CGA 48108	lower than -1.0	0.6	50

[a]Conditions: Digitonin fragmented chloroplasts with about 0.5 mg chlorophyll in a total vol. of 2 ml; 25 mM MES buffer, pH 5.5. Incubations were performed in 15 ml Fernbach flasks (with rubber stopper) in a photo-thermostat (20°C, 35 Klux). After 1 hr illumination, chlorophyll and ethane were determined.

chloroplasts is stimulated by low potential compounds of the bipyridylium type (Ciba-Geigy series, CGA, generously donated by Drs.H.P. Fischer and E. Ebert), with redox potentials between -0.45 V and -0.51 V. Compounds of E_o' of lower than -0.7 V are not active in these particles. As shown in Table 2 the above

TABLE 2

EFFECTS OF ASCORBATE, SUPEROXIDE DISMUTASE AND CATALASE ON PARAQUAT ENHANCED ETHANE FORMATION AND CHLOROPHYLL BLEACHING IN DIGITONIN FRAGMENTED CHLOROPLASTS[a]

Additions	Concentration	Ethane formed (pmol/mg chl)	chlorophyll bleached (%)
Control	-	3	13
PQ	3×10^{-4} M	8	30
PQ + catalase	195 units added	8.4	35
PQ + SOD	50 units added	3.5	30
PQ + ascorbate	3×10^{-3} M	1.6	12

[a]Conditions as in Table 1.

ethane forming reaction is partially inhibited by SOD and ascorbate and not influenced by catalase, while chlorophyll bleaching is only inhibited by ascorbate and not influenced by either SOD or catalase. We assume that light activated PS I, in the absence of the reduced PET chain or of exogenous electron donors oxidizes chlorophyll molecules and eventually unsaturated fatty acids of the chloroplast fragments:

(h)

PS I - PA I $\xrightarrow{h\ x\ \nu}$ PS I" - PA I (PS I " = excited PS I)

PS I" - PA I ------> PS I^{+} - PA $I^{\cdot}$

PS I^{+} - PA $I^{\cdot}$ + O_2 ------> PS I^{+} - PA I + O_2^{-}

PS I^{+} - PA $I^{\cdot}$ + PQ ------> PS I^{+} - PA I + $PQ^{\cdot}$

$PQ^{\cdot}$ + O_2 ------> PQ + O_2^{-}; 2 O_2^{-} $\xrightarrow[2H^{+}]{SOD}$ H_2O_2 + O_2

PS I^{+} - PA I + α-lin ------> PS I-PA I + α-lin·

α-$lin^{\cdot}$ + O_2 ---> α-linOO·

α-lin OO· + chl -----> chl^{+} + α-lin OOH; α-linOOH ----> ethane + other products

PS I^{+} - PA I + chl ------> chl^{+} + PS I - PA I

chl^{+} --------> bleached products; ascorbate reduces PS I^{+}-PA I.

As outlined here, we can differentiate between DCMU-induced photooxidations, involving only PS II with chl-PS II^{+} as oxidizing species possibly including $^{1}O_2$, and low potential dye-induced photooxidations depending on both PS II and PS I at low light intensities. After PET is damaged to a certain extent, a new PS I reaction may take over including both PS I^{+} and O_2^{-} as destructive species, oxidizing unsaturated fatty acids and cooxidizing chlorophyll.

(4) Commercial and phytopathological aspects.

(I) Commerical aspects. As discussed at the "Biocide-Workshop" held in spring 1979 at the Univeristy of Konstanz (to be published as a special issue of *Z.Naturforsch.*), herbicides of the PQ and DCMU type are of great commercial importance. Since several years, resistance against certain herbicides (e.g. Atrazin) has been observed. It is necessary, therefore, to continuously look for new herbicides of the above types, study the mechanisms of their function and elaborate simple test systems. Gas chromatographuc determination of ethane formation may represent a simple test system for screening both DCMU and PQ types (Refs.41-43; see also Table 1). Another aspect is the toxicity of PQ type herbicides;[9] the study of the reaction mechanisms of PQ activities is the prerequisite for studying possible antidotes as well as for the elaboration of new "low". or even "lower" potential herbicides which might not be toxic for animals since they do not possess electronegative compounds as the reducing side of PS I (lower than -0.55 V).

(II) Phytopathological aspects. Kato[62] and Montalbini et al.[71] reported on enhanced unsaturated fatty acid oxidations connected with necrotic reactions or (as in the case of certain systemic diseases) leaf yellowing after viral infections of higher plants. Leaf yellowing and/or necrotisation drastically reduce the yields of certain crops during and after epidemic diseases caused by these viruses. Since cooxidations of leaf pigments is a well known reaction of lipoxygenases[90,91] implicating "activated oxygen", the above diseases are problems of "oxygen toxicity". With simple indicator reactions, either present or absent in more or less resistant lines, the plant breeder should be able to differentiate between these lines within weeks instead of months. In order to provide these tools, we have to study the role of oxygen in host-plant parasite relationships.

REFERENCES

1. Allen, J.F. and Hall, D.O. (1973) Biochem. Biophys. Res. Commun. 52, 856-862.
2. Allen, J.F. (1975) Nature, 256, 599-600.
3. Allen, J.F. (1977) in Superoxide and Superoxide Dismutases, Michelson, A.M., McCord, J.M., and I. Fridovich., eds., Academic Press, New York, pp. 417-436.
4. Asada, K., Urano, M. and Takahashi, M. (1973) Eur. J. Biochem. 36, 257-266.
5. Adams, D.O. and Yang, S.F. (1979) Proc. Nat. Acad. Sci. USA, 76, 170-174.
6. Andrews, T.J. and Lorimer, G.H. (1978) FEBS Lett. 90, 1-9.
7. Arntzen, C.J. (1979) Z. Naturforsch., in press.
8. Asami, S. and Akazawa, F. (1978) Plant Physiol. 62, 98--986.
9. Autor, A.P. (1977) Biochemical Mechanisms of Paraquat Toxicity, Academic Press, New York.
10. Bartsch, R.G., Meyer, T.E. and Robinson, (1967) in Structure and Function of Cytochromes, Okunuki, K., Kamen, M.D. and Sekuzu, I., eds., University Park Press, Baltimore, pp. 443-451.
11. Böhme, M., Pfleiderer, W. and Elstner, E.F. (1979): Submitted for publication.
12. Böger, P. and Kunert, K.J. (1978) Z. Naturforsch. 33c, 688-694.
13. Bloch, K. (1969) Account. Chem. Res. 2, 193-200.
14. Beauchamp, C. and Fridovich, I. (1970) J. Biol. Chem. 245, 4641-4646.
15. Cobern, D., Hobbs, J.S., Lucas, R.A. and Mackenzie, D.J. (1966) J. Chem. Soc., pp. 1897-1902.
16. Doll, S., Lutz, C. and Ruppel, H.G. (1976) Z. Pflanzenphysiol. 80, 166-176.
17. Dewey, V.C. and Kidder, G.W. (1966) Arch. Biochem. Biophys. 115, 401-406.
18. Dickes, J.W. and Friend, J. (1967) Phytochemistry, 6, 1193-1202.
19. Dodge, A.D. (1977) in Herbicides and Fungicides. Factors Affecting their Activity, McFarlane, N.M., ed., The Chemical Society, London, pp.7-21.
20. Drawert, F., Heimann, W., Emberger, R. and Tressl, R. (1966) Ann. Chem. 694, 200-208.
21. Dumelin, E.E. and Tappel, A.L. (1977) Lipids, 12, 894-900.
22. Egneus,H.,Heber, U., Mathiessen, U. and Kirk, M. (1975) Biochim. Biophys. Acta, 408. 252-268.
23. Elstner, E.F., Pistorius, E., Böger, P. and Trebst, A. (1968) Planta, 79, 146-161.
24. Elstner, E.F., Heupel, A. and Vaklinova, S. (1970) Z. Pflanzenphysiol. 62, 184-200.
25. Elstner, E.F. and Kramer, R. (1973) Biochim. Biophys. Acta, 314, 340-353.
26. Elstner, E.F. and Heupel, A. (1974) Z. Naturforsch. 29c, 564-571.
27. Elstner, E.F. and Heupel, A. (1974) Z. Naturforsch. 29c, 559-563.
28. Elstner, E.F. and Konze, J.R. (1974) FEBS Lett. 45, 18-21.
29. Elstner, E.F. and Heupel, A. (1975) Planta, 123, 145-154.
30. Elstner, E.F., Stoffer, C. and Heupel, A. (1975) Z. Naturforsch. 30 c, 53-56.
31. Elstner, E.F. and Heupel, A. (1976) Anal. Biochem. 70, 616.620.
32. Elstner, E.F. and Heupel, A. (1976) Arch. Biochem. Biophys. 173, 614-622.
33. Elstner, E.F., Wildner, G.F. and Heupel, A. (1976) Arch. Biochem. Biophys. 173, 623-630.

34. Elstner, E.F. and Konze, J.R. (1976) Nature, 263, 351-352.
35. Elstner, E.F., Konze, J.R., Selman, B.R. and SToffer, C. (1976) Plant Physiol. 58, 163-168.
36. Elstner, E.F. and Frommeyer, D. (1978) FEBS Lett. 86, 143-146.
37. Elstner, E.F. and Frommeyer, D. (1978) Z. Naturforsch. 33c, 276-279.
38. Elstner, E.F., Saran, M., Bors, W. and Lengfelder, E. (1978) Eur. J. Biochem. 89, 61-66.
39. Elstner, E.F. and Zeller, H. (1978) Plant Sci. Lett. 13, 15-20.
40. Elstner, E.F. and Youngman, R. (1978) Ber. Deutsch. Bot. Ges. 91, 569-577.
41. Elstner, E.F. (1979) in Encyclopedia of Plant Physiology, Vol. 6, No. 2, Gibbs, M. and Latzko, E., eds., Springer Verlag, Berlin, in press.
42. Elstner, E.F. and Pils, I. (1979) Z. Naturforsch., in press.
43. Elstner, E.F. (1979) Z. Naturforsch., in press.
44. Epel, B.L. and Neumann, J. (1973) Biochim. Biophys. Acta, 325, 520-529.
45. Firl, J., Frommeyer, D. and Elstner, E.F. (1979):In preparation.
46. Foote, C.S. (1970) Ann. N.Y. Acad. Sci. 171, 139-148.
47. Fridovich, I. (1975) Annu. Rev. Biochem. 44, 147-159.
48. Fuller, R.C., Kidder, G.W., Nugent, N.A., Dewey, V.C. and Rigopoulos, N. (1971) Photochem. Photobiol. 14, 359-371.
49. Govindjiee and Van Rensen, J.J.S (1978) Biochim. Biophys. Acta, 505, 183-213.
50. Glidewell, S.M. and Raven, J.A. (1975) J. Exp. Bot. 26, 579-588.
51. Halliwell, B. (1978) progr. Biophys. Mol. Biol. 33, 1-54.
52. Harris, N. and Dodge, A.D. (1972) Planta, 104, 201-209.
53. Harris, N. and Dodge, A.D. (1972) Planta, 104, 210-219.
54. Hatanaka, A. and Harada, T. (1973) Phytochemistry, 12, 2341-2346.
55. Heath, R.L. and Packer, L. (1968) Arch. Biochem. Biophys. 125, 189-198.
56. Heath, R.L. and Packer, L. (1968) Arch. Biochem. Biophys. 125, 850-857.
57. Heber, U. (1976) J. Bioenerg. Biomembranes, 8, 157-172.
58. Jackson, C., Dench, J., Moore, A.L., Halliwell, B., Foyer, C.H. and Hall, D.O. (1978) Eur. J. Biochem. 91, 339-344.
59. Ilan, Y.A., Czapski, G. and Meisel, D. (1976) Biochim. Biophys. Acta, 430, 209-224.
60. Iwai, K., Kobashi, M., Suzuki, T. and Bunno, M. (1979) in Chemistry and Biology of Pteridines, Kisliuk, R.L. and Brown, G.M., eds., Elsevier/North Holland, New York, pp. 111-116.
61. Kaiser, W. (1976) Biochim. Biophys. Acta, 440, 476-482.
62. Kato, S. (1977) Ann. Phytopathol. Soc. Japan, 43, 472-480.
63. Kaufman, S. (1979) in Chemistry and Biology of Pteridines, Kisliuk, R. and Brown, G.M., eds., Elsevier/North Holland, New York, pp. 117-124.
64. Konze, J.R. and Elstner, E.F. (1976) FEBS Lett. 66, 8-11.
65. Konze, J.R. and Elstner, E.F. (1978) Biochim. Biophys. Acta, 528, 213-221.
66. Krause, G.H. and Heber, U. (1976) in The Intact Chloroplast, Barber, J., ed., Elsevier, Amsterdam, pp. 171-124.
67. Lengfelder, E., Saran, M. and Bors, W. (1977) Int. Conf. on Singlet Oxygen and Related Species, August 21-26, 1977, Pinawa, Canada, Abstr. No. L-3.
68. Lengfelder, E. and Elstner, E.F. (1978) Hoppe-Seylers Z. Physiol. Chem. 359, 751-757.
69 Lengfelder, E., Fuchs, C., Younes, M. and Weser, U. (1979) Biochim. Biophys. Acta, 567, 492-502.
70. McCord, J.M. and Fridovich, I. (1969) J. Biol. Chem. 244, 6049-6055.
71. Montalbini, P., Koch, F., Burba, M. and Elstner, E.F. (1978) Physiol. Plant. 12, 211-223.
72. Nakajima, R., Sano, H. and Yamazaki, I. (1969) Biochim. Biophys. Acta, 172, 578-581.
73. Nishikimi, M. (1975) Arch. Biochem. Biophys. 166, 273-279.
74. Osmond, C.B. and Björkman, O. (1972) Carnegie Inst. Washington Yearbook 1971-72, pp. 141-148.
75. Ottolenghi, A. (1959) Arch. Biochem. Biophys. 79, 355-363.
76. Patterson, C.O.P. and Myers, J. (1973) Plant Physiol. 51, 104-109.
77. Pallet, K.E. and Dodge, A.D. (1976) Proc. Brit. Crop. Protection Conf. Weeds, pp. 235-240.

78. Peiser, G.D. and Yang, S.F. (1978) Phytochemistry, 17, 79-84.
79. Radmer, R.J. and Kok, B. (1976) Plant Physiol. 58, 336-340.
80. Rembold, H. (1975) in Chemistry and Biology of Pteridines, Pfleiderer, W., ed., de Gruyter, Berlin-New York, pp. 359-371.
81. Riely, C.A., Cohen, G. and Lieberman, M. (1974) Science, 183, 208-210.
82. Ridley, S.M. (1977) Plant Physiol. 59, 724-732.
83. Schobert, B. and Elstner, E.F. (1979): Submitted for publication.
84. Siedow, J., Yoccum, C.F. and San Pietro, A. (1973) in Current Topics in Bioenergetics, Academic Press, New York, pp. 107-122.
85. Singh, A. and Petkau, A. (1978) Singlet Oxygen and Related Species in Chemistry and Biology (Photochem.Photobiol. 28, 429 ff).
86. Terao, J. and Matsushita, S. (1977) J. Amer. Oil Chem. Soc. 54, 234-238.
87. Trebst, A. (1974) Annu. Rev. Plant Physiol. 25, 423-458.
88. Trebst, A. (1970) Ber. Deutsch. Bot. Ges. 83, 373-398.
89. Trebst, A., Pistorius, E. and Elstner, E. (1966) in Currents in Photosynthesis, A.D. Donker, Rotterdam, pp. 409-419.
90. Veldink, G.A., Vliegenthart, F.G. and Boldingh, J. (1977) Progr. Chem. Fats Lipids, 15, 131-166.
91. Weber, F. and Grosch, W. (1976) Z. Lebensm. Unters. Forsch. 161, 223-230.

ACTIVE OXYGEN SPECIES IN HERBICIDE ACTION

R.J. YOUNGMAN, K.E. PALLETT* AND A.D. DODGE,
School of Biological Sciences, University of Bath, Clavertown Down, Bath, Avon, BA2 7AY, England

INTRODUCTION

Of the large number of herbicides currently available for use throughout the world, a major percentage are thought to have a primary site of action in the chloroplast and more specifically on some portion of the photosynthetic electron transport chain. Herbicides in this area fall into two classes: (i) those which prevent electron flow, and (ii) those which cause a diversion of electron transport and result in the generation of certain toxic species. Recent work described here has suggested that active oxygen species play a major role in the phytotoxic action of both groups of herbicides.

RESULTS AND DISCUSSION

(1) Photosynthetic inhibitor herbicides. This is a large group of compounds and includes the ureas, uracils, hydroxybenzonitriles, triazines and triazinones. After application to the plant, the herbicide enters the chloroplast and binds to a proteinaceous component of the thylakoid membrane. This results in an interruption of electron flow between Q and plastoquinone.[18,20,22] A recent study with four of these photosynthetic inhibitors: bromacil, ioxynil, metribuzin and monuron (CMU),** revealed that in general, the toxic effects elicited by these compounds were similar.[16] In detached flax (*Linum usitatissimum* var. Reina) cotyledons floated on herbicide solutions, a rapid inhibition of CO_2 fixation was followed by a loss of chloroplast integrity.[16,17] The first visible symptom of herbicide activity was a bleaching of the chloroplast pigments in which carotene breakdown preceded that of chlorophyll (Figure 1). Electron microscope studies of treated cotyledons showed that the bleaching was accompanied by a swelling of the chloroplasts and was followed by a progressive disorganisation of the pigment containing thylakoid membranes.

*Present address: ARC, Weed Research Organisation, Yarnton, Oxford, England.
**Abbreviations: Bromacil : 5-bromo-6-methyl-3-(1-methyl-n-propyl)uracil; Ioxynil : 3,5-diiodo-4-hydroxybenzonitrile; Metribuzin : 4-amino-6-t-butyl-3 (methylthio)-1,2,4-triazin-5(4H)-one; Monuron, CMU : 3-(4-chlorophenyl)-1,1-dimethylurea; Paraquat : 1,1'-dimethyl-4,4'-bipyridylium dichloride; Diquat : 1,1'-ethylene-2,2'-bipyridylium dibromide; DABCO : 1,4-diazobicyclo(2,2,2)-octaine; DBMIB : 2,5-dibromo-3-ethyl-6-isopropyl-p-benzoquinone

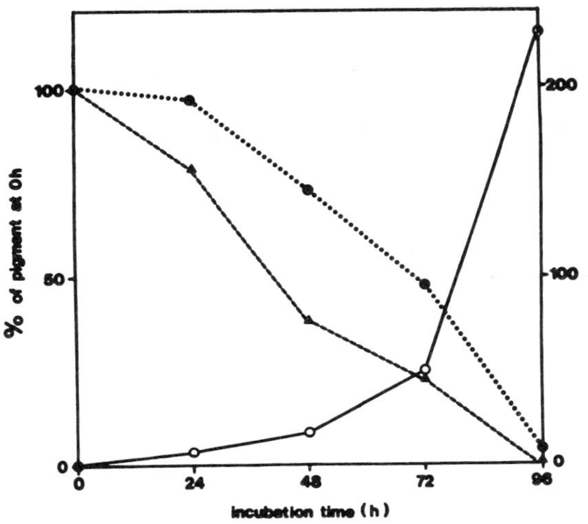

Fig. 1. Effect of $1x10^{-3}$ M monuron on ethane generation (○———○), chlorophyll (●·········●), and α- and β-carotene levels (▲----▲) in flax cotyledons illuminated at 30 W/m^2.

Ethane generation, which is thought to be an indicator of membrane lipid peroxidation,[9,12,21,24] increased immediately after application of the herbicide. The high rate of release of ethane after 72 hours corresponded with a period of rapid chloroplast membrane disorganisation and a total loss of pigment (Figure 1).

It is suggested that when photosynthetic electron transport is prevented by these inhibitor herbicides, excitation energy still absorbed by the pigment system is dissipated by the carotenoids.[13] The preferential breakdown of the carotene pigments would suggest that this protective system was overloaded. Unquenched triplet excited chlorophyll may pass excitation energy to triplet (ground state) oxygen, thus forming excited singlet oxygen. Table 1 shows that the destruction of certain chloroplast pigments was retarded by incubation under argon instead of air. By use of the singlet oxygen quencher DABCO in addition to the herbicide monuron under aerobic conditions, the breakdown of chlorophyll was retarded (Table 2). Singlet oxygen, generated when the pigment protective systems are overloaded, is believed to interact with unsaturated fatty acids which constitute a large proportion of the chloroplast membrane lipids, leading to the formation of lipid hydroperoxides.[19] These, on further breakdown would give rise to ethane. The accompanying membrane changes would be visible under the electron microscope.

A convenient model of the <u>in vivo</u> inhibited system might be the incubation of chloroplast fragments in the light in the absence of an electron acceptor.

TABLE 1

PIGMENT CONTENT OF DETACHED FLAX COTYLEDONS TREATED WITH MONURON (CMU)[a]

Pigment	0 Hours	72 Hours	
		CMR/Air	Air/Argon
Chlorophyll	1.669	0.956	1.620
α- and β-Carotenes	0.066	0.015	0.061
Xanthophylls	0.249	0.123	0.231

[a]After 0 hours and 72 hours illumination at 30 $W3m^2$. Monuron (CMU) concentration $1x10^{-3}M$. Herbicide solutions were degassed before use.

TABLE 2

EFFECT OF DABCO ON THE CHLOROPHYLL CONTENT OF DETACHED FLAX COTYLEDONS TREATED MONURON[a]

Treatment	mg chlorophyll/g fresh wt
Water	1.735
Monuron	1.184
DABCO	1.610
Monuron + DABOC	1.598

[a]Cotyledons were treated with $1x10^{-3}M$ monuron and illuminated for 96 hours at 5.25 W/m^2. DABCO ($1x10^{-3}M$) was made up in aqueous solution.

Experiments with such a system showed that chlorophyll bleaching was promoted by increasing light intensity (Figure 2). Chlorophyll destruction was retarded when chloroplasts were incubated under argon instead of air (Figure 3). Ethane generation in the model system, as an indicator of lipid peroxidation, was promoted by light (Figure 4). Further confirmation of a possible role for singlet oxygen in promoting membrane destruction was obtained by the use of the quencher DABCO. In this case, chlorophyll breakdown was reduced to some extent (Table 3).

(2) Photosynthetic energy deviators. The bipyridylium herbicides, paraquat and diquat lead to a rapid bleaching of green plant tissue upon illumination. In addition to a requirement for light, early experiments demonstrated that the phytotoxic effects were dependent upon the presence of oxygen.[5,15] In the chloroplast, bipyridyls accept electrons from the reducing side of photosystem I and become univalently reduced in the process.[10] Although this is in direct competition with the reduction of NADP and subsequent CO_2 fixation, plant starvation due to a lack of elaborated carbon is not the primary factor in the action of the

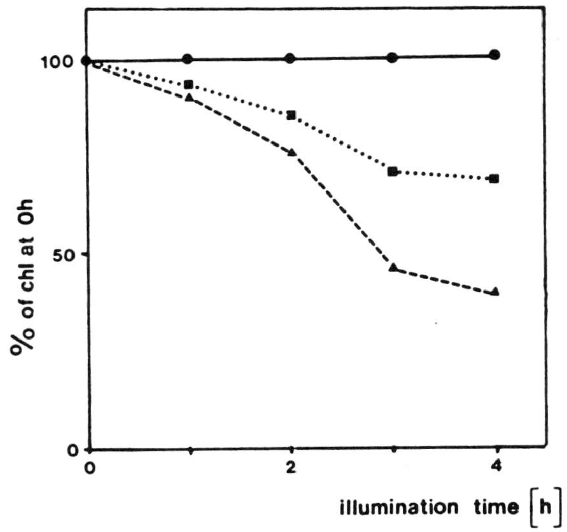

Fig. 2. Effect of light intensity on chlorophyll content of pea chloroplast fragments. ●——● dark incubation, ▲---▲ 275 W/m^2, ■·····■ 132 W/m^2. Reaction mixture contained in 3 ml: 50 mM tricine-NaOH buffer pH 7.5, 0.3 M NaCl, 3 mM $MgCl_2$.

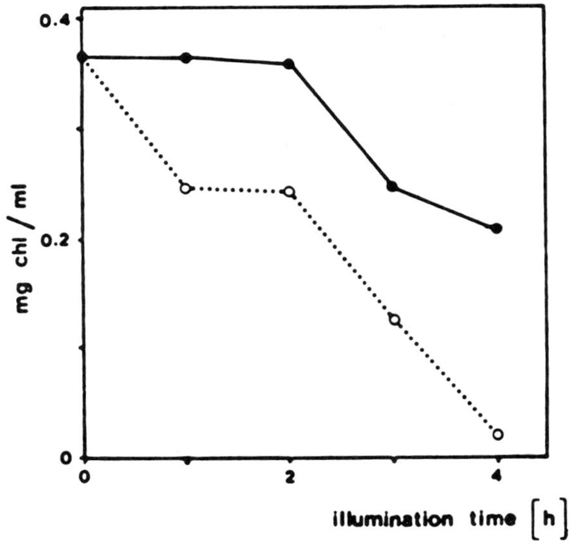

Fig. 3. Effect of argon on chlorophyll levels of pea chloroplast fragments illuminated with 275 W/m^2 light. Conditions as for Fig. 2. O-----O air, ●——● argon.

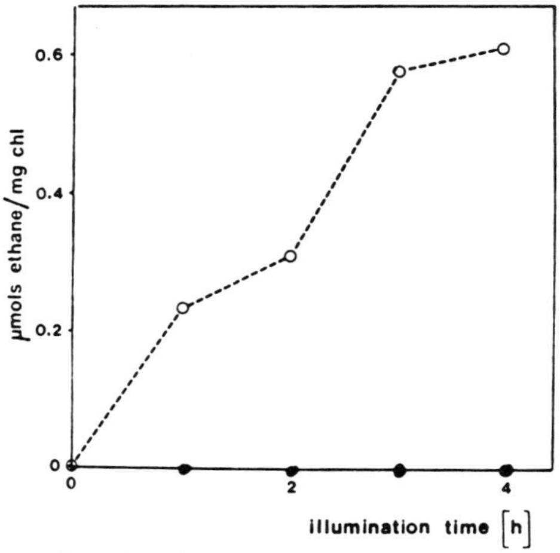

Fig. 4. Effect of illumination on ethane generation from pea chloroplast fragments. Conditions as for Fig. 2. ●——● dark control, ○-----○ illuminated with 275 W/m^2 light.

TABLE 3

EFFECT OF DABCO ON CHLOROPHYLL LEVELS OF PEA CHLOROPLAST FRAGMENTS[a]

Treatment	mg chlorophylls/ml
0 hours illumination	0.400
2 hours illumination	0.304
2 hours illumination + DABCO	0.353

[a]Conditions as for Fig. 2; light intensity 275 W/m^2.

herbicide.[5] The primary mechanism of action derives from the fate of the reduced paraquat radical. It was originally thought that the initial causative toxic species in bipyridyl action was H_2O_2, formed by the interaction of oxygen with the reduced bipyridyl.[4,5] More recent work has established that the superoxide anion radical is an earlier, more reactive intermediate.[10]

The ability of isolated chloroplasts to generate superoxide in the presence of paraquat has been demonstrated using the oxidising potential of the radical to form nitrite from hydroxylamine.[2,9] Figure 5 shows the results of this experiment, together with the inhibition of nitrite formation when SOD was also present in the reaction mixture. Further evidence for the dependence of superoxide formation on electron flow was obtained through the use of the inhibitors CMU and DBMIB (Table 4). Paraquat also stimulated the production of H_2O_2 by isolated chloroplasts, but SOD had no effect on this formation, indicating that paraquat increased the production of superoxide which dismutated (enzymatically and/or non-enzymatically to form H_2O_2.

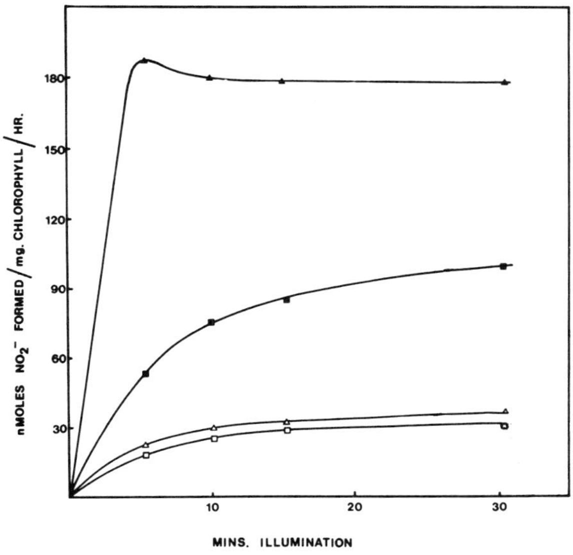

Fig. 5. Nitrite formation from hydroxylamine. Reaction mixture contained in 3 ml . 50 mM phosphate buffer pH 7.8, 1 µmole NH_2OH, chloroplasts with 75 µg chlorophyll and where indicated, 6.6 µM paraquat, 50 Units SOD. Illuminated at 275 W/m^2 at 22°C. ■——■ control, ▲——▲ + paraquat, □——□ +SOD, △——△ + paraquat and SOD.

TABLE 4

EFFECT OF PHOTOSYNTHETIC TRANSPORT INHIBITORS ON NITRITE FORMATION FROM HYDROXYLAMINE[a]

Treatment	-CMU or DBMIB	+CMU or DBMIB
	(nmoles nitrite/mg chlorophyll/hr)	
Control	100	17
+ Paraquat	195	28

[a]Conditions as for Fig. 5; illumination for 15 min.

The damage to chloroplast fragments after treatment with paraquat was followed by monitoring the generation of ethane from the reaction mixture. Ethane release was considerably stimulated by paraquat and -linolenic acid which is thought to be one of the primary fatty acid substrates in the membrane for lipid peroxidation[9,12] (Table 5).

TABLE 5

ETHANE GENERATION IN CHLOROPLAST FRAGMENTS[a]

Treatment	Ethane (nmol/g fresh wt)
Control	0
Paraquat	0.6
α-Linolenic acid	8.5
Paraquat + α-linolenic acid	15.2

[a]Reaction mixture contained in 3 ml : chloroplast fragments with 100 μg chlorophyll, 75 μmol phosphate buffer pH 7.8, 5 μmol NH_4Cl, 5 μmol NaN_3, 50 mg BSA and where indicated, 1 μmol α-linolenic acid and 10^{-6}M paraquat. Reaction time was 30 min.

The application of paraquat to detached flax cotyledons showed that the herbicide greatly promoted chlorophyll and carotenoid pigment breakdown.[9,23,24] In this case, as with the photosynthetic inhibitor herbicides, the carotene pigments were preferentially destroyed. When the superoxide scavenger, copper-penicillamine (PA-Cu) was present,[14] the breakdown of photosynthetic pigments was significantly retarded.[9,23] Furthermore, PA-Cu markedly reduced the generation of ethane from paraquat treated flax cotyledons[9,24] (Figure 6). Additional experiments in which the fatty acid content of chloroplasts isolated from paraquat treated flax cotyledons was assessed, showed a large decrease in the levels of the unsaturated linoleic and linolenic acids (Figure 7).

It is suggested that superoxide formed as a result of bipyridyl herbicide interaction with photosynthetic electron transport is produced in excess of the endogenous SOD and ascorbate defence mechanisms. It could then participate in further reactions which would lead to the formation of more reactive species such as singlet oxygen and $Fe^{2+}\cdot H_2O_2$ (Refs. 3, 7). The instigation of lipid peroxidation by these toxic oxygen species (produced as a result of bipyridyl or inhibitor herbicide action) would lead to the destruction of cellular membranes and ethane generation. The disruption of the tonoplast and plasmalemma have been shown to be early effects after paraquat treatment[11] and this is followed by complete cellular destruction and plant death. The biochemical retarding action of PA-Cu is also manifest in

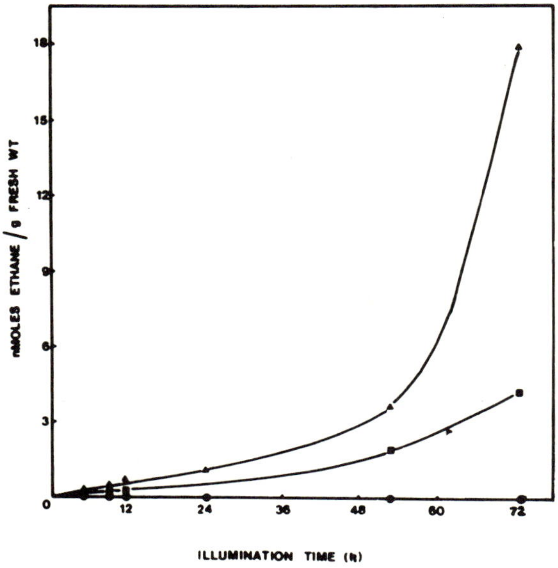

Fig. 6. Ethane generation from flax cotyledons illuminated with light of 5,500 lux at 20°C. Twenty cotyledons (ca. 0.2 g) used per treatment. When present, paraquat concentration was 10^{-6} M; PA-Cu concentration was equivalent to 50 SOD Units. ▲—▲ + paraquat, ■—■ + paraquat and PA-Cu, ●—● control.

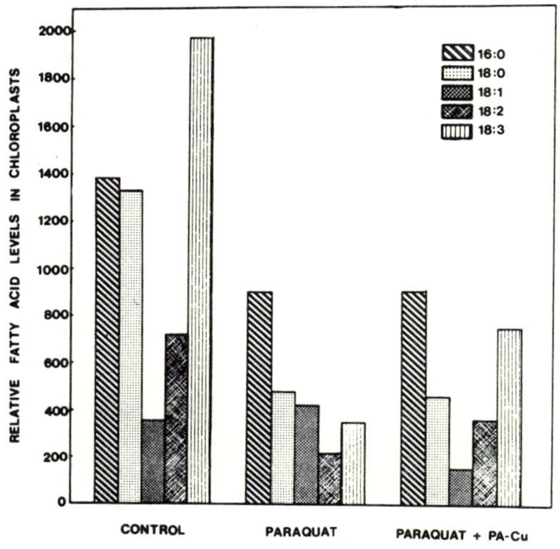

Fig. 7. Fatty acid levels of chloroplasts isolated from treated flax cotyledons. Conditions as for Fig. 6. Fatty acid extraction and identification as desribed in Ref. 23.

the ultrastructural changes detected by electron microscopy.

In summary, the generation and fate of superoxide is of primary importance in the mechanism of action of the bipyridyl herbicides. Hydrogen peroxide is less important in this process, but can probably participate in reactions leading to damage, once the sequence has been initiated.[25]

Conclusions. The illuminated chloroplast is particularly susceptible to photodynamic damage. The presence of excited pigment systems and electron transport in close association with a high oxygen tension is a potential hazard to the plant. To cope with the problems of excess excitation energy, activated oxygen species and other free radicals, the plant has developed a complex system of defences including SOD, ascorbate and α-tocopherol which are generally sufficient to deal with the normal conditions of photosynthesis. The action of the two groups of herbicides discussed here, is largely as a result of an overloading of the endogenous protection mechanisms by an increased generation of activated oxygen species.

The prevention or inhibition of certain protective systems in the chloroplast offers potential targets for new herbicides. It is of interest that some herbicides such as metfluorazone (Sandoz 6706) (Refs. 1,2), appear to act by primarily inhibiting the biosynthesis of carotenoids. In this instance, the herbicidal effects are promoted by increased light intensity.

AKNOWLEDGEMENTS

This work was financed by grants from the U.K. Science Research Council and ICI Ltd., Plant Protection Division to R.J.Y. and a University of Bath Research Studentship awarded to K.E.P. The authors are grateful to Professor E.F. Elstner for helpful advice and discussions and to Dr. E. Lengfelder for a gift of penicillamine-copper complex.

REFERENCES

1. Bartels, P.G. and Hyde, A. (1970) Plant Physiol. 45, 807-810.
2. Bartels, P.G. and McCullogh, C. (1972) Biochem. Biophys. Res. Commun. 48, 16-22.
3. Bors, W., Saran, M., Lengfelder, E., Spöttl, R. and Michel, C. (1974) Curr. Top. Radiat. Res. 9, 247-309.
4. Davenport, H.E. (1963) Proc. Roy. Soc. B, 157, 332-345.
5. Dodge, A.D. (1971) Endeavour, 30, 130-135.
6. Elstner, E.F. and Heupel, A. (1976) Anal. Biochem. 70, 616-620.
7. Elstner, E.F., Saran, M., Bors, W. and Lengfelder, E. (1978) Eur. J. Biochem. 89, 61-66.
8. Elstner, E.F., Stoffer, C. and Heupel, A. (1975) Z. Naturforsch. 30c, 53-57.
9. Elstner, E.F. and Youngman, R. (1978) Ber. Deutsch. Bot. Ges. 91, 569-577.
10. Farrington, J.A., Ebert, M., Land, E.J. and Fletcher, K. (1973) Biochim. Biophys. Acta, 314, 372-381.
11. Harris, N. and Dodge, A.D. (1972) Planta, 104, 201-209.
12. Konze, J.R. and Elstner, E.F. (1978) Biochim. Biophys. Acta, 528, 213-221.
13. Krinsky, N.I. (1971) in Carotenoids, Isler, O., ed., Birkhäuser Verlag, Basel, pp. 669-716.

14. Lengfelder, E. and Elstner, E.F. (1978) Hoppe-Seylers Z. Physiol. Chem. 359, 751-757.
15. Mees, G. (1960) Ann. App. Biol. 48, 601-612.
16. Pallett, K.E. (1978) Ph.D. Thesis, University of Bath.
17. Pallett, K.E. and Dodge, A.D. (1976) Proc. Brit. Crop Protection Conf. Wees, pp. 235-240.
18. Pallett, K.E. and Dodge, A.D. (1979) Pest. Sci., in press.
19. Rawls, H.R. and Van Santen, P. (1970) J. Amer. Oil Chem. Soc. 47, 121-125.
20. Render, G. (1976) Biochim. Biophys. Acta, 440, 287-300.
21. Riely, C.A., Cohen, G. and Liebermann, M. (1974) Science, 183, 208-210.
22. Trebst, A. and Draber, W. (1979) Adv. Pest. Sci., Int. Cong. Pest. Chem., Zurich, 1978, Part 2, pp. 223-234.
23. Youngman, R.J. and Dodge, A.D. (1979) Z. Naturforsch. (Special issue), in press.
24. Youngman, R.J., Dodge, A.D., Lengfelder, E. and Elstner, E.F. (1979) Experientia, in press.
25. Youngman, R.J., Dodge, A.D., Lengfelder, E. and Elstner, E.F. (1979) Plant Physiol. 63 Suppl. Abstr. 230.

Index